T0339566

Energy and Behaviour

Energy and Behaviour
Towards a Low Carbon Future

Edited by

Marta Lopes

Carlos Henggeler Antunes

Kathryn B. Janda

Academic Press is an imprint of Elsevier
125 London Wall, London EC2Y 5AS, United Kingdom
525 B Street, Suite 1650, San Diego, CA 92101, United States
50 Hampshire Street, 5th Floor, Cambridge, MA 02139, United States
The Boulevard, Langford Lane, Kidlington, Oxford OX5 1GB, United Kingdom

Notices
Knowledge and best practice in this field are constantly changing. As new research and experience
broaden our understanding, changes in research methods, professional practices, or medical treatment
may become necessary.

Practitioners and researchers must always rely on their own experience and knowledge in evaluating
and using any information, methods, compounds, or experiments described herein. In using such
information or methods they should be mindful of their own safety and the safety of others, including
parties for whom they have a professional responsibility.

To the fullest extent of the law, neither the Publisher nor the authors, contributors, or editors, assume
any liability for any injury and/or damage to persons or property as a matter of products liability,
negligence or otherwise, or from any use or operation of any methods, products, instructions, or ideas
contained in the material herein.

Library of Congress Cataloging-in-Publication Data
A catalog record for this book is available from the Library of Congress

British Library Cataloguing-in-Publication Data
A catalogue record for this book is available from the British Library

ISBN 978-0-12-818567-4

For information on all Academic Press publications
visit our website at https://www.elsevier.com/books-and-journals

Publisher: Candice Janco
Acquisition Editor: Graham Nisbet
Editorial Project Manager: Michelle W. Fisher
Production Project Manager: Kamesh Ramajogi
Cover Designer: Christian J. Bilbow

Typeset by SPi Global, India

Working together
to grow libraries in
developing countries

www.elsevier.com • www.bookaid.org

We dedicate this book to the memory of Hal Wilhite (1946–2019) who was a colleague, friend, and founding member of eceee (European Council for an Energy Efficient Economy). He was tirelessly interested in people, energy, and a better world, and we hope this book will inspire future generations to follow his example.

Contents

PART 2   Energy behaviour across sectors

**CHAPTER 2.6   Energy, human activity, and knowledge:
Addressing smart city challenges**............................**237**

Sarah J. Darby

PART 3 Modelling energy behaviour

CHAPTER 3.1 Energy and enjoyment: The value of household electricity consumption **263**

Phil Grunewald and Marina Diakonova

CHAPTER 3.2 Developing quantitative insights on building occupant behaviour: Supporting modelling tools and datasets **283**

Tianzhen Hong, Jared Langevin, Na Luo, and Kaiyu Sun

PART 4 Promoting behaviour change

List of figures

List of tables

Contributors

Wokje Abrahamse
Victoria University of Wellington, Wellington, New Zealand

Eeva-Lotta Apajalahti
University of Helsinki, Helsinki, Finland

Katrin Arning
Chair of Communication Science/Human Computer Interaction Center, RWTH Aachen University, Aachen, Germany

Carlos Henggeler Antunes
INESC Coimbra, DEEC, University of Coimbra, Coimbra, Portugal

Hermano Bernardo
School of Technology and Management, Polytechnic Institute of Leiria, Leiria; INESC Coimbra, DEEC, University of Coimbra, Coimbra, Portugal

Paolo Bertoldi
European Commission, Joint Research Centre, Ispra, Italy

Seraja Bock
Potsdam Institute for Climate Impact Research (PIK), Potsdam, Germany

Sylvia Breukers
Duneworks, Eindhoven, The Netherlands

Émile J.L. Chappin
Faculty of TPM, Delft University of Technology, Delft, The Netherlands

Sarah J. Darby
Environmental Change Institute, University of Oxford, Oxford, United Kingdom

Marina Diakonova
Environmental Change Institute, University of Oxford, Oxford, United Kingdom

Luís C. Dias
CeBER, Faculty of Economics; INESC Coimbra, DEEC, University of Coimbra, Coimbra, Portugal

Elisabeth Dütschke
Fraunhofer Institute for Systems and Innovation Research ISI, Karlsruhe, Germany

Maria J. Figueroa
Department of Management Society and Communication, Copenhagen Business School, Frederiksberg, Denmark

Joachim Globisch
Fraunhofer Institute for Systems and Innovation Research ISI, Karlsruhe, Germany

Phil Grunewald
Environmental Change Institute, University of Oxford, Oxford, United Kingdom

Eva Heiskanen
University of Helsinki, Helsinki, Finland

Tianzhen Hong
Building Technology and Urban Systems Division, Lawrence Berkeley National Laboratory, Berkeley, CA, United States

Kathryn B. Janda
Energy Institute, Bartlett School of Environment Energy and Resources, University College London, London, United Kingdom

Senja Laakso
University of Helsinki, Helsinki, Finland

Oliver Lah
Mobility and International Cooperation, Wuppertal Institute for Climate, Environment and Energy, Berlin, Germany

Jared Langevin
Building Technology and Urban Systems Division, Lawrence Berkeley National Laboratory, Berkeley, CA, United States

Wiebke Lass
Potsdam Institute for Climate Impact Research (PIK), Potsdam, Germany

Marta Lopes
Polytechnic Institute of Coimbra—ESAC; INESC Coimbra, DEEC, University of Coimbra, Coimbra, Portugal

Na Luo
Building Technology and Urban Systems Division, Lawrence Berkeley National Laboratory, Berkeley, CA, United States

Loren Lutzenhiser
Portland State University, Portland, OR, United States

António Gomes Martins
INESC Coimbra, DEEC, University of Coimbra, Coimbra, Portugal

Kaisa Matschoss
University of Helsinki, Helsinki, Finland

Mithra Moezzi
QQForward, Marin County, CA, United States

Ruth M. Mourik
Duneworks, Eindhoven, The Netherlands

Luis Mundaca
International Institute for Industrial Environmental Economics at Lund University, Lund, Sweden

Luís Neves
School of Technology and Management, Polytechnic Institute of Leiria, Leiria; INESC Coimbra, DEEC, University of Coimbra, Coimbra, Portugal

Igor Nikolic
Faculty of TPM, Delft University of Technology, Delft, The Netherlands

Gabriela Oliveira
CeBER, Faculty of Economics; INESC Coimbra, DEEC, University of Coimbra, Coimbra, Portugal

Jenny Palm
IIIEE, International Institute for Industrial Environmental Economics, Lund University, Lund, Sweden

Jane S. Peters
Opinion Dynamics Corporation, Portland, OR, United States

Fritz Reusswig
Potsdam Institute for Climate Impact Research (PIK), Potsdam, Germany

Margaret Samahita
International Institute for Industrial Environmental Economics at Lund University, Lund, Sweden

Geertje Schuitema
University College Dublin, Dublin, Ireland

Roman Seidl
Öko-Institut, Freiburg, Germany

Jonas Sonnenschein
International Institute for Industrial Environmental Economics at Lund University, Lund, Sweden

Kaiyu Sun
Building Technology and Urban Systems Division, Lawrence Berkeley National Laboratory, Berkeley, CA, United States

Patrik Thollander
Department of Management and Engineering, Division of Energy Systems, Linköping University, Linköping; Department of Building, Energy and Environment Engineering, University of Gävle, Gävle, Sweden

Luc F.M. van Summeren
School of Innovation Sciences, Eindhoven University of Technology, Eindhoven, The Netherlands

Anna J. Wieczorek
School of Innovation Sciences, Eindhoven University of Technology, Eindhoven, The Netherlands

Neil Yorke-Smith
Faculty of EEMCS, Delft University of Technology, Delft, The Netherlands;
Olayan School of Business, American University of Beirut, Beirut, Lebanon

Barbara Zaunbrecher
Chair of Communication Science/Human Computer Interaction Center, RWTH
Aachen University, Aachen, Germany

Foreword

Putting the brakes on climate change is perhaps the most critical global challenge facing this generation, and reducing the consumption of fossil fuels offers the largest opportunity for meeting it. New technologies will almost surely be required, but we know that it's not true that 'if you build it, they will come'. One reason is that new technologies bring risks and benefits. To be widely adopted, they must be acceptable to those who would use them. Even when a technology seems highly promising to its proponents, acceptance isn't assured. This is one of the lessons of the history of nuclear power. Understanding the conditions under which technologies are acceptable and ways to address their risks and proponents' concerns is a critical need (NRC, 1996).

Public policies that require or financially incentivize change will also almost surely be required, but they also have limitations. Policies such as regulatory requirements or carbon taxes, which look ideal to their proponents, often fail to be adopted because of public or industry opposition. And even when they are adopted, implementation can fall far short of expectations because of exceptions, evasions, and failures to attend fully to the barriers the intended adopters face. For both technologies and policies to achieve their technical potential, the agents of change must to take into account the concerns, interests, and social and behavioural contexts of the targets of change. To do this, change agents must be informed by what I have called the science of human-environment interactions (Stern, 1993).

Social and behavioural scientists have been engaged in such work for over 40 years, as the contributions in this book acknowledge, though often out of the field of vision of policy makers and technologists. This is evident in the recent call from the Intergovernmental Panel on Climate Change (IPCC, 2018) for lifestyle and behavioural and societal change to hold global warming to no more than 1.5°C and in increasing support from several governments for social and behavioural science research on limiting climate change.

The editors and authors of this book recognise, though, that, although such research has a long history, the overall corpus is not yet adequately broad or integrated. Psychologists, for example, have focused mainly on easily measurable, frequent consumer behaviours by individuals and households that can have immediate impact but have too often overlooked less frequent actions, such as acquisition of more energy-efficient or renewable energy producing household equipment that can have much greater impact on a longer timescale (Dietz et al., 2009; Kastner and Stern, 2015). They have generally looked at nonregulatory, nonfinancial interventions alone, even though these interventions may well have their greatest impact when combined with incentives. The most effective interventions to change household consumer behaviours seem to be ones that combine the value of incentives in getting consumers' attention with marketing and implementation features that take advantage of social and behavioural science insights to ease consumers' physical, social, and cognitive constraints (Gardner and Stern, 1996; Wolske and Stern, 2018). Such interventions involve change in the behaviours of the agents of change—often organisations that supply energy technologies—with the behaviours of the consumers who are the targets of change.

And there are further frontiers for social and behavioural sciences to contribute by examining how individuals, outside their roles as consumers, can help limit climate change: by acting within formal organisations and in informal communities to limit their climate footprints, as citizens and within social movements to influence public policies, and as parents and teachers to influence the next generation. These sciences also need to examine social processes beyond the individual. This work includes studies of organisational actors, both as energy users and as influences on other parts of energy systems (e.g. Stern et al., 2016) and of social movements and other cultural forces that can facilitate or impede transformational changes, especially on generational timescales. They also examine decision processes that can better inform societal choices about technologies or policies that present benefits and risks that may be distributed unequally amongst social groups (NRC, 1996, 2008).

Achieving goals for reducing fossil fuel consumption will thus require the integration of thinking from many social science disciplines, of the social and natural sciences, and of science and practice. This volume provides an important entrée into this critical domain of research and practice. Like the literature as a whole, some of the chapters in this volume focus on household consumer behaviour on relatively short timescales. But other chapters also venture beyond this, examining the actions of business organisations, collective efforts at renewable energy provision, matters of building and urban design, and other broader issues. It includes a section on modelling to emphasise the necessary integration of the engineering and physical sciences with the social sciences. By doing these things, it helps make connections between different types of science and scientists that are needed to build the integrated effort required for meeting the monumental challenge of limiting climate change.

Paul C. Stern

References

Dietz, T., Gardner, G.T., Gilligan, J., Stern, P.C., Vandenbergh, M.P., 2009. Household actions can provide a behavioral wedge to rapidly reduce US carbon emissions. Proc. Natl. Acad. Sci. 106, 18452–18456.

Gardner, G.T., Stern, P.C., 1996. Environmental Problems and Human Behavior. Allyn & Bacon.

Intergovernmental Panel on Climate Change (IPCC). (2018). Global warming of 1.5°C: an IPCC special report on the impacts of global warming of 1.5°C above pre-industrial levels and related global greenhouse gas emission pathways, in the context of strengthening the global response to the threat of climate.

Kastner, I., Stern, P.C., 2015. Examining the decision-making processed behind household energy investments: a review. Energy Res. Soc. Sci. 10, 72–89.

National Research Council (NRC), 1996. Understanding Risk: In forming Decisions in a Democtatic Society. National Academy Press, Washington DC.

National Research Council (NRC), 2008. Public Participation in Environmental .... National Academy Press, Washington DC.

Stern, P.C., 1993. A second environmental science: Human-environment interactions. Science 260, 1897–1899.

Stern, P.C., Janda, K.B., Brown, M.A., Steg, L., Vine, E.L., Lutzenhiser, L., 2016. Opportunities and insights for reducing fossil fuel consumption by individuals and organizations. Nat. Energy 1, 16043.

Wolske, K.S., Stern, P.C., 2018. Contributions of psychology to limiting climate change: Opportunities through consumer behavior. In: Clayton, S., Manning, C. (Eds.), Psychology and Climate Change: Human Perceptions, Impacts, and Responses. Elsevier, Amsterdam, pp. 127–160.

Energy and behaviour: Challenges of a low-carbon future

1

Marta Lopes[a,b], Carlos Henggeler Antunes[b], Kathryn B. Janda[c]

[a]Polytechnic Institute of Coimbra—ESAC, Coimbra, Portugal
[b]INESC Coimbra, DEEC, University of Coimbra, Coimbra, Portugal
[c]Energy Institute, Bartlett School of Environment Energy and Resources,
University College London, London, United Kingdom

1 Introduction

A global effort is currently underway to combat climate change and to adopt actions for a sustainable low-carbon future. Under the United Nations Framework Convention on Climate Change (UNFCCC), the 2015 Paris Agreement has set a goal to limit the rise in average global temperatures to 1.5°C above preindustrial levels (UNFCCC, 2019). All over the world, governments, national authorities, and other stakeholders from civil society and the private sector are being called to implement measures to limit greenhouse gas (GHG) emissions and adapt to climate change.

The Intergovernmental Panel on Climate Change (IPCC) has recognised that anthropogenic GHG emissions are mainly driven by the population size, economic activity, human lifestyle, energy use, land use patterns, technology, and climate policy (IPCC, 2014). Many measures to limit GHG relate to the energy system, which can be viewed at different scales from global to local. These measures include efforts to change both how societies *produce* energy (called the 'supply side') and how societies *use* energy ('demand side'). Most future projections focusing on energy systems forecast that roughly half of the low-carbon solution will come from decarbonising energy supply and half can be created by reducing energy demand through a combination of energy efficiency and behaviour change (OECD/IEA, 2018).

Energy efficiency is often construed as using better technology to deliver the same level of energy service with less primary energy. It is thought to be easy to quantify, and there are a number of studies outlining the energy efficiency 'gap' between what is and what could be (Hirst and Brown, 1990; Dietz, 2010). Behaviour change, however, is tricky. What a 'better' behaviour is depends on many factors,

Energy and Behaviour. https://doi.org/10.1016/B978-0-12-818567-4.00030-2

including whether one is trying to avoid waste through 'conservation' (c.f. Black et al. (1985), Gardner and Stern (2002), Karlin et al. (2014), Peters (2019)), define appropriate limits through 'sufficiency' (Princen, 2005; Darby and Fawcett, 2018; Bertoldi, 2019), or pursue fairness and equity for vulnerable socioeconomic groups (Hall et al., 2013).

Europe has committed to lead on global climate action, and the European Commission adopted a strategic vision to achieve net-zero GHG emissions by 2050 (EC, 2018). All economic sectors have a significant role to play in the energy transition, and the European strategy recognises and strengthens the central role of engaging the public, as end users, consumers, and citizens (EC, 2018). The transition towards a decentralised power system based on renewables will require a smarter and more flexible energy system, built on end users' individual and collective involvement as energy consumers and producers (Geelen et al., 2013). In addition to being increasingly important in the energy supply system, people and organisations have always been and will continue to be vital players in energy demand, which is the *raison d'être* for this book. Ultimately the European strategy acknowledges that moving to a net-zero GHG society will require people to change their daily lives, the way they work, move, consume, and interact socially (EC, 2018).

In this book the authors investigate the evolving relationship between energy and behaviour to move towards a low-carbon future. However, as the title of this introduction emphasises, there are some significant challenges in moving a field of thought from one that focuses mainly on technological substitution to one that pursues an integrative approach to fundamental social change. For those readers with a quantitative orientation, we point to ecological economist Herman Daly's insight: what really counts is often not countable (Daly, 1991). We hope a book that acknowledges and explores this ongoing tension in the field and that will be an important resource for researchers, graduate students, practitioners, and policy-makers.

This introduction focuses the reader's attention on the moving target of what low-carbon energy behaviours are and could be, in not only European countries but also worldwide. It sets the scene for the book by providing a brief history of energy and behaviour research and defining what the editors hope to achieve in the book. Next, it delivers a guided tour for readers. Finally, it concludes with a call for action to move beyond the conventional boundaries of energy and behaviour research.

2 The role of energy and behaviour in moving towards a low-carbon future

To provide some background for the book, we tackle two important topics. First, we provide a brief history of the energy and behaviour literature, identifying the paths that have been travelled thus far and noting challenges that have yet to be addressed. Second, we outline the book's major contributions to this literature.

2.1 A brief history of energy and behaviour

Although people are central to the energy system, the idea of an 'energy behaviour' is still evolving and somewhat contested. How societies are motivated to use or conserve energy has been a topic addressed sporadically by social scientists for more than a century (see Rosa et al. (1988) for a review of the historical literature). A complete review of the energy and social science literature since its resurgence due to the oil crises of the 1970s is beyond the scope of this introduction. However, a recent meta-review of the literature from 1999 to 2013 shows that the field has dominantly focussed on the supply side (Sovacool, 2014).

When energy demand is investigated, a variety of research patterns emerge. Wilson et al. (2015), for example, investigated energy demand from four different viewpoints, including a physical, technical, and economic model (PTEM); an energy service approach; social practice theories; and sociotechnical transitions theory. These authors found that each approach provided a window onto a complex landscape, but none provided a complete explanation for historical trends nor a comprehensive basis for predicting the future. Together, however, the evidence shows that understanding energy demand is a sociotechnical problem rather than one that is either social or technical.

This introduction singles out two strands in the existing energy demand literature for further discussion: the physical-technical-economic model (Lutzenhiser, 1993) and individualisation (Maniates, 2002).

2.1.1 The PTEM and its limitations

According to Lutzenhiser (1993), energy efficiency research has long been dominated by the physical-technical-economic model (PTEM) that assumes people make economically rational decisions about the physical and technical environment they live in. Thus they will automatically invest in more efficient technologies when it is cost effective to do so, according to predefined (statistically characterised) patterns. Research in this vein assumes that people are fully rational utility maximisers, given their budget constraints (Lutzenhiser, 1992; Wilson and Dowlatabadi, 2007).

However, there is strong evidence of behavioural inconsistencies due to cognitive limitations and psychological biases that individuals face when making decisions, which are explored by behavioural economists (Mundaca et al., 2019) and psychologists. Psychology contributes to the understanding of personal determinants (e.g. intentions, attitudes, norms, beliefs, and values) and contextual influences that shape energy behaviours, facilitate or hinder the adoption of energy-efficient technologies, and predict people's acceptability of energy policies and infrastructural changes (Abrahamse and Schuitema, 2019). Beyond psychological factors, people may seek to produce and reproduce other qualities in the world around them such as comfort, cleanliness, or convenience (Shove, 2003). These qualities can be seen as social or cultural constructs, rather than attributes belonging to a particular individual or technology.

2.1.2 Individualisation and a counterpoint

Energy behaviour is often characterised as a set of individual actions that influence energy consumption and production (Lopes et al., 2015). For example, the International Energy Agency describes energy behaviour as '*all human actions that affect the way that fuels (electricity, gas, petroleum, coal, etc.) is utilised to achieve desired services, including the acquisition or disposal of energy-related technologies and materials, the ways in which these are used, and the mental processes that relate to these actions*' (IEA-DSM, 2015). In terms of energy consumption, for example, this may include people purchasing new energy-consuming goods and services; whether they repair, maintain, and improve energy-consuming devices (or don't); and how they use equipment, buildings, and transportation. Moreover, climate conscious consumer choices will drive the market towards products and services with low-carbon footprints (Stern et al., 2016).

Definitions focussed on individuals have been considered narrow, and a broad interest in the behaviour of actors in all segments of energy system is emerging (Stephenson et al., 2015). These include, but are not limited to, organisations, communities, and middle actors (Janda and Parag, 2013; Janda, 2014; Parag and Janda, 2014). Another counterpoint to individualisation is social practice theories. These hold that people have social practices that are not determined exclusively by individual choices (whether economic or noneconomic). Instead, social practices are framed by sociotechnical assemblages. In the case of cleanliness, for example, Shove (2003) has shown that social norms for showering have increased dramatically over the last few generations. This is not only the result of individual choice but also a confluence in the relationship between indoor plumbing, hot water provision, the number of bathrooms, and a host of social discourses about sanitation and social order, pleasure, duty, and nature.

The definition of an 'energy behaviour', therefore, depends on who is studying this phenomenon, how they study it, and what they hope to achieve. Much of the current literature defaults to assuming this term means individual actions in homes, leaving groups, and their dynamics in businesses or industry comparatively understudied (Moezzi and Janda, 2014; Stern et al., 2016). This book addresses individuals at home and also moves beyond them, as discussed in the succeeding text.

2.2 Contributions to the literature

Energy and Behaviour: Towards a Low Carbon Future presents an international and multifaceted approach to understanding the interface of people and energy, which involves both engaging people in the energy system and changing the energy system to better accommodate people. Contributors (1) introduce the major disciplinary and interdisciplinary approaches to this field; (2) provide cross-sectoral perspectives on the challenges in residential and nonresidential buildings, industry, transport, cities, and energy communities; (3) critically review evolving modelling approaches to address this sociotechnical challenge; and (4) address new areas where additional evidence is required for interventions and policy-making.

There are five ways that this book contributes to the literature and to the field more broadly.

First, it supports a broader definition of 'behaviour' than the individualised one. In this book, 'energy and behaviour' is broadly understood as the role of people, organisations, and technology in energy use. As the authors in Part I show, 'people' may be individuals, groups, or society.

Second, the book explicitly and deliberately seeks to expand the field beyond studies that tend to focus on people in their homes. Part 2 of the book contains six chapters that were commissioned to move beyond the usual focus on homes and homeowners to look at people and decision-making in nonresidential buildings, transportation systems, industrial organisations, and cities.

Third, the book includes an explicit focus on modelling in Part 3, including quantifying energy behaviour and decision modelling. The integration of quantitative approaches to energy behaviour is notable because other recent initiatives have fostered a more qualitative approach (e.g. Foulds and Robison, 2018; Hui et al., 2018; Fahy et al., 2019). We believe there is complementary value in modelling tools to gain insights about problems, design methodological approaches to derive effective solutions, and assess the upscaling of interventions.

Fourth, the book challenges the ways in which policies have been designed to be both generalisable and targeted to specific barriers. Authors in Part 4 of the book address these issues and begin to chart a way towards integrative policies that incorporate both behaviour change and technological change, rather than one or the other.

Finally, for readers new to the field, we have developed what we hope is a useful Appendix of resources, such as most relevant international conferences, programmes, and journals in the 'energy and behaviour' field. We provide this information both as a historical resource and one for future growth. From a historical perspective, we have included the initiation dates of various conferences and journals to show that much of the early energy and behaviour research appeared in conference proceedings (e.g. the American Council for an Energy-Efficient Economy summer study started in 1980). The proceedings are available online but not indexed on the Web of Science. In a world dominated by the Internet searches, we wish to remind readers that, prior to 2006, there was no journal specifically dedicated to energy efficiency and that the premier journal focussed broadly on energy and social science began only in 2014. The energy and behaviour field is still evolving, and we invite readers who are interested in its development to visit the Appendix. We hope the Appendix conveys a further sense of the depth and breadth of this landscape and will help readers navigate their contributions to its evolution.

3 A guided tour to the book

This book is structured in four parts, plus an Appendix. **Part I** introduces disciplinary approaches on energy and behaviour with views from psychology, behavioural economics, and the social sciences. It provides an introduction to the study of energy and behaviour, as well as showing there is not a single accepted approach to this topic. **Part II** explores different aspects of energy and behaviour research across different energy end-use sectors ranging from households, nondomestic buildings, industry, transport, and smart cities. Because it is important to recognise that energy and behaviour research can be quantitative, qualitative, or mixed methods, **Part III** presents an overview of several quantitative approaches to energy and behaviour, including modelling. **Part IV** reflects on opportunities and challenges to implementing energy and behaviour change in the real-world including policies and programmes.

3.1 Part I—Understanding energy behaviour: Disciplinary approaches and beyond

Part I contains three chapters, each of which takes a different theoretical approach to energy and behaviour. It starts with a psychological approach from Abrahamse and Schuitema (2019) and then moves to Mundaca et al. (2019)'s economics-based assessment of behavioural strategies. It concludes with an interdisciplinary call to action from Moezzi and Lutzenhiser (2019), which expands the frame beyond disciplines focussed on individual actions (psychology and economics) to a broader sociotechnical spectrum, looking at the relationship between technologies and society at different levels. Even though they are not located in this section of the book, other chapters also consider individual behaviours in the context with other institutional factors. These include Figueroa and Lah (2019), Mourik et al. (2019), Darby (2019), Oliveira et al. (2019) and Heiskanen et al. (2019).

Abrahamse and Schuitema (2019) provide an overview of research from psychology on household energy consumption. Their chapter describes the motivations for people to save energy at home and the main drivers of adoption rates of new energy-saving technologies. Then, it explores how to evaluate the effectiveness of interventions to encourage energy conservation behaviours and discusses research on the acceptability of energy policies and infrastructural changes.

Mundaca et al. (2019) review, analyse, and discuss various policies and research encompassing economic and nonmonetary factors affecting energy use and decarbonisation efforts from a household consumption-based perspective in Scandinavia. The chapter introduces behavioural economics and its applications in this context, critically analyses policy mixes targeting sustainable energy use and decarbonisation in Scandinavia, presents case studies that investigate economic and noneconomic factors affecting energy use and decarbonisation activities, and derives policy implications.

Moezzi and Lutzenhiser (2019) provide a counterpoint to the first two chapters. They discuss the limitations of the individual behaviour frame for multidisciplinary work and point to the need to see intersections of people, energy, and technology at multiple levels, especially given the scale and scope of climate policy goals. They remind us that people do not only purchase, use, and maintain technologies but also '*design, produce, sell, maintain, regulate, and collectively shape technologies*'.

The chapter proposes developing more formal interdisciplinary methods for energy research and development, especially for engineers and social scientists working together.

Questions we hope this section will raise for the reader is what is energy behaviour? Who is 'behaving' and why? Finally, where are the actions occurring? Abrahamse and Schuitema (2019) and Mundaca et al. (2019) both focus their research on households rather than business, industry, or transport. Moezzi and Lutzenhiser (2019)'s approach could be applied to households, businesses, transport, or industry. In the next section, we address places of energy consumption that go beyond households.

3.2 Part II—Energy behaviour across sectors

As noted earlier, much of the traditional energy and behaviour research has directed towards individual people at home (Moezzi and Janda, 2014; Janda, 2014; Stern et al., 2016). This section contains six chapters that stress the multiple locations where energy and behaviour matter, not only in housing but also in nondomestic buildings (Bernardo and Martins (2019); Arning et al. (2019)), industry (Palm and Thollander, 2019), transportation (Figueroa and Lah, 2019), community level energy production (Mourik et al., 2019), and smart cities (Darby, 2019).

Bernardo and Martins (2019) discuss the intertwined relation of behaviours and technology in nondomestic buildings. The chapter provides a broad context from the perspective of buildings' life cycle, explores the behavioural issues affecting the buildings' energy performance, and discusses the role of building automation and control systems and also the design requirements of human-technology interaction. It concludes that interdisciplinary cooperation among members of research teams is required for energy-efficient building design, coping simultaneously with technological and behavioural issues and their interrelations and privileging occupant-centred approaches.

Arning et al. (2019) present an empirically based decision-making process model of energy-efficient refurbishment investments, which compares the decision process of investors in residential and in nonresidential buildings. The chapter highlights the impact of intermediaries and their role in renovation decisions concerning (non)residential buildings to identify the most promising levers for policies supporting a more rapid transition towards a sustainable building stock. Recommendations are derived about how to promote the diffusion of low-carbon energy-efficient refurbishment measures in the building sector.

Palm and Thollander (2019) outline the major principles for energy efficiency and energy management in industry and discuss how to develop the energy management perspective with theories from the social sciences focusing on how situated actions, tacit knowledge, and social networks can influence decision-making. The aim is to develop new insights contributing to existing research on energy efficiency in industry exploiting the synergies between different fields, which can also be used by practitioners. The chapter proposes an energy management model, which should be adjusted to the local context, focusing on the importance of knowledge and communication, offering guidelines on how to implement sustainable values, routines, and behaviours in a way that energy efficiency becomes embedded in industrial organisations.

Figueroa and Lah (2019) provide an overview on the policies and actions that can facilitate a collective transition and supportive human response to the adoption of changes towards a decarbonized transport sector. The chapter discusses the increasing energy demand for transport, reviews the opportunities and cobenefits that facilitate transport decarbonisation, outlines changing conditions affecting human dimensions and energy for transport, discusses synergies between actors and collective community approaches and the role these may play to facilitate scaling up policies and solutions, and derives recommendations.

Mourik et al. (2019) investigate how institutional arrangements at EU and national and local levels influence business models for community energy projects. The chapter presents community energy projects, such as community virtual power plants, discussing the underlying community logic, providing case studies, and exploring how energy project business models can be community centred. It further examines how existing policy, regulation, and ways of doing influence these energy community business models and how they may influence the energy system restructuring in the long run.

Darby (2019) discusses the challenges brought by the spread of ICT into everyday city life. The chapter presents the 'smart city' concept and reviews relevant lessons from energy studies, regarding not only opportunities but also challenges associated with affordable energy services, security and privacy, control, accountability, and resilience. Questions to research and to smart city initiatives are derived.

3.3 Part III—Modelling energy behaviour

The third part of this book contains four chapters focussed on various aspects of quantitative energy and behaviour research, which this book presents as complementary to the qualitative approaches taken by other authors in this volume and elsewhere (e.g. Foulds and Robison (2018), Hui et al. (2018), Fahy et al. (2019)). Grunewald and Diakonova (2019) use mixed methods (high-resolution electricity use patterns coupled with time-use information) to understand household demand in a social context; Hong et al. (2019) demonstrate several modelling tools and datasets to explore the influence of occupants in energy performance of buildings, Chappin et al. (2019) use agent-based tools to model the social dynamics of energy end use, and Oliveira et al. (2019) discuss different preference elicitation approaches to support energy-related decisions.

Grunewald and Diakonova (2019) expand the focus of household analysis from mere electricity use patterns to include underlying social activity patterns, since units of energy and money alone are insufficient to understand the dynamics of energy demand. The chapter presents a new type of time-use data collection method (interactive app rather than paper diary) to understand the relationship between activities and their electricity footprint, which can help to develop more effective means to engage energy users and to reshape their demand. Reported enjoyment of activities offer insights into strategies to reshape demand. The authors present electricity consumption patterns for the most common activities and how these differ at times of electricity system peaks.

Hong et al. (2019) introduce state-of-the-art methods, tools, and datasets for quantifying occupant impacts on building energy use and occupant comfort. The chapter offers an overview of how occupants can influence building environments and energy performance and highlights gaps in the abilities of building energy simulation programmes to represent these influences. State-of-the-art methods and modelling tools that enable more sophisticated occupant behaviour simulation are reviewed, along with the most prominent datasets available to support quantitative behaviour model development. An overview of application areas for occupant behaviour modelling tools and datasets across the building life cycle is presented, with case studies. The chapter identifies opportunities and challenges associated with occupant behaviour simulation to support the design and operation of low-energy buildings that foster greater occupant satisfaction.

Chappin et al. (2019) review the opportunities and challenges offered by agent-based model (ABM) approaches for understanding the social dynamics of energy end use. The chapter discusses the ability of ABM to capture heterogeneous, boundedly rational, and imperfectly informed behavioural factors at the individual, household or businesses, and neighbourhood levels, including the use of microgeneration and local storage. Three archetypal case studies of ABM are reviewed: energy efficiency in domestic heating, electric vehicle adoption, and energy management in smart grids. The importance of ABM in the context of multidisciplinary studies of energy behaviours is emphasised.

From a decision science perspective, Oliveira et al. (2019) present three preference elicitation approaches used in energy decision contexts: multicriteria decision analysis (MCDA), conjoint analysis, and problem structuring methods. The potential of these approaches for structuring and supporting energy-related decisions is illustrated through several applications, namely, the assessment of initiatives to promote energy efficiency using a combination of soft systems methodology (a problem structuring method) and ELECTRE TRI (an MCDA method based on the exploitation of an outranking relation devoted to the sorting problem). Other applications reviewed include the evaluation of policies to foster the development of smart grids in Brazil, using the Delphi method and ELECTRE TRI, and the modelling of consumer preferences for electric vehicles, using choice-based conjoint analysis.

3.4 Part IV—Promoting behaviour change

The fourth part of this book reflects on what it takes to change energy behaviours to promote a more sustainable world. It contains four chapters. Heiskanen et al. (2019) review behaviour change interventions highlighting the confounding influence of context to generalise behaviour change efforts, Reusswig et al. (2019) present results from a behaviour change intervention in Berlin, Bertoldi (2019) provides an overview of the market barriers to energy efficiency and European Union policies designed to overcome them, and Peters (2019) presents a historically grounded overview of US energy behaviour programmes.

Heiskanen et al. (2019) review behaviour change interventions aiming to support energy conservation and evaluate how context is taken into account in the most popular intervention types, such as convenience, information, feedback, and social influence. The authors also explore how context, namely, organisational, geographical, and practice, is likely to influence intervention outcomes. Based on this exploration the authors suggest how to better address contextual factors in sustainable energy interventions.

Reusswig et al. (2019) explore the interplay between consumers and citizens in the transition to low-carbon cities. The chapter reports the results of a 1-year urban real-world lab experiment in the capital of Germany, Berlin, with households reducing their personal carbon footprints by using various types of feedback, including a weekly carbon tracker. The authors consider the ethical and political dimensions of the experiment and make policy recommendations.

Bertoldi (2019) provides a summary of the main energy-efficient barriers and discusses policies to address energy efficiency and energy conservation, such as energy and carbon taxes, personal carbon allowances, and energy-saving feed-in tariffs. European Union policies for sustainable behaviours in energy end use are presented and conclusions are provided.

Peters (2019) provides an overview of the US energy programme landscape outlining the role of government, states, utilities, and private sector market actors in seeking to influence behaviour change by end users. The chapter discusses how behaviour has been integrated into the three most common types of energy programmes offered to end users in the United States: energy efficiency, demand management and pricing, and market transformation programmes. She concludes with recommendations for the future of behaviour change in energy efficiency programmes in the United States.

4 Conclusions and next steps

This book shows that energy behaviour can be influenced at both the individual and societal levels, but effectively addressing their influence on the energy system requires making the underlying theoretical frameworks explicit (MEN, IPEEC, and IEA, 2018). The contributions in this book show that different disciplinary fields address energy behaviour differently, through approaches that are sometimes complementary and sometimes competing (Lopes et al., 2012; Moezzi and Janda, 2014; Moezzi and Lutzenhiser, 2019).

Economics and psychology share a common focus on individual choices and provide policy prescriptions seeking to influence those choices (Sorrell, 2015). Going beyond the individual perspective of energy behaviours, social sciences such as sociology and anthropology argue that energy is a means to provide useful services that enable normal and socially acceptable activities to be carried out as part of the daily life (Wilhite, 2008; Strengers, 2012). Hence, energy demand is not a consequence of individual decisions but a reflection of the social organisation in which rules, practices, and routines are embedded (Moezzi and Lutzenhiser, 2010; Shove and Walker, 2010).

Many policy strategies for promoting energy efficiency and reducing energy demand have focussed on technology development, regulation, financial incentives, and information provision (Stern et al., 2016), which are strongly influenced by the PTEM framework and an individual perspective. Decades of behaviour change interventions in Europe have not yet succeeded in achieving long-term and significant improvements in energy efficiency and energy demand reduction (Gynther et al., 2011; EEA, 2013). In fact, emerging research is disclosing that, when behavioural interventions are focussed narrowly on individual actions, the effects are uncertain and only deliver marginal short-term benefits (Mundaca et al., 2019). When aiming to reduce energy demand and GHG emissions, the most promising actions are those having higher impacts when considering both technical potential (i.e. the amount of reduction) and behavioural plasticity (i.e. capability of delivering effective behaviour change) (Stern et al., 2016).

Going through this book, relevant conclusions can be drawn on the interplay between energy and behaviours towards a low-carbon future, which should inform future work in this field. These include but are not limited to the following:

- the need to move beyond the usual scale of disciplinary problem solving and redesign the sociotechnical energy system (rather than redesigning individual technologies or expecting to change people's behaviours);
- the importance of interdisciplinary work and close cooperation between all stakeholders (e.g. researchers, policy-makers, industry, businesses, middle actors, interest groups, and other organisations). This should be developed in real-world practice rather than idealised in theory;
- the value of modelling tools to gain insights about problems, design methodological approaches to derive effective solutions, and assess the upscaling of interventions;
- the opportunity for reengineering regulations and integrating policies while maximising cobenefits to society, leveraging citizen participation, and fostering interactivity.

Transitioning to a low-carbon energy system is not simply a matter of changing fuel sources. It implies profound and large-scale transformations in the sociotechnical energy system and how people fit in and interact with it and each other (Sorrell, 2015; Darby, 2019). Thus finding alternative paths will require innovative interdisciplinary work, at all scales, bringing together engineering, economics, environmental, social, and political sciences (Stern et al., 2016; Darby, 2019; Abrahamse and Schuitema, 2019). More than a challenge that lies ahead, interdisciplinary work for a more sustainable low-carbon future offers the opportunity to design more participative and effective development strategies on a global scale. This book is a contribution to better understanding each other's perspectives to reach this end. Our hope is that the book as whole will serve both as a primer for new entrants in the energy efficiency and behaviour field and a critical resource for researchers, postgraduate students, practitioners, and policy-makers.

Acknowledgements

We would like to thank all the people who helped to make this edited book possible, in particular all the authors for their contributions, Inês F. Reis for her assistance during the revision process, and the supportive team at Elsevier. We also would like to acknowledge the support of the project grants UID/Multi/00308/2019, ESGRIDS (POCI-01-0145-FEDER-016434), MAnAGER (POCI-01-0145-FEDER-028040), and Learn2Behave (02/SAICT/2016-023651), from the Energy for Sustainability Initiative of the University of Coimbra and from the Bartlett School of Environment Energy and Resources at University College London.

References

Abrahamse, W., Schuitema, G., 2019. Psychology and energy conservation: Contributions from theory and practice. In: Lopes, M., Antunes, C.H., Janda, K.B. (Eds.), Energy and Behaviour: Towards a Low-Carbon Future. Academic Press, Elsevier.

Arning, K., Dütschke, E., Globisch, J., Zaunbrecher, B., 2019. The challenge of improving energy efficiency in the building sector—taking an in-depth look at decision-making on investments in energy-efficient refurbishments. In: Lopes, M., Antunes, C.H., Janda, K.B. (Eds.), Energy and Behaviour: Towards a Low-Carbon Future. Academic Press, Elsevier.

Bernardo, H., Martins, A.G., 2019. Resource-efficient non-domestic buildings: Intertwining Behaviours and technology. In: Lopes, M., Antunes, C.H., Janda, K.B. (Eds.), Energy and Behaviour: Towards a Low-Carbon Future. Academic Press, Elsevier.

Bertoldi, P., 2019. Overview of the European Union policies to promote more sustainable Behaviours in energy end-users. In: Lopes, M., Antunes, C.H., Janda, K.B. (Eds.), Energy and Behaviour: Towards a Low-Carbon Future. Academic Press, Elsevier.

Black, J.S., Stern, P.C., Elworth, J.T., 1985. Personal and contextual influences on household energy adaptations. Appl. Psychol. 70, 3–21.

Chappin, É.J.L., Nikolic, I., Yorke-Smith, N., 2019. Agent-based modelling of the social dynamics of energy end-use. In: Lopes, M., Antunes, C.H., Janda, K.B. (Eds.), Energy and Behaviour: Towards a Low-Carbon Future. Academic Press, Elsevier.

Daly, H.E., 1991. Steady-State Economics: Second Edition with New Essays. Island Press, Washington DC.

Darby, S.J., 2019. Energy, human activity and knowledge: Addressing Smart City challenges. In: Lopes, M., Antunes, C.H., Janda, K.B. (Eds.), Energy and Behaviour: Towards a Low-Carbon Future. Academic Press, Elsevier.

Darby, S., Fawcett, T., 2018. Energy Sufficiency: An Introduction. Concept Paper, Eceee European Council For An Energy Efficient Economy.

Dietz, T., 2010. Narrowing the us energy efficiency gap. Proc. Natl. Acad. Sci. 107, 16007–16008.

EC, 2018. A clean planet for all—a European strategic long-term vision for a prosperous, modern, competitive and climate neutral economy. Com(2018) 773—communication from the commission to the European Parliament, the European council, the council. In: The European Economic and Social Committee. European Commission, The Committee Of The Regions And The European Investment Bank Brussels.

EEA, 2013. Achieving Energy Efficiency through Behaviour Change: What Does It Take? European Environment Agency, Copenhagen.

Fahy, F., Goggins, G., Jensen, C. (Eds.), 2019. Energy Demand Challenges in Europe— Implications for Policy, Planning and Practice. Palgrave Pivot.

Figueroa, M.J., Lah, O., 2019. What do we know about the role the human dimension plays in shaping a sustainable low-carbon transport transition? In: Lopes, M., Antunes, C.H., Janda, K.B. (Eds.), Energy and Behaviour: Towards a Low-Carbon Future. Academic Press, Elsevier.

Foulds, C., Robison, R. (Eds.), 2018. Advancing Energy Policy—Lessons On The Integration Of Social Sciences And Humanities. Palgrave Pivot.

Gardner, G.T., Stern, P.C., 2002. Environmental Problems and Human Behavior. Pearson Learning Solutions, Boston.

Geelen, D., Reinders, A., Keyson, D., 2013. Empowering the end-user in smart grids: Recommendations for the design of products and services. Energy Policy 61, 151–161.

Grunewald, P., Diakonova, M., 2019. Energy and enjoyment—The value of household electricity consumption. In: Lopes, M., Antunes, C.H., Janda, K.B. (Eds.), Energy and Behaviour: Towards a Low-Carbon Future. Academic Press, Elsevier.

Gynther, L., Mikkonen, I., Smits, A., 2011. Evaluation of European energy behavioural change Programmes. Energy Efficiency, 1–16.

Hall, S.M., Hards, S., Bulkeley, H., 2013. New approaches to energy: Equity, justice and vulnerability. Introduction to the special issue. Local Environ. 18, 413–421.

Heiskanen, E., Matschoss, K., Laakso, S., Apajalahti, E.-L., 2019. A critical review of energy behaviour change: The influence of context. In: Lopes, M., Antunes, C.H., Janda, K.B. (Eds.), Energy and Behaviour: Towards a Low-Carbon Future. Academic Press, Elsevier.

Hirst, E., Brown, M., 1990. Closing the efficiency gap: Barriers to the efficient use of energy. Resour. Conserv. Recycl. 3, 267–281.

Hong, T., Langevin, J., Luo, N., Sun, K., 2019. Developing quantitative insights on building occupant behavior: Supporting modeling tools and datasets. In: Lopes, M., Antunes, C.H., Janda, K.B. (Eds.), Energy and Behaviour: Towards a Low-Carbon Future. Academic Press, Elsevier.

Hui, A., Day, R., Walker, G. (Eds.), 2018. Demanding Energy—Space, Time and Change. Palgrave Macmillan.

IEA-DSM, 2015. Task 24 phase I—Closing the loop: Behaviour change in Dsm—From theory to practice. International Energy Agency—Demand Side Management—Energy Efficiency Task 24.

IPCC 2014. Climate Change 2014—Ar5 Synthesis Report—Summary For Policymakers.

Janda, K.B., 2014. Building communities and social potential: Between and beyond organizations and individuals in commercial properties. Energy Policy 67, 48–55.

Janda, K.B., Parag, Y., 2013. A middle-out approach for improving energy performance in buildings. Building Research & Information 41, 39–50.

Karlin, B., Davis, N., Sanguinetti, A., Gamble, K., Kirkby, D., Stokols, D., 2014. Dimensions of conservation: Exploring differences among energy behaviors. Environ. Behav. 46, 423–452.

Lopes, M.A.R., Antunes, C.H., Martins, N., 2012. Energy Behaviours as promoters of energy efficiency: A 21st century review. Renew. Sust. Energ. Rev. 16, 4095–4104.

Lopes, M.A.R., Antunes, C.H., Martins, N., 2015. Towards more effective Behavioural energy policy: An integrative modelling approach to residential energy consumption in Europe. Energy Res. Soc. Sci. 7, 84–98.

Lutzenhiser, L., 1992. A cultural model of household energy consumption. Energy 17, 47–60.

Lutzenhiser, L., 1993. Social and behavioral aspects of energy use. Annu. Rev. Energy Environ. 18, 247–289.

Maniates, M., 2002. Individualization: Plant a tree, buy a bike, save the world? In: Princen, T., Maniates, M., Conca, K. (Eds.), Confronting Consumption. MIT Press, Cambridge, MA.

MEN, IPEEC & IEA, 2018. Behaviour Change for Energy Efficiency: Opportunities for International Cooperation in the G20 and beyond. Argentinian Ministry of Energy, International Partnership For Energy Efficiency Cooperation, International Energy Agency.

Moezzi, M., Janda, K.B., 2014. From "if only" to "social potential" in schemes to reduce building energy use. Energy Res. Soc. Sci. 1, 30–40.

Moezzi, M., Lutzenhiser, L., 2019. Beyond energy behaviour: A broader way to see people for climate change technology planning. In: Lopes, M., Antunes, C.H., Janda, K.B. (Eds.), Energy and Behaviour: Towards a Low-Carbon Future. Academic Press, Elsevier.

Moezzi, M. & Lutzenhiser, L. What's missing in theories of the residential energy user. In: ACEEE, Ed. 2010 ACEEE Summer Study On Energy Efficiency In Buildings—The Climate For Efficiency Is Now, 2010 Pacific Grove, CA.

Mourik, R.M., Breukers, S., Summeren, L.F.M.V., Wieczorek, A.J., 2019. The impact of the institutional context on the potential contribution of new business models to Democratising the energy system. In: Lopes, M., Antunes, C.H., Janda, K.B. (Eds.), Energy and Behaviour: Towards a Low-Carbon Future. Academic Press, Elsevier.

Mundaca, L., Samahita, M., Sonnenschein, J., Seidl, R., 2019. Behavioural economics for energy and climate change policies and the transition to a sustainable energy use—a Scandinavian perspective. In: Lopes, M., Antunes, C.H., Janda, K.B. (Eds.), Energy and Behaviour: Towards a Low-Carbon Future. Academic Press, Elsevier.

OECD/IEA, 2018. Energy Efficiency 2018—Analysis and Outlooks to 2040. International Energy Agency, p. 2018.

Oliveira, G., Dias, L.C., Neves, L., 2019. Preference elicitation approaches for energy decisions. In: Lopes, M., Antunes, C.H., Janda, K.B. (Eds.), Energy and Behaviour: Towards a Low-Carbon Future. Academic Press, Elsevier.

Palm, J., Thollander, P., 2019. Reframing energy efficiency in industry—A discussion of definitions, rationales and management practices. In: Lopes, M., Antunes, C.H., Janda, K.B. (Eds.), Energy and Behaviour: Towards a Low-Carbon Future. Academic Press, Elsevier.

Parag, Y., Janda, K.B., 2014. More than filler: Middle actors and socio-technical change in the energy system from the "middle-out". Energy Res. Soc. Sci. 3, 102–112.

Peters, J., 2019. A brief history of behaviour in United States energy Programmes: Landscape, integration, and future opportunities. In: Lopes, M., Antunes, C.H., Janda, K.B. (Eds.), Energy and Behaviour: Towards a Low-Carbon Future. Academic Press, Elsevier.

Princen, T., 2005. The Logic of Sufficiency. The MIT Press, Cambridge, MA.

Reusswig, F., Lass, W., Bock, S., 2019. Urban low-carbon futures: Results from real-world lab experiment in Berlin. In: Lopes, M., Antunes, C.H., Janda, K.B. (Eds.), Energy and Behaviour: Towards a Low-Carbon Future. Academic Press, Elsevier.

Rosa, E.A., Machlis, G.E., Keating, K.M., 1988. Energy and society. Annu. Rev. Sociol. 14, 149–172.

Shove, E., 2003. Comfort, Cleanliness & Convenience. Berg, Oxford, New York.

Shove, E., Walker, G., 2010. Governing transitions in the sustainability of everyday life. Res. Policy 39, 471–476.

Sorrell, S., 2015. Reducing energy demand: A review of issues, challenges and approaches. Renew. Sust. Energ. Rev. 47, 74–82.

Sovacool, B.K., 2014. What are we talking about? Analyzing fifteen years of energy scholarship and proposing a social science research agenda. Energy Res. Soc. Sci. 1, 1–29.

Stephenson, J., Barton, B., Carrington, G., Doering, A., Ford, R., Hopkins, D., Lawson, R., Mccarthy, A., Rees, D., Scott, M., Thorsnes, P., Walton, S., Williams, J., Wooliscroft, B., 2015. The energy cultures framework: Exploring the role of norms, practices and material culture in shaping energy behaviour in New Zealand. Energy Res. Soc. Sci. 7, 117–123.

Stern, P.C., Janda, K.B., Brown, M.A., Steg, L., Vine, E.L., Lutzenhiser, L., 2016. Opportunities and insights for reducing fossil fuel consumption by households and organizations. Nat. Energy 1, 16043.

Strengers, Y., 2012. Peak electricity demand and social practice theories: Reframing the role of change agents in the energy sector. Energy Policy 44, 226–234.

UNFCCC, 2019. Understanding the un climate change regime. In: United Nations Framework Convention on Climate Change.

Wilhite, H., 2008. New thinking on the agentive relationship between end-use technologies and energy-using practices. Energy Efficiency 1, 121–130.

Wilson, C., Dowlatabadi, H., 2007. Models of decision making and residential energy use. Annu. Rev. Environ. Resour. 32, 169–203.

Wilson, C., Janda, K.B., Bartiaux, F., & Moezzi, M., 2015. Technical, economic, social, and cultural perspectives on energy demand. In: Ekins, P., Bradshaw, M., & Watson, J. (Eds.), Oxford Scholarship Online.

Understanding energy behaviour: Disciplinary approaches and beyond

Psychology and energy conservation: Contributions from theory and practice 1.1

Wokje Abrahamse[a], Geertje Schuitema[b]
[a]Victoria University of Wellington, Wellington, New Zealand
[b]University College Dublin, Dublin, Ireland

1 Introduction: What can psychology contribute to energy research?

Globally, domestic energy consumption accounts for a substantial proportion of greenhouse gas emissions, primarily through the combustion of fossil fuels. In a recent report, the Intergovernmental Panel on Climate Change (IPCC, 2018) argues that the mitigation of emissions requires the use of new technologies, clean energy sources, reduced deforestation, improved sustainable agricultural methods, and, importantly, changes in individual and collective behaviour. Many developed nations currently have plans to reduce carbon emissions from domestic energy consumption, referred to as the 'energy transition'. This transition includes a move from fossil fuel-based energy to renewable energy and also a reduction in overall energy consumption levels.

A transition to a low-carbon future implies a change in how individual households interact with the energy system. Households can make substantive contributions to reducing carbon emissions from domestic energy consumption via changes in individual behaviours and through the adoption of energy-efficient technologies, such as solar panels and heat pumps. In some cases, households have become producers and consumers of energy; they are referred to as energy prosumers (Toffler, 1980, p. 174). Households can produce energy via microgeneration technologies, which are small-scale systems that generate energy and/or heat for domestic usage (e.g. the installation of solar panels). Some consider energy storage (e.g. via electric vehicles) and the use of automated control systems (which help balance the supply and demand of electricity on the grid) as part of the role of prosumers too (e.g. Schuitema et al., 2017).

Households play an increasingly 'active' role in the energy system. In this chapter, we examine the interplay between individual households and the energy system (broadly defined). We use empirical research findings from psychology to illustrate the following questions:

(i) What motivates people to save energy at home?
(ii) What are the main drivers of adoption rates of new energy-saving technologies?

Energy and Behaviour. https://doi.org/10.1016/B978-0-12-818567-4.00001-6

(iii) How effective are behaviour change interventions to reduce energy consumption?

(iv) How acceptable are energy policies and infrastructural changes to people?

Interventions and policies that aim to substantially reduce carbon emissions from the domestic sector need to incorporate an understanding of individual motivations, perceptions, and behaviours. Psychological research can provide insights into factors that explain energy conservation behaviours, it can help improve the effectiveness of behaviour change interventions, and it can provide a better understanding of public support for (pricing) policies, regulations, technology, and infrastructural changes to facilitate the 'energy transition'. Collaborations between different academic disciplines are necessary to achieve this aim, and in this chapter, we elaborate on the role psychologists can play in such interdisciplinary research collaborations.

Our chapter is divided into six sections. In Section 1, we discuss the factors that can explain energy conservation behaviours, using insights from psychological theories and empirical research findings. In Section 2, we explore the question of how to evaluate the effectiveness of interventions to encourage energy conservation behaviours and then move on to discuss some of the most commonly used behaviour change interventions in this field. Sections 3 and 4 describe the main drivers of adoption rates of new energy-saving technologies and discuss which factors are related to the acceptability of energy policies and infrastructural changes. In Section 5, we argue for more interdisciplinary energy research, whereby research methods and theories used in psychology are complemented with those from other relevant disciplines, such as sociology, economics, building sciences, and environmental sciences. In Section 6, we provide some suggestions for future research in this area.

2 Explaining energy conservation behaviours

What motivates people to save energy at home? Judging by information campaigns that aim to promote energy conservation, the general assumption appears to be that people are mainly (and solely) motivated by money. Many energy conservation campaigns emphasise the monetary aspects of energy-saving actions and do not necessarily mention environmental benefits (De Groot and Steg, 2009). The findings from various psychological studies indicate that people are not always and not only motivated by price signals or financial motivations. Psychological research suggests that while financial motivations play an important role in encouraging people to save energy at home, it is by no means the only motivator (Abrahamse and Steg, 2009, 2011; Bolderdijk et al., 2013). Psychological energy research sometimes examines one or two motivation in isolation, or it uses theoretical frameworks, encompassing multiple variables and the relationships between these variables, to better understand what motivates people to save energy at home. In this section, we first describe the role of financial and environmental motivations and then move on to describe two theoretical frameworks that are applied often in the energy domain. The key concepts from these two theories are summarised in Table 1.

Table 1 Overview of the variables from the theory of planned behaviour and the value–belief–norm theory applied to energy conservation

	Examples	Key concepts	Description of concept	Sample survey statements used to measure concept
Theory of planned behaviour (TPB)	Abrahamse and Steg (2009), Chen, Xu, and Day (2017), and Pals and Singer (2015)	Intentions	People's intention to perform a given behaviour	'How likely are you to save energy at home in the next month?'
		Attitude towards the behaviour	People's positive or negative evaluation of engaging in a given behaviour	'I like the idea of conserving energy at home'
		Subjective norm	The perception that other people would (not) endorse a given behaviour	'Most of my neighbours are trying to save energy'
		Perceived behavioural control	Extent to which people believe they are capable of doing a given behaviour	'It is difficult for me to save energy at home'
Value–belief–norm theory	Abrahamse and Steg (2011) and Yeboah and Kaplowitz (2016)	Personal norms	A moral obligation to act	'I feel guilty when I waste a lot of energy'
		Ascription of responsibility	Extent to which people assume responsibility for energy-related problems	'I feel responsible for the environmental consequences of my energy consumption'
		Awareness of consequences	Extent to which people are aware that energy use contributes to environmental problems	'Saving energy helps to mitigate the effects of climate change'
		Environmental concern	Extent to which people are concerned about environmental problems	'The balance of nature is easily upset'
		Values	Guiding principles in people's lives	'It is important to me to care for nature'

In psychological research on energy conservation behaviours, a distinction is commonly made between efficiency behaviours and curtailment behaviours (Gardner and Stern, 2008). Efficiency behaviours refer to actions that result in less energy being used for a given level of energy service (e.g. replacing conventional light bulbs with energy-efficient lighting and putting in cavity wall insulation). Efficiency behaviours come at a financial cost as they involve a one-off purchase or installation (e.g. energy-efficient bulbs are more expensive than conventional bulbs; having insulation installed can be very expensive). Curtailment behaviours, on the other hand, are low-cost actions that need to be done regularly to achieve energy savings (e.g. lowering thermostat settings and switching off the lights) and can involve the loss of convenience and comfort (less heating and less lighting). Efficiency behaviours can help reduce environmental impacts more significantly compared with curtailment behaviours (Gardner and Stern, 2008).

Research findings indicate that efficiency behaviours and curtailment behaviours are associated with different motivations (e.g. Karlin et al., 2014; Lillemo, 2014). For example, a study with a broadly representative sample of households from the United States found that financial motivations were associated with efficiency behaviours, but not with curtailment behaviours (Karlin et al., 2014). In this study, households were asked to indicate whether they had ever implemented an energy efficiency measure, such as installing insulation or purchasing an energy-efficient appliance. Households were also asked how often ('never' to 'always') they did a range of curtailment behaviours, such as switching off lights when leaving a room. The survey asked people about their level of awareness of the cost of their energy bill and their levels of awareness of environmental problems (which could be considered proxies for financial and environmental motivations) and a range of sociodemographic characteristics, such as income and home ownership. The study findings indicate that people with higher incomes, people who owned their homes, and people with a higher awareness of the cost of their energy bill were significantly more likely to have implemented an efficiency behaviour. Environmental concern was unrelated to the uptake of efficiency behaviours. Conversely, people with higher levels of environmental concern were significantly more likely to have undertaken curtailment behaviours. Other work has also found that efficiency behaviours are related to financial considerations and that curtailment behaviours are related to environmental concerns (for an example from Sweden, see Lillemo, 2014).

Several scenario-based studies have examined the extent to which financial and environmental concerns could motivate changes in energy conservation intentions and behaviours. A study from the United States directly compared the effect of an energy conservation campaign that was framed in terms of financial benefits or environmental benefits to people's intentions to conserve energy in a hypothetical scenario (Schwartz et al., 2015). Participants in this study were randomly assigned to one of three groups and received a message that advertised a residential energy programme. One message emphasised only the financial benefits of joining this programme ('reduce your energy bill'), one message emphasised only the environmental benefits ('reduce your environmental impact'), and a third message emphasised both ('reduce your energy bill and your environmental impact'). The study findings indicate that people expressed a higher willingness to enrol in the energy conservation programme when they had been told about the environmental benefits of saving energy compared with people who read the 'financial' message, or the combined message. A message framed in terms of environmental benefits was associated with a significantly stronger willingness to enrol in the programme, suggesting that energy conservation campaigns might benefit from emphasising the environmental benefits of saving energy. This might be even more fruitful when a campaign aims to encourage the uptake of curtailment behaviours, as these behaviours tend to be motivated by environmental concerns.

Several researchers have used theoretical frameworks from psychology to examine the factors related to energy conservation intentions and behaviours. Here, we focus on two psychological theories commonly used in this domain: the theory of planned behaviour (TPB) and the value–belief–norm (VBN) theory. The theory of planned behaviour (Ajzen, 1991) provides a framework for understanding the determinants of specific behaviours. According to the TPB, behaviours are determined by behavioural intentions, and these intentions, in turn, are guided by attitudes towards the behaviour, subjective norms, and perceived behavioural control. Attitudes refer to a positive or negative evaluation of a specific behaviour (e.g. 'I like the idea of lowering the thermostat setting to conserve energy'). Subjective norms refer to the extent to which other people (would) endorse the behaviour (e.g. 'my friends would not approve it if I were to lower the thermostat setting'). Lastly, perceived behavioural control refers to the extent to which people feel able to do the behaviour (e.g. 'I'm liable to forget to lower the thermostat setting').

Researchers have found that the theory of planned behaviour is a useful framework for studying energy-saving intentions and behaviours. A study of Dutch households (Abrahamse and Steg, 2011) found that perceived behavioural control and attitudes towards energy conservation were positively associated with intentions to reduce energy consumption. People who felt more control over their ability to save energy and who expressed more positive attitudes towards energy conservation also expressed a stronger intention to conserve energy. A study from the United States (Pals and Singer, 2015) examined the determinants of several energy-related behaviours using the TPB, such as unplugging appliances when not in use (a curtailment behaviour) and installing new insulation (an efficiency behaviour). When sociodemographic variables were controlled for, perceived behavioural control and attitudes were the strongest predictors of intentions to reduce energy consumption. A study among lower income households in the United States (Chen et al., 2017) found that intentions to save energy were largely predicted by behavioural control and attitudes. These studies seem to indicate that perceived behavioural control and attitudes are predictive of intentions and/or behaviours, but that endorsement of energy conservation by other people (i.e. subjective norms) is not necessarily associated with intentions and behaviours. This could have something to do with the fact that many energy conservation behaviours may not be visible to important others (e.g. friends). In addition, the ways in which the concept of subjective norms is measured in surveys may not necessarily capture the complexities of inter-household dynamics and social influence.

A second theory that is applied in this area is the value–belief–norm theory (VBN; Stern, 2000). The value–belief–norm theory describes a causal chain of variables that ultimately predict behaviour. At the start of this chain of variables are values, which are generally described as guiding principles in people's lives (such as altruism or care for nature). Values are motivational goals that are assumed to guide behaviours across different situations. For example, people with strong environmental values are assumed to hold beliefs that are in line with these values. According to VBN, values are associated with general environmental concern. For example, people with stronger altruistic values are assumed to have higher levels of environmental concern, whereas people with stronger egoistic values are assumed to have lower levels of environmental concern. Environmental concern, in turn, predicts people's awareness of the consequences of their behaviour (e.g. that energy consumption contributes to climate change via the use of fossil fuels). Awareness of consequences, then, predicts the extent to which people feel responsible for these consequences. Levels of ascription of responsibility are associated with variable called personal norms, which reflects the extent to which people feel morally obliged to change their behaviour (or feel guilty when they do not change their behaviour). Personal norms, in turn, are predictive of behaviour. The VBN theory proposes that a general set of values informs behaviour-specific beliefs and norms, which then predict behaviour.

The VBN theory has also been applied in the field of domestic energy consumption, albeit it to a lesser degree than the TPB. Yeboah and Kaplowitz (2016) examined the relationships between the variables from the VBN and energy conservation behaviours in a sample of students, faculty, and staff from Michigan State University. They asked people to indicate how often (from 'never' to 'always') they did the following: turn off computers and printers overnight, unplug electrical appliances when not in use, turn off lights in unoccupied rooms, and purchase energy-efficient appliances. These four behaviours were then added to form a composite measure of energy conservation behaviours. They found that the variables from the VBN predicted engagement in energy conservation behaviour. Personal norms (the extent to which people felt morally obliged to conserve energy), ascription of responsibility (the extent to which people felt responsible for the problems associated with energy use), and awareness of consequences (the extent to which people thought their energy conservation could reduce the effect of climate change) were positively associated with engagement in energy conservation behaviours. Their analysis also suggests that people with stronger self-transcendence values (e.g. altruism and care for nature) were more likely to engage in energy conservation behaviours, whereas participants with stronger self-enhancement values (e.g. achievement, power) engaged in such behaviours less often. A qualitative study (Mirosa et al., 2013) of households in New Zealand found that values related to achievement (i.e. self-enhancement) and hedonism (pleasure) influenced how people used energy at home. The authors also note that the same value 'can promote energy-efficient behaviours or act as obstacles to change' (p. 455). Hedonism ('pleasure') values, for example, were associated not only with energy-efficient actions such as hanging laundry on the line but also to inefficient behaviours, like taking longer showers.

3 Design and evaluate interventions to change behaviour

Research on the effectiveness of interventions to encourage energy conservation in households is voluminous (see, e.g. Abrahamse et al., 2005; Delmas et al., 2013 for reviews). Michie et al. (2011) define behaviour change interventions as 'coordinated sets of activities designed to change specified behaviour patterns' (p. 42). For example, an energy conservation campaign that discusses energy-saving tips can be considered an intervention because it is designed to encourage the uptake of energy conservation behaviours. In this section, we first briefly describe three research designs that are used in intervention research and then move onto commonly used behaviour change intervention in energy research.

3.1 Design of intervention research

The effectiveness of behaviour change interventions is generally assessed via field experiments, whereby the prevalence of a particular behaviour is measured before and after an intervention and, ideally, compared with a control group. Different research designs are employed to examine the effect of behaviour change interventions.

A randomised experiment involves randomly assigning individual households (or people) to one or more treatment groups or a no-treatment control group. Random assignment helps ensure that there are no systematic differences between groups taking part in the study other than whether they receive an intervention or not. This allows researchers to be more confident in claiming causation (i.e. that the intervention resulted in any observed changes between groups and not some other factor).

An example of the use of a randomised experiment is provided by a study by Asensio and Delmas (2015). The researchers randomly assigned households to one of three groups. One treatment group received information and feedback about their energy consumption, and this was framed in terms of how much money a household had saved by conserving energy (or money they had spent by using more energy). A second treatment group of households received information and feedback about their energy consumption, which was framed in terms of public health benefits of saving energy. These households were told what they had achieved in terms of reductions (or increases) in harmful emissions (e.g. air pollution). A control group did not receive any intervention. The results indicate that relative to the control group, households who had received the public health feedback saved energy, whereas households in the money saving group slightly increased their energy consumption. Because the researchers used a randomised experiment and compared the effects with a control group, they can be reasonably confident that the changes between the groups are attributable to the intervention. The results indicate that the use of price signals (telling households they have saved money) is less effective in encouraging energy conservation compared with an appeal to altruistic motivations (public health).

In applied settings, the use of a randomised experiment is not always feasible because of certain constraints. For example, it would not be feasible to randomly assign households to different treatment groups in cases where households voluntarily sign up to receive an intervention (e.g. a home retrofit scheme). This is known as 'self-selection'. It would still be possible to recruit households to be part of a control group. This research design is referred to as a quasi-experimental design. Households are part of different groups (treatment vs control group), but assignment to these groups does not occur at random, which means some caution is warranted in claiming causation. For example, people who sign up to an intervention may already be more motivated to change their behaviour compared with people who did not sign up, which may affect the findings. In a quasiexperimental design, researchers can collect information from households about other variables of interest to examine any systematic differences between groups prior to the implementation of the intervention (e.g. income and environmental concern).

In other cases, it is not possible to include a control group of households. This occurs when in principle everybody is exposed to an intervention, such as a nation- or statewide information campaign to encourage the uptake of insulation (for an example, see Staats et al., 1996). If the aim is to examine the effect of such an intervention, it is possible to measure energy consumption behaviours (or opinions and beliefs) before the intervention and after the intervention. This will give insight into changes in behaviours and beliefs over time. But because there is no control group, it is more difficult to claim causation.

3.2 **Types of interventions**

A distinction can be made between informational and structural interventions (Steg and Vlek, 2009). Informational interventions aim to strengthen people's motivations to conserve energy through, for example, increasing knowledge or awareness, changing attitudes, or changing social norms. Commonly used informational approaches include the provision of information, the provision of feedback about energy savings, commitment making, and the use of social norms. Structural interventions change the context in which households make decisions, without necessarily changing people's motivations. Examples include the use of financial incentives and disincentives and the availability of efficient technologies. Here, we focus on two commonly used interventions in the energy use literature: information and feedback provision.

Two types of information are generally provided to households. General information, often conveyed via media campaigns, communicates the link between human behaviour and environmental problems. This can include, for example, the impacts of household energy consumption on carbon emissions through the combustion of fossil fuels. Specific information includes energy-saving tips about how households can reduce their energy consumption. The provision of information is assumed to increase people's knowledge or their awareness of energy-related problems—also known as the 'knowledge-deficit' approach (Moser and Dilling, 2011). But the provision of information is generally not very effective in encouraging energy conservation. While information provision can result in increased knowledge, it does not necessarily lead to concomitant changes in behaviour (Abrahamse et al., 2005; Schultz, 2014).

Several researchers have started to examine the role of social norms in encouraging energy conservation, in an effort to enhance the effectiveness of information provision. Social norms refer to what society expects of us. According to focus theory of normative conduct, put forward by Robert Cialdini and colleagues (e.g. Cialdini et al., 1990), social norms guide our behaviour when these norms are made salient. Information provision can make a descriptive social norm apparent, which refers to what other people are doing ('a majority of your neighbours are saving energy'); it can also make an injunctive social norm salient, which refers to what other people expect of us ('people in your neighbourhood approve of energy conservation'). The strength and appeal of this social norm approach are that it is grounded in theory. Many studies from the energy conservation field suggest that this theory-based intervention is successful in practice (e.g. Allcott, 2011; Allcott and Mullainathan, 2010; Nolan et al., 2008; Schultz et al., 2007).

In one of these studies, Californian households received a door hanger encouraging them to use fans instead of air conditioning to conserve energy (Nolan et al., 2008). The information on these door hangers was framed differently for different households. One group of households were asked to save energy because of the financial benefits ('save money by conserving energy'); money, as we saw previously, is an important driver of energy decisions at home. A second group of households received an environmental plea ('protect the environment by conserving energy'), also an important determinant of energy conservation behaviours. A third group received a message that appealed to people's sense of social responsibility ('do your part to conserve energy for future generations'). A fourth group received a social norm message. In this message, households were told that a majority of residents in their city were conserving energy ('join your neighbours in conserving energy'). A fifth group of households were the control group; they received neutral information ('you can save energy by using fans instead of air conditioning'). The results indicate that households in the social norm group significantly reduced their electricity consumption, compared with the other groups combined.

Feedback is another intervention that is used frequently in energy research. Feedback conveys to households how much gas and/or electricity they have saved and is often displayed in financial units (e.g. dollars saved) and/or energy units (e.g. kilowatt-hour). According to feedback intervention theory (Kluger and DeNisi, 1996), feedback can enhance an individual's motivation to perform well. A recent study reports the findings of two randomised control trials on the effectiveness of feedback provision on energy conservation (Ayres et al., 2013). In one of these trials, 85,000 customers of Californian utility company Sacramento Municipal Utility District (SMUD) took part, with SMUD providing electricity consumption data for these customers. Customers were randomly assigned to a treatment group or a control group. Households in the treatment group received a home energy report, which included feedback about their monthly electricity consumption compared with the same month of the previous year and a range of comparisons with similar neighbours (i.e. appealing to descriptive and injunctive norms). The energy report also included energy-saving tips. The results indicate that relative to the control group, households in the treatment group reduced electricity consumption by 1.2% and that this reduction was sustained over a period of 7 months.

Feedback appears to be more effective when it is given frequently and when it includes some sort of comparison (e.g. a historical comparison or a comparison with 'similar' neighbours). Reviews suggest feedback provision can effectively encourage households to conserve energy, with energy savings ranging between 5% and 15% (e.g. Abrahamse et al., 2005; Karlin et al., 2015). Other intervention research in the domain of domestic energy consumption includes the use of commitment making, goal setting, and nudging (see Box 1). For more extensive reviews of the intervention literature, see, for example, Abrahamse et al. (2005) and Delmas et al. (2013) (Table 2).

Box 1 The use of green defaults

Nudging, or choice architecture, is an increasingly popular intervention to change behaviour. A nudge involves changing the choice set that people have, without reducing or changing the number of choices. For example, nudges can imply changing the default, which is the option people receive unless they explicitly request something else (i.e. unless they opt out). In research by Pichert and Katsikopoulos (2008), nudging was used to encourage consumer uptake of green electricity. In one of their studies, a laboratory experiment, participants were asked to imagine they had moved house and were choosing a new electricity provider. The researchers used two scenarios. In one scenario, the 'green' utility company was presented as the default option, and in a second scenario, the 'grey' utility company was presented as the default. The findings indicate that people were more likely to choose the green utility company when it was presented as the default, compared with when the green utility company was not the default option.

Table 2 Examples of three interventions and their theoretical basis used in energy research

Intervention	Theoretical basis	It works how...	Example of studies
Information provision	Knowledge-deficit model	Provision of information fills gap in knowledge or changes attitudes, which in turn alters behaviour	Staats et al. (1996)
Social norm information and feedback	Focus theory of normative conduct	Making a social norm salient can encourage behaviour change, because people are guided by what other people do	Nolan et al. (2008) and Schultz et al. (2007)
Feedback provision	Feedback intervention theory	Provision of feedback enhances people's motivation to do well	Ayres et al. (2013) and Abrahamse et al. (2007)

4 Drivers of technology adoption

As explained earlier, efficiency behaviours refer to actions that result in less energy being used for a given level of energy service. This includes the adoption of energy-efficient technologies, such as energy-efficient (smart) appliances, smart meters, or heat pumps. But it also refers to the installation of microgeneration technologies, which are small-scale systems that can generate or store electricity. These small-scale generation systems are often based on photovoltaic (PV) rooftop panels, which can be used for self-consumption and charging batteries (e.g. in electric vehicles) or delivered to the grid. In some countries, households can receive financial compensation for delivering electricity to the grid, but this is not the case everywhere. As the adoption of efficiency behaviour can lead to significant reductions of the environmental impact of households, the questions that need to be addressed are as follows: when and why do people adopt such technologies? What are their motivations and concerns?

Two main motivations are important for the adoption of microgeneration and smart automation technologies, that is, financial and environmental motives (Gangale et al., 2013; Hargreaves et al., 2010; Jager, 2006). People indicate that they are willing to consider the adoption of technologies if they believe that they will financially benefit. As a result, technologies are often promoted as being financially attractive. However, as microgeneration technologies often also require large investments, it is questionable whether the investments outweigh the costs in a reasonable amount of time. Also, if there are no or low investment costs, as is, for example, the case with smart meters, raising expectations that there are financial benefits may backfire. People who expected financial gains from the installation of smart meters reacted disappointed or frustrated when they learned the financial benefits were very small (Hargreaves et al., 2010).

Environmental motivations are mentioned as frequently as financial motivations for the adoption of microgeneration technologies and smart technologies. For example, the option to analyse and monitor energy consumption patterns is an important motivation for people to adopt smart meters (Hargreaves et al., 2010). Also, people who are aware and concerned about their energy consumption are more likely to adopt microgeneration technologies. They see microgeneration technologies as a means to reduce their environmental impact. For example, Jager (2006) found that buyers of solar panels generally had a higher-than-average awareness of environmental problems. This is related to their values, that is, if people strongly endorse environmental values, they are more inclined to support renewable energy sources (Perlaviciute and Steg, 2015). This implies that if people are confronted with the negative consequences of their energy use, they are more likely to adopt microgeneration technologies and support policies that allow more renewable energy in general. For example, after the nuclear accident in Fukushima, public support for nuclear energy decreased significantly (Latré, Perko and Thijssen, 2017). This effect was larger in countries that are geographically close to Japan. Also, some scholars have linked the accident in Fukushima to increased political and public support for renewable energy sources in Germany (Renn and Marshall, 2016).

In addition to financial and environmental motives, more technology-specific motivations for adoption have also been examined. For example, the technology adoption model (TAM) is a theoretical framework that aims to explain the adoption of new technologies (Davis, 1989; Davis et al., 1989). According to this framework, the adoption of new technologies depends on two factors. Firstly, people are more likely to adopt new technologies when they believe that the technology will improve their performance and help them to achieve their goals (referred to as perceived usefulness). This is related to perceived benefits; if people believe technologies will benefit them, they are more likely to adopt them. This provides an explanation for the slow uptake of various microgeneration technologies, as the perceived benefits of many technologies are small for households, whereas the costs are usually relatively large.

Another theoretical framework that is used to explain the adoption of technologies is the diffusion of innovations framework proposed by Rogers (1995, first edition 1962). Rogers argues that the adoption of technologies spreads through social networks. People have different needs and expectations and vary in the level of risks they find acceptable. Also, they perceived the five dimensions of technologies differently. As a result, some people will adopt technologies in different stages of the diffusion process. 'Innovators' are the first to adopt technologies, willing to accept relatively large risks, followed by 'early adopters', who are typical block leaders and opinion leaders. Next, 'early majorities' follow, who are rarely leaders but adopt new technologies before the average person. Then 'late majority' and 'laggards' follow in the adoption chain, who are generally sceptical of change and only willing to adopt technologies when a majority has tried it out. Therefore, when promoting a technological change, it is important to understand the characteristics of the target segment that will help or hinder adoption of the innovation.

The key characteristic that determines in which stage people adopt new technologies is their level of innovativeness, which is defined as the degree to which an individual is likely to adopt new products relative to other people. Innovativeness is related to consumer novelty seeking, which refers to the desire to seek out new product information (Manning et al., 1995). For example, interest in the technology of smart meters (as another 'gadget') was found to be an important motivator for people to participate in a smart meter trial (Hargreaves et al., 2010). Innovativeness is also associated with opinion leadership, which reflects people's ability to influence others' attitudes or behaviours on a specific topic (Grewal et al., 2000). Hence, opinion leaders are often actively seeking information about the technology they are interested in.

In addition, the likelihood that a technology will be adopted depends on people's perception of the technology on five dimensions. The first dimension is the relative advantage of the technology and is defined by Rogers as the degree to which a technology or innovation is perceived as being better than the idea it supersedes. The relative advantage of technologies is very similar to the 'perceived usefulness' in the TAM and refers how much a user expects to benefit from the technology. Indeed, people who expected to benefit from smart grid technologies were more likely to adopt them (Broman-toft et al., 2014).

The second dimension is the value compatibility, referring to how much the technology aligns with the user's values, needs, and experiences, in other words, how much is the technology a fit with its user. However, this remains rather vague, and to address this, Karahanna et al. (2012) identified and empirically tested 15 different components of compatibility with technologies. They concluded that there are four distinct aspects of compatibility, namely, compatibility with people's values, prior experiences, existing routines, and practices, the latter referring to how a new technology fits with one's current lifestyle.

Complexity is the third dimension and refers to how difficult or easy it is to understand or use the technology. This is comparable with the 'ease of use' in the TAM. The importance of this dimension is illustrated in a study by Yang and Newman (2013) who interviewed users on how they interacted with various types of smart thermostats. They found that some users were reluctant to change the temperature if they found a thermostat 'very confusing to use' (p. 96). Instead, they opted for adjusting their clothing to the temperature. Smart thermostats are partially 'smart' because they 'learn' users' preferences. However, smart thermostats were reported to misinterpret users' intentions and thus make unwanted changes to the temperature. One person was reported to call the thermostat 'arrogant' (p. 97) and attributing human characteristics to it. It reveals frustration and referring the feeling that the thermostat would act independently of somebody's attempts to control it.

Tribality is the fourth dimension, referring to the extent to which a technology can be tested with limited costs. This is important, as it gives people the opportunity to get positive experiences with technologies. For example, people who participated in a trial with electric vehicles reported that things like charging was not as complicated as they had thought, although there were also things people had to get used to (Graham-Rowe et al., 2012). All in all, trials and test periods give people a chance to explore the use of technologies, which may take away some of their worries and concerns that may stop people from adopting these technologies. However, the opposite may happen too: people may feel less positive and more concerned after a trial period. Therefore, offering trials need to be carefully considered.

The last dimension is observability, which Rogers defines as the degree to which the results of an innovation are visible to others. This is related to symbolic functions of technologies, which refers to the notion that people use technologies to signal their identity to other people. Many people assume that microgeneration technologies are merely adopted for functional reasons: an electric vehicle is to drive from A to B, and a solar PV panel is installed to generate electricity. However, such technologies mean more to people, and they have an affective and symbolic function as well. Affective functions refer to emotional experiences derived from using new technologies, such as joy or pleasure (Dittmar, 1992). New technologies can lead to both negative and positive emotions when people first learn to use an innovation, which can influence their expectations and evaluations about these technologies (King and Slovic, 2014; Wood and Moreau, 2006). Symbolic functions refer to a sense of self or social identity that is reflected by, or built from, the possession of new technologies (Dittmar, 1992). In other words, we can shape a positive self-image by buying, owning, and displaying products (Belk, 1988). For example, the owner of an electric vehicle or solar panels may signal to others that he or she cares about the environment (Noppers et al., 2014; Schuitema et al., 2013).

The adoption of many technologies is largely market driven. Their adoption tends to be voluntary in nature and is often encouraged by subsidies, information and awareness campaigns, and trials. However, there are also technologies that are introduced as part of government regulations, such as smart meters and automated control systems. Such technologies are often rolled out by utilities as a result of these regulations. The roll-out of such technologies may start with a voluntary phase where people can choose whether or not they are willing to adopt them, but sooner or later, the adoption will get a more mandatory nature. If adoption is no longer voluntary, perceived fairness of procedures and regulations (referred to as procedural fairness) and trust in authorities and industry responsible for the roll-out become more important for adoption decisions.

Procedural fairness refers to the perceived fairness of the procedures used in decision-making processes (Tyler, 2000). This includes not only the way decisions were made but also how they were communicated to the public. Decisions are generally seen as fair if they were made honestly, impartially, and objectively and not based on personal values and biases. The party who makes the decision needs to be trusted (e.g. perceived to be benevolent and caring), needs to be concerned about the situation and the needs and wishes of the public, needs to be open to arguments, and needs to try to find the best and fairest solution. Finally, in the communication with the public, people value it when respect is shown to their rights and status and when they are treated politely and with respect. Fairness and trust issues played an important role for the backlash that arose against smart meters in the Netherlands in 2011. Hoenkamp et al. (2011, p. 269) state that 'The Dutch experience shows that smart metering is up for failure when the technical and commercial aspects are considered to be more important than the interests of the end-users'.

Particularly, the fear of lack of privacy is driving backlashes of smart meters: 'By measuring energy consumption by appliance every fifteen minutes, they capture details on how you spend your days and nights. When you turn off the lights. When you take a shower. When you leave home. When your electric vehicle is charged. But who is watching you? Who has access to all this information?' (Balough, 2011). The lack of privacy was the underlying reason for the backlash against smart meters in, for example, the Netherlands, Texas, and California.

5 Public acceptance of infrastructure and policies

Large structural and infrastructural changes are needed, such as large-scale renewable energy plants, transmission and distribution lines, and storage capacity. As a result, changes in the landscape are inevitable as this means that large visible infrastructural changes will appear, such as pylons and transmission lines, geothermal power plants, or wind and solar farms. Particularly, local communities are likely to face drastic changes in their neighbourhoods. In addition, policies and regulations are needed to facilitate an energy transition too. For example, in the European Union, some energy-inefficient technologies are forbidden (e.g. the sale of incandescent light bulbs was banned from 2009), while other energy-efficient technologies are supported (e.g. solar energy and heat pumps), and smart meters, which are an important part of the energy transition, are becoming mandatory in more and more countries. Also, the introduction of demand-side management (DSM) programmes aimed at reducing or balancing the energy demand is common. These often include financial policies in combination with information and feedback on household energy use.

Whether people accept and support such infrastructural changes and policies is arguably one of the most critical factors that will lead to a successful energy transition. Without it, the much-needed changes will be delayed and potentially abolished if opposition is very strong. Wüstenhagen et al., (2007) conceptualise social acceptance as a construct that has three different acceptance dimensions, namely, socio-political, community, and market (see Fig. 1). 'Socio-political acceptance' is the broadest and most general form of acceptance and refers to the acceptance of policies and technologies in general. Community acceptance refers to how communities that are close to specific projects evaluate them. This refers primarily to the response of communities to decisions made with regard to these specific projects. The last dimension is market acceptance, which refers to the process of markets adopting new technologies. This relates to questions about whether the market is willing to invest in new technologies and innovations.

Psychologists mainly research socio-political acceptance and community acceptance. For example, people who have strong environmental values and feel a moral obligation to contribute to the solution of environmental problems are more likely to accept policies aimed at reducing energy conservation (Steg et al., 2006). Also, the type of policy is important for their acceptance levels: people are generally more

FIG. 1

The triangle of social acceptance of renewable energy innovation.

Modified from Wüstenhagen, R., Wolsink, M., and Bürer, M.J., 2007. Social acceptance of renewable energy innovation: an introduction to the concept. Energy Policy 35, 2683–2691.

likely to accept measures aimed to increase energy efficiency compared with policies aimed at curtailment behaviour (Steg et al., 2005). Also, so-called pull measures (i.e. measures aimed at making behaviour change more attractive) are generally more acceptable than push measures (i.e. measures aimed at making the undesired behaviour less attractive) (Steg et al., 2005). However, the acceptance of push and pull measures depends on how many other people find a policy acceptable. De Groot and Schuitema (2012) observed that people found a tax increase much more acceptable when a majority supported this policy (implying a strong descriptive social norm) compared with when a minority supported this tax increase (implying a weak descriptive social norm).

The lack of community acceptance for specific infrastructural projects was long associated with NIMBYism (not in my backyard). However, NIMBYism is now generally seen as too simplistic and erroneous (Devine-Wright, 2005, 2007; Ek, 2005; Wolsink, 2007). NIMBYism is often used as an excuse to ignore public concerns and objections to projects. People do not resist changes per se, but they often have valid reasons for their objections, such as increased exposure to risks and changes to their local environment and community (Dent and Galloway Goldberg, 1999; Devine-Wright and Howes, 2010). A common assumption from project planners and authorities is that this resistance is caused by a 'knowledge gap', which tends to lead to a flow of information to communities. However, it is often not a lack of knowledge that is driving objections as many people would be well informed about the issues. Knowing about a project and its consequences does not necessarily mean that people agree with it. Also, information provision may not be directed at people's concerns at all. A more effective communication strategy is a two-way communication that is open and allows an exchange of information between parties, with the possibility to transform decisions and opinions on both sides (Frewer, 2004; Rowe and Frewer, 2000).

Another common assumption is that communities need to be given something in return, which is often done in the form of compensation and the provision of benefits. Indeed, providing community compensation has led to increased acceptance (Walker et al., 2014; Warren and McFadyen, 2010); however, if compensation is not tailored to the specific needs of a community, such well-intended initiatives may well backfire and increase criticism. For example, it can lead to distrust in the agenda of the developer (Eirgrid, 2016), feelings of not being taken seriously or even seen as bribes to ensure community acceptance (Ter Mors et al., 2012). To avoid this, an in-depth analysis of the local history and context is needed to develop tailored strategies that may lead to genuine community acceptance of certain projects (Perlaviciute et al., 2018). An example of project that was well embedded and widely accepted by the local community in Northern Ireland is described by Devine-Wright (2011). This case study analysis found that a local tidal energy project was received very positively, because the symbolic meaning of the local community (i.e. what the place meant for people who lived in the local community) and the project were a good fit.

In the literature, the terms acceptance and support are often used interchangeably. However, Batel et al., (2013) argue that acceptance and support are two different constructs. They define acceptance as 'the act to accept' that includes responding affirmative or tolerating something as well as enduring something without protest. Support, on the other hand, refers to approving or subscribing to something as well as encouraging, upholding, or defending something as being the right thing to do. Although there is overlap between the terms, people can accept something (i.e. tolerate or endure without protest), while they do not support (approve or subscribe) it (Batel et al., 2013). As such, one may see support as a more internalised form of acceptance. That there is a difference between acceptance and support is illustrated through the findings of a study by Dreyer and Walker (2013), who studied public acceptance and support for a carbon tax in Australia. In their survey, 21.4% of participants indicated to be 'neutral' when asked to evaluate the tax and give their opinion about it (an indication of acceptance). However, when forced to choose whether or not they supported the carbon tax, one-third of participants indicated they supported the tax, whereas two-thirds did not support it. Also, they found that people who supported the tax thought it was more effective and fairer than people who did not support it. The Australian carbon tax was abolished a few years after implementation, partly as a result of opposition from the public and industry.

6 Psychology's contributions to energy research in interdisciplinary settings

Interdisciplinary research is defined as the synthesis of two or more disciplines, leading to the establishment of a new level of discourse and integration of knowledge (Klein, 1990; Schuitema and Sintov, 2017). Interdisciplinary research is especially needed to solve wicked societal problems, as to optimally better understand the complex reality of such problems and design optimal solutions for them. How to meet society's energy demand in a sustainable, reliable, and affordable way is an example of a wicked societal problem that needs interdisciplinary research. An integration across many different disciplines, including but not limited to engineering, environmental sciences, computer sciences, mathematics, geoscience, economics, philosophy, law, anthropology, business, sociology, and psychology, is needed in energy research (for a complete overview, see Sovacool, 2014).

What is the role of psychology in energy research? The current changes in the energy grid and power systems affect everybody, including individuals and households. These changes require that people more actively participate in the energy system, adopt new technologies, and support infrastructural changes and policies that are needed for an energy transition. Psychologists can help to understand how and why people respond to these changes and how they can be encouraged to change their behaviour. Specifically, psychology can contribute to energy research in two distinct ways.

Firstly, psychologists study motivations of human behaviour. Psychologists ask: why do people behave the way they do; what are the (psychological) drivers and barriers to specific behaviours? These include a broad range of constructs, such as financial and environmental motivations, and social norms. Moreover, psychologists try to understand individual differences, that is, not everybody behaves the same. Psychologists not only do study behaviour but also look at underlying reasons for people's acceptance of, for example, infrastructural changes and policies, the adoption of technologies. Understanding differences and similarities between people is an important part of understanding people's motivations. As we explained earlier, energy conservation, technology adoption, and public acceptance of infrastructure are very important aspects of the current energy transition. Therefore, psychologists have an important role to play.

Secondly, psychologists use their understanding of behavioural motivations to inform, develop, and evaluate interventions aimed at changing behaviour. As illustrated earlier, psychologists are heavily involved in the design and evaluation of interventions to encourage energy conservation and technology adoption (for an example, see Box 2). 'Interventions' include a broad range of measures that can be used to change behaviour, including but not limited to education or media campaigns aimed at increasing knowledge, information, and feedback provision that targets social norms or the use of defaults ('nudging').

For a successful energy transition, research should not only be interdisciplinary in terms of including various academic disciplines but also involve stakeholders such as industry, businesses, politics, interest groups, and other organisations. This is typically referred to as transdisciplinary research (Sintov and Schuitema, 2018; Spreng, 2014). All stakeholders form an important part of the complex puzzle that needs to be solved. Psychologists can play an important role here, especially as these stakeholders often deal directly with individual consumers and communities. Therefore, psychologists can advise on issues arising in this context. One area where collaboration can be strengthened is in the evaluation of interventions. Often, practitioners develop behaviour change interventions, but these are not necessarily grounded in existing

---

Box 2 Interdisciplinary energy research: An example

Thermal imaging is a technology that uses infrared radiation that makes the flow of heat visible in and around the home. In energy research, it is used to visualise where heat is escaping from a home. The images display different temperatures in different colours to indicate areas of heat loss. For example, if a home is not double glazed or does not have insulation, heat will escape through these surfaces, and thermal imaging makes this heat loss visible. The image can make visible how people might be able to conserve energy (e.g. through draught stripping or putting in double glazing). A group of researchers from the University of Plymouth, in the United Kingdom, set out to test the effectiveness of thermal imaging to encourage energy conservation (Goodhew et al., 2015). This is a good example of an interdisciplinary study, as it integrates theories and methods from the physical sciences (engineering) and the social sciences (psychology). In a first study, households either received a thermal image or an energy audit. Households who received a thermal image has significantly increased the uptake of energy-saving behaviours and reduced their energy consumption, compared with a control group and the audit group. In a second study, householders were more likely to install draught proofing measures after seeing a thermal image. The findings of these two studies suggest that making heat loss visible via thermal imaging can encourage energy conservation. It also highlights that research can be particularly useful when it combines insights on new technologies with evidence on how people use these technologies for energy conservation.

research and are rarely evaluated systematically (i.e. by including control groups or creating a baseline to compare the results with). As a result, many interventions are poorly evaluated. As such, it would be beneficial if more psychologists were to advise on the design and evaluation of interventions.

In addition, psychologists understand communication processes and are therefore well equipped to study communication strategies and processes. In the current energy transition, there are many questions about how to engage the public and communities and how and what should be communicated to them. Scientists, politicians, industry, and other organisations struggle to effectively communicate with the public. Psychologists who study communication and decision-making processes can therefore make contributions to help improve our understanding of which message should be communicated and how the message should be communicated and by whom (Pidgeon and Fischhoff, 2011).

In interdisciplinary research, each discipline has a unique contribution. For example, researchers in the energy domain model energy demand and supply, flexibility in the energy system, or potential for renewable energy sources. Often, such models are based on assumptions about how people will respond. Psychologists can challenge some of these underlying assumptions and help develop more realistic ones. Vice versa, psychologists have a lot to learn from other disciplines; for example, psychologists are more likely to target 'low-impact' curtailment behaviours and would need to gain a better understanding of technical, physical, and economic constraints to encouraging more impactful behaviour changes, such as the uptake of efficiency measures.

Finally, interdisciplinary research is very much needed, yet there are many obstacles to be a successful interdisciplinary researcher (Schuitema and Sintov, 2017). To illustrate the dynamics of interdisciplinary research, one of us has reflected on an interdisciplinary experience to illustrate issues and lessons learned (see Box 3).

7 Further research: What are next steps?

Psychological research on energy consumption contributes to understanding the factors that shape energy consumption, the factors that facilitate or hinder the adoption of energy-efficient technologies, and the factors that predict people's acceptability of energy policies and infrastructural changes. It also explores how households can be encouraged to take up efficiency and curtailment behaviours through behaviour change interventions such as information and feedback provision. Specifically, psychology should focus on contributing to four main researches in energy research.

Firstly, there is a clear imperative for substantially reducing emissions from the domestic energy sector, as highlighted, for example, by the recent IPCC special report mentioned at the start of this chapter. Psychological research tends to focus mainly on understanding and changing curtailment behaviours, whereas research on how to encourage efficiency behaviours is relatively scarce. Here, there is scope for (more) interdisciplinary research, perhaps in collaboration with engineering, building sciences, or

> ### Box 3 Working in an interdisciplinary setting. Personal reflections from Geertje Schuitema
>
> Bridging the gap between engineers and psychology: My first week in a new job
>
> I started my current position in 2014. In my first week, I was introduced to many engineers as I was cohired by an energy research institute that is heavily engineering based. As is the case in new jobs, I was frequently asked what my research was about. I learned from previous experience in interdisciplinary research projects that it's best not to assume any deep pre-existing knowledge about the work psychologists do and therefore it's best not to explain in too much detail what my research is about. Hence, I answered generally 'Well, I look at the role of consumers in energy systems'. The first response to that simple sentence was overwhelmingly positive: 'Very good. You are just what we need! When I model the energy demand of buildings, people tend to open a door or a window or turn on the stove, which really ruins my model. Can you help with that?'. There is a short silence while my brain is processing what this means: people open their window, why is that weird? They are cold and turn on a stove, so what? Then I realised what the problem was: a good model was the aim, not a comfortable building for its occupants. In my first week there, I had similar request, which all came down to the question if I could please 'fix' people, because their behaviour was most inconvenient for various reasons. 'Fixing' people could mean anything from making them stop opening their doors or to tell them to accept wind turbines. I see this example as illustrative of a deeper underlying view: many disciplines I work with have a technology-centred energy system as underlying assumption. This means that the technology needs to be perfect and models need to be optimal. This realisation made me aware of my own view: I have a user-centred view, implying that I think that energy systems are designed to meet the needs of people. And if meeting the needs of individual people means that technologies are not perfect and models are suboptimal, so be it. Of course, I realise that there are physical laws that constrains an energy system and that people cannot always be at the centre of it. And I'm not even speaking of the market, policy, cultural, geological, and legal context of energy systems, which is important too. But for me, an important reminder of my first week in my new job is that interdisciplinary collaboration begins with understanding each other's assumptions and perspectives. So, let me finish by answering those who asked me to 'fix' people: no, I cannot fix them, and I don't want to either. However, what I can do is to try and understand why they behave the way they do and think of interventions to encourage them to change based on their motivations.—*Geertje Schuitema*

environmental sciences (see also Box 2). For example, engineers can identify and implement features of the building envelope that promote energy efficiency (e.g. eaves that act as shade/cooling), environmental scientists can quantify the environmental impact or carbon footprint of energy-saving features of the home and individual household behaviours, and psychologists can provide insights into how to encourage behaviour changes so that the features of building design are used in a way to maximise the environmental impact (in terms of emission reductions). Some of these collaborations are well established (for an overview, see Abrahamse and Shwom, 2018), but more interdisciplinary energy research is warranted given the current climate change challenge.

Secondly, psychological research highlights that it is important to understand how people use and interact with energy-efficient technologies and infrastructure. For example, energy-efficient light bulbs can theoretically help reduce energy use, and they are widely available, so why do people still use incandescent bulbs? And do people have the skills and knowledge required to make efficient use of energy-saving features? Some studies, for example, have shown that people alter their behaviour after they have installed energy-efficient lighting (e.g. leaving lights on for longer) counteracting initial energy savings (Schleich et al., 2014). This is also referred to as the rebound effect. If the adoption of new technologies does not necessarily lead to the commensurate reductions in energy consumption levels, it suggests that individual preferences and behaviours need to be taken into consideration.

A third contribution of psychologists lies in the design and evaluation of behavioural interventions and interventions that may enhance public support for changes. Many intervention studies are conducted via (online) scenario studies or lab experiments, while there is a huge need for more field experiments: how do people respond to interventions in real life. We acknowledge that field experiments are time intensive, difficult to control, and sometimes frustrating to run. However, we make a plea for an increase of field experiments as they give more realistic information of how and why people respond to interventions. Setting up such field experiments requires not only interdisciplinary research teams that can involve other scientific disciplines but also industry or other practitioners. In addition, there are ample opportunities for psychologists to evaluate interventions without setting them up. Industry, city councils, and other practitioners often initiate interventions aimed at behaviour change. However, these initiatives are rarely systematically evaluated. Psychologists could more actively seek opportunities to study initiatives taken, whereas practitioners could seek more advice from psychologists on how to evaluate them (see also Sintov and Schuitema, 2018).

Finally, public support is a crucial precondition to implement policies that support sustainable energy developments or build new energy infrastructure (e.g. wind farms, pylons, and geothermal energy plants), which is typically studied by psychologists. Issues around public resistance of energy policies or infrastructural projects are often largely underestimated by the involved parties and, as a result, often addressed too late. Hence, public support should be taken into account from the start of projects (Perlaviciute et al., 2018), implying that psychologists should be involved in planning and implementation processes of policies and other energy projects to evaluate, analyse, and address the dynamics of each specific process.

However, there are many challenges for energy research to become truly interdisciplinary (see for a full discussion, Schuitema and Sintov, 2017). These challenges include knowledge and skills of researchers, as well as institutional barriers that hinder successful interdisciplinary research. Institutional barriers occur at primarily in the area of funding, university structures, and academic reward systems.

With respect to funding, obtaining interdisciplinary funding is difficult, as funding schemes focus very much on specific disciplinary areas. And if interdisciplinary research is funded, schemes tend to cover disciplinary areas that are closely related, such as science, technology, engineering, and mathematics (STEM) research or social sciences and humanities. Crossovers between science and social science funding are still relatively rare. If funding bodies fund such crossover projects, they typically face the challenge to find reviewers who are equipped to evaluate all aspects of such interdisciplinary proposals (Nightingale and Scott, 2007).

Although universities increasingly acknowledge the need for interdisciplinary research, they have no mechanisms in place to reward this. As a result, interdisciplinary researchers are often evaluated against monodisciplinary standards, which will be harder for them to meet. This is specifically problematic for early career researchers, especially those on fixed contracts. As a result, many researchers feel that interdisciplinary research will harm their career prospects, even though they feel the need to contribute to societal problems (Rhoten and Parker, 2004). Related to the last point are barriers that lie in the academic publication: interdisciplinary research is more difficult to publish in high-ranked journals. As long as interdisciplinary research is not valued as much as monodisciplinary research in academia, it will be very difficult to motivate researchers to fully engage in this type of collaboration.

In sum, psychologists have much to offer to energy research, and there is a large range of interesting questions that still need to be addressed. This work must be done in interdisciplinary settings to fully reach its potential impact. However, there are many institutional barriers that currently hinder interdisciplinary research, and it is important that they are addressed in the near future. If the aim is to significantly reduce carbon emissions, more interdisciplinary empirical studies are needed to elucidate how this can be achieved most effectively.

References

Abrahamse, W., Steg, L., Vlek, C., Rothengatter, T., 2005. A review of intervention studies aimed at household energy conservation. J. Environ. Psychol. 25 (3), 273–291.

Abrahamse, W., Steg, L., Vlek, C., Rothengatter, T., 2007. The effect of tailored information, goal setting, and tailored feedback on household energy use, energy-related behaviors, and behavioral antecedents. J. Environ. Psychol. 27 (4), 265–276.

Abrahamse, W., Shwom, R., 2018. Domestic energy consumption and climate change mitigation. Wiley Interdiscip. Rev. Clim. Chang. 9, 525–534.

Abrahamse, W., Steg, L., 2009. How do socio-demographic and psychological factors relate to households' direct and indirect energy use and savings? J. Econ. Psychol. 30, 711–720.

Abrahamse, W., Steg, L., 2011. Factors related to household energy use and intention to reduce it: The role of psychological and socio-demographic variables. Hum. Ecol. Rev. 18, 30–40.

Ajzen, I., 1991. The theory of planned behavior. Organ. Behav. Hum. Decis. Process. 50, 179–211.

Allcott, H., 2011. Social norms and energy conservation. J. Public Econ. 95, 1082–1095.

Allcott, H., Mullainathan, S., 2010. Behavior and energy policy. Science 327, 1204–1205.

Asensio, O.I., Delmas, M.A., 2015. Nonprice incentives and energy conservation. Proc. Natl. Acad. Sci. 112, E510–E515.

Ayres, I., Raseman, S., Shih, A., 2013. Evidence from two large field experiments that peer comparison feedback can reduce residential energy usage. J. Law Econ. Organ. 29, 992–1022.

Balough, C.D., 2011. Privacy Implications of Smart Meters. Chic. Kent Law Rev. 86, 161–191.

Batel, S., Devine-Wright, P., Tangeland, T., 2013. Social acceptance of low carbon energy and associated infrastructures: a critical discussion. Energy Policy 58, 1–5.

Belk, R.W., 1988. Possessions and the extended self. J. Consum. Res. 15, 139–168.

Bolderdijk, J.W., Steg, L., Geller, E.S., Lehman, P.K., Postmes, T., 2013. Comparing the effectiveness of monetary versus moral motives in environmental campaigning. Nat. Clim. Chang. 3, 413–416.

Broman-toft, M., Schuitema, G., Thøgersen, J., 2014. Responsible technology acceptance: model development and application to consumer acceptance of smart grid technology. Appl. Energy 134, 392–400.

Chen, C.F., Xu, X., Day, J.K., 2017. Thermal comfort or money saving? Exploring intentions to conserve energy among low-income households in the United States. Energy Res. Soc. Sci. 26, 61–71.

Cialdini, R.B., Reno, R.R., Kallgren, C.A., 1990. A focus theory of normative conduct: recycling the concept of norms to reduce littering in public places. J. Pers. Soc. Psychol. 58, 1015–1026.

Davis, F.D., 1989. Perceived usefulness, perceived ease of use, and user acceptance. MIS Q. 13, 319–340.

Davis, F.D., Bagozzi, R.P., Warshaw, P.R., 1989. User acceptance of computer technology: A comparison of two theoretical models. Manag. Sci. 35, 982–1003.

De Groot, J.I.M., Schuitema, G., 2012. How to make the unpopular popular? Policy characteristics, social norms and the acceptability of environmental policies. Environ. Sci. Pol. 19–20, 100–107.

De Groot, J.I., Steg, L., 2009. Mean or green: Which values can promote stable pro-environmental behavior? Conserv. Lett. 2, 61–66.

Delmas, M.A., Fischlein, M., Asensio, O.I., 2013. Information strategies and energy conservation behavior: a meta-analysis of experimental studies from 1975 to 2012. Energy Policy 61, 729–739.

Dent, E.B., Galloway Goldberg, S., 1999. Challenging "resistance to change". J. Appl. Behav. Sci. 35, 25–41.

Devine-Wright, P., 2005. Beyond NIMBYism: towards an integrated framework for understanding public perceptions of wind energy. Wind Energy 8, 125–139.

Devine-Wright, P., 2007. Reconsidering public attitudes and public acceptance of renewable energy technologies: a critical review. School of Environment and Development, University of Manchester, Manchester.

Devine-Wright, P., 2011. Place attachment and public acceptance of renewable energy: a tidal energy case study. J. Environ. Psychol. 31, 336–343.

Devine-Wright, P., Howes, Y., 2010. Disruption to place attachment and the protection of restorative environments: a wind energy case study. J. Environ. Psychol. 30, 271–280.

Dittmar, H., 1992. The social psychology of material possessions: to have is to be. Harvester Wheatsheaf and St. Martin's Press, Hemel Hempstead.

Dreyer, S., Walker, I., 2013. Acceptance and support of the Australian carbon policy. Soc. Justice Res 26, 343–362.

Eirgrid, 2016. Reviewing and improving our public consultation process. Eirgrid, Dublin.

Ek, K., 2005. Public and private attitudes towards "green" electricity: the case of Swedish wind power. Energy Policy 33, 1677–1689.

Frewer, L., 2004. The public and effective risk communication. Toxicol. Lett. 149, 391–397.

Gangale, F., Mengolini, A., Onyeji, I., 2013. Consumer engagement: an insight from smart grid projects in Europe. Energy Policy 60, 621–628.

Gardner, G.T., Stern, P.C., 2008. The short list: The most effective actions US households can take to curb climate change. Environ. Sci. Policy Sustain. Dev. 50, 12–25.

Goodhew, J., Pahl, S., Auburn, T., Goodhew, S., 2015. Making heat visible: promoting energy conservation behaviors through thermal imaging. Environ. Behav. 47, 1059–1088.

Graham-Rowe, E., Gardner, B., Abraham, C., Skippon, S., Dittmar, H., Hutchins, R., Stannard, J., 2012. Mainstream consumers driving plug-in battery-electric and plug-in hybrid electric cars: a qualitative analysis of responses and evaluations. Transp. Res. A Policy Pract. 46, 140–153.

Grewal, R., Mehta, R., Kardes, F.R., 2000. The role of the social-identity function of attitudes in consumer innovativeness and opinion leadership. J. Econ. Psychol. 21, 233–252.

Hargreaves, T., Nye, M., Burgess, J., 2010. Making energy visible: a qualitative field study of how householders interact with feedback from smart energy monitors. Energy Policy 38, 6111–6119.

Hoenkamp, R., Huitema, G.B., De Moor-Van Vugt, A.J.C., 2011. The neglected consumer: the case of the smart meter rollout in the Netherlands. Renew. Energy Law Pol. Rev. 4, 269–282.

IPCC, 2018. Global warming of 1.5°C. Retrieved from: Intergovernmental Panel on Climate Change, https://www.ipcc.ch/sr15/.

Jager, W., 2006. Stimulating the diffusion of photovoltaic systems: a behavioural perspective. Energy Policy 34, 1935–1943.

Karahanna, E., Agarwal, R., Angst, C.M., 2012. Reconceptualizing compatibility beliefs in technology acceptance research. MIS Q. 30, 781–804.

Karlin, B., Davis, N., Sanguinetti, A., Gamble, K., Kirkby, D., Stokols, D., 2014. Dimensions of conservation: exploring differences among energy behaviors. Environ. Behav. 46, 423–452.

Karlin, B., Zinger, J.F., Ford, R., 2015. The effects of feedback on energy conservation: a meta-analysis. Psychol. Bull. 141, 1205.

King, J., Slovic, P., 2014. The affect heuristic in early judgments of product innovations. J. Consum. Behav. 13, 411–428.

Klein, J.T., 1990. Interdisciplinarity: History, Theory, and Practice. Wayne State University Press, Detroit, MI.

Kluger, A.N., DeNisi, A., 1996. The effects of feedback interventions on performance: a historical review, a meta-analysis, and a preliminary feedback intervention theory. Psychol. Bull. 119, 254–284.

Latré, E., Perko, T., Thijssen, P., 2017. Public opinion change after the Fukushima nuclear accident: the role of national context revisited. Energy Policy 104, 124–133.

Lillemo, S.C., 2014. Measuring the effect of procrastination and environmental awareness on households' energy-saving behaviours: an empirical approach. Energy Policy 66, 249–256.

Manning, K.C., Bearden, W.O., Madden, T.J., 1995. Consumer innovativeness and the adoption process. J. Consum. Psychol. 4, 329–345.

Michie, S., Van Stralen, M.M., West, R., 2011. The behaviour change wheel: a new method for characterising and designing behaviour change interventions. Implement. Sci. 6, 42–50.

Mirosa, M., Lawson, R., Gnoth, D., 2013. Linking personal values to energy-efficient behaviors in the home. Environ. Behav. 45, 455–475.

Moser, S.C., Dilling, L., 2011. Communicating climate change: closing the science-action gap. In: The Oxford Handbook of Climate Change and Society. Oxford University Press, Oxford, pp. 161–174.

Nightingale, P., Scott, A., 2007. Peer review and the relevance gap: ten suggestions for policymakers. Sci. Public Policy 34, 543–553.

Nolan, J.M., Schultz, P.W., Cialdini, R.B., Goldstein, N.J., Griskevicius, V., 2008. Normative social influence is underdetected. Personal. Soc. Psychol. Bull. 34, 913–923.

Noppers, E.H., Keizer, K., Bolderdijk, J.W., Steg, L., 2014. The adoption of sustainable innovations: driven by symbolic and environmental motives. Glob. Environ. Chang. 25, 52–62.

Pals, H., Singer, L., 2015. Residential energy conservation: the effects of education and perceived behavioral control. J. Environ. Stud. Sci. 5, 29–41.

Perlaviciute, G., Steg, L., 2015. The influence of values on evaluations of energy alternatives. Renew. Energy 77, 259–267.

Perlaviciute, G., Schuitema, G., Devine-Wright, P., Ram, B., 2018. At the heart of becoming more sustainable: the public acceptability of energy projects. IEEE Power Energy Mag. 16, 49–55.

Pichert, D., Katsikopoulos, K.V., 2008. Green defaults: information presentation and pro-environmental behaviour. J. Environ. Psychol. 28, 63–73.

Pidgeon, N., Fischhoff, B., 2011. The role of social and decision sciences in communicating uncertain climate risks. Nat. Clim. Chang. 1, 35–41.

Renn, O., Marshall, J.P., 2016. Coal, nuclear and renewable energy policies in Germany: from the 1950s to the "Energiewende". Energy Policy 99, 224–232.

Rhoten, D., Parker, A., 2004. Risks and rewards of an interdisciplinary research path. Science 306, 2046.

Rogers, E.M., 1995. Diffusion of Innovations, fourth ed. Free Press, New York.

Rowe, G., Frewer, L.J., 2000. Public participation methods: a framework for evaluation. Sci. Technol. Hum. Values 25, 3–29.

Schleich, J., Mills, B., Dütschke, E., 2014. A brighter future? Quantifying the rebound effect in energy efficient lighting. Energy Policy 72, 35–42.

Schuitema, G., Sintov, N.D., 2017. Should we quit our jobs? Challenges, barriers and recommendations for interdisciplinary energy research. Energy Policy 101, 246–250.

Schuitema, G., Anable, J., Skippon, S., Kinnear, N., 2013. The role of instrumental, hedonic and symbolic attributes in the intention to adopt electric vehicles. Transp. Res. A Policy Pract. 48, 39–49.

Schuitema, G., Ryan, L., Aravena, C., 2017. The consumer's role in flexible energy systems: an interdisciplinary approach to changing consumers' behavior. IEEE Power Energy Mag. 15, 53–60.

Schultz, P., 2014. Strategies for promoting proenvironmental behavior: lots of tools but few instructions. Eur. Psychol. 19, 107–115.

Schultz, P.W., Nolan, J.M., Cialdini, R.B., Goldstein, N.J., Griskevicius, V., 2007. The constructive, destructive, and reconstructive power of social norms. Psychol. Sci. 18, 429–434.

Schwartz, D., Bruine de Bruin, W., Fischhoff, B., Lave, L., 2015. Advertising energy saving programs: the potential environmental cost of emphasizing monetary savings. J. Exp. Psychol. Appl. 21, 158–166.

Sintov, N.D., Schuitema, G., 2018. Odd couple or perfect pair? Tensions and recommendations for social scientist-industry partnerships in energy research. Energy Policy 117, 247–251.

Sovacool, B.K., 2014. What are we doing here? Analyzing fifteen years of energy scholarship and proposing a social science research agenda. Energy Res. Soc. Sci. 1, 1–29.

Spreng, D., 2014. Transdisciplinary energy research – reflecting the context. Energy Res. Soc. Sci. 1, 65–73.

Staats, H.J., Wit, A.P., Midden, C.Y.H., 1996. Communicating the greenhouse effect to the public: evaluation of a mass media campaign from a social dilemma perspective. J. Environ. Manag. 46, 189–203.

Steg, L., Vlek, C., 2009. Encouraging pro-environmental behaviour: an integrative review and research agenda. J. Environ. Psychol. 29 (3), 309–317.

Steg, L., Dreijerink, L., Abrahamse, W., 2005. Factors influencing the acceptability of energy policies: a test of VBN theory. J. Environ. Psychol. 25, 415–425.

Steg, E.M., Dreijerink, L., Abrahamse, W., 2006. Why are energy policies acceptable and effective? Environ. Behav. 38, 92–111.

Stern, P.C., 2000. New environmental theories: toward a coherent theory of environmentally significant behavior. J. Soc. Issues 56, 407–424.

Ter Mors, E., Terwel, B.W., Daamen, D.D.L., 2012. The potential of host community compensation in facility siting. Int. J. Greenh. Gas Con. 11, 130–138.

Toffler, A., 1980. The Third Wave. William Morrow, New York.

Tyler, T.R., 2000. Social Justice: outcome and procedure. Int. J. Psychol. 35, 117–125.

Walker, B.J.A., Wiersma, B., Bailey, E., 2014. Community benefits, framing and the social acceptance of offshore wind farms: an experimental study in England. Energy Res. Soc. Sci. 3, 46–54.

Warren, C.R., McFadyen, M., 2010. Does community ownership affect public attitudes to wind energy? A case study from south-west Scotland. Land Use Policy 27, 204–213.

Wolsink, M., 2007. Wind power implementation: the nature of public attitudes: Equity and fairness instead of 'backyard motives'. Renew. Sust. Energ. Rev. 11, 1188–1207.

Wood, S.L., Moreau, C.P., 2006. From fear to loathing? How emotion influences the evaluation and early use of innovations. J. Mark. 70, 44–57.

Wüstenhagen, R., Wolsink, M., Bürer, M.J., 2007. Social acceptance of renewable energy innovation: an introduction to the concept. Energy Policy 35, 2683–2691.

Yang, R., Newman, M.W., 2013. Learning from a learning thermostat: lessons for intelligent systems for the home. In: Proceedings of the 2013 ACM International Joint Conference on Pervasive and Ubiquitous Computing, Zurich, 8–12 September.

Yeboah, F.K., Kaplowitz, M.D., 2016. Explaining energy conservation and environmental citizenship behaviors using the value-belief-norm framework. Hum. Ecol. Rev. 22, 137–160.

Further reading

Attari, S.Z., DeKay, M.L., Davidson, C.I., De Bruin, W.B., 2010. Public perceptions of energy consumption and savings. Proc. Natl. Acad. Sci. 107, 16054–16059.

Baird, J.C., Brier, J.M., 1981. Perceptual awareness of energy requirements of familiar objects. J. Appl. Psychol. 66, 90–96.

IEA, 2018. Global EV Outlook 2018: Towards Cross-Modal Electrification. International Energy Agency, Paris.

Behavioural economics for energy and climate change policies and the transition to a sustainable energy use—A Scandinavian perspective

1.2

Luis Mundaca[a], Margaret Samahita[a], Jonas Sonnenschein[a], Roman Seidl[b]

[a]*International Institute for Industrial Environmental Economics at Lund University, Lund, Sweden*
[b]*Öko-Institut, Freiburg, Germany*

1 Introduction

Scandinavian countries (i.e. Denmark, Norway, and Sweden) are often listed as world leaders when it comes to clean energy, innovation policy and environmental protection, and top multiple international rankings (e.g. the Global Green Economy Index, the Legatum Prosperity Index, and the Global Cleantech Innovation Index). When it comes to energy and climate change policies, Scandinavian countries have also adopted a number of ambitious targets that aim to support or guide their national efforts, in line with the 'EU Energy Roadmap 2050'[a] (IEA and Nordic Energy Research, 2013). They currently seek pathways leading to a carbon-neutral energy-economy system in 2050 (IEA and Nordic Energy Research, 2016), and policy efforts are expected to play an important role in the 'climate-neutral Europe by 2050' strategy.[b]

Whereas the main focus of the literature and policy debate has been on technology options and innovation policies, less attention has been paid to behavioural aspects affecting or shaping the transition. For instance, major studies in the region indicate that infrastructure and technology (e.g. long-distance transport systems, investment in electricity grids, and development of wind power generation) are a major challenge

---

[a] For details, see https://ec.europa.eu/energy/en/topics/energy-strategy-and-energy-union/2050-energy-strategy.
[b] For details, see https://ec.europa.eu/energy/en/topics/energy-strategy-and-energy-union/2050-long-term-strategy.

Energy and Behaviour. https://doi.org/10.1016/B978-0-12-818567-4.00004-1

in encouraging a climate-neutral economy (IEA and Nordic Energy Research, 2013, 2016). Energy efficiency potentials are often investigated, and accelerated renovation efforts in the building sector and the construction of zero-energy buildings have been identified as other critical success factors (IEA and Nordic Energy Research, 2016; Seljom et al., 2017). These studies claim that various technological measures can support sustainable energy-economy systems, while behavioural change is limited to an afterthought—either in the context of stressing core assumptions behind technological solutions or confined to the need to alter behavioural patterns. Perceptions from Scandinavian stakeholders about the challenges of low-carbon energy transitions confirm the dominance of technology-oriented solutions (e.g. the integration of renewables) (Sovacool et al., 2018), with behavioural aspects receiving far less attention.

Despite various policy efforts targeting sustainable energy behaviour, there has been relatively little research in this area. This includes, for instance, knowledge gaps related to the motivation to adopt renewable energy in Sweden (Bergek and Mignon, 2017), psychological barriers affecting energy efficiency in Norway (Lillemo, 2014), or the potential for behavioural change via information policy schemes in Denmark (Ölander and Thøgersen, 2014). At the same time, evidence from the region indicates that, for example, social norms represent a barrier to increased energy efficiency (Throne-Holst et al., 2008) and an opportunity for mitigation actions (Lindman et al., 2013), cultural values can limit low-carbon energy technologies (LCET) choices in the building sector (Lindkvist et al., 2014), and commitment from individuals is a critical factor affecting the performance of energy-efficient and mitigation initiatives at the local level (Kasa et al., 2012). While there is a clear consensus that the low-carbon energy futures are dependent on ambitious technological innovations and policies (IEA and Nordic Energy Research, 2016; Sovacool, 2017; Sovacool et al., 2018), fragmented knowledge remains with respect to behavioural aspects that frame or affect this transition.

Within this context, we explore the interplay between policies (e.g. economic incentives and information schemes) and behavioural aspects (e.g. cognitive, motivational, and contextual issues) that shape or influence the adoption and use of LCET in Scandinavia. To this end, we review and analyse various issues associated with energy, behaviour, and mitigation action. Our study is guided by a behavioural economics framework (details in the next section); it seeks to complement existing knowledge and support policymaking. This framework is motivated by growing evidence that suggests that interventions addressing cognitive, motivational, and contextual aspects can play an important role in promoting sustainable energy behaviour (e.g. Allcott, 2011; de Coninck et al., 2018; Frederiks et al., 2015; Pollitt and Shaorshadze, 2011). With due limitations, this chapter provides a review, analysis, and discussion of various policy and research efforts encompassing economic and nonmonetary factors affecting energy use and decarbonisation efforts from a consumption-based perspective in Scandinavia. We focus on the individual/household level.

Overall, we argue that greater attention needs to be given to behavioural issues in policy design, implementation, and evaluation. Emerging research suggests that there are various opportunities to complement existing policy portfolios. However, caution is needed as behavioural interventions do not offer a panacea that can steer sustainable energy use, and ambitious price and nonprice interventions remain relevant. Thus, various complementarities and synergies need to be analysed in more detail. Ethical considerations also need to be taken into account, and policymakers need to embrace new directions for sustainable energy use. We conclude that a more stringent, integrated, behavioural, and technological policy approach is needed.

The chapter is organised as follows. We start by providing a short introduction to behavioural economics and its applications in the context of energy use and decarbonisation (Section 2). Section 3 offers a critical analysis of policy mixes targeting sustainable energy use and decarbonisation in Scandinavia. Section 4 presents four case studies that investigate economic and noneconomic factors, including the role of subsidies, prosocial behaviour, and peer effects in the adoption of solar photovoltaic (PV) technologies; the choice of different carbon pricing mechanisms and its effects on the willingness to pay for carbon emission reductions; the potential role of personal carbon trading given different socioecological and economic motivations; and the role of loss aversion and information framing for the effectiveness of smart metering. Policy implications resulting from these case studies are drawn in Section 5. Finally, our conclusions are presented in Section 6.

2 Behavioural economics for energy use and climate mitigation

Broadly speaking, behavioural economics (BE) explores the cognitive, motivational, and contextual factors that affect people's preferences, decision-making processes, and resulting choices. As Thaler (2015) argued, BE is economic analysis combined with key insights from psychology and knowledge from other social sciences. The following sections provide a brief introduction to this field in line with the scope and orientation of this chapter.

2.1 Basic concepts and foundations

The main aim of BE is twofold. First, it aims to provide a more *realistic* understanding of human economic behaviour by stressing the psychological and sociological foundations of economic analysis (Camerer, 1999). Second and with a strong orientation towards consumers (Reisch and Zhao, 2017), it aims to improve the *accuracy* of economic analysis, notably in three areas: theories, predictions, and policy (Camerer et al., 2011).

The development of BE was driven by growing dissatisfaction with neoclassical economics, particularly its disconnection from psychology and other social sciences (Camerer et al., 2011). Firstly, questions were raised about *rational choice theory*, which is the standard decision-making model in neoclassical economics. It is based on the *Homo economicus*, a rational agent with clearly defined preferences and unbounded optimisation skills, who never makes systematic mistakes and always seeks to maximise his/her utility. Secondly, there was particular discontent with the critical assumptions supporting rational choice theory, namely, rationality, self-control, and self-interest (Tomer, 2017). According to Mullainathan and Thaler (2000), BE underlines the fact that humans are rationally bounded (Simon, 1979, 1986) (i.e. have limited ability to process information and make rational, optimal choices), lack self-control (Thaler and Shefrin, 1981) (i.e. the inability to regulate one's feelings, emotions, or behaviour when confronted with a given task, activity, temptation, or choice), and can be selfless (i.e. individuals are concerned about other people, too) and thus express social preferences (e.g. altruism). Thirdly, dissatisfaction also grew with respect to the epistemological positioning of neoclassical economics. The fundamental arguments include a rather positivistic approach, notably the development of factual knowledge, devotion to quantifiable aspects, deductive reasoning, and the dismissal of subjective, qualitative aspects (Camerer et al., 2011; Tomer, 2017).

BE focuses on two, interwoven areas, namely, the analysis of intrinsic aspects (e.g. heuristics and biases) and extrinsic factors (e.g. the decision context) that can better explain economic behaviour (Reisch and Zhao, 2017; Thaler and Sunstein, 2008). On the one hand, BE investigates and understands systematic divergences from neoclassical economics. These intrinsic factors, or so-called *behavioural anomalies* (Gillingham and Palmer, 2014), are often described as patterns or deviations that explain why humans fail to behave as the rational theory model would predict (more details later) (Shogren and Taylor, 2008). When behavioural anomalies lead to a systematic difference between decision utility (i.e. expected or intended utility at the time of choice) and experienced utility (i.e. utility experienced after the choice) (Kahneman, 1994; Kahneman and Thaler, 2006), they are categorised as *behavioural failures* (Gillingham and Palmer, 2014) and provide better grounds for policy interventions. Within the context of sustainable energy use and climate change mitigation, examples of behavioural anomalies are heuristics, limited attention, loss aversion, and status quo bias (Frederiks et al., 2015; Gillingham and Palmer, 2014; Gowdy, 2008; Lopes et al., 2012; Steg et al., 2015; Tietenberg, 2009) (see examples in Section 2.2).

Heuristics (or 'rules of thumb') are simplified or intuitive decision-making rules that often lead to immediate but suboptimal choices or incorrect outcomes (Slovic et al., 2002). When confronted with multiple options (i.e. choice overload), heuristics can drive individuals to resort to *satisficing* (a mix of 'satisfying' and 'sufficing' as opposed to 'maximising' (Gigerenzer and Goldstein, 1996)), in other words, a decision-making strategy that achieves a satisfactory or acceptable ('good enough') result, rather than finding the 'best' or 'optimal' solution among different alternatives (Kahneman, 2003a; Simon, 1979). Limited attention is a cognitive constraint that is often defined as the rate at which the brain can process information (Dukas, 2004). Cognitive mechanisms restrict the brain to focus (only) on the most important information, and limited attention has already been identified as a scarce resource in the new era of information overload (Falkinger, 2008). Loss aversion is the tendency to place greater value on relative losses than gains (Kahneman, 2003b). Perceived gains and losses are dependent on and relative to a 'reference point' (Kahneman, 2003b; Kahneman et al., 1991). Loss aversion plays a role in *status quo bias* (Samuelson and Zeckhauser, 1988), which reflects the preference of individuals to do nothing and stick to default or given options (Hausman and Welch, 2010; Reisch and Zhao, 2017). Loss aversion has multiple, conceptual ramifications in BE (e.g. reference dependence, risk aversion, and status quo bias). For a detailed review, see Kahneman and Tversky (2000).

In addition to behavioural anomalies, BE also focuses on the specific effects of the decision context or *contextual factors* affecting decision-making processes and outcomes (Reisch and Zhao, 2017). In the domain of sustainable energy use and climate change mitigation, relevant contextual factors include, for example, decision framing, peer effects, and social norms (Abrahamse et al., 2005; Frederiks et al., 2015; Gillingham and Palmer, 2014; Lindman et al., 2013; Lopes et al., 2012). The framing of choices (e.g. via defaults and information settings) has been consistently highlighted as critical in multiple decision settings (Gillingham and Palmer, 2014; Hausman and Welch, 2010; Sunstein, 2014). As Hoeffler and Ariely (1999) argue, preferences are constructed and can be subject to contextual influencers, particularly when consumers are exposed to a new choice or task. Peer effects are usually examined in the context of economic analyses of social interactions (Manski, 2000). According to Bursztyn et al. (2014), the underlying (economic) mechanisms of peer effects are 'social learning' (i.e. peers learn from the choice made by other(s))and 'social utility' (i.e. ownership of a good/service directly affects others' utility of possessing the same good/service). Social norms are (in)formal social considerations or patterns that govern or influence the behaviour of a given group, community, or society at large (Abrahamse and Steg, 2013). The literature (Cialdini et al., 1990, 1991) differentiates two types of social norms: (i) descriptive norms that reflect the degree to which behaviour is perceived as collective and (ii) injunctive norms that denote the degree to which behaviour is supposed to be normally accepted or rejected.

From a methodological point of view, BE is *empirical* and *inductive* in nature (Camerer et al., 2011; Thaler, 2015). This is reflected in the methods that are used in BE research, which include experiments, surveys, and computer simulations (Tomer, 2017). Experiments are divided into three main categories: in the lab, in the field, and natural experiments. Borrowing heavily from medical research, BE experiments are often designed in a way that participants are randomly allocated to treatment and control groups (Gandhi et al., 2016; Hahn and Metcalfe, 2016). Combined with the use of statistical or econometric techniques (Thaler, 2015), they aim to test the internal and external validity of the study and its procedures (List and Price, 2016; Price, 2014). Given the growing interest of neuroscientists in economics, magnetic resonance imaging is also becoming part of the BE methodological toolkit (Camerer et al., 2005; Shogren and Taylor, 2008).

Finally, we acknowledge that it is difficult to draw a clear line between the multiple strands, tools, and ideas that BE encompasses. Given this caveat and based on the aforementioned, we propose a conceptual framework that encompasses and distinguishes several critical domains (see Fig. 1). By no means exhaustive, it is anchored in four main areas (Camerer et al., 2011; Thaler, 2015; Tomer, 2017): (cumulative)

- *Endowment effect/reference* (Kahneman et al, 1990, 1991; Thaler, 1981; Knetsch, 1989; Dinner et al, 2011)
- *Status-quo bias/reference* (Kahneman et al, 1991; Samuelson & Zeckhauser, 1988; Ritov & Baron, 1992; Camerer & Lovalo, 1999; Terrell, 1994)
- *Value function* (Tversky & Kahneman, 1992; Kahneman & Tversky, 1984; Tversky & Kahneman, 1981)
- *Loss aversion/reference* (Kahneman & Tversky, 1979; Shogren & Taylor, 2008)

- *Discounting* (Hyperbolic/Implicit) (Loewenstein & Thaler, 1989; Thaler, 1981; Shane, Loewenstein & O'Donoghue , 2002; Coller & Williams, 1999)
- *Risk (aversion) and time-varying decision* (Camerer & Loewenstein, 2004; Frederick et al, 2004; O'Donogue and Rabin, 2000; Loewenstein et al. 2003; Bell, 1985; Thaler & Shefrin, 1981)
- *Value commitment* (Ashraf et al, 2006; Green & Myerson, 1994; Della & Malmendeir, 2006)

Prospect Theory   **Intertemporal Choice**

Behavioural Economics

Conceptual framework for sustainable energy use and decarbonisation

Norms and Moral Behaviour

- *Fairness* (Kahneman et al., 1986; Cardenas & Carpenter, 2008; Fehr & Schmidt, 1999; Falk et al, 2008; Forsythe et al, 1994)
- *Cooperation* (conditional) (Ostrom, 1998; Frey & Meier, 2004; Fischbacher et al., 2001)
- *Reciprocity* (Croson et al, 2005; Fehr & Gächter, 2000; Gouldner, 1960; Falk & Fishbacher, 2006; Berg et al, 1995)
- *Warm-glow effect* (Andreoni, 1990; Crumpler & Grossman, 2008; Isen & Levin, 1972; Menges et al., 2005; Gneezy & Rustichini, 2000)
- *Norm-based motivation* (Andreoni et al, 2009; Brekke et al, 2003; Nyborg et al, 2006; Biel & Thogersen, 2007; Goldsmith 2011)

Cognitive Science & Bounded Rationality

- *Choice overload* (Schwartz, 2004; Iyengar & Lepper, 2000; Scheibehenne et al, 2010; Reed et al, 2011; Hogarth & Reder 1987; Smith, 1991; Fehr & Rangel, 2011)
- *Heuristics* (sub-optimal) methods (Simon, 1947; 1957; Camerer & Loewenstein, 2004; Thaler, 1991; Heath & Soll, 1996; Tversky & Kahneman, 1981; Tversky & Shafir, 1992)
- *Salience* (Kahneman, 2003; Avineri, 2012)
- *Satisficing behaviour* (Simon, 1947, 1972, 1979; March & Simon, 1963; Winter, 2000; Augier & March, 2002)
- *Self-deception* (Mijovetic & Prelec, 2010; Mazar & Ariely, 2006)

FIG. 1

Behavioural economics conceptual framework applicable to sustainable energy use and decarbonisation activities.

prospect theory (Kahneman and Tversky, 1979; Tversky and Kahneman, 1992), intertemporal choice (Loewenstein and Thaler, 1989; Loewenstein, 1988), norms and moral behaviour (Ajzen and Fishbein, 1972; Cialdini et al., 1991), and cognitive science and bounded rationality (Simon, 1947, 1986; Tversky and Kahneman, 1974). In the field of energy use and climate mitigation, Brekke and Johansson-Stenman (2008) and Pollitt and Shaorshadze (2011) make similar distinctions. Some of these aspects are developed in more detail in the following sections.

2.2 Applications in the field of sustainable energy use and climate action: A snapshot

Globally, residential sector energy use is responsible for a large part of carbon dioxide (CO_2) emissions. Despite growing concerns people express about climate change, a value–action gap persists between these concerns and households' actual energy use. While carbon pricing is one alternative to close this gap, BE research has repeatedly shown that people do not always behave rationally, meaning that price interventions are insufficient or can even backfire. Softer policy interventions, commonly known as *nudges*, have been proposed as complementary ways to reduce household energy use. This section outlines several recent studies that have applied BE to sustainable energy use. We examine the behavioural biases addressed and the interventions tested. We conclude with examples of applications in Scandinavia.

Interventions based on social norms have, so far, been the most popular type of nudge (Andor and Fels, 2018). People are often uncertain about their best course of action in different contexts, including energy use, and knowing what their neighbours do (descriptive norm) can reduce this uncertainty. Furthermore, providing information about social norms can trigger competition among households, which can also reduce energy use (Abrahamse et al., 2005). For high-intensity energy users, social comparison corrects their biased belief that their energy use is at an acceptable level and creates the notion that energy-saving practices are socially desirable. On the other hand, low-intensity energy users may increase their energy use, the so-called *boomerang effect*. One way to combat this is through an injunctive norm, which informs users that conserving energy is desirable (Allcott, 2011). A related concern is *moral licensing*, whereby households who learn that their energy use is low feel 'licensed' to do less in other proenvironmental areas, for example, by consuming more water (Tiefenbeck et al., 2013). It is therefore important to consider these unintended side effects when using nudges based on social norms.

OPOWER conducted a well-known, large-scale field experiment with 600,000 households. This study illustrated the effectiveness of including social norms in home energy reports (Allcott, 2011). Providing information about a household's energy use compared with similar neighbours, combined with tips for saving energy, yielded an average treatment effect of 2.03%, equivalent to 0.62 kWh per day. It was assessed to be at least as cost-effective as similar programmes and shows that running a large-scale behavioural intervention has become increasingly feasible and can overcome the common problem of underpowered samples (Allcott and Mullainathan, 2010).

Interventions based on commitment and goal setting aim to address bias in energy users who discount future payment streams compared with immediate ones. This bias results in underinvestment in energy-efficient appliances, or procrastination in making such investments, as the immediate cost is much more salient than the long-term benefits (Fuerst and Singh, 2018; Lillemo, 2014). A typical commitment nudge asks households to promise (publicly) to conserve energy, potentially with a set goal of a certain amount and deadline (Abrahamse et al., 2005). Such interventions create extra motivation for households to achieve the goal of conserving energy and simultaneously appeal to their self-image (for internal commitments) or public image (for external commitments). While there are early indications of positive effects, more substantial research is required as many studies suffer from underpowered samples, and it is often difficult to single out the effect of commitment or goal setting in experiments that study combined interventions (Abrahamse et al., 2005; Andor and Fels, 2018).

The type of information and how it is presented have also been identified as a low-cost intervention that can influence behaviour. Humans are limited in how much information they can process, and cognitive overload can lead to satisficing rather than optimising behaviour, hence the need to communicate information in a simple manner (Frederiks et al., 2015). This can take the format of easily understood energy labels on appliances or feedback on energy usage, as illustrated in the OPOWER example earlier (Abrahamse et al., 2007; Allcott, 2011; Andor et al., 2017; Ayres et al., 2013;

Faruqui et al., 2010). Loss aversion and the sunk-cost fallacy also contribute to households underinvesting in energy-efficient appliances. This can be addressed by framing information such that there is a focus on the cost of continuing energy-wasting practices and reducing the salience of the (sunk) cost of existing appliances (Frederiks et al., 2015; Pollitt and Shaorshadze, 2011).

Cognitive overload can also cause status quo bias, whereby households stick to preset options even if there are better alternatives. A good example is choosing energy contracts. Providing an energy-efficient default can increase the number of households choosing this option. The idea is to relieve them of the responsibility of having to actively make this choice, since a default can be perceived as a recommendation by the provider or energy experts (Momsen and Stoerk, 2014; Sunstein and Reisch, 2014). Similarly, household appliances can be preset to the energy-efficient setting by manufacturers to encourage households to use energy-efficient programmes on dishwashers or washing machines (Frederiks et al., 2015) or printing double sided (Egebark and Ekström, 2016).

In Scandinavia, BE interventions addressing sustainable energy use and climate action are limited, but growing in popularity. For example, one study in Sweden showed that greater attention needs to be given to proenvironmental behaviour and social norms (Lindman et al., 2013). It was found that willingness to purchase carbon allowances from the European Union Emission Trading Scheme (EU-ETS)—and permanently retire them to reduce emissions—is driven not only by price mechanisms but also by proenvironmental motivations and social norms. That is, people who feel personally responsible for reducing the impacts of climate change appear to be more inclined to buy carbon allowances when they know that others are doing the same. In Denmark, another study explored default settings and reference dependence (Ölander and Thøgersen, 2014). It addressed the acceptance rate of installing smart metering (SM) technology based on default settings. It found that the acceptance rate was significantly higher (80% approx.) if the installation of the SM was presented as the default (i.e. customers had to opt-out) than the same option (60%) being framed as opt-in (i.e. customers had to actively choose the SM installation) (Ölander and Thøgersen, 2014). In Norway, research addressing behavioural anomalies and policy interventions found that individuals with a higher predisposition to procrastinate were less willing to participate in energy-saving programmes (Lillemo, 2014). The reviewed literature suggests that further research is needed to determine whether identified behavioural anomalies lead to systematic differences between decision utility and experienced utility.

3 Policies for sustainable energy use and decarbonisation in Scandinavia

The aim of this section is to provide a brief review of policy mixes targeting sustainable energy use and decarbonisation in Scandinavia. We first use different indicators and trends to question the effectiveness of previous policy interventions from a consumption perspective. After collecting diagnostic evidence about energy use emissions, we characterise policy portfolios and discuss to what extent policies address technology uptake versus behavioural change and behavioural anomalies versus market failures. We address the potential 'divide' between conventional policy instruments and behavioural interventions and examine how a lack of behavioural interventions might explain trends regarding emissions associated with consumption.

3.1 Energy use and carbon emissions: Do we need further policy interventions?

Here, we briefly investigate the need to foster sustainable energy use and emission reductions by looking at some key indicators in the context of energy use and consumer behaviour. By no means exhaustive, we use figures and trends in three domains: energy use in buildings, consumption-based carbon emissions, and emissions from aviation.

First, it is important to acknowledge that many factors explain energy use (e.g. end-use technology, income, population growth, quality of building stock, and energy pricing) and that it is not limited to behavioural aspects and policy (Grubler et al., 2012). This said, average electricity consumption per capita was relatively high across Scandinavian countries (14.3 MWh) compared with the OECD region (8 MWh), OECD Europe (5.9 MWh), and the world (3.05 MWh) in 2015 (IEA, 2017a) (see also Fig. 2). For the particular case of energy use in buildings, high consumption

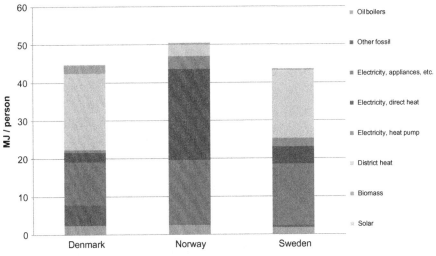

FIG. 2

Energy demand per capita in urban buildings in 2013.

Data from IEA, Nordic Energy Research, 2016. Nordic Energy Technology Perspectives 2016: Cities, Flexibility and Pathways to Carbon-Neutrality. Nordic Energy Research, International Energy Agency, Oslo, Paris.

is explained by high demand for heating in winter, the expansion of floor area per person, and building stocks badly in need of retrofitting (IEA and Nordic Energy Research, 2016). For instance, in Sweden, studies show that energy use per floor area has increased since 1995 (Nässén and Holmberg, 2005) and that energy use for heating is twice as high in new buildings compared with the best-performing buildings constructed in the 1980s (Nässén et al., 2008). Driven by deficiencies in public policy, similar aspects are found in Norway—including a lack of innovation in the building industry (Ryghaug and Sørensen, 2009). In Denmark and Sweden, per capita appliance intensity (kilowatt-hour/appliance unit/per person) has grown or remained unchanged since 2000, and no consistent downward trends are observed (see Fig. 3).

As the population in the region has become wealthier, increases in floor area and transport demand per capita are also observed (IEA and Nordic Energy Research, 2016), which, in turn, has offset important efficiency gains in technology. Inertia in efficiency improvements is also explained by relatively low energy prices (IEA and Nordic Energy Research, 2013; Nässén et al., 2008) particularly in Norway and Sweden. While average heating intensity (kWh/m^2) improved by 0.5% in the Nordic region (i.e. Scandinavia plus Finland and Iceland) for the period 1990–2013, modelling studies exploring net zero emission pathways indicate that this is insufficient and that improvements in the range of 2.5% per year are needed (IEA and Nordic Energy Research, 2016). Together with the extensive retrofitting of existing buildings (IEA and Nordic Energy Research, 2016; Nässén and Holmberg, 2005), it has been argued

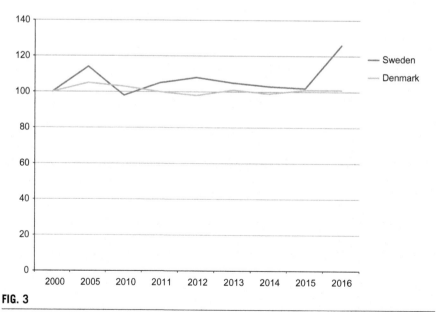

FIG. 3

Per capita appliance energy intensity (index 2000) in the residential sector.

Data from IEA, 2017. Energy Efficiency Indicators 2017 Database. OECD/IEA, Paris. Note: Data not available for Norway.

that behavioural change is needed to foster stringent mitigation pathways in the building sector (IEA and Nordic Energy Research, 2016).

Second, whereas production-based emission accounting is currently more common in methodological and policy terms (Davis et al., 2011; Peters, 2008), consumption-based approaches can reveal a different picture. In fact, the use of both approaches may help in reducing international carbon leakage by providing a more appropriate and less misleading picture of countries' emission reduction policies (Davis and Caldeira, 2010; Larsen and Hertwich, 2009; Peters, 2008). For Scandinavia, quantitative studies show very-high-carbon emissions embodied in imports that satisfy domestic consumption (Mundaca et al., 2015; Peters and Hertwich, 2006), resulting in a negative emission trading balance (see Fig. 4). As a whole, net imports of emissions have grown by nearly 95% since 1990, an average of 4.1% per year. However, figures show large discrepancies among countries: net imports grew by 362% in Denmark, 169% in Norway, and 38% in Sweden for the period 1990–2015.[c]

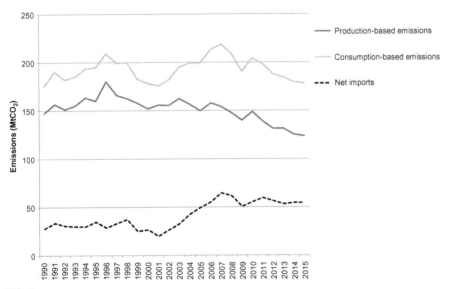

FIG. 4

Production- and consumption-based CO_2 emissions in Scandinavia and resulting net imports.

Data from Peters, G.P., Minx, J.C., Weber, C.L., Edenhofer, O., 2011. Growth in emission transfers via international trade from 1990 to 2008. Proc. Natl. Acad. Sci. 108, 8903–8908; Marland, G., Boden, T., Andres, R., 2017. Global, Regional, and National Fossil-Fuel CO2 Emissions. Carbon Dioxide Information Analysis Center, Oak Ridge National Laboratory, U.S. Department of Energy, Oak Ridge; UN, 2017. National Inventory Submissions 2017|UNFCCC [WWW Document]. https://unfccc.int/process/transparency-and-reporting/reporting-and-review-under-the-convention/greenhouse-gas-inventories-annex-i-parties/submissions/ national-inventory-submissions-2017.

---

[c] The Global Carbon Atlas was used as dataset for the specific case of territorial consumption and CO_2 emissions. Available at http://globalcarbonatlas.org/en/CO2-emissions.

When carbon emissions related to the final consumption of goods and services are estimated, some similarities can still be identified. For instance, carbon footprints for mobility, services, and manufactured goods are relatively higher than other categories (e.g. clothing and construction), and their contributions to household emissions are in the range of 28%–34%, 18%–23%, and 10%–14%, respectively, across countries (Hertwich and Peters, 2009). As Scandinavian nations have become wealthier, travel frequency and distances travelled have also increased (IEA and Nordic Energy Research, 2013, 2016). While progress in reducing production-based (or territorial) emissions can be discerned (IEA and Nordic Energy Research, 2016; Mundaca et al., 2015), these trends suggest that greater policy efforts are needed to reduce consumption-based emissions.

Third, accounting for international aviation by apportioning its emissions to national or even subnational jurisdictions can help to point out respective contributions and responsibilities and highlight the need for specific policy interventions (Wood et al., 2010). National accounting of greenhouse gas (GHG) emissions in Scandinavia (and all other countries) disregards emissions from international shipping and aviation. CO_2 from international transport is growing at a higher rate than the average for all other sectors (Bows-Larkin, 2015). Projections for the aviation sector foresee emissions growth of 140% between 2013 and 2050 (Kuramochi et al., 2018). The contribution of aviation to global CO_2 emissions may reach 22% by 2050 (Cames et al., 2015). Historical trends in Scandinavia are similar. Swedish residents' international air travel emissions increased by 61% between 1990 and 2014, when they reached the same level as emissions from Swedish car traffic (Larsson et al., 2018), which is another illustration of the accounting gap if a national, production-based emission approach is chosen. The United Nations Framework Convention on Climate Change (UNFCCC) GHG inventory data reveal even stronger growth in emissions from international aviation bunkers in Denmark (+63% between 1990 and 2016) and Norway (+141%). Air travel in Scandinavia is expected to continue to increase in the future, for example, by 3.6% annually in Sweden until 2022 (Andersson and Falck, 2017).

This brief analysis suggests that policy mixes have been unable to effectively drive sustainable patterns of energy use and decarbonisation across Scandinavian countries, particularly from a consumption point of view. To explain why policy mixes have not driven sustainable energy use in the past, it is useful to investigate the composition of policy portfolios and examine the extent to which these policies have addressed behavioural change.

3.2 Policy portfolios: Development and focus

From a methodological point of view, this section is based on a directed content analysis (Hsieh and Shannon, 2005), which is understood as an approach that examines and compares documentary evidence in an inductive manner (Elo and Kyngäs, 2008). The review covers the period 1990–2015 and follows the taxonomy of policy instruments proposed by the Intergovernmental Panel on Climate Change (IPCC) (Kolstad et al., 2014). To carry out the content analysis, four categories were defined: (1) geographical coverage, (2) policy instruments at the national level, (3) orientation of policy design and mitigation actions (e.g. market failures, behavioural anomalies, and demand vs. supply side), and (4) policy status. For category 2, this includes the following policy interventions (coding in parenthesis): (1) economic incentives (e.g. taxes, subsidies, and emission trading), (2) regulatory approaches (e.g. building codes and minimum performance standards), (3) information schemes (e.g. education, public awareness campaigns, and labelling programmes), (4) voluntary actions (e.g. GHG emission reduction pledges and sectoral energy efficiency targets), and (5) provision of public goods and procurement (e.g. investment in public transport and procurement rules). The IEA Policies and Measures Database was our main dataset.[d] Policies were coded as 'implemented' (1) or 'ended' (0) for the period under analysis. The IEA Energy Efficiency Indicators Database[e] was used to address energy use in the residential sector.

Four main issues emerge when analysing the evolution of policy mixes in Scandinavia. First, there has been significant growth in interventions that address energy use and decarbonisation. While the number of implemented policies was at a minimum before the mid-1990s, a peak of 127 was reached in 2012 (see Fig. 5). The average number of policies implemented on an annual basis was estimated to be approximately 4.5 for the period 1990–2015. By the end of the period, 118 policies

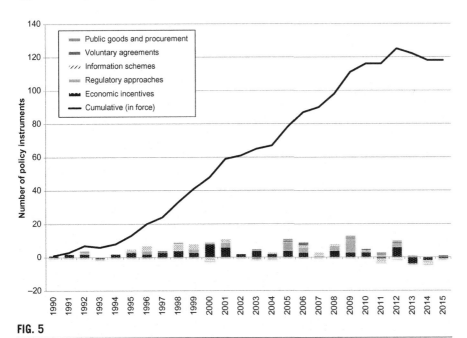

FIG. 5

Evolution of policy interventions in Scandinavia for the period 1990–2015. Bars show net annual additions (i.e. implemented policies minus terminated policies). Cumulative (in force) policy interventions are shown by the *solid line*.

---

[d] Available at https://www.iea.org/policiesandmeasures/.
[e] Available at https://www.iea.org/statistics/efficiency/.

had been in force; numbers for Norway and Sweden were relatively similar (43 and 46, respectively), while Denmark lagged behind (29).

Second, economic instruments have dominated the development of policy portfolios (see Fig. 6). By 2015, nearly 50% of interventions that promoted sustainable energy use and decarbonisation efforts operated via market-based incentives (almost 40% in Denmark and Sweden and 63% in Norway). The 2012 peak in active policies was concomitant with the highest number of effective economic instruments (65) in the region.

Third, the foci of policy portfolios vary over time, with a growing trend for demand-side interventions. Unlike Denmark and Sweden, which have increased demand-side efforts, Norway has maintained a relatively equal balance between supply and demand-side interventions. For the entire period under analysis, the share of all demand-side policies (implemented and withdrawn) reached 67% in Denmark and 71% in Sweden, while it was only 55% in Norway.

Fourth, policy portfolios tend to address market failures, while much less attention is given to behavioural anomalies and contextual factors affecting energy use and decarbonisation measures. The content analysis reveals a traditional policy mix that is restricted to the internalisation of external costs, the reduction of financial constraints, and the moderation of information. Cognitive, motivational, and contextual aspects were almost always consigned to information schemes. The latter includes public awareness campaigns, labelling of products, training, and advice programmes.

From a city-level perspective, Scandinavian cities tend to focus their policy portfolios on LCET adoption and associated infrastructure. In Sweden's capital, for instance, the 'Strategy for a fossil-fuel free Stockholm by 2040' outlines technological solutions for sustainable energy use, transportation, and resource-efficiency, but does not explicitly consider behavioural aspects (City of Stockhom, 2016). In Denmark, Copenhagen's climate mitigation policy is also focused on LCET solutions, although some behavioural interventions are explicitly mentioned in the city's short-term roadmap (2017–20), notably the promotion of cycling to work and energy-saving behaviours for municipal employees (City of Copenhagen, 2016). Overall, there are only a few, rare examples of the explicit consideration of behavioural change, let alone behavioural anomalies in municipal climate portfolios (Sonnenschein, 2016). Moreover, sectors that are particularly relevant in the context of behavioural change, such as the climate footprint of consumption or travel (including air travel), are typically outside the scope of municipal climate policy.

As indicated in Section 2.2, the evidence that we found of policy interventions that explicitly address behavioural aspects in Scandinavia seems to be confined to experimental settings, pilot studies, and/or small-scale utility-driven initiatives. Although at the global level there seems to be a growing institutionalisation of behavioural insights informing policymaking (OECD, 2017), our review seems to question this for the region. This is despite the fact that we found significant growth in policy instruments and a relatively balanced mix between supply and demand-side interventions. However, both a traditional policymaking approach and a resulting policy mix focused on technology markets and related market failures

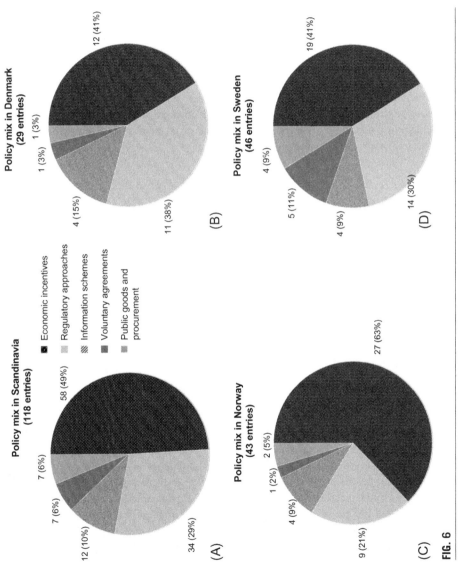

FIG. 6

Absolute and relative composition of implemented policy mixes for Scandinavia (A) and per country (B–D) in 2015.

appear to be apparent. Our findings are consistent with recent studies addressing global policy portfolios (Mundaca et al., 2019) and demand-side policies across Norway and Sweden (Moberg et al., 2018). The findings seem to support the information-deficit model (Blake, 1999; Owens, 2000) in policymaking. Here, a central assumption is that the provision of information per se supposedly enables end-users to make more rational decisions about their energy use and/or decarbonisation measures (Bager and Mundaca, 2017). This approach is, however, likely to be deficient because it ignores how information is actually understood, used, or put into practice.

While further work is clearly needed, our analyses of emissions from energy use and the review of policy portfolios suggest that the traditional policymaking approach may be inadequate and that innovative and ambitious policy approaches that target behavioural aspects more explicitly should be given more attention in policy design, implementation and evaluation.

4 Case studies

In the following sections, we elaborate on four case studies that explore economic and noneconomic factors affecting energy use and decarbonisation activities. The section stresses the need to better understand and integrate behavioural aspects in economic and technology assessments. This is critical to provide more comprehensive analyses and better guidance to policymakers.

4.1 Untangling the drivers of solar PV adoption

Interest in solar energy and PV cell systems has grown steadily. In Sweden, 79.2MW were installed in 2016, representing an increase of 63% compared with the amount of power installed in 2015 (Lindahl, 2016).[f] Despite this fast growth, however, solar-powered electricity makes up only 0.09% of Sweden's total annual electricity production.[g]

Realising that investment costs were a key barrier to household adoption of PV, the Swedish government instituted two major policy changes. In 2009, households were offered a subsidy that contributed to the installation cost of solar PV systems. In 2010, the government started to promote net metering, by removing a fee levied on households that sold excess electricity to electricity companies, thus making it more profitable to install PV. However, uncertainty remains regarding the subsidy scheme, with changes often made at short notice. The high number of applications combined with an upper cap, which varies depending on the government's budget, has led to

---

[f] The International Renewable Energy Agency estimates the capacity at 140MW in 2016 (IRENA, 2017).

[g] For details see https://www.scb.se/hitta-statistik/sverige-i-siffror/miljo/energi/.

long waiting times (Lindahl, 2016). These uncertainties only serve to magnify the cost barrier and have been identified as a key issue for further PV development in Sweden (Palm, 2015).[h]

On top of financial factors, peer influence has been identified as an important driver of PV diffusion both in Sweden and abroad (Palm, 2016, 2017; Rai et al., 2016). Additionally, prosocial behaviour theory predicts that increased visibility will increase agents' prosociality; however, introducing monetary rewards may result in crowding-out effects (Bénabou and Tirole, 2006). The visibility of PV within the neighbourhood and households' proenvironmentalism are, thus, two factors that need to be considered when forming policies that seek to encourage proenvironmental behaviour. However, a lack of quantitative insights hinders a better understanding of the decision-making processes in the solar PV context in Sweden.

Our study sought to understand and quantify the effects of subsidies and non-monetary factors affecting the likelihood to adopt solar PV in Sweden ($N = 208$). It consisted of a survey sent to 208 Swedish house owners who do not currently have PV.[i] The key outcome variable was respondents' likelihood of installing PV given: the percentage of investment costs covered by the subsidy, the visibility of the PV installation (main residence or holiday home), and peer effects, in terms of seeing or hearing about PV from (un)familiar sources. These explanatory variables were all hypothesised to positively affect the likelihood of adopting PV. The latter ranged from 1 to 4, where 1 is very unlikely and 4 is very likely. Each respondent was randomly assigned to a treatment group with a specific subsidy percentage, which varied from 0 to 100 in increments of 10. Respondents answered the survey with respect to their primary or holiday home (visibility equals 1 or 0, respectively). This was randomly assigned for respondents who owned both types of home. Peer effects were measured as the total frequency of seeing or hearing about PV in the last month, ranging from 0 (never), to 1 (at least once), to 2 (at least weekly), to 3 (at least daily). The survey ended with questions about respondents' reasons for not yet installing PV, their level of environmentalism, and demographics.

First-order effects were investigated with a correlation analysis between our dependent variable *likelihood* and the explanatory variables *visibility*, *peer effect*, and *subsidy*.[j] Our hypotheses regarding monetary incentives and peer effects appear to be confirmed, as both were positively and significantly correlated with likelihood (*P*-value = 0.48 and 0.22, respectively). However, the effect of visibility was not statistically significant. No cross-correlation coefficients were significant, indicating few interaction effects, if any. These results were confirmed by an ordered logistic

---

[h] Other studies have commonly found that cost is a barrier to adoption and that providing financing solutions can lead to a substantial increase in installations (Arvizu et al., 2011; Crago and Chernyakhovskiy, 2017; Hughes and Podolefsky, 2015; Kirkpatrick and Bennear, 2014; Mitchell et al., 2011).
[i] A complete description of the survey, including descriptive statistics, is available in the working paper (Samahita and Mundaca, 2018).
[j] Full results are provided in the working paper (Samahita and Mundaca, 2018).

regression of the likelihood of adoption for *visibility*, *peer effect*, *subsidy*, and their interactions. The model with only main effects explains 31% (Nagelkerke R^2) of the variance of PV adoption, though visibility is not significant. A unit increase in *peer effect* increases the odds of a higher likelihood of installing PV by 1.48, while a 10% increase in the subsidy increases the odds of a higher likelihood by 1.04. When interactions are included, none are found to be significant, and their additions do not substantially increase the explanatory power of the model in terms of R^2 values.

Main effects are robust after controlling for other relevant, demographic variables, while odds ratios for *peer effect* and *subsidy* are of similar magnitude (1.46 and 1.04, respectively). The inclusion of control variables raises Nagelkerke R^2 to 35.9%. Higher levels of environmentalism increase the likelihood of adoption, while older people are marginally less likely to adopt.

Our findings reveal that both the subsidy scheme and peer effects are important underlying mechanisms affecting the likelihood to adopt solar PV. However, the visibility of solar PV installations is insignificant. Contrary to indications from the reviewed literature, interactions among main predictors do not matter, and crowding-out effects are not identified. Peer effects come mostly from hearing about PV installations. Nevertheless, both seeing and hearing have an impact on the likelihood to adopt, if the source of information is known to the potential adopter. All in all, we conclude that the national subsidy scheme and, by extension, its long-term certainty are critical to supporting the adoption of solar PV in the country. Measures to strengthen peer effects should also be considered by policymakers, as will be elaborated on in Section 5.

4.2 Exploring payment mechanisms and their effects on willingness to pay for carbon emission reductions

Environmental economic theory suggests that a uniform carbon price is the most efficient policy instrument to mitigate climate change. In practice, carbon pricing covers only 15% of global GHG emissions, and implemented prices are around 15 EUR/tCO_2 (World Bank et al., 2018), which is clearly below what is needed to reach the goals of the Paris Agreement (Stiglitz et al., 2017). In theory, efficient and explicit carbon tax rates are determined by equalising marginal climate change damages with marginal (technical) abatement costs (MAC). In practice, however, various market and behavioural failures imply that technology choices and energy consumption do not follow the marginal abatement cost curve (Gillingham and Palmer, 2014). Sustainable energy use and incentives for climate action depend on contextual factors such as personal influences (attitudes, awareness, and beliefs), attributes of the required behavioural change (comfort or service level), and the value people give to associated environmental benefits (Clayton et al., 2015; Wilson et al., 2015).

Perceived environmental benefits of consumption decisions can be seen as complementary to MACs. In the context of climate change, they are commonly quantified by eliciting people's willingness to pay (WTP) for mitigation. The current literature provides data about WTP for mitigation in very specific settings and sectors, including air travel (Brouwer et al., 2008; MacKerron et al., 2009), green electricity (Adaman et al., 2011; Nomura and Akai, 2004; Sagebiel et al., 2014; Wiser, 2007), national climate policy (Kotchen et al., 2013; Uehleke, 2016), and offsetting emissions by purchasing and retiring emission allowances (EUA) from the EU emission trading system (Diederich and Goeschl, 2013; Löschel et al., 2017). These studies have resulted in WTP estimates that range from around $5\,EUR/tCO_2$ to more than $100EUR/tCO_2$, suggesting low external validity and making these estimates of little use in the Swedish policymaking context.

To address differences across WTP studies and the lack of policy-relevant data in the Swedish context, we conducted a contingent valuation study ($N = 500$) that compared four payment vehicles to elicit WTP for climate mitigation: climate surcharges on short-distance flights, on long-distance flights, and motor vehicle fuels and voluntary offsetting by buying and cancelling EUAs (Sonnenschein and Mundaca, 2019; Sonnenschein and Smedby, 2019). Respondents were presented with all of these payment vehicles (in random order), and their WTP was elicited per flight for the surcharge on short- and long-distance flights and per litre for the fuel surcharge. These values were converted into WTP/tCO_2 to have a common unit for comparing estimates.[k]

When measured per tonne of CO_2, mean WTP is significantly higher for the surcharge on short-distance flights than for any other payment vehicle, while all surcharges are associated with significantly higher WTPs than voluntary offsetting with EUAs (see Fig. 7 and Table 1 later). The low WTP for EUA offsetting can be largely explained by the fact that merely 29% of respondents were willing to pay for climate change mitigation via this payment vehicle in principle, while only 5% had actually purchased offsets before. This finding supports existing empirical evidence that coercive payment vehicles are preferred to voluntary payment vehicles and elicit higher WTP values in a sustainable energy context (Ek and Söderholm, 2008; Menges and Traub, 2009; Segerstedt and Grote, 2016; Wiser, 2007). Moreover, the significantly higher WTP/tCO_2 for short- compared with long-haul flights supports the low-cost hypothesis (Blasch and Farsi, 2014; Diekmann and Preisendörfer, 2003). The latter claims that in a context with lower absolute costs for environmentally conscious behaviour, WTP for the respective good is higher. The absolute cost of paying a surcharge is much higher for long-haul than short-haul flights because CO_2 emissions from long-haul flights are about four times higher.

When investigating the drivers of WTP for a plane ticket surcharge by means of regression analysis (Sonnenschein and Smedby, 2019), several policy-relevant factors were identified. Respondents' WTP for an air ticket surcharge was higher: as their income increased, if they had left-leaning political views, if they felt responsible

---

[k] See Sonnenschein and Mundaca (2019) for details regarding survey design, sampling procedure, WTP elicitation process, and conversion factors from WTP/flight and WTP/L to WTP/tCO_2.

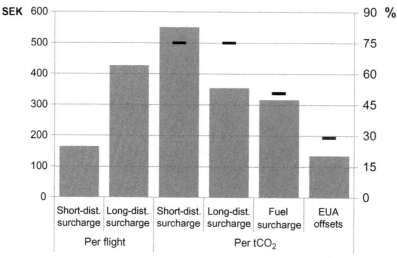

FIG. 7

Mean WTP (columns, left axis) and the percentage of people willing to pay in principle (dashes, right axis) for the four payment vehicles. Mean WTP for car travel before transformation to SEK/tCO$_2$ is 0.79 SEK/L of fuel (The average exchange rate at the time of study (2017/18) was 10 SEK/EUR (source: xe.com)).

Based on Sonnenschein, J., Mundaca, L., 2019. Is one carbon price enough? Effects of payment vehicle choice on willingness to pay. Energy Res. Soc. Sci. 52: 30–40. http://dx.doi.org/10.1016/j.erss.2019.01.022.

for their emissions, and if they wanted revenue from the surcharge to be earmarked for climate mitigation and sustainable transport projects. In contrast, being a frequent flyer significantly reduced average WTP for an air ticket surcharge, which is a first indication of free-riding behaviour and an argument in favour of mandatory interventions.

Table 1 Differences between mean WTP for the four payment vehicles (in SEK/tCO$_2$)

Difference in mean WTP	Long-distance surcharge	Fuel surcharge	EUA offsets
Short-distance surcharge	196***	234***	415***
Long-distance surcharge		39**	220***
Fuel surcharge			181***

Based on Sonnenschein, J., Mundaca, L., 2019. Is one carbon price enough? Effects of payment vehicle choice on willingness to pay. Energy Res. Soc. Sci. 52: 30–40. http://dx.doi.org/10.1016/j. erss.2019.01.022.
All differences are significant at the 1% level in Wilcoxon signed-rank tests. For two-tailed t-tests
**** indicate significance at the 1% level and ** at the 5% level.*

4.3 **Analysing personal carbon trading via an agent-based model**

It is argued that one of the more ambitious carbon pricing interventions targeting the household sector is personal carbon trading (PCT). Although it has been repeatedly debated, it has never been implemented on a large scale (Eyre, 2010; Fawcett et al., 2010; Fawcett and Parag, 2010). The basic feature of a PCT scheme is that households are given a CO_2 allowance that they may use to cover their carbon-based energy use at home or for mobility. This amount is reduced over time, giving households an increasing incentive to limit their carbon footprint, for example, by implementing energy efficiency measures.

Current empirical support for the effectiveness of PCT is scarce (Parag and Fawcett, 2014). Although some studies have been conducted (Capstick and Lewis, 2010; Kempener, 2009; Li et al., 2015; Tang et al., 2015), it remains unclear how households would respond to different PCT schemes and what this would mean in terms of emission reduction potential. Because real-world implementations are difficult, computer modelling approaches with artificial populations, such as agent-based modelling (ABM) and social simulation (Bonabeau, 2002; Grimm et al., 2010; Squazzoni et al., 2014), offer a methodology to analyse behavioural aspects. This is consistent with methodological approaches used in BE as described in Section 2 (see Tomer, 2017).

We ran an ABM simulation[l] to explore PCT dynamics by focusing on *heterogeneity*. Differences in households' energy-related behaviour differ with respect to (among others) their income class, household size, and relative socioecological and economic motivation. These distinctions are unusual in the current PCT modelling literature; however, we argue that the model is more realistic if heterogeneous (Coad et al., 2009) rather than homogeneous rational agents are considered.

Our model considers the role of household size, income group, and heterogeneity in motivation (cf. the *consumat* approach (Janssen and Jager, 2002)). Each agent is given a specific value for both socioecological and economic motivations. The simulation aims to help us understand social learning effects and the relative impact of socioecological[m] versus economic concerns. According to the available literature, different possible behaviours in response to a PCA scheme are more or less likely to be adopted (Wallace et al., 2010). Our simulation examined investment in carbon reduction actions such as private transport, energy-saving products/technology, and energy usage, following Zanni et al. (2013).

---

[l] Details of the model can be found in the report (Seidl, 2018), and a basic version is accessible at OpenABM via https://www.comses.net/codebases/14ed682a-a0ed-4b55-8e86-ad99df4d2187/releases/1.0.0.

[m] For instance, one study found that protecting the environment was the strongest driver for photovoltaic investments (Braito et al., 2017). As people differ with respect to this motivation, assuming a uniform population would likely overestimate the potential for investments.

Our population of agents represented households of different sizes (from 1 to 5 and over 5 members[n]) and 10 monthly income classes. It was applied (and roughly calibrated) to the situation in Denmark's residential sector, and initial values were collected from available Danish data. Denmark's residential sector consumes most of the country's energy, accounting for 32% of total final energy consumption in 2015 (International Energy Agency, 2017). This is despite the introduction of various carbon and energy taxes, which made up 67.8% of the final price of energy in 2016 (Energy, transport and environment indicators, 2017). The model estimates the price of one allowance (i.e. 1kg of carbon) to be 0.033 Euro.[o] This pricing considers CO_2 to be independent of the fuel source (Burgess, 2016).[p] We initialized a population of 10,000 agents, and each simulation ran 120 time steps (10years). The initial total allocation of allowances matched current average emissions and was divided equally among the population (i.e. on a per capita basis). This meant that some households received more and some received less allowances than they would need. Allowances were reduced by 2% per year over the simulation period.

Although the ABM exercise is clearly experimental, preliminary results can be summarised as follows. In general, there is an initial *status quo bias*, as households resist change even if this would be economically rational. Wallace et al. (2010) argue that it is cognitively demanding to understand certain, key aspects of PCT, notably related to 'cap and trade' and the declining cap. Our results suggest that the households differ with respect to their learning, that is, how fast they adapt or react to the trading scheme. As hypothesised, those with relatively higher economic motivation are more active; however, others adapt via social learning (Bandura, 1969). In fact, the results show that the likelihood of households selling their allowances is higher if neighbouring households also sell, which suggest a degree of normative behaviour. This can be seen in the spatial pattern that emerges (Fig. 8), which resembles a camouflage design (darker-coloured households are more active). As expected and as a function of the allowance price (Fig. 9A), some, more economically driven households, also invest in carbon mitigation measures that reduce their demand below the allowed limit, which enables them to sell their excess allowance to the market (Fig. 9B).

In general, our simulation shows that higher percentages of 'annual allowance reduction' result in more year-on-year trading activity as more households need more allowances, followed by sharp decreases (producing a sawtooth pattern, see Fig. 9D). These annual 'reminders' prompt agents to reduce their CO_2 emissions by investing, suggesting that annual reductions 'motivate' their green consciousness. However,

---

[n] Neither children nor age were explicitly considered.
[o] 32.89 Euro cents per kWh (2017)=0.033 in the model. See https://www.statista.com/statistics/596329/electricity-household-price-denmark.
[p] In 2015, Denmark's CO_2 electricity intensity roughly accounted for 345 g/kWh. See https://ens.dk/sites/ens.dk/files/Statistik/figures2016.xlsx.

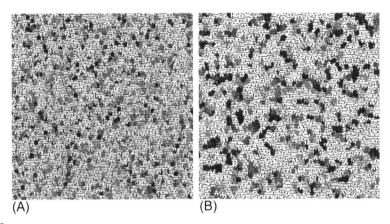

FIG. 8

The visual pattern after initialization (A) and after 120 time steps (B). Households (depicted as a house shape) selling few *(light colour)* or many *(dark colour)* allowances are not arbitrarily distributed across the virtual settlement area, but develop into clumps (although initialized randomly).

Fig. 9C also shows that households that are motivated by socioecological concerns mitigate more that those motivated by economic concerns. This finding is consistent with previous evidence from ABM (Kempener, 2009), from experimental simulation (Capstick and Lewis, 2010), and with the experience from voluntary carbon rationing action groups (Howell, 2012), which showed that people with proenvironmental attitudes tend to be more supportive of PCT.

Further behavioural aspects that were identified in previous studies of PCT interventions include carbon budgeting effects and fairness considerations. The carbon budgeting effect associated with PCT means that people may adapt their behaviour simply by being aware of their carbon budget and its limits. It could be shown that this carbon budgeting effect occurred in addition to the financial incentive effect of carbon pricing via PCT or carbon taxes (Capstick and Lewis, 2010; Zanni et al., 2013). Regarding the acceptance of PCT, which was shown to be substantial in several surveys (Jagers et al., 2010; Wallace et al., 2010), it appears to be important that the allocation of allowances is carried out according to rules that are perceived as fair, including aspects such as household size, children, and available income (Bristow et al., 2010; Jagers et al., 2010).

With due limitations, these results suggest that behavioural aspects add a dynamic layer that cannot be observed in traditional, cost-effective analyses of carbon trading schemes. In turn, they also suggest that more consideration needs to be given to understanding the changes in behaviour targeted by policies (like PCT) that aim to encourage the widespread transformation of daily energy use behaviour (cf. Steg, 2018).

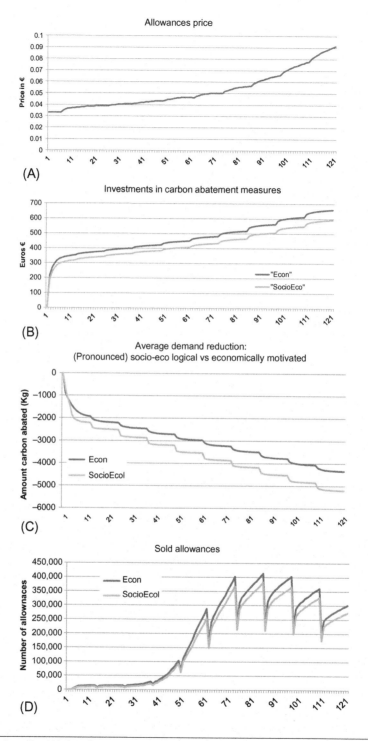

FIG. 9

Selected results: (A) the allowance price trend, (B) investments in abatement measures, (C) average demand reduction for 'motivated' agents (socioecological/economic motivation values exceed other dimensions by 0.3), and (D) allowances sold by more economically or socioecologically motivated households.

4.4 Investigating salient loss aversion via smart metering

In the European Union (EU), the introduction of smart metering (SM) is expected to result in a 10% reduction in energy consumption in the residential sector (EC, 2011). A prime postulate in the policy discourse is that the information provided by SM enables energy users to make more rational decisions about their energy service demands (e.g. lighting).[q]

Despite criticism of the effectiveness of SM, the literature highlights that little is known about the behavioural factors that may affect how energy users interact with SM technology, notably the role of cognitive limitations, norms, and beliefs regarding SM technology (Darby, 2010; Gangale et al., 2013; Gerpott and Paukert, 2013; Kaufmann et al., 2013; Krishnamurti et al., 2012; Nachreiner et al., 2015; van der Werff and Steg, 2016). In fact and in line with traditional policy-making, behavioural aspects have consistently been overlooked. Implementation efforts have focused heavily on technical and economic issues such as investment needs, cost–benefit analyses, grid connectivity, participation of network operators, and the responsibilities of utilities (Depuru et al., 2011; EC, 2014; Faruqui et al., 2010a; Giordano et al., 2013). The latter is consistent with the emerging literature on the application of BE to energy use, which concludes that cognitive biases and motivational factors are frequently ignored by both policymakers and practitioners (Frederiks et al., 2015). Gillingham and Palmer (2014, p. 13) highlight this knowledge gap and argue that 'more research is needed that closely examines both the information provision context and the behavioural failure'. To the best of our knowledge, there has been no investigation of the effect of loss-framed salient information provided by SM in a Scandinavian country. We developed a case study to address this gap.

A pilot field experiment[r] was conducted to investigate the potential impact of salient loss aversion on electricity use via a nonprice intervention for a group of SM users in Copenhagen, Denmark.[s] We hypothesised that consumers would underestimate potential gains and overestimate potential financial losses resulting from electricity use. The experiment used a so-called 'add-on' SM. These SM are added to an analogue meter to collect, store, and transmit data. Users accessed information via a software installed on their smartphones, tablets, or computers. Initially, 85 potential participants were identified. Of these, 65 gave written consent. However, only 16 had installed the SM technology when the experiment began (18.8% participation rate). Consequently, the intervention group consisted of 11 households, and the reference

---

[q] The EU Energy Services Directive (Directive 2006/32/EC) accelerated the development of SM technology, inter alia by requiring suppliers to offer consumers competitively priced individual meters that could provide time-of-use information (Art. 13, Sub 1) and accurate bills based on actual consumption (Art. 13, Sub 2). Its implementation prompted the rollout of SMs in many member states.

[r] As the reader may know, a pilot study is a small-scale, preparatory version of a major study (Baker, 1998; Moore et al., 2011).

[s] See Bager and Mundaca (2017) for full details.

group consisted of five households. In this context, our sample size and participation rate were similar to those found in the literature.[t]

In a nutshell, the intervention was as follows. The control group received 'standard SM' consumption information given in kilowatt-hours (kWh) and Danish krone (DKK)[u] on an hourly, daily, weekly, monthly, or yearly basis (see Fig. 10A). They could also evaluate their current consumption with respect to a predetermined annual budget (in DKK)—typically their electricity costs in the previous calendar year. This information was unframed; the cost of electricity for a day, month, and year was simply stated. The intervention group received the same data, plus additional information in the form of the estimated cost of daily and weekly standard use and the daily and annual cost of standby use (see Fig. 10B). This information was framed as a salient loss by its presentation in the software interface, which read 'money lost from electricity consumption' followed by the monetary value. The SM displayed the amount spent per day as a running total; this figure was updated every few seconds, every day.

A difference-in-differences approach (Abadie, 2005) addressed the between-group effect. Two baselines were used. The first approach (baseline 1) tracked absolute electricity consumption and measured the relative change from the first to the last week of the intervention. The second approach (baseline 2) calculated average consumption during the experiment based on the deviation of the data from a sample of 3000 Danish households. From a statistical point of view, a paired-sample t-test was used to compare the significance of mean values for the reference and intervention groups. We also estimated the eta-squared (η^2) statistic to calculate the effect size.

The findings indicated that electricity use decisions were affected by the way information was designed, framed, and presented. For daily energy use, the first baseline approach found that the reference group reduced their consumption by an average of 7%. However, the intervention group reduced their daily consumption by 18%. Using the second baseline, an increase in daily electricity consumption of 2% for the reference group was estimated, compared with a decrease of 5% for the

---

[t] Given that this is a pilot study, it is important to note that small samples (and short-lived interventions) predominate in SM and environmental studies (Darby, 2006; Ehrhardt-Martinez et al., 2010; Faruqui et al., 2010b; Gans et al., 2013; Osbaldiston and Schott, 2011; Owen and Ward, 2007). For example, Bradley et al. (2016, p. 112) reports 10 households in the intervention group and a participation rate of 8%. In their analysis of smart grids (including SMs) in Denmark, Hansen and Hauge (2017) used a sample of 20 households in a pilot experiment that addressed perceived and enacted control. Given inherent limitations, the American Journal of Neuroradiology suggests that statistical power figures and sample size calculations should be required only for large-scale field studies and not pilot experiments (Trout et al., 2007). It is argued that small samples are also a consequence of the intrusive nature of SM technology, and there is growing literature that stresses ethical and privacy concerns among SM customers (Bohli et al., 2010; Cuijpers and Koops, 2013; Kaufmann et al., 2013). Negative perceptions and a lack of trust seem to be critical factors (Gangale et al., 2013; Gerpott and Paukert, 2013; Krishnamurti et al., 2012; Raimi and Carrico, 2016).
[u] 1 DKK=0.134 euro (15 September 2018) (www.oanda.com).

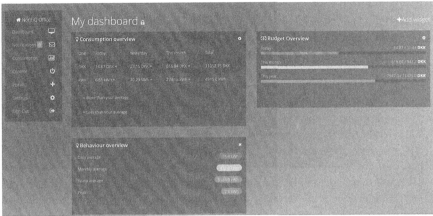

(A)

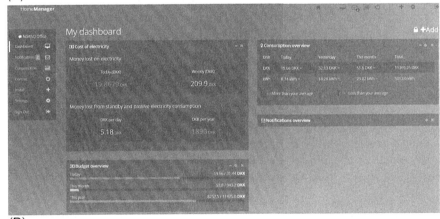

(B)

FIG. 10

Screenshot of SM technology and related information available to (A) the control group (top panel) and (B) the intervention group (bottom panel).

Reproduced from Bager, S., Mundaca, L., 2017. Making 'smart meters' smarter? Insights from a behavioural economics pilot field experiment in Copenhagen, Denmark. Energy Res. Soc. Sci. 28, 68–76.

intervention group. Thus, a differential effect of −11% (baseline 1) and −7% (baseline 2) was found for daily electricity consumption. Reductions in standby consumption were more pronounced, with a differential effect of 16%–25%.

From a statistical point of view, we found a significant decrease in electricity use when a 30% significance level ($\alpha = 0.30$) and a 70% confidence interval were used. These values seem to be consistent with significance thresholds suggested for pilot studies in medical research (Lee et al., 2014; Schoenfeld, 1980; Trout et al., 2007). For daily electricity use, there was a significant decrease between the reference group

($M=5.00$, $SD=1.09$) and the intervention group ($M=4.75$, $SD=0.65$): $t(26)=1.17$, P-value $= .25$. The mean decrease in the intervention group was 0.24, with a confidence interval ranging from 0.025 to 0.471. Following the guidelines proposed by Cohen (Cohen, 1988), the estimated η^2 statistic (0.05) is in the range of a 'moderate' effect. For standby consumption, there is a statistically significant decrease in electricity use between the reference group ($M=0.91$, $SD=0.13$) and the intervention group ($M=0.87$, $SD=0.13$): $t(25)=1.11$, P-value $= .27$. The mean decrease in the intervention group was 0.03, with a confidence interval ranging from 0.0021 to 0.0771. Here again, the η^2 statistic (0.05) was in the range of a moderate effect.

The relatively larger reductions in electricity use by the intervention group confirmed, initially, our hypothesis that framing reduced consumption as avoiding a loss would prove to be a greater incentive than viewing it as gaining a benefit. The results are also consistent with the literature that underscores the relevance of salience in SM studies (Price, 2014). However, compared with the current literature on 'standard' feedback that is provided via enabling technologies (e.g. SM, in-home displays, or web portals) (Darby, 2006; Faruqui et al., 2010b; Schleich et al., 2013), our results show a relatively similar level of effectiveness.

Taking a BE perspective, our results suggest that how SM information is shown to households has an impact on how it is perceived and acted upon. Despite the limitations inherent in a pilot study, notably the small sample size, our findings suggest that policies that address SM technology need to go beyond the simple provision of information and also consider how it is designed and presented to users. While our findings suggest the usefulness of the intervention, they also indicate the need for additional research. This pilot study paves the way for a full-scale field experiment. In any event, we argue that the level of effectiveness of similar, nonpricing policy interventions is very likely to be context and method specific.

5 Policy implications

Based on our case studies, a number of policy implications can be drawn in relation to the design, choice, and potential implementation of behaviour-oriented policy interventions. However, caution is needed as such interventions do not offer a panacea that can steer sustainable energy use.

Our case study on households' likelihood to install solar PV found that both monetary and nonmonetary factors play a role. Concerning monetary factors, we found that, unsurprisingly, higher subsidies increase the likelihood that households will adopt the technology. While installation subsidies have been in place in Sweden since 2009, the scheme continues to be plagued by uncertainty. Its coverage has been changed multiple times, often at short notice. The same applies to the cap on its amount. Additionally, some households have waited up to 4years for a decision regarding the subsidy. Taken together, these factors raise adoption barriers for risk-averse decision-makers. Clearly, addressing these problems is a first step towards a more behavioural approach in policymaking.

While financial rewards or subsidies have been found to crowd out prosocial actions in some settings, especially when visibility is high (Ariely et al., 2009; Mellström and Johannesson, 2008), findings from our study on solar PV adoption do not support this case. Our findings showed no significant interaction between subsidies and visibility. The implication is that policymakers can implement the installation subsidy, without concerns that higher subsidies will crowd out the intrinsic motivation of Swedish households.

Our finding concerning peer effects highlights the importance of considering nonmonetary factors in policymaking. Moreover, the peer-learning channel matters. *Hearing* about PV significantly increases the likelihood of adoption, while *seeing* it does not. A potential explanation is that hearing about PV typically occurs in social settings or through traditional media, facilitating a more detailed discussion of its benefits. Seeing PV, for example, while driving past a neighbour's house, does not. Local policymakers could consider holding information-sharing events or promoting the formation of solar community organisations, where thorough and guided discussions about the benefits and challenges of PV adoption take place, thus enhancing peer-learning effects.

With respect to carbon pricing policies, our findings need to be put in wider context, particularly in relation to outcomes from BE studies addressing similar aspects. First, counting on individuals to make voluntary payments, for example, via emission offsetting, is vastly insufficient as only some people are (in principle) willing to purchase offsets. WTP is low, and very few people have actually offset their emissions (Sonnenschein and Mundaca, 2019). Although nudging measures (e.g. by making it the default when booking travel) may increase the share, the results of small trials are mixed (Araña and León, 2013; Löfgren et al., 2012). In contrast, (additional) mandatory carbon pricing interventions appear to be viable, as WTP for climate change mitigation is significant and positive if a coercive mechanism is applied (Sonnenschein and Mundaca, 2019). WTP for climate mitigation differs, however, across areas of consumption and cost contexts (Blasch and Farsi, 2014; Sonnenschein and Mundaca, 2019). These differences can be seen as a pragmatic argument against uniform carbon pricing that is argued to be economically efficient (Aldy and Stavins, 2012; Weitzman, 2014) and thus an argument in favour of differentiated tax rates.

Furthermore, the design and framing of a carbon pricing intervention have a large influence on its acceptability. Calling an intervention a 'tax' lowers acceptance compared with framing it as a 'surcharge' or 'contribution' (Brannlund and Persson, 2012). Moreover, earmarking revenue for climate change mitigation can significantly increase acceptance (Sælen and Kallbekken, 2011; Sonnenschein and Smedby, 2019). It appears particularly important to stress that revenues are used for climate change mitigation in cases where carbon pricing is unlikely to be effective (Carattini et al., 2017), for example, when it is insufficient to steer behaviour away from carbon-intensive activities. Moreover, small financial incentives may run the risk of crowding-out intrinsic environmental behaviour (Bolderdijk and Steg, 2015; Gneezy et al., 2011), as previously discussed—albeit not applicable—in the case of solar PV.

PCT is an ambitious policy alternative to carbon taxes. Although the economic incentives may appear to be similar, the former may be preferable from a behavioural perspective. This is because, in contrast to taxation, it ensures that some level of mitigation occurs and makes carbon emissions (and action) salient. This, in turn, may increase an individual's interest in the level of their own emissions and fosters a 'carbon budgeting' effect among participants. This suggests that, for the same price level, a PCT scheme can be more effective than a carbon tax (Capstick and Lewis, 2010; Zanni et al., 2013). While in Sweden carbon taxation has greater support than a potential PCT scheme (Jagers et al., 2010), evidence suggests that the acceptability of PCT can be significantly increased by assuring the fair allocation of allowances (Bristow et al., 2010). PCT seems to be particularly viable among people with proenvironmental attitudes, confirmed by evidence from an experimental simulation (Capstick and Lewis, 2010), the field experience of voluntary carbon rationing action groups (Howell, 2012), and by the ABM exercise (see Section 4.3 earlier and Kempener, 2009).

Finally, our pilot field experiment with SM investigated the relationship between loss aversion and energy use behaviour in a nonprice policy intervention. The results underscore the need in policymaking to give greater attention to motivational and psychological drivers that can reduce electricity demand, which is a key gap in the literature (Abrahamse et al., 2005; Steg et al., 2015). In particular, the findings stress the extent to which salient loss aversion information may affect energy users and whether effects can be maintained in the long term. Our experiment suggests that policies that address SM technology need to have a more comprehensive understanding of the design of the feedback. Our findings were in line with the emerging literature in the field, which highlights that policies that frame incentives as losses are more effective than those that frame them as rewards (Gillingham and Palmer, 2014; Price, 2014).[v] The study advocates for closer collaboration with policymakers and further experimentation and evaluation of the choice of policy instruments. To that end, the experiment suggests that SM policy needs to consider not only the provision of information but also *how* it is framed. It supports claims that SM feedback, per se, is unlikely to be sufficient to promote energy efficiency and conservation (Ehrhardt-Martinez et al., 2010; Nachreiner et al., 2015). Behavioural change can be better facilitated via the provision of information that is carefully framed (Abrahamse et al., 2005; Allcott and Mullainathan, 2010; Frederiks et al., 2015).

6 Conclusions

Taken together, these analyses suggest that Scandinavia needs a more integrated behavioural and technological approach to policymaking if sustainable energy use and deep decarbonisation efforts are to be pursued. Our growing knowledge of energy

---

[v] From a policy point of view, 'persistence' of behavioural interventions in the long run is a critical challenge for scholars and policymakers (Abrahamse et al., 2005; Ehrhardt-Martinez et al., 2010; Owen and Ward, 2007).

use, decarbonisation, and behaviours are an opportunity to complement the existing mix of policy instruments with measures that go beyond technology. Currently, most decarbonisation policy portfolios are dominated by market-based instruments and a traditional policymaking approach. The assumption is that the 'right' price signals or 'correct' information drives energy users to make 'rational' decisions about energy use and decarbonisation. However, the results of our analyses question this assumption. The failure to acknowledge behavioural issues and consumption-based policy approaches may generate a path dependency in which policy recommendations are confined to technology, production-based incentives, or markets.

While it is clear that further research is needed, energy use and carbon emission consumption patterns suggest that policy mixes have been ineffective or insufficient. Given the various 'irrationalities' or 'anomalies' that exist, it is clear that much more attention needs to be given to decision-making processes, the *use* of energy technologies, and the context of carbon-intensive consumption patterns. Nevertheless, pricing mechanisms must not be ruled out as they incentivise energy users, and research shows that they complement behavioural interventions. Exploiting complementarities and synergies between price and nonprice policy interventions—both of which have to become much more stringent to be effective—is of prime importance when designing innovative policy approaches targeting behavioural change.

By no means can it be claimed that BE interventions are the ultimate solution to encouraging sustainable energy behaviour and deep decarbonisation. Our understanding of the effectiveness of dedicated, consumption-based behavioural change and nonprice interventions in the context of sustainable energy use and climate action remains limited. Furthermore, emerging research shows that behavioural interventions can also be uncertain and ineffective. Marginal improvements and short-term benefits may mask a lack of effectiveness in the long run. Effective, persistent behavioural policy approaches are therefore critical. Moreover, the growing literature on behavioural insights shows the importance of ethical issues (e.g. are individuals manipulated into making unwanted choices?), so choice settings need careful evaluation and thought before implementation.

These issues underscore the value of evidence-based information and the comprehensive evaluation of policy to support the design, choice, and implementation of instruments. An example comes from our study on solar PV adoption, where evidence of crowding out detected in experimental research may be contradicted by experiments run in another context. BE research can increase our knowledge of microeconomic decision-making processes and value-based choices and can unearth the motivational, psychological, and contextual factors that affect energy demand or mitigation actions. Empirical data can inform policymakers about, for example, self-control (why do we use too much electricity even when it can undermine our own welfare?), limited attention (why do we miss important information included on energy labels?), or satisficing (what are the key elements driving mental shortcuts when buying appliances?). Although the results obtained so far tend to be highly context dependent, the experience in Scandinavia stresses that rigorous assessments are needed to bridge the gap between policy and science. Behavioural scientists and policymakers need to work much more closely together, as doing so will help to accelerate the transition towards sustainable energy use and a net zero carbon economy. Given the enormous challenges posed by the Paris Agreement and its aspirational target, policymakers need to embrace new directions for sustainable energy use and decarbonisation.

References

Abadie, A., 2005. Semiparametric difference-in-differences estimators. Rev. Econ. Stud. 72, 1–19.

Abrahamse, W., Steg, L., 2013. Social influence approaches to encourage resource conservation: a meta-analysis. Glob. Environ. Chang. 23, 1773–1785.

Abrahamse, W., Steg, L., Vlek, C., Rothengatter, T., 2005. A review of intervention studies aimed at household energy conservation. J. Environ. Psychol. 25, 273–291. https://doi.org/10.1016/j.jenvp.2005.08.002.

Abrahamse, W., Steg, L., Vlek, C., Rothengatter, T., 2007. The effect of tailored information, goal setting, and tailored feedback on household energy use, energy-related behaviors, and behavioral antecedents. J. Environ. Psychol. 27, 265–276. https://doi.org/10.1016/j.jenvp.2007.08.002.

Adaman, F., Karalı, N., Kumbaroğlu, G., Or, İ., Özkaynak, B., Zenginobuz, Ü., 2011. What determines urban households' willingness to pay for CO2 emission reductions in Turkey: a contingent valuation survey. Energy Policy, Special Section on Offshore Wind Power Planning, Economics and Environment. 39, 689–698. https://doi.org/10.1016/j.enpol.2010.10.042.

Ajzen, I., Fishbein, M., 1972. Attitudes and normative beliefs as factors influencing behavioral intentions. J. Pers. Soc. Psychol. 21, 1–9. https://doi.org/10.1037/h0031930.

Aldy, J.E., Stavins, R.N., 2012. The promise and problems of pricing carbon theory and experience. J. Environ. Dev. 21, 152–180. https://doi.org/10.1177/1070496512442508.

Allcott, H., 2011. Social norms and energy conservation. J. Public Econ. 95, 1082–1095. https://doi.org/10.1016/j.jpubeco.2011.03.003.

Allcott, H., Mullainathan, S., 2010. Behavior and energy policy. Science 327, 1204–1205. https://doi.org/10.1126/science.1180775.

Andersson, M., Falck, M., 2017. Skatt på flygresor (Lagrådsremiss). Finansdepartementet, Stockholm.

Andor, M.A., Fels, K.M., 2018. Behavioral economics and energy conservation—a systematic review of non-price interventions and their causal effects. Ecol. Econ. 148, 178–210. https://doi.org/10.1016/j.ecolecon.2018.01.018.

Andor, M.A., Gerster, A., Sommer, S., 2017. Consumer inattention, heuristic thinking and the role of energy labels (No. 671). Ruhr Econ. Papers. https://doi.org/10.2139/ssrn.2874905.

Araña, J.E., León, C.J., 2013. Can defaults save the climate? Evidence from a field experiment on carbon offsetting programs. Environ. Resour. Econ. 54, 613–626. https://doi.org/10.1007/s10640-012-9615-x.

Ariely, D., Bracha, A., Meier, S., 2009. Doing good or doing well? Image motivation and monetary incentives in behaving prosocially. Am. Econ. Rev. 99, 544–555.

Arvizu, D., Balaya, P., Cabeza, L., Hollands, T., Jäger-Waldau, A., Kondo, M., Konseibo, C., Melesh, V., Ko, W., Stein, Y., Tamaura, H., Xu, R.Z., 2011. Direct solar energy. In:

Edenhofer, O., Pichs-Madruga, R., Sokona, Y., Seyboth, K., Matschoss, P., Kadner, S., … von Stechow, C. (Eds.), IPCC Special Report on Renewable Energy Sources and Climate Change Mitigation. Cambridge University Press, pp. 333–400. https://doi.org/10.1017/CBO9781139151153.007.

Ayres, I., Raseman, S., Shih, A., 2013. Evidence from two large field experiments that peer comparison feedback can reduce residential energy usage. J. Law Econ. Org. 29, 992–1022. https://doi.org/10.1093/jleo/ews020.

Bager, S., Mundaca, L., 2017. Making 'smart meters' smarter? Insights from a behavioural economics pilot field experiment in Copenhagen, Denmark. Energy Res. Soc. Sci. 28, 68–76.

Baker, T., 1998. Doing Social Research, third ed. McGraw-Hill Humanities/Social Sciences/Languages, Boston.

Bandura, A., 1969. Social-learning theory of identificatory processes. In: Goslin, D.A. (Ed.), Handbook of Socialization Theory and Research. Rand McNally & Co.

Bénabou, R., Tirole, J., 2006. Incentives and prosocial behavior. Am. Econ. Rev. 96, 1652–1678.

Bergek, A., Mignon, I., 2017. Motives to adopt renewable energy technologies: evidence from Sweden. Energy Policy 106, 547–559.

Blake, J., 1999. Overcoming the 'value-action gap' in environmental policy: tensions between national policy and local experience. Local Environ. 4, 257–278.

Blasch, J., Farsi, M., 2014. Context effects and heterogeneity in voluntary carbon offsetting—a choice experiment in Switzerland. J. Environ. Econ. Policy 3, 1–24. https://doi.org/10.1080/21606544.2013.842938.

Bohli, J.M., Sorge, C., Ugus, O., 2010. A privacy model for smart metering. In: Presented at the 2010 IEEE International Conference on Communications Workshops, Capetown, pp. 1–5. https://doi.org/10.1109/ICCW.2010.5503916.

Bolderdijk, J.W., Steg, L., 2015. Promoting sustainable consumption: the risks of using financial incentives. In: Reisch, L., Thøgersen, J. (Eds.), Handbook of Research on Sustainable Consumption. Edward Elgar Publishing. https://doi.org/10.4337/9781783471270.

Bonabeau, E., 2002. Agent-based modeling: methods and techniques for simulating human systems. Proc. Natl. Acad. Sci. U. S. A. 99 (Suppl 3), 7280–7287.

Bows-Larkin, A., 2015. All adrift: aviation, shipping, and climate change policy. Clim. Pol. 15, 681–702. https://doi.org/10.1080/14693062.2014.965125.

Bradley, P., Coke, A., Leach, M., 2016. Financial incentive approaches for reducing peak electricity demand, experience from pilot trials with a UK energy provider. Energy Policy 98, 108–120.

Braito, M., Flint, C., Muhar, A., Penker, M., Vogel, S., 2017. Individual and collective socio-psychological patterns of photovoltaic investment under diverging policy regimes of Austria and Italy. Energy Policy 109, 141–153. https://doi.org/10.1016/j.enpol.2017.06.0.

Brannlund, R., Persson, L., 2012. To tax, or not to tax: preferences for climate policy attributes. Clim. Pol. 12, 704–721. https://doi.org/10.1080/14693062.2012.675732.

Brekke, K., Johansson-Stenman, O., 2008. The behavioural economics of climate change. Oxf. Rev. Econ. Policy 24, 280–297.

Bristow, A.L., Wardman, M., Zanni, A.M., Chintakayala, P.K., 2010. Public acceptability of personal carbon trading and carbon tax. Ecol. Econ. 69, 1824–1837. https://doi.org/10.1016/j.ecolecon.2010.04.021.

Brouwer, R., Brander, L., Van Beukering, P., 2008. "A convenient truth": air travel passengers' willingness to pay to offset their CO2 emissions. Clim. Chang. 90, 299–313. https://doi.org/10.1007/s10584-008-9414-0.

Burgess, M., 2016. Personal carbon allowances: a revised model to alleviate distributional issues. Ecol. Econ. 130, 316–327.

Bursztyn, L., Ederer, F., Ferman, B., Yuchtman, N., 2014. Understanding mechanisms underlying peer effects: evidence from a field experiment on financial decisions. Econometrica 82, 1273–1301. https://doi.org/10.3982/ECTA11991.

Camerer, C., 1999. Behavioral economics: reunifying psychology and economics. Proc. Natl. Acad. Sci. 96, 10575–10577. https://doi.org/10.1073/pnas.96.19.10575.

Camerer, C., Loewenstein, G., Prelec, D., 2005. Neuroeconomics: how neuroscience can inform economics. J. Econ. Lit. 43, 9–64.

Camerer, C.F., Loewenstein, G., Rabin, M., 2011. Advances in Behavioral Economics. Princeton University Press.

Cames, M., Graichen, J., Siemons, A., Cook, V., 2015. Emission Reduction Targets for International Aviation and Shipping. European Parliament. Directorate General for Internal Policies, Freiburg.

Capstick, S.B., Lewis, A., 2010. Effects of personal carbon allowances on decision-making: evidence from an experimental simulation. Clim. Pol. 10, 369–384. https://doi.org/10.3763/cpol.2009.0034.

Carattini, S., Baranzini, A., Thalmann, P., Varone, F., Vöhringer, F., 2017. Green taxes in a post-Paris world: are millions of nays inevitable? Environ. Resour. Econ. 68, 97–128. https://doi.org/10.1007/s10640-017-0133-8.

Cialdini, R.B., Reno, R.R., Kallgren, C.A., 1990. A focus theory of normative conduct: recycling the concept of norms to reduce littering in public places. J. Pers. Soc. Psychol. 58, 1015.

Cialdini, R.B., Kallgren, C.A., Reno, R.R., 1991. A focus theory of normative conduct: a theoretical refinement and reevaluation of the role of norms in human behavior. In: Advances in Experimental Social Psychology. Elsevier, pp. 201–234.

City of Copenhagen, 2016. CPH 2025 Climate Plan. Roadmap 2017–2020. Technical and Environmental Administration, Copenhagen.

City of Stockhom, 2016. Strategy for a Fossil-Fuel Free Stockholm by 2040 (No. 134–175/2015).

Clayton, S., Devine-Wright, P., Stern, P.C., Whitmarsh, L., Carrico, A., Steg, L., Swim, J., Bonnes, M., 2015. Psychological research and global climate change. Nat. Clim. Chang. 5, 640. https://doi.org/10.1038/nclimate2622.

Coad, A., de Haan, P., Woersdorfer, J.S., 2009. Consumer support for environmental policies: an application to purchases of green cars. Ecol. Econ. 68, 2078–2086.

Cohen, J., 1988. Statistical Power Analysis for the Behavioral Sciences, second ed. Lawrence Erlbaum Associates, Hillsdale, NJ.

Crago, C.L., Chernyakhovskiy, I., 2017. Are policy incentives for solar power effective? Evidence from residential installations in the Northeast. J. Environ. Econ. Manag. 81, 132–151. https://doi.org/10.1016/J.JEEM.2016.09.008.

Cuijpers, C., Koops, B.J., 2013. Smart metering and privacy in Europe: lessons from the Dutch case. In: Gutwirth, S., Leenes, R., Hert, P., Poullet, Y. (Eds.), European Data Protection: Coming of Age. Springer, Netherlands, pp. 269–293.

Darby, S., 2006. The Effectiveness of Feedback on Energy Consumption, a Review for DEFRA of the Literature on Metering, Billing and Direct Displays, Environmental Change Institute, University of Oxford, Oxford.

Darby, S., 2010. Smart metering: what potential for householder engagement? Build. Res. Inf. 38, 442–457.

Davis, S., Caldeira, K., 2010. Consumption-based accounting of CO2 emissions. Proc. Natl. Acad. Sci. 107, 5687–5692. https://doi.org/10.1073/pnas.0906974107.

Davis, S., Peters, G., Caldeira, K., 2011. The supply chain of CO2 emissions. Proc. Natl. Acad. Sci. 108, 18554–18559.

de Coninck, H., Revi, A., Babiker, M., Bertoldi, P., Buckeridge, M., Cartwright, A., Dong, W., Ford, J., Fuss, S., Hourcade, J.C., Ley, D., Mechler, R., Newman, P., 2018. Strengthening and implementing the global response. In: Masson-Delmotte, V., Zhai, P., Pörtner, H.O., Roberts, D., Skea, J., Shukla, P.R., … Waterfield, T. (Eds.), Global Warming of 1.5°C. An IPCC Special Report on the Impacts of Global Warming of 1.5°C above Pre-Industrial Levels and Related Global Greenhouse Gas Emission Pathways, in the Context of Strengthening the Global Response to the Threat of Climate Change, Sustainable Development, and Efforts to Eradicate Poverty. In Press.

Depuru, S., Wang, L., Devabhaktuni, V., 2011. Smart meters for power grid: challenges, issues, advantages and status. Renew. Sust. Energ. Rev. 15, 2736–2742.

Diederich, J., Goeschl, T., 2013. Willingness to pay for voluntary climate action and its determinants: field-experimental evidence. Environ. Resour. Econ. 57, 405–429. https://doi.org/10.1007/s10640-013-9686-3.

Diekmann, A., Preisendörfer, P., 2003. Green and greenback: the behavioral effects of environmental attitudes in low-cost and high-cost situations. Ration. Soc. 15, 441–472. https://doi.org/10.1177/1043463103154002.

Dukas, R., 2004. Causes and consequences of limited attention. Brain Behav. Evol. 63, 197–210.

EC, 2011. Next Steps for Smart Grids: Europe's Future Electricity System Will Save Money and Energy—Press Summary. Brussels.

EC, 2014. Benchmarking Smart Metering Deployment in the EU-27 With a Focus on Electricity (No. COM (2014) 356 Final). Brussels.

Egebark, J., Ekström, M., 2016. Can indifference make the world greener? J. Environ. Econ. Manag. 76, 1–13. https://doi.org/10.1016/j.jeem.2015.11.004.

Ehrhardt-Martinez, K., Donnelly, K.A., Laitner, S., et al., 2010. Advanced Metering Initiatives and Residential Feedback Programs: A Meta-Review for Household Electricity-Saving Opportunities. American Council for an Energy-Efficient Economy, Washington, DC.

Ek, K., Söderholm, P., 2008. Norms and economic motivation in the Swedish green electricity market. Ecol. Econ. 68, 169–182. https://doi.org/10.1016/j.ecolecon.2008.02.013.

Elo, S., Kyngäs, H., 2008. The qualitative content analysis process. J. Adv. Nurs. 62, 107–115. https://doi.org/10.1111/j.1365-2648.2007.04569.x.

EU, 2017. Energy, transport and environment indicators, 2017 edition. Publications Office of the European Union, Luxembourg.

Eyre, N., 2010. Policing carbon: design and enforcement options for personal carbon trading. Clim. Pol. 10, 432–446.

Falkinger, J., 2008. Limited attention as a scarce resource in information-rich economies. Econ. J. 118, 1596–1620. https://doi.org/10.1111/j.1468-0297.2008.02182.x.

Faruqui, A., Sergici, S., Sharif, A., 2010. The impact of informational feedback on energy consumption—a survey of the experimental evidence. Energy 35, 1598–1608. https://doi.org/10.1016/j.energy.2009.07.042.

Faruqui, A., Harris, D., Hledik, R., 2010a. Unlocking the€ 53 billion savings from smart meters in the EU: How increasing the adoption of dynamic tariffs could make or break the EU's smart grid investment. Energy Policy 38, 6222–6231.

Faruqui, A., Sergici, S., Sharif, A., 2010b. The impact of informational feedback on energy consumption—a survey of the experimental evidence. In: Energy, Demand Response

Resources: the US and International Experience Demand Response Resources: the US and International Experience. vol. 35, pp. 1598–1608.

Fawcett, T., Parag, Y., 2010. An introduction to personal carbon trading. Clim. Pol. 10, 329–338.

Fawcett, T., Hvelplund, F., Meyer, N.I., 2010. (Chapter 4). Making it personal: per capita carbon allowances. In: Sioshansi, F.P. (Ed.), Generating Electricity in a Carbon-Constrained World. Academic Press, Boston, MA, pp. 87–107.

Frederiks, E.R., Stenner, K., Hobman, E.V., 2015. Household energy use: applying behavioural economics to understand consumer decision-making and behaviour. Renew. Sust. Energ. Rev. 41, 1385–1394. https://doi.org/10.1016/J.RSER.2014.09.026.

Fuerst, F., Singh, R., 2018. How present bias forestalls energy efficiency upgrades: a study of household appliance purchases in India. J. Clean. Prod. 186, 558–569. https://doi.org/10.1016/j.jclepro.2018.03.100.

Gandhi, R., Knittel, C., Pedro, P., Wolfram, C., 2016. Running randomized field experiments for energy efficiency programs: a practitioner's guide. Econ. Energy Environ. Policy 5, 7–26.

Gangale, F., Mengolini, A., Onyeji, I., 2013. Consumer engagement: an insight from smart grid projects in Europe. Energy Policy 60, 621–628.

Gans, W., Alberini, A., Longo, A., 2013. Smart meter devices and the effect of feedback on residential electricity consumption: evidence from a natural experiment in Northern Ireland. Energy Econ. 36, 729–743.

Gerpott, T.J., Paukert, M., 2013. Determinants of willingness to pay for smart meters: an empirical analysis of household customers in Germany. Energy Policy 61, 483–495.

Gigerenzer, G., Goldstein, D., 1996. Reasoning the fast and frugal way: models of bounded rationality. Psychol. Rev. 103, 650–669.

Gillingham, K., Palmer, K., 2014. Bridging the energy efficiency gap: policy insights from economic theory and empirical evidence. Rev. Environ. Econ. Policy 8, 18–38.

Giordano, V., Fulli, G., Jiménez, M., Mengolini, A., Meletiou, A., Covrig, C., Ardelean, M., Filiou, C., 2013. Smart Grid Projects in Europe: Lessons Learned and Current Developments. Joint Research Centre of the European Commission, Petten.

Gneezy, U., Meier, S., Rey-Biel, P., 2011. When and why incentives (don't) work to modify behavior. J. Econ. Perspect. 25, 191–210. https://doi.org/10.1257/jep.25.4.191.

Gowdy, J., 2008. Behavioral economics and climate change policy. J. Econ. Behav. Organ. 68, 632–644.

Grimm, V., Berger, U., DeAngelis, D.L., Polhill, J.G., Giske, J., Railsback, S.F., 2010. The ODD protocol: a review and first update. Ecol. Model. 221, 2760–2768.

Grubler, A., Johansson, T., Mundaca, L., Nakicenovic, N., Pachauri, S., Riahi, K., Rogner, H., Strupeit, L., 2012. Energy primer. In: Johansson, T.B., Patwardhan, A., Nakicenovic, N., Gomez-Echeverri, L. (Eds.), Global Energy Assessment—Toward a Sustainable Future. Cambridge University Press/the International Institute for Applied Systems Analysis, Cambridge, New York/Laxenburg, pp. 99–150.

Hahn, R., Metcalfe, R., 2016. The impact of behavioral science experiments on energy policy. Econ. Energy Environ. Policy 5, 27–44.

Hansen, M., Hauge, B., 2017. Scripting, control, and privacy in domestic smart grid technologies: insights from a Danish pilot study. Energy Res. Soc. Sci. 25, 112–123.

Hausman, D., Welch, B., 2010. Debate: to nudge or not to nudge. J. Polit. Philos. 18, 123–136. https://doi.org/10.1111/j.1467-9760.2009.00351.x.

Hertwich, E.G., Peters, G.P., 2009. Carbon footprint of nations: a global, trade-linked analysis. Environ. Sci. Technol. 43, 6414–6420. https://doi.org/10.1021/es803496a.

Hoeffler, S., Ariely, D., 1999. Constructing stable preferences: a look into dimensions of experience and their impact on preference stability. J. Consum. Psychol. 8, 113–139. https://doi.org/10.1207/s15327663jcp0802_01.

Howell, R.A., 2012. Living with a carbon allowance: the experiences of carbon rationing action groups and implications for policy. Energy Policy, Modeling Transport (Energy) Demand and Policies. 41, 250–258. https://doi.org/10.1016/j.enpol.2011.10.044.

Hsieh, H.-F., Shannon, S.E., 2005. Three approaches to qualitative content analysis. Qual. Health Res. 15, 1277–1288. https://doi.org/10.1177/1049732305276687.

Hughes, J.E., Podolefsky, M., 2015. Getting green with solar subsidies: evidence from the California solar initiative. J. Assoc. Environ. Resour. Econ. 2, 235–275. https://doi.org/10.1086/681131.

IEA, 2017a. World Energy Statistics [WWW Document]. http://www.iea.org/statistics/statisticssearch/.

IEA, Nordic Energy Research, 2013. Nordic Energy Technology Perspectives: Pathways to a Carbon Neutral Energy Future. Nordic Energy Research, International Energy Agency, Oslo, Paris.

IEA, Nordic Energy Research, 2016. Nordic Energy Technology Perspectives 2016: Cities, Flexibility and Pathways to Carbon-Neutrality. Nordic Energy Research, International Energy Agency, Oslo, Paris.

International Energy Agency, 2017. Energy Policies of IEA Countries: Denmark 2017. IEA, Paris.

IRENA, 2017. Renewable Energy Statistics. Abu Dhabi.

Jagers, S., Löfgren, Å., Stripple, J., 2010. Attitudes to personal carbon allowances: political trust, fairness and ideology. Clim. Pol. 10, 410–431. https://doi.org/10.3763/cpol.2009.0673.

Janssen, M.A., Jager, W., 2002. Stimulating diffusion of green products. J. Evol. Econ. 12, 283–306.

Kahneman, D., 1994. New challenges to the rationality assumption. J. Inst. Theor. Econ./Zeitschrift Für Gesamte Staatswiss, 18–36.

Kahneman, D., 2003a. Maps of bounded rationality: psychology for behavioral economics. Am. Econ. Rev. 93, 1449–1475.

Kahneman, D., 2003b. A perspective on judgment and choice: mapping bounded rationality. Am. Psychol. 58, 697–720.

Kahneman, D., Thaler, R.H., 2006. Anomalies: utility maximization and experienced utility. J. Econ. Perspect. 20, 221–234.

Kahneman, D., Tversky, A., 1979. Prospect theory: an analysis of decision under risk. Econom. J. Econ. Soc., 263–291.

Kahneman, D., Tversky, A., 2000. Choices, Values, and Frames. Cambridge University Press, New York.

Kahneman, D., Knetsch, J., Thaler, R., 1991. Anomalies: the endowment effect, loss aversion, and status quo bias. J. Econ. Perspect. 5, 193–206.

Kasa, S., Leiren, M., Khan, J., 2012. Central government ambitions and local commitment: climate mitigation initiatives in four municipalities in Norway and Sweden. J. Environ. Plan. Manag. 55, 211–228. https://doi.org/10.1080/09640568.2011.589649.

Kaufmann, S., Künzel, K., Loock, M., 2013. Customer value of smart metering: explorative evidence from a choice-based conjoint study in Switzerland. Energy Policy 53, 229–239.

Kempener, R., 2009. Simulating personal carbon trading: an agent-based model. In: Sci. Technol. Policy Res. Electron. Work. Pap. Ser. Univ. Sussex UK.

Kirkpatrick, A.J., Bennear, L.S., 2014. Promoting clean energy investment: an empirical analysis of property assessed clean energy. J. Environ. Econ. Manag. 68, 357–375. https://doi.org/10.1016/J.JEEM.2014.05.001.

Kolstad, C., Urama, K., Broome, J., Bruvoll, A., Cariño Olvera, M., Fullerton, D., Gollier, C., Hanemann, W.M., Hassan, R., Jotzo, F., Khan, M.R., Meyer, L., Mundaca, L., 2014. Social, Economic and Ethical Concepts and Methods. In: Edenhofer, O., Pichs-Madruga, R., Sokona, Y., Farahani, E., Kadner, S., Seyboth, K., … Minx, J.C. (Eds.), Climate Change 2014: Mitigation of Climate Change. Contribution of Working Group III to the Fifth Assessment Report of the Intergovernmental Panel on Climate Change. Cambridge University Press, Cambridge, New York, NY.

Kotchen, M.J., Boyle, K.J., Leiserowitz, A.A., 2013. Willingness-to-pay and policy-instrument choice for climate-change policy in the United States. Energy Policy, Special Section: Long Run Transitions to Sustainable Economic Structures in the European Union and Beyond. 55, 617–625. https://doi.org/10.1016/j.enpol.2012.12.058.

Krishnamurti, T., Schwartz, D., Davis, A., Fischhoff, B., de Bruin, W., Lave, L., Wang, J., 2012. Preparing for smart grid technologies: a behavioral decision research approach to understanding consumer expectations about smart meters. Energy Policy, Modeling Transport (Energy) Demand and Policies. 41, 790–797.

Kuramochi, T., Höhne, N., Schaeffer, M., Cantzler, J., Hare, B., Deng, Y., Sterl, S., Hagemann, M., Rocha, M., Yanguas-Parra, P.A., Mir, G.-U.-R., Wong, L., El-Laboudy, T., Wouters, K., Deryng, D., Blok, K., 2018. Ten key short-term sectoral benchmarks to limit warming to 1.5°C. Clim. Pol. 18, 287–305. https://doi.org/10.1080/14693062.2017.1397495.

Larsen, H., Hertwich, E., 2009. The case for consumption-based accounting of greenhouse gas emissions to promote local climate action. Environ. Sci. Pol. 12, 791–798. https://doi.org/10.1016/j.envsci.2009.07.010.

Larsson, J., Kamb, A., Nässén, J., Åkerman, J., 2018. Measuring greenhouse gas emissions from international air travel of a country's residents methodological development and application for Sweden. Environ. Impact Assess. Rev. 72, 137–144. https://doi.org/10.1016/j.eiar.2018.05.013.

Lee, E., Whitehead, A., Jacques, R., Julious, S., 2014. The statistical interpretation of pilot trials: should significance thresholds be reconsidered? BMC Med. Res. Methodol. 14, 41.

Li, J., Fan, J., Zhao, D., Wang, S., 2015. Allowance price and distributional effects under a personal carbon trading scheme. J. Clean. Prod. 103, 319–329.

Lillemo, S.C., 2014. Measuring the effect of procrastination and environmental awareness on households' energy-saving behaviours: an empirical approach. Energy Policy 66, 249–256. https://doi.org/10.1016/j.enpol.2013.10.077.

Lindahl, J., 2016. National Survey Report of PV Power Applications in Sweden 2016.

Lindkvist, C., Karlsson, A., Sørnes, K., Wyckmans, A., 2014. Barriers and Challenges in nZEB Projects in Sweden and Norway. Energy Procedia, Renewable Energy Research Conference, RERC 2014. 58, 199–206. https://doi.org/10.1016/j.egypro.2014.10.429.

Lindman, Å., Ek, K., Söderholm, P., 2013. Voluntary citizen participation in carbon allowance markets: the role of norm-based motivation. Clim. Pol. 13, 680–697. https://doi.org/10.1080/14693062.2013.810436.

List, J.A., Price, M., 2016. The use of field experiments in environmental and resource economics. Rev. Environ. Econ. Policy 10, 206–225.

Loewenstein, G.F., 1988. Frames of mind in intertemporal choice. Manag. Sci. 34, 200–214. https://doi.org/10.1287/mnsc.34.2.200.

Loewenstein, G., Thaler, R., 1989. Anomalies: intertemporal choice. J. Econ. Perspect. 3, 181–193.

Löfgren, Å., Martinsson, P., Hennlock, M., Sterner, T., 2012. Are experienced people affected by a pre-set default option—results from a field experiment. J. Environ. Econ. Manag. 63, 66–72. https://doi.org/10.1016/j.jeem.2011.06.002.

Lopes, M.A.R., Antunes, C.H., Martins, N., 2012. Energy behaviours as promoters of energy efficiency: a 21st century review. Renew. Sust. Energ. Rev. 16, 4095–4104.

Löschel, A., Sturm, B., Uehleke, R., 2017. Revealed preferences for voluntary climate change mitigation when the purely individual perspective is relaxed – evidence from a framed field experiment. J. Behav. Exp. Econ. 67, 149–160. https://doi.org/10.1016/j.socec.2016.12.007.

MacKerron, G.J., Egerton, C., Gaskell, C., Parpia, A., Mourato, S., 2009. Willingness to pay for carbon offset certification and co-benefits among (high-)flying young adults in the UK. Energy Policy 37, 1372–1381. https://doi.org/10.1016/j.enpol.2008.11.023.

Manski, C.F., 2000. Economic analysis of social interactions. J. Econ. Perspect. 14, 115–136. https://doi.org/10.1257/jep.14.3.115.

Mellström, C., Johannesson, M., 2008. Crowding out in blood donation: was Titmuss right? J. Eur. Econ. Assoc. 6, 845–863. https://doi.org/10.1162/JEEA.2008.6.4.845.

Menges, R., Traub, S., 2009. An experimental study on the gap between willingness to pay and willingness to donate for green electricity. Finanz. Public Finance Anal. 65, 335–357. https://doi.org/10.1628/001522109X477804.

Mitchell, C., Sawin, J.L., Pokharel, G.R., Kammen, D., Wang, Z., Jaccard, M., Langniss, O., Lucas, H., Blanco, R.T., Usher, E., Verbruggen, A., Wuestenhagen, R., Yamaguchi, K., 2011. Policy, financing and implementation. In: Edenhofer, O., Pichs-Madruga, R., Sokona, Y., Seyboth, K., Matschoss, P., Kadner, S., … von Stechow, C. (Eds.), IPCC Special Report on Renewable Energy Sources and Climate Change Mitigation. Cambridge University Press, pp. 865–950. https://doi.org/10.5860/CHOICE.49-6309.

Moberg, K., Aall, C., Dorner, F., Reimerson, E., Ceron, J.-P., Sköld, B., Sovacool, B., Piana, V., 2018. Mobility, food and housing: responsibility, individual consumption and demand-side policies in European deep decarbonisation pathways. Energy Effic. https://doi.org/10.1007/s12053-018-9708-7.

Momsen, K., Stoerk, T., 2014. From intention to action: can nudges help consumers to choose renewable energy? Energy Policy 74, 376–382. https://doi.org/10.1016/j.enpol.2014.07.008.

Moore, C., Carter, R.E., Nietert, P., Stewart, P., 2011. Recommendations for planning pilot studies in clinical and translational research. Clin. Transl. Sci. 4, 332–337.

Mullainathan, S., Thaler, R.H., 2000. Behavioral Economics (SSRN Scholarly Paper No. ID 245828). Social Science Research Network, Rochester, NY.

Mundaca, L., Román, R., Cansino, J.M., 2015. Towards a green energy economy? A macroeconomic-climate evaluation of Sweden's CO_2 emissions. Appl. Energy 148, 196–209.

Mundaca, L., Sonnenschein, J., Steg, L., Höhne, N., Ürge-Vorsatz, D., 2019. The global expansion of climate mitigation policy interventions, the Talanoa Dialogue and the role of behavioural insights. Environ. Res. Commun 1 (6), 1–14.

Nachreiner, M., Mack, B., Matthies, E., Tampe-Mai, K., 2015. An analysis of smart metering information systems: a psychological model of self-regulated behavioural change. Energy Res. Soc. Sci., Special Issue on Smart Grids and the Social Sciences. 9, 85–97.

Nässén, J., Holmberg, J., 2005. Energy efficiency—a forgotten goal in the Swedish building sector? Energy Policy 33, 1037–1051. https://doi.org/10.1016/j.enpol.2003.11.004.

Nässén, J., Sprei, F., Holmberg, J., 2008. Stagnating energy efficiency in the Swedish building sector—economic and organisational explanations. Energy Policy 36, 3814–3822. https://doi.org/10.1016/j.enpol.2008.07.018.

Nomura, N., Akai, M., 2004. Willingness to pay for green electricity in Japan as estimated through contingent valuation method. Appl. Energy 78, 453–463. https://doi.org/10.1016/j.apenergy.2003.10.001.

OECD, 2017. Behavioural insights and public policy: lessons from around the world. OECD Publishing, Paris.

Ölander, F., Thøgersen, J., 2014. Informing versus nudging in environmental policy. J. Consum. Policy 37, 341–356.

Osbaldiston, R., Schott, J., 2011. Environmental Sustainability and Behavioral Science: Meta-Analysis of Pro-environmental Behavior Experiments. Environ. Behav. 0013916511402673.

Owen, G., Ward, J., 2007. Smart meters in Great Britain: the next steps. Sustain. First, 1–68.

Owens, S., 2000. "Engaging the public": information and deliberation in environmental policy. Environ. Plan. A 32, 1141–1148.

Palm, A., 2015. An emerging innovation system for deployment of building-sited solar photovoltaics in Sweden. Environ. Innov. Soc. Transit. 15, 140–157. https://doi.org/10.1016/J.EIST.2014.10.004.

Palm, A., 2016. Local factors driving the diffusion of solar photovoltaics in Sweden: a case study of five municipalities in an early market. Energy Res. Soc. Sci. 14, 1–12. https://doi.org/10.1016/J.ERSS.2015.12.027.

Palm, A., 2017. Peer effects in residential solar photovoltaics adoption—a mixed methods study of Swedish users. Energy Res. Soc. Sci. 26, 1–10. https://doi.org/10.1016/J.ERSS.2017.01.008.

Parag, Y., Fawcett, T., 2014. Personal carbon trading: a review of research evidence and real-world experience of a radical idea. Energy Emiss. Control Technol. 2014, 23–32.

Peters, G.P., 2008. From production-based to consumption-based national emission inventories. Ecol. Econ. 65, 13–23.

Peters, G.P., Hertwich, E.G., 2006. Pollution embodied in trade: the Norwegian case. Glob. Environ. Chang. 16, 379–387.

Pollitt, M.G., Shaorshadze, I., 2011. The Role of Behavioural Economics in Energy and Climate Policy (No. 1165). EPRG Working Paper. https://doi.org/10.17863/CAM.1140.

Price, M.K., 2014. Using field experiments to address environmental externalities and resource scarcity: major lessons learned and new directions for future research. Oxf. Rev. Econ. Policy 30, 621–638.

Rai, V., Reeves, D.C., Margolis, R., 2016. Overcoming barriers and uncertainties in the adoption of residential solar PV. Renew. Energy 89, 498–505. https://doi.org/10.1016/J.RENENE.2015.11.080.

Raimi, K., Carrico, L., 2016. Understanding and beliefs about smart energy technology. Energy Res. Soc. Sci. 12, 68–74.

Reisch, L.A., Zhao, M., 2017. Behavioural economics, consumer behaviour and consumer policy: state of the art. Behav. Public Pol. 1, 190–206. https://doi.org/10.1017/bpp.2017.1.

Ryghaug, M., Sørensen, K., 2009. How energy efficiency fails in the building industry. Energy Policy 37, 984–991. https://doi.org/10.1016/j.enpol.2008.11.001.

Sælen, H., Kallbekken, S., 2011. A choice experiment on fuel taxation and earmarking in Norway. Ecol. Econ., Special Section—Earth System Governance: Accountability and Legitimacy. 70, 2181–2190. https://doi.org/10.1016/j.ecolecon.2011.06.024.

Sagebiel, J., Müller, J.R., Rommel, J., 2014. Are consumers willing to pay more for electricity from cooperatives? Results from an online choice experiment in Germany. Energy Res. Soc. Sci. 2, 90–101. https://doi.org/10.1016/j.erss.2014.04.003.

Samahita, M., Mundaca, L., 2018. Visibility, Peer Effects and Monetary Incentive in Solar PV Adoption in Sweden—Evidence From a Survey Experiment (Working Paper). International Institute for Industrial Environmental Economics (IIIEE), Lund, Sweden.

Samuelson, W., Zeckhauser, R., 1988. Status quo bias in decision making. J. Risk Uncertain. 1, 7–59.

Schleich, J., Klobasa, M., Gölz, S., Brunner, M., 2013. Effects of feedback on residential electricity demand—findings from a field trial in Austria. Energy Policy 61, 1097–1106.

Schoenfeld, D., 1980. Statistical considerations for pilot studies. Int. J. Radiat. Oncol. 6, 371–374.

Segerstedt, A., Grote, U., 2016. Increasing adoption of voluntary carbon offsets among tourists. J. Sustain. Tour. 24, 1541–1554. https://doi.org/10.1080/09669582.2015.1125357.

Seidl, R., 2018. Assessing a personal carbon trading scheme with an agen-based model (Working Paper). Öko Institut, Freiburg.

Seljom, P., Lindberg, K., Tomasgard, A., Doorman, G., Sartori, I., 2017. The impact of zero energy buildings on the Scandinavian energy system. Energy 118, 284–296. https://doi.org/10.1016/j.energy.2016.12.008.

Shogren, J.F., Taylor, L.O., 2008. On behavioral-environmental economics. Rev. Environ. Econ. Policy 2, 26–44.

Simon, H., 1947. Administrative behaviour. Aust. J. Public Adm. 9, 241–245. https://doi.org/10.1111/j.1467-8500.1950.tb01679.x.

Simon, H., 1979. Models of Thought. Yale University Press, New Haven, CT.

Simon, H., 1986. Rationality in psychology and economics. J. Bus. 59, S209–S224.

Slovic, P., Finucane, M., Peters, E., MacGregor, D.G., 2002. Rational actors or rational fools: implications of the affect heuristic for behavioral economics. J. Socio-Econ. 31, 329–342. https://doi.org/10.1016/S1053-5357(02)00174-9.

Sonnenschein, J., 2016. Rapid decarbonisation of cities by addressing behavioural failures? A critical review of policy interventions. In: Presented at the BEHAVE—4th European Conference on Behaviour and Energy Efficiency, Coimbra, Portugal.

Sonnenschein, J., Mundaca, L., 2019. Is one carbon price enough? Effects of payment vehicle choice on willingness to pay. Energy Res. Soc. Sci. 52, 30–40. https://doi.org/10.1016/j.erss.2019.01.022.

Sonnenschein, J., Smedby, N., 2019. Designing air ticket taxes for climate change mitigation: insights from a Swedish valuation study. Clim. Pol. https://www.tandfonline.com/doi/full/10.1080/14693062.2018.1547678.

Sovacool, B., 2017. Contestation, contingency, and justice in the Nordic low-carbon energy transition. Energy Policy 102, 569–582. https://doi.org/10.1016/j.enpol.2016.12.045.

Sovacool, B., Kester, J., de Rubens, G., Noel, L., 2018. Expert perceptions of low-carbon transitions: investigating the challenges of electricity decarbonisation in the Nordic region. Energy 148, 1162–1172. https://doi.org/10.1016/j.energy.2018.01.151.

Squazzoni, F., Jager, W., Edmonds, B., 2014. Social simulation in the social sciences: a brief overview. Soc. Sci. Comput. Rev. 32, 279–294.

Steg, L., 2018. Limiting climate change requires research on climate action. Nat. Clim. Chang. 8, 759–761. https://doi.org/10.1038/s41558-018-0269-8.

Steg, L., Perlaviciute, G., van der Werff, E., 2015. Understanding the human dimensions of a sustainable energy transition. Front. Psychol. 6, 1–17.

Stiglitz, J.E., Stern, N., Duan, M., Edenhofer, O., Giraud, G., Heal, G., La Rovere, E.L., Morris, A., Moyer, E., Pangestu, M., 2017. Report of the high-level commission on carbon prices. World Bank.

Sunstein, C., 2014. Choosing not to choose. Duke Law J. 64, 1.

Sunstein, C.R., Reisch, L., 2014. Automatically green: behavioral economics and environmental protection. Harv. Environ. Law Rev. 38, 127–158. https://doi.org/10.2139/ssrn.2245657.

Tang, L., Wu, J., Yu, L., Bao, Q., 2015. Carbon emissions trading scheme exploration in China: a multi-agent-based model. Energy Policy 81, 152–169.

Thaler, R., 2015. Misbehaving: The Making of Behavioral Economics. Norton, Grand Haven, MI.

Thaler, R.H., Shefrin, H.M., 1981. An economic theory of self-control. J. Polit. Econ. 89, 392–406.

Thaler, R., Sunstein, C., 2008. Nudge: Improving Decisions About Health, Wealth, and Happiness, Revised & Expanded edition. Penguin Books, New York.

Throne-Holst, H., Strandbakken, P., Stø, E., 2008. Identification of households' barriers to energy saving solutions. Manag. Environ. Qual. Int. J. 19, 54–66. https://doi.org/10.1108/14777830810840363.

Tiefenbeck, V., Staake, T., Roth, K., Sachs, O., 2013. For better or for worse? Empirical evidence of moral licensing in a behavioral energy conservation campaign. Energy Policy 57, 160–171. https://doi.org/10.1016/j.enpol.2013.01.021.

Tietenberg, T., 2009. Reflections—energy efficiency policy: pipe dream or pipeline to the future? Rev. Environ. Econ. Policy 3, 304–320.

Tomer, J., 2017. Advanced Introduction to Behavioral Economics. Edward Elgar Publishing, Cheltenham; Northampton, MA.

Trout, A., Kaufmann, T., Kallmes, D., 2007. No significant difference … says who? Am. J. Neuroradiol. 28, 195–197.

Tversky, A., Kahneman, D., 1974. Judgment under uncertainty: heuristics and biases. Science 185, 1124–1131.

Tversky, A., Kahneman, D., 1992. Advances in prospect theory: cumulative representation of uncertainty. J. Risk Uncertain. 5, 297–323.

Uehleke, R., 2016. The role of question format for the support for national climate change mitigation policies in Germany and the determinants of WTP. Energy Econ. 55, 148–156. https://doi.org/10.1016/j.eneco.2015.12.028.

van der Werff, E., Steg, L., 2016. The psychology of participation and interest in smart energy systems: comparing the value-belief-norm theory and the value-identity-personal norm model. Energy Res. Soc. Sci. 22, 107–114.

Wallace, A.A., Irvine, K.N., Wright, A.J., Fleming, P.D., 2010. Public attitudes to personal carbon allowances: findings from a mixed-method study. Clim. Pol. 10, 385–409.

Weitzman, M.L., 2014. Can negotiating a uniform carbon price help to internalize the global warming externality? J. Assoc. Environ. Resour. Econ. 1, 29–49. https://doi.org/10.1086/676039.

Wilson, C., Crane, L., Chryssochoidis, G., 2015. Why do homeowners renovate energy efficiently? Contrasting perspectives and implications for policy. Energy Res. Soc. Sci. 7, 12–22. https://doi.org/10.1016/j.erss.2015.03.002.

Wiser, R.H., 2007. Using contingent valuation to explore willingness to pay for renewable energy: a comparison of collective and voluntary payment vehicles. Ecol. Econ. 62, 419–432. https://doi.org/10.1016/j.ecolecon.2006.07.003.

Wood, F.R., Bows, A., Anderson, K., 2010. Apportioning aviation CO_2 emissions to regional administrations for monitoring and target setting. Transp. Policy 17, 206–215. https://doi.org/10.1016/j.tranpol.2010.01.010.

World Bank, Ecofys, Vivid Economics, 2018. State and Trends of Carbon Pricing 2018. World Bank, Washington, DC.

Zanni, A.M., Bristow, A.L., Wardman, M., 2013. The potential behavioural effect of personal carbon trading: results from an experimental survey. J. Environ. Econ. Policy 2, 222–243. https://doi.org/10.1080/21606544.2013.782471.

Further reading

IEA, 2017b. Energy Efficiency Indicators 2017 Database. OECD/IEA, Paris.

Marland, G., Boden, T., Andres, R., 2017. Global, Regional, and National Fossil-Fuel CO2 Emissions. Carbon Dioxide Information Analysis Center, Oak Ridge National Laboratory, U.S. Department of Energy, Oak Ridge.

Peters, G.P., Minx, J.C., Weber, C.L., Edenhofer, O., 2011. Growth in emission transfers via international trade from 1990 to 2008. Proc. Natl. Acad. Sci. 108, 8903–8908.

UN, 2017. National Inventory Submissions 2017|UNFCCC [WWW Document]. https://unfccc.int/process/transparency-and-reporting/reporting-and-review-under-the-convention/greenhouse-gas-inventories-annex-i-parties/submissions/national-inventory-submissions-2017.

Beyond energy behaviour: A broader way to see people for climate change technology planning

1.3

Mithra Moezzi[a], Loren Lutzenhiser[b]

[a]*QQForward, Marin County, CA, United States*
[b]*Portland State University, Portland, OR, United States*

1 Introduction

It is now easier than ever before to find research and arguments that jointly address people and energy technology. Most of this work is quite specialised, with little high-level guidance available as to how overall discourse on people and energy use is organised. For those without a social sciences background, this has led to some confusion. In particular, 'behaviour' is often used to shorthand how people relate to energy systems. This term puts the emphasis on what people purchase and how to get them to change their behaviour with respect to specific devices. Those topics are aligned with straightforward programme interventions and technology deployment goals. But if one thinks instead of how societies use energy, why this differs from one locale to another, how use has changed over history, and how it might change in the future, the behavioural viewpoint on its own is much too limited and stylised to answer much. It cannot capture energy use systems.

The orientation to behaviour reflects the 40-plus year tradition of seeing energy efficiency as a property of machines. Consideration of the humans who use these machines has mostly come as an afterthought, even in the past 15 years as climate change concerns came to the forefront of energy work. A recent study found that less than 3% of the academic literature sample on 'energy efficiency' publications over 1909–2018 could be classified as social sciences (Dunlop, 2019). Taking behaviour explicitly into account represents progress over a purely machine-oriented view of efficiency, but without more, it presents a narrow and static view of how people matter.

This weakness was evident in the energy efficiency era of energy planning (Lutzenhiser, 2014; Shove, 2018). Now that climate change problems are the main concern for much of energy planning, the limitations are more evident and more consequential. But there also may be a higher probability that they can be overcome, which is why we summarise it here for those who might not be familiar with it. We imagine the readers of this chapter to include engineers, technology designers, policymakers, policy planners, and students in these fields, all with interest in better understanding how to conceptually integrate people into their projects on climate-friendly energy technologies. This paper offers some tools, ideas, and examples as to how that might happen.

1.1 Climate change is a different energy problem than energy efficiency

The problem of climate change brings together a huge and diverse set of researchers and policymakers, more so than for any policy problem in history. Even for the energy aspects of climate change alone, the scope, scale, and dimensions of the questions being asked, and the goals pursued by governments, are much grander than they were when the goal was promoting relative energy efficiency. The problem is no longer saving incremental energy on a device by device basis relative to a counterfactual baseline, but instead contributing to absolute emissions reductions at a very large scale and in the meantime trying to assure a reliable energy system that can handle the new demands made of it in a hotter, more complex world. Energy efficiency programmes have had major effects on how energy is used. But much bigger and better outcomes will be needed for 'smarter' and otherwise better devices, buildings, and grids to be put in place and to truly provide the benefits promised of them.

Currently, achieving greenhouse gas (GHG) emissions reductions is usually seen as an outcome of a completely refreshed set of technical efficiencies to help reduce and control energy demand (buildings, appliances, transportation systems, etc.) and a mammoth overhaul of energy supply (e.g. renewable rather than fossil fuel energy sources, prosumers, and a smart grid). This must be accomplished not just for one jurisdiction but, in some form, worldwide. In this technology deployment view, the problem is to get the right technologies developed and into place—a 'build it and they will come' strategy. But what are the right technologies? And what effects do they have? How do they change the sociotechnical system of energy?

2 From technology to people and back again

The basic policy stance on energy technology for climate change starts with the idea of a technology as a neutral material object with energy saving properties. But socially, no technology is neutral, and savings are not a property of a technology but of its use. This section steps back to reinterpret how technology and people fit together, interact, and affect each other. To set the stage for a deeper dive into social thinking about people and technologies, we identify two issues related to the contributions of different scientific disciplines and areas of professional practice.

The first problem involves disciplinary divisions of labour. The conceptual model of technical solutions that can be actualised by the right behavioural elements delegates the behavioural and social sciences to an end-of-pipe role. That is, people are instruments of these technological solutions. But people are much more complex (and interesting) than this strategy would suggest.

Second, even if the issue was promoting one single technology, a conceptual blind spot that sees people as only buyers and users would still need correcting. People play much deeper parts in the story: they design, produce, sell, maintain, regulate, and collectively shape technologies. These multiple roles are even more consequential when the goals are to introduce multiple efficiencies, and even completely new technologies, in short time frames. As we will argue later, thinking about the problem as involving inherently complex and dynamic 'sociotechnical system' of people, energy, and material world, rather than starting from an idealised technological arsenal, already starts to illuminate some of the more hidden challenges of large-scale technological change and to invite in a new set of expertise.

2.1 Learning from history: Technologies to systems

Technology is more than machines or devices. Work in the history of technology illustrates how devices only function as parts of larger systems that have structures and dynamics that shape design and deployment (Hughes, 1986; Kranzberg, 1986; Marcus and Segal, 1989). By *dynamics*, we mean the ongoing interplay of system elements such as sources of materials supply, transport, energy, and production processes that work together to sustain system operations. These dynamics have critical implications for how a technology works and the degree to which it meets policy and engineering expectations.

By *shaping*, we mean the influences and constraints that the system dynamics exert on technical innovation and system expansion—along paths that result from mutual relationships among technologies, policies, and stakeholders and not in directions that are blocked. For example, in the United States, affordable compressor-based air conditioning enabled home builders to build large tracts of simpler houses that were less able to provide passive cooling than earlier designs, contributing to the need for using mechanical air conditioning (Cooper, 1998). Once air conditioning is well-established, efforts to help encourage efficiency, such as energy efficiency rating schemes, can discourage home designs that could afford dropping air conditioning despite being a lower-energy design (Kordjamshidi and King, 2009).

When trying to imagine ways that technologies could change in the future, these historical sequences and the direction they force cannot be overlooked. For example, historians have documented how technology system arrangements exert resistance to change. As case in point being, Thomas Hughes' (1993) studies of how electric grids developed quite differently in different countries. He also identified ways in which technical systems exhibit inertias, momentum, and punctuated rates of change (Hughes, 1987). The 'path dependencies' that industries and technologies exhibit that are sometimes referenced by economists are a related phenomenon (Vergne and Durand, 2010).

The history of technology is also full of stories about little technical things making big sociotechnical changes (e.g. moveable type and knowledge, the compass and empire, or the transistor and data) and also big things making change difficult (e.g. railroad track system geographies or dams on rivers preventing passage, 'locked-in' patterns of trade relationships, and effects of highway locations on commercial development). Looking towards the future, we can see how the entire range of imaginable and even critically necessary technological change will be caught up in similar system dynamics as past histories are projected into the future. Engineers who are reading this will appreciate what a systems framework may imply for design—both in terms of how things must be made and how their use may open new possibilities for innovation.

2.2 Technologies as social and society as technological

Building on insights from historians, in the past 25 years, sociologists interested in technological systems and technology change begun to incorporate knowledge about physical devices and systems into their thinking about society. Early modern social theory (e.g. Parsons, 1952) proposed a very conservative view of society as organised in subsystems that interacted and functioned to bind people, groups, and institutions together. Unfortunately, this view was found wanting when societies were torn by social change in the 1960s and 1970s. Sociology, anthropology, and political science all began to take more of a political economy view of society, in which conflicts and power were seen as key features. As a result, social theory turned its back on systems thinking for several decades, with a few notable exceptions (e.g. Luhmann, 1982).

However, the large-scale failure of technologies (Perrow, 1984) and environmental crises (e.g. nuclear accidents, civil engineering failures, and widespread toxic exposures) beginning in the 1970s led other social scientists to turn their attention to technological systems, to the practice of science and engineering, and to the political processes through which decisions about technologies are made. All of these turned out to be areas in which understanding of technologies depends on knowledge about the workings of humans involved with those technologies.

The work of sociologists and other social scientists focused on technology has merged with scholarship in science and technology studies (STS), and the result has been a robust view of technological systems as deeply social and of a society deeply embedded in a technological matrix. We now speak of the two sides of the coin (devices vs people) less simplistically and see both as bound up in multiscale, complex, and dynamic *sociotechnical systems*. We also recognise that these systems are actually 'heterogeneous', being made up of complexly related people, groups, ideas, material devices, software, technological subsystems, energy and fuels, rules (e.g. engineering standards, and regulatory codes), institutions (e.g. the law, accounting, engineering education, manufacturing firms, retailers, etc.), and so on. This view has then led to questions about how technologies in these systems are *shaped* and *formed* (e.g. Bijker et al., 1987). Research has shown that any and all of the heterogeneous elements of a sociotechnical system can play a role in stabilisation and change at various scales. Put in a context of large-scale technological change, with considerable climate change innovation and policy significance, this work in STS is also relevant to more recent studies of sweeping sociotechnical *transitions* (e.g. from coal to electricity, though also including transitions in technologies of demand) that have taken place in the past and could unfold in the future (Smil, 2010; Geels, 2004). A transitions perspective allows us to see that technologies and systems have changed substantially in the past and have always been changing, contra an assumption of equilibriums. So 'transition management' has become a topic of interest in environmental policy (Meadowcroft, 2009).

2.3 Climate change and technological change beyond a behaviour vantage point

How can this work at much larger scales be applied to the problem at hand and our thinking about 'behaviour'? We started with a simple problem of how to get a handle on how consumer behaviour might be changed so that they adopt the 'right stuff' and use it in the 'right ways'. The reader might reasonably wonder how to get back down to earth.

Let's start by going back to the conventional framing and seeing if we can now reinterpret. Since bringing 'behaviour' into the equation means that we have come to think about a 'device side' and a 'people side' of a system, let's put that into a concrete example, say that of a household using energy and generating emissions through the use of devices. We can now think about even the homeliest example in a very different way in the future. A sociotechnical systems view enables us to begin seeing modern life as made up of people and devices that are inextricably connected and interdependent. One cannot exist without the other.[a]

If we recognise that these sociotechnical systems are highly organised and have some continuity and stability (which allows predictable everyday life to go on), we should be able to identify the parts that are stabilising. In a heterogeneous system, some reside in persons' routine habits and practices in using devices. But these are connected to household dynamics, work and home life, cultural understandings and beliefs, and technical ensembles that go with the dwelling and are understood to be normal and proper—these are not usually questioned nor would a behavioural perspective really allow this. The power grid and the grid supplier, the prices charged, the reliability of the energy service, weather, the surrounding environment, and neighbourhood context are all implicated in technology use, energy flows, and emissions and in their stabilisation and resistance to change. At the same time, change does take place, and even the most isolated and local parts of sociotechnical systems are under some continuous pressures for change. The question is, however, can we

---

[a]This has always been true of *Homo sapiens* and earlier hominids. They refashion bits of their natural environments and use those, along with language and social life, to make a living, reproduce, and so forth. The differences in modern societies are the sheer scale and complexity of the technologies, the dependence of all of us on their smooth workings, and the fact that we can be almost completely ignorant of those dependencies, which is a luxury that our ancestors could not afford.

understand transitions and systems well enough to intelligently intervene in them (Meadows, 1999) for a more hospitable future for humans and the planet? Or is another approach than direct intervention needed?

Turning back to the argument to better integrate the social sciences in technology systems, what if designers of devices and planners of policy interventions could see those systems, from top to bottom, as filled up with people playing critical roles? People invent and construct systems of technologies. They manage, direct, and operate systems (and subsystems on multiple scales). They service and modify; they regulate and shape; they buy goods and services, consume energy and natural resources, and travel and connect with each other, producing 'demands' on systems. *So changes in hardware mean changes in systems (with likely both hoped-for and unintended changes).* And that means changing people (and changes for people) as well as hardware and software. But it also means going beyond 'changing behaviour' to understand what people are doing, why they are doing it that way, which understandings, beliefs, and values are involved, and how interactions with devices and machines and infrastructures produce social value and stabilise patterns of action and forms of consumption. And it involves respect for what people know and do and are willing to engage about.

2.4 **Behaviour in the energy efficiency world**

Climate action plans and strategies conventionally identify *energy efficiency* as a crucial ingredient. Energy efficiency has nearly a 50-year history in Europe and the United States that involves a combination of government energy conservation rules and laws, public and private research and development to create more efficient devices, processes and buildings, grassroots citizen action, and utility programmes to purchase, induce, and support replacement of less efficient equipment with forms that use less energy. Because this has grown up as in the context of regulated energy monopolies in the United States (Lutzenhiser, 2014), it has taken on a highly specialised and technical character, which scripts activities narrowly and takes up goals of only modest changes in energy use—and as Elizabeth Shove notes "undermines that which it is expected to achieve" (Shove, 2018).

But with the advent of concerns about climate change and policies requiring that larger efficiencies be gathered, an interest in behaviour change has emerged that builds upon earlier energy conservation interests (discussed in the following text). In the current policy scene in California, for example, the category of behaviour change activities titled 'BROs' (behaviour, retro-commissioning, and operations) is now a large fraction of proposed energy utility efficiency expenditures (Navigant, 2017). However, as noted earlier, in the behaviour change framework, people are first and foremost seen as instruments, or at least as instrumental entities interacting with devices. The prescription is obvious enough. People need to change their behaviour to fit the optimal technologies, starting with buying the approved target device (e.g. Axsen and Kurani, 2012).

There is a substantial literature on 'rebound effects' of energy efficiency technology interventions in which the planned-for savings do not materialise. If there is blame to be had, it is often applied to the users who 'take back' savings by 'increasing their comfort' or acting in other more mysterious ways (for reviews, see Herring and Sorrell, 2009). The term rebound is often used in poorly articulated ways that end up confusing purely economic explanations (which people buy more because it costs less) with the many other things that could be at play, such as poor installed performance or coinciding technological changes. In short, failure to understand actual events and outcomes through a 'people as instrument' lens has been widespread in the energy efficiency world. But, again, as long as the goals of energy efficiency have been modest, and the savings estimated to be adequate, these weaknesses could be overlooked.

2.5 **Behaviour change movement**

Beginning with the energy crises of the 1970s, the need to conserve energy and reduce waste enlisted social and behavioural scientists to the energy cause. A more social perspective on energy and social groups that was reflected in early studies (e.g. Newman and Day, 1975) was supplemented by insights by psychologists and other behavioural scientists and what might be called a 'big tent' review by the US National Research Council gathered many different strands of people-related energy work in a volume titled *Energy Use: The Human Dimension* (Stern and Aronson, 1984). But by the late 1980s the institutionalisation of energy *efficiency* in government programmes, appliance regulations, building codes, R&D agendas, engineering practice, and utility programmes took a nearly exclusive hardware (if sometimes also software) focus, pushing the energy users to the sidelines (Lutzenhiser, 2014).

Early in the 2000s, however, a behaviour change movement became evident in European and US energy research and policy (Owens and Driffill, 2008). Work in this tradition generally identifies people's behaviour with respect to energy as problematic—indeed sometimes as the chief problem—and seeks to find ways to change this behaviour to actions considered more desirable with respect to specific ideals. This is a natural step to better incorporate people when originated within a technology-centred framework for energy savings, since behaviour is the ingredient that determines the efficacy of technological solutions. If people don't buy the desired technology, that technology cannot save anything, and if they use it differently than predicted, the modelled savings will not be achieved.

This move put people more prominently back on the energy policy map and permitted partial accounting for behavioural energy conservation, which had been nearly invisible in mainstream energy programming throughout the 1990s. A variety of modes of behaviour change interventions developed, for example, applying behavioural economics principals such as setting opt-out defaults rather than relying on opt-in choices (Frederiks et al., 2015), social comparison (e.g. Carrico and Riemer, 2011), gamification in domestic (Johnson et al., 2017) or workplace, information-centred strategies such as real-time feedback through home energy monitoring (Hargreaves et al., 2013), and others (Abrahamse et al., 2005). These methods provide deliberate, concrete ways of relating people to devices, making them appealing for intervention-style programmes.

From a social science perspective, such pursuits of behaviour change are troublesome or at least an unnecessarily limited way of viewing the potential for changing energy use. They set 'solutions' in terms of individual behaviour alone, as noted earlier. Behaviour itself is usually seen as a consequence of 'values and attitudes' (Shove, 2010) interacting with level of 'information' that a person has or lacks (Owens and Driffill, 2008). As detailed in the next section, this misses a lot. But even within the behaviour change paradigm, there are some logical and practical difficulties. Ideal behaviours are often arbitrarily and statically defined, and sometimes without even a realistic assessment of what really saves energy and how much, at the same time privileging energy saving behaviours over everything else. And applying behaviour change principles often relies on a conceptual model of correct behaviour that has little ability to incorporate the degrees of freedom commensurate with the wide variety of different circumstances, people, and practices. For example, people may behave much more conservatively than the models allow (e.g. Boucher, 2017; Deumling et al., 2019; Hagbert and Bradley, 2017), and this variety is ignored. Most importantly perhaps, it is not clear that behaviour change programmes have long-term benefits.

2.6 Social science ambivalence

For many, this new attention to people invited by the 'behaviour change' movement was good news, as it seemed to promise a way of looking more holistically at real energy use and designing strategies that took better advantage of the intricate links among the social and the technical—the questions behind, for example, the observation that some of the most technologically advanced countries use energy vastly differently than others. Ideally, this would lead to sociotechnical transformations to lower emissions versus the classic 'technical swap-out' perspective. Most of the new attention, however, was confined to just two angles on thinking about people: convincing individuals that they should buy specific products and that they should use energy differently in a set of prescribed ways.

This perspective treats the social (people) elements and the technical (device) elements separately. It furthermore privileges specific technical solutions often with little more than an elementary understanding of what people do and why—as in the example of programmable thermostats mentioned earlier and illustrated in the box in the following text. This fits relatively neatly into engineering work and into semi-classical scientific methods, as it generates quantitative and quasiquantitative data that can be easily modelled or used to describe a population. But most importantly in terms of the reasons for this chapter, it fails to give engineers very much to do in terms of designing efficient systems of energy use.

To complicate matters further, in the energy policy field, 'behaviour' has come to be a catch-all for what individuals, households, companies, or other collectives do: how they act in physical space. This may at first seem to be an adequate basis for talking about people's role in energy use, because it connects so closely to actions. Actions are the endpoint in engineering–economic models of purchase decisions, for example; they can be used to describe relationships between humans and the physical infrastructure; they can be modelled (crudely—in part because empirical data on behaviour is so poor and hard to sample) in various more detailed quantitative models, such as stochastic expression of occupancy or of envelope interaction in building simulation models, or decisions as expressed in agent-based models of technology adoption and use.

2.7 Thinking in social science terms about people and technologies

We believe that insights from the history and social studies of technology on socio-technical systems broadens the view and points to a host of features of the social and technological landscape that strongly affect action and social processes that are either not captured or severely distorted in the 'behaviour change' amendments to the conventional device-centred energy efficiency view.

To begin, consider how persons organise and act through social practices that are structured across scales, from the micro-habitual to the macro-historical. Over the past decade, a rich vein of argument has been tapped by *social practice theory* (Shove et al., 2012). Seeing social practices as simultaneously composed of technical/artefact elements, human skill at use/interaction with devices, and bodies of knowledge/understanding that give the actions meaning and social form, social practices can range from food/cooking/eating in particular ways to securing incomes, educating children, sheltering, securing cleanliness and comfort, etc. All of these can vary, and all are bundles of device–body–meaning performances. As noted, they are simultaneously local and situated, as well as embedded in/constitutive of larger social networks and forms, institutions, matrices of larger sociotechnical arrangements, etc.[b]

This may seem esoteric for our intended readers. But an example or two should be helpful. Let's say that one has learned (from their elders) and have chosen (based on study and prejudice) to eat vegetables very often and as fresh as possible. Their spouse reinforces this when she cooks, and the grown children signal approval and disapproval as suits them. This business of vegetable eating involves growing things in the summer (which requires a garden) and doing some preserving and, when more is needed, trips to farmers markets and grocery stores. This, in turn, requires some walking and driving, which requires some walkways and a car. The food needs some cold storage, so a refrigerator is needed to check that box. And if they want to cook

---

[b]These are our representations of social practice theory, particularly as applied to our interests in households and mundane technologies caught up in large-scale systems implicated in climate change. But we are not social practice theorists, so our summary certainly lacks nuance and detail. We strongly suggest that interested readers refer to two original works: Shove (2003) *Comfort, Cleanliness and Convenience: The Social Organization of Normality* and Shove, Pantzar, and Watson (2012) *The Dynamics of Social Practice: Everyday Life and How it Changes.*

it, they'll need to keep some pans and a stove. And then there's clean up. And on, and on. The particular eating practice involves a lot of social agreement and division of labour, resources, devices and tools, places to store and use them, and ways to get about. This is not to mention vegetable magazine subscriptions, being on the lookout for garden pests and larcenous children, and more. It's a social practice, and persons can only pull it off with a lot of help and a lot of highly interconnected bits of the world, including people and machines, as well as nature and society.

Let's say another person lives in an apartment and works at coding in their office cube from 10:00 a.m. until 10:00 p.m. They forget to eat. There is a microwave at work and a vending machine in the hallway. They can order takeout on their phone and either pick it up on the way to the bus or have it delivered by someone driving their own car (thankfully, the tip can be paid through the app). That's another practice; another set of people; another set of habits and understandings; another set of devices.

Back in the climate change context, there are energy flows involved in both of these practice patterns. We could estimate them with some accounting, and some scholarly traditions do this in generalised form through the notion of urban energy metabolism (e.g. Kennedy et al., 2011). But that's complicated, and the practitioners rarely are interested or able to study their energy demands. There are also carbon and other emissions produced, and these can be calculated and compared (and will also vary a good deal depending on sources of fuels and energy conversion technologies). In addition, a stream of 'life cycle' carbon, energy, and pollution implications can be investigated for every product or process used (which is done even less frequently).

Engineers are already familiar with systems thinking, so these simple depictions of practice patterns should resonate and contrast with the narrower behaviour-centred view. This shifts attention from energy use as a bunch of independent end uses, each to be directly tackled, to one where the whole life cycle of energy use is the object at issue. Taking a *systems view* doesn't necessarily simply reveal a world that is hopelessly complicated. Instead it is an important starting point to understanding what's going on physically and ecologically, and grounds inquiry and design in an appreciation that the social and technological are intrinsically bound up in other another. As a result, it will take some work and care and decent information to make a reasonable attempt to change anything. It also builds on the understanding that the standard 'behaviour change' interventions with respect to sociotechnical systems often fall short. Simply saying "get more efficient devices" or "figure out what the 'users' or 'consumers' are doing wrong" or "nudge these people to do the right thing (which is that thing we want them to do)" very clearly is not a starting point. If anything, it's a dead end.

The rise and fall of programmable thermostats

Energy efficiency programmes have often seen a commitment to a particular technology as a highly promising energy saving technology, with little attention to the expected users and their existing practices. The latter are usually substituted by theoretical logic models that can readily be designed for, while trying to determine and express what people are actually doing is a daunting statistical and practical problem (though less so now than 15 years ago). Programmable thermostats in the United States are one example. Early programmes estimated that programmable thermostats could provide large heating and cooling energy savings for households based on a series of assumptions about what people did and what they would do with programmable thermostats. Actual experience revealed, eventually, that users often did not use the programming features—for which users were mocked. Eventually better interfaces were designed and learning thermostats, which did not require user intervention, were promoted as the answer. But the issues go beyond user interfaces. Savings assumptions were often modelled on the basis that people did not turn off or set back (or set up, in the case of cooling) their thermostat at night or when they were away—whereas empirical data show that most do. Plus, programmable thermostats reinforce the message that heating and cooling is by default 'on' and ready to correct deviations, already part of early engineering assumptions about home air conditioning (Cooper, 1998, 2008).

For years, programmable thermostats were listed as approved Energy Star Devices by the US Environmental Protection Agency (EPA) and Department of Energy. But the specification was removed in 2009, with the explanation given: "While EPA recognizes the potential for programmable thermostats to save significant amounts of energy, there continue to be questions concerning the net energy savings and environmental benefits achieved under the previous ENERGY STAR programmable thermostat specification." (EPA, 2019a). The EPA site does still contain very detailed instructions for programming and use of these thermostats in order to use them to save money (EPA, 2019b), but has now included a 'smart thermostat' (web connected with energy use feedback information and utility communications) specification—slightly less detailed instructions about how the user should be configured in order for the thermostat to help save energy (EPA, 2019c).

2.8 The realm of quasiproblems

In summary, people are much more complicated and adept and highly organised in social groups than simple behavioural representations claim. We argued this narrowness invites the pursuit of narrow 'solutions' to climate change that can be totally unrealistic or that fall far short of the benefits expected of them. And we suggested that in focusing on these narrow solutions, more efficient and less risky approaches are not considered or are locked out. Getting this wrong not only leads to limited understanding but also leads to poor outcomes in policy and business and incorrect interpretations of those outcomes—the problems involved with self-referential conceptual framings, projecting organisational aims on people, and the problem of limited tools creating their own representations and solutions to nonreal quasiproblems.

3 From behaviour to people in practice

So far, we have outlined some difficulties of relying on 'behaviour' as a proxy for people's role in energy use. But restricting attention to this slice has been paralleled by a dominance of highly instrumental and often unrealistic, unsympathetic views of people, failing to take advantage of the fact that technology is imminently social. The remainder of this chapter starts to tackle how some of this 'missing half' could be brought to bear on energy technology planning.

3.1 **Reasonable protests**

Regarding the social scientific views of technology outlined earlier, readers may have reservations about how these broader and more intricate interpretations of technology could be made generally actionable and useful in the practical realms of climate change–related technology development and planning, which are already complicated enough. We are hardly at the place where these complexities can be systematically expressed. Uncertainties and potentially relevant decision criteria (Bhardwaj et al., 2019) are overwhelming. At the same time, the simple causal story (Stone, 1989) of how people are wasting energy and how a refresh with efficient versions of the same technologies will reverse this is highly engrained in institutions, regulations, models, habits, etc. So, there is a lot of inertia.

Moving to better ways of seeing people in climate change technology planning requires accepting that much of social sciences strength lies in articulating new and mostly true 'stories' to colleagues, including those in other fields. These stories should encourage re-conceptualisations of established assumptions and methods, more pertinent sets of questions, and overall a recast of how the world as a technology canvas is seen, away from stable uniform problems of efficiently fulfilling needs. There is experimental evidence that this can happen (Bodin et al., 2016). With this vision in mind, there are some actionable pathways available, and they are not that hard. The remainder of this paper covers three angles on doing so.

3.2 **Vocabulary and other tools**

Table 1 provides a sample list of terms that are routinely used in market and policy documents for representing how people fit into energy use. These words are the US vocabulary and interpretations that will be somewhat different than, for example, European ones. Terms in the first column link to conceptual models dominant in the narrowly technology-centric view of how people figure into energy efficiency criticised earlier. For example, 'consumers' are purchasers, who then easily fit into policy instruments that aim to get these consumers to buy differently. There is little hint that those purchasers are affected other than receiving a clean set of product benefits. There is an assumption that these consumers make deliberate choices in a gamut of products available to them. And there is an assumption that with the right intervention methods, consumers will follow through. The third column provides an option for a more social scientific term that partially escapes these rather sterile models. For example, the term 'people' instead of 'consumer' at least opens the possibility of seeing subjects as moving through the world with bodies and in contexts, well beyond the pocketbook and, though patterned from a population view, in infinite variety. And people can be referred to in terms of social groups—such as households, facilities teams, classrooms, unions, etc.—that do not 'behave' (or 'prefer', 'think', or 'learn') like individuals, a fundamental principle of social sciences.

Swapping one term for another is not progress on its own. But recognising the hidden assumptions and loading of these usual terms, signalled by a new vocabulary, can point to and legitimise a creative space—for example, for mutual conversations

Table 1 Revamping energy efficiency vocabulary.

Usual term	Rationale	More social scientific term as possible substitute
Consumer	'Consumers' render people as utility-maximising purchasers of goods and resources, i.e. investor logic. This has little to do with why people do things nor any other dimensions of personhood or motivation.	Person—suggests a social actor who is more than someone who merely buys things being delivered by suppliers of goods and services.
Preference	'Preferences' are what individual consumers have; things that they select for purchase over other alternatives.	Value—suggests that choices are made on the basis of both individual likes and dislikes, but also social and cultural meanings.
Customer	'Customer' renders people as the source of revenue stream of companies, relative to which companies can manage offerings so as to increase revenue or to serve other company interests. In some discussions, 'customer' is appropriate, but it is a loaded substitute when used to refer to more general situations.	Person—in another sense someone that has a life outside of the market place.
Occupant	'Occupant' renders people as interchangeable passive objects who are served indoor conditions. People are not seen as social actors living and working together in dynamic and diverse systems and circumstances, routinely adapting to and affecting these circumstances.	Inhabitant—someone who has developed habits of living (and living with others) in a place that they have furnished and that holds meaning for them.
Behaviour (as a discrete action)	'Behaviour' is a very general term that invites a view where people are quite separate from technology and from each other and even from their own minds and bodies.	Actions, interface interactions, purchases.
Behaviour (as what people do)	As mentioned earlier.	Practice—shared social patterns that incorporate actions, devices and knowledge and that are organised across scales (from individual habit to living group, community, and society). Habits—generally patterned and recurrent actions and interactions with technical devices that are learned, often optimally efficient and usually unconscious.

Continued

Table 1 Revamping energy efficiency vocabulary—cont'd

Usual term	Rationale	More social scientific term as possible substitute
Barriers	'Barriers' refer to presumed set of factors that prevent people from behaving according to some top-down model of ideal energy behaviour (e.g. purchasing a more efficient model when cost-effective or not believing in climate change). The policy goal is then to overcome barriers so that idealised target behaviours can be achieved. This invites a rigid and self-serving view of what should happen, rather than one that tries to more fully understand why people act as they do, to adapt conceptual models and offerings, or to accept limitations of policy instruments.	Circumstances—the complex realities of everyday life operating at multiple scales. The realm in which all of the social, cultural, habitation, practice, and other sociotechnical interactions mentioned earlier play out and make sense to people. While 'barriers' are bits of the world picked out through the lens of some change or product being promoted ('barriers TO' some action satisfying program goals), circumstances explain what people actually do and value.

between engineers and social scientists—that extends beyond the simple market behaviour models.

In addition to working on a better vocabulary, one could imagine a tool kit of basic notions and questions for policymakers and engineers to ask in designing devices, systems, and deployments, as well as designing the original imaginaries and policy goals from which these concrete goals derive. Imagine, for example, a "Very Short Introduction to Social Sciences for Climate Change Technology Planning" as in the Oxford University Press series. Whatever this set of tools, it would be applied at different layers and time points, due to all that is unpredictable in figuring out how a technology will evolve in society (see, e.g. Walker et al., 2013).

So in a more unified paradigm, we see the people and device interactions with a much richer and nuanced understanding of both social and technical parts of intrinsically bound up sociotechnical systems/arrangements/dynamics. This should mean that when facing any energy technology design, development, or implementation problem, we can ask new questions and pursue new answers (including new ways of seeing people and devices and new ways to design devices and make them relevant and adaptable). For example, if not 'occupants' but 'inhabitants', then what does the former miss? What does the latter bring in? What are the questions to ask in order to uncover and assess issues for design/deployment?

3.3 Multi-, trans-, and interdisciplinary approaches and beyond

Finally, over the same period as the climate change problem came to prominence, there has been a transformation in research and communications enabled by the

Internet. There is far easier access to research, data, opinions, and tools than there was two decades ago, making it much easier to be informed as well as easier to be shallowly informed. The nature of the problems, research teams, and empirical data now at play beg for integration and coordination of key concepts and methods across disciplinary approaches. But disciplines, researchers, and institutions have developed historically to do what they do, and these conventions remain inscribed in powerful ways (e.g. through the need to maintain funding streams, through systems of academic credit, etc.). Given the historically situated positions of disciplines, researchers, and institutions, this sort of coordination, still in the early stages, will take time, sustained effort, and strategy. But given the scope of the problems related to climate change that humanity and the planet are facing, and given the thoroughly intertwined human–machine–environment nature of the sociotechnical systems that will need to be altered on a vast scale, there is little choice but to make the attempt. There are useful precedents, for example, in disaster research and public health and recent developments along these lines in environmental research that are encouraging (One Earth, 2019).

One of the tensions is misunderstanding about the 'sciences' aspect of social sciences, which are quite different from the experimental and controlled observational methods that are taught as idealised scientific methods in school. Even when these methods can be applied, they are usually far from ideal statistical and logical validity (e.g. Amrhein et al., 2019), because (for one thing) the real world is too messy to make this practical. But the social scientists and bureaucrats and other applied types (e.g. with journalism or business degrees) also don't understand the technical/technological side, and they had better not have to start with a full engineering undergraduate curriculum to get there. This has been a failing of engineering: the insularity and inaccessibility on campus and among the knowledge disciplines. This now gives the engineers a unique mission to educate and interpret/translate engineering information and understandings (e.g. thermodynamics, heat transfer, cooling technologies and alternatives, etc.) to social scientists and to the public.

3.4 Bigger questions, alternative approaches

All of this could help energy professionals work better together on improving technology development and planning towards lower energy use and lower greenhouse gas emission without harming their well-being. But there are also possibilities beyond top-down technology planning. A main one is to recognise people as innovators. Here, we do not mean 'innovators' as the first to adopt a new technology on the market. Rather, throughout history, laypeople have been developing and modifying technology, integrating it in their special projects and into their everyday lives of social practices. These creative and adaptive capabilities are reflected in the hugely diverse ways that people manage hot weather, how they organise in the face of disasters (Haque and Etkin, 2007), or how they rushed to develop apps for smartphones that soon became a main reason for these phones. This social potential (Moezzi and Janda, 2014) has implications for technology development.

Arguably, efficiency measures may be trending towards more rigidity (e.g. in office buildings, people in the building may not be able to control the lights), to more outside control via smartness, and to lower diversity (e.g. towards elimination of nonelectric energy sources). Imposed rigidity can impede ability to adapt and improve technologies and technology use to local conditions and preferences. Instead, sponsored technology development could continue to take inspiration from local innovations (e.g. in lower-impact cooling technologies) and to routinely consider the risks of rigid technologies. This changes the perspective from a top-down one where humans comply with technology to a more fluid one.

4 Conclusions

People are integral parts of the energy system. To capture their roles in shaping the energy system and its environmental and social outcomes requires going beyond the normal boundaries of energy behaviour frameworks. Otherwise, problem solving is likely to remain confined to efforts to control small pieces of human–technology interaction. Those steps have a place, but they do not allow for how the energy system evolves or could evolve with the right designs.

Social scientists are good at identifying and framing problems, and engineers can be very good and creative at solving them. Our main takeaway message is that there is an opportunity to do more than is being done and that this can start by better understanding each other's problems versus remaining superficial if convenient overlaps. Given deliberate attention, this could lead to finding ways that the skills and insights of different disciplines complement each other and can jointly move towards re-designed systems rather than re-designed technologies or the hope of deliberately 'correcting' people.

References

Abrahamse, W., Steg, L., Vlek, C., Rothengatter, T., 2005. A review of intervention studies aimed at household energy conservation. J. Environ. Psychol. 25 (3), 273–291.

Amrhein, V., Greenland, S., McShane, B., 2019. Scientists rise up against statistical significance. Nature 567 (7748), 305.

Axsen, J., Kurani, K.S., 2012. Social influence, consumer behavior, and low-carbon energy transitions. Annu. Rev. Environ. Resour. 37 (1), 311–340.

Bhardwaj, A., Joshi, M., Khosla, R., Dubash, N.K., 2019. More priorities, more problems? Decision-making with multiple energy, development and climate objectives. Energy Res. Soc. Sci. 49 (March), 143–157.

Bijker, W., Hughes, T., Pinch, T. (Eds.), 1987. The Social Construction of Technological Systems: New Directions in the Sociology and History of Technology. MIT Press.

Bodin, R., Chermack, T.J., Coons, L.M., 2016. The effects of scenario planning on participant decision-making style: a quasi-experimental study of four companies. J. Futures Stud. 20, 21–40. https://doi.org/10.6531/JFS.2016.20(4).A21.

Boucher, J.L., 2017. The logics of frugality: reproducing tastes of necessity among affluent climate change activists. Energy Res. Soc. Sci. 31 (September), 223–232.

Carrico, A.R., Riemer, M., 2011. Motivating energy conservation in the workplace: an evaluation of the use of group-level feedback and peer education. J. Environ. Psychol. 31 (1), 1–13.

Cooper, G., 1998. Air Conditioning America: Engineers and the Controlled Environment, 1900-1960. Johns Hopkins University Press.

Cooper, G., 2008. Escaping the house: comfort and the California garden. Build. Res. Inf. 36 (4), 373–380.

Deumling, R., Poskanzer, D., Meier, A., 2019. 'Everyone has a peer in the low user tier': the diversity of low residential energy users. Energy Efficiency 12 (1), 245–259.

Dunlop, T., 2019. Mind the gap: a social sciences review of energy efficiency. Energy Res. Soc. Sci. 56.

U.S. Environmental Protection Agency, 2019a. Thermostat Specs. https://www.energystar.gov/index.cfm?c=archives.thermostats_spec.

U.S. Environmental Protection Agency, 2019b. Proper Use Guidelines. https://www.energystar.gov/products/heating_cooling/programmable_thermostats/proper_use_guidelines.

U.S. Environmental Protection Agency, 2019c. Smart Thermostats. https://www.energystar.gov/products/heating_cooling/smart_thermostats.

Frederiks, E.R., Stenner, K., Hobman, E.V., 2015. Household energy use: applying behavioural economics to understand consumer decision-making and behaviour. Renew. Sust. Energ. Rev. 41 (January), 1385–1394.

Geels, F., 2004. Processes and patterns in transitions and system innovations: refining the co-evolutionary multi-level perspective. Technol. Forecast. Soc. Chang. 72, 681–696.

Hagbert, P., Bradley, K., 2017. Transitions on the home front: a story of sustainable living beyond eco-efficiency. Energy Res. Soc. Sci. 31 (September), 240–248. https://doi.org/10.1016/j.erss.2017.05.002.

Haque, C.E., Etkin, D., 2007. People and community as constituent parts of hazards: the significance of societal dimensions in hazards analysis. Nat. Hazards 41 (2), 271–282.

Hargreaves, T., Nye, M., Burgess, J., 2013. Keeping energy visible? Exploring how householders interact with feedback from smart energy monitors in the longer term. Energy Policy 52 (January), 126–134.

Herring, H., Sorrell, S., 2009. Energy Efficiency and Sustainable Consumption: The Rebound Effect. Springer.

Hughes, T.P., 1986. The seamless web: technology, science, etcetera, etcetera. Soc. Stud. Sci. 16, 281–292. SAGE.

Hughes, T.P., 1987. The evolution of large technological systems. In: Bijker, W., Hughes, T.P., Pinch, T. (Eds.), The Social Construction of Technological Systems: New Directions in the Sociology and History of Technology. MIT Press, pp. 51–82.

Hughes, T.P., 1993. Networks of Power: Electrification in Western Society, 1880-1930. JHU Press.

Johnson, D., Horton, E., Mulcahy, R., Foth, M., 2017. Gamification and serious games within the domain of domestic energy consumption: a systematic review. Renew. Sust. Energ. Rev. 73 (June), 249–264.

Kennedy, C., Pincetl, S., Bunje, P., 2011. The study of urban metabolism and its applications to urban planning and design. Environ. Pollut. 159 (8), 1965–1973.

Kordjamshidi, M., King, S., 2009. Overcoming problems in house energy ratings in temperate climates: a proposed new rating framework. Energy Build. 41 (1), 125–132.

Kranzberg, M., 1986. Technology and history: 'Kranzberg's Laws'. Technol. Cult. 27 (3), 544–560.

Luhmann, N., 1982. The world society as a social system. Int. J. Gen. Syst. 8 (3), 131–138.

Lutzenhiser, L., 2014. Through the energy efficiency looking glass. Energy Res. Soc. Sci. 1 (March), 141–151.

Marcus, A.I., Segal, H.P., 1989. Technology in America: A Brief History. Palgrave.

Meadowcroft, J., 2009. What about the politics? Sustainable development, transition management, and long term energy transitions. Policy Sci. 42, 323–340. https://doi.org/10.1007/s11077-009-9097-z.

Meadows, D.H., 1999. Leverage Points: Places to Intervene in a System. Academy for Systems Change. http://donellameadows.org/archives/leverage-points-places-to-intervene-in-a-system/.

Moezzi, M., Janda, K.B., 2014. From 'if only' to 'social potential' in schemes to reduce building energy use. Energy Res. Soc. Sci. 1 (March), 30–40.

Navigant, 2017. Energy efficiency potential and goals study for 2018 and beyond. Consultant Report to the California Public Utilities Commission, http://docs.cpuc.ca.gov/PublishedDocs/Efile/G000/M194/K614/194614840.PDF.

Newman, D.K., Day, D., 1975. The American Energy Consumer. Ballinger, Cambridge, MA.

One Earth, 2019. A new journal from Cell Press, *One Earth* publishes research across the spectrum of natural, social, and applied sciences, with a particular interest in integrated, transdisciplinary studies. https://www.cell.com/one-earth/home.

Owens, S., Driffill, L., 2008. How to change attitudes and behaviours in the context of energy. Energy Policy 36 (12), 4412–4418.

Parsons, T., 1952. The Social System. Routledge.

Perrow, C., 1984. Normal Accidents. Basic Books.

Shove, E., 2010. Beyond the ABC: climate change policy and theories of social change. Environ Plan A 42 (6), 1273–1285.

Shove, E., 2018. What is wrong with energy efficiency? Build. Res. Inf. 46 (7), 779–789.

Shove, E., Pantzar, M., Watson, M., 2012. The Dynamics of Social Practice: Everyday Life and How It Changes. Sage.

Smil, V., 2010. Energy Transitions: History, Requirements, Prospects. Praeger.

Stern, P.S., Aronson, E., 1984. Energy Use: The Human Dimension. National Academies Press.

Stone, D.A., 1989. Causal stories and the formation of policy agendas. Polit. Sci. Q. 104 (2), 281–300.

Vergne, J.-P., Durand, R., 2010. The missing link between the theory and empirics of path dependence: conceptual clarification, testability issue, and methodological implications. J. Manag. Stud. 47, 727–759.

Walker, W., Haasnoot, M., Kwakkel, J.H., 2013. Adapt or perish: a review of planning approaches for adaptation under deep uncertainty. Sustainability 5, 955–979.

Energy behaviour across sectors

Resource-efficient nondomestic buildings: Intertwining behaviour and technology

2.1

Hermano Bernardo[a,b], António Gomes Martins[a,c]

[a]INESC Coimbra, DEEC, University of Coimbra, Coimbra, Portugal
[b]School of Technology and Management, Polytechnic Institute of Leiria, Leiria, Portugal
[c]Department of Electrical and Computer Engineering, University of Coimbra, Coimbra, Portugal

1 Perspective on the life-cycle path of buildings

The concept of energy neutral building (Thomas and Duffy, 2013) has been specified and disseminated under several forms, including in regulation and in several geographies (California Energy Commission, 2007) (European Commission, 2018).

A variation of this concept consists of the neutral buildings regarding greenhouse gas (GHG) emissions (Sørensen et al., 2017), which represents a step forward compared with the more classical approach to solely improve building energy use. In fact, there are several possible ways to reduce energy consumption with different impacts on GHG emissions.

Improving energy performance of buildings corresponds to implement some changes, either technological or managerial or both, leading to a lower energy consumption whilst keeping the same comfort level. It mainly addresses the operating phase of the whole life cycle of a building. The ambition of achieving energy neutrality of a building goes beyond the improvements of efficient energy use. Energy neutrality requires that the efforts towards a better use of energy resources are complemented with the possibility of in situ renewable primary energy conversion. This sustainable in situ energy supply aims at the satisfaction of the residual energy consumption that remains after ensuring an (near)optimal inherent building energy performance (European Commission, 2018).

It has been shown that the operational phase of a building represents the majority of energy use during the whole life cycle of the building (Soares et al., 2017). Nevertheless, a cradle to grave approach (Khasreen et al., 2009) has been pointed out as the most comprehensive and rational regarding a sustainable use of resources. All activities preceding construction, the construction itself, and the dismantling of a building have significant impact on the use of energy and other resources that should

Energy and Behaviour. https://doi.org/10.1016/B978-0-12-818567-4.00006-5

be taken into account as well. Therefore, the nearly zero-energy building (NZEB) concept (European Commission, 2018), virtuous as it can be, is limited in scope.

Several well-known factors influence the future energy performance of a building since the very early stages up to the end of its life cycle: location, exposure, orientation, (passive) design, construction, test and commissioning, operation, demolition, reutilization of components, and residue management (Soares et al., 2017).

Each one of these factors, up to the operation phase, has a specific influence on the resulting use of energy until the end of the useful life of the building. The building structures, including its envelope characteristics, together with all the active systems that provide energy services for internal comfort, are the material basis that set the limits to the quality of the building's performance. Examples of such energy services are heating and air conditioning, lighting, internal transportation, and the building automation and control system (BACS) that controls energy service provision (Tian et al., 2018). The set of physical and technical characteristics of a building defines the limits within which the behaviour of building occupants influences the actual performance of the whole system—the building itself and the equipment.

A more generalised view on the influence of behavioural factors on the building impact on the use of resources should also take into account the behaviour of all the agents that develop some kind of activity throughout the building's life cycle, and not only occupants: designers, contractors, construction workers, facilities management (FM) staff, building managers, and visitors. The scope of this chapter is limited, though, to the operational phase of buildings and to the role of those agents that have some influence on the building's energy performance during this phase: occupants, FM staff, and building managers.

The importance of the operational phase has to be seen from a broader perspective both scales, space and time. Whatever is learned from systematic data gathering for the characterisation of the building's operational phase is very important also for the constant improvement of design approaches and methods. This configures an endless iterative process where design tools lead to increasingly more efficient buildings and, sequentially, these originate new data streams on systems and behaviours that lead to the improvement of design tools. Software simulation platforms, abundantly used for efficient building design, are one such example, benefiting from permanent improvements that are possible through data collection on buildings' use (Huang and Niu, 2016). This sequential closed loop process has a potential positive effect on the evolution of the quality of the performance of buildings: intelligent design for maximising comfort and minimizing use of resources, consistent measurement of relevant data towards performance indicators, systematic monitoring of occupants' behaviour, level of satisfaction and quality of the interaction with buildings' systems, research for interpreting the data towards model improvement, updating/reformulation of design assumptions, models and methods, and back to the beginning in an endless iterative, closed loop path (Tian et al., 2018).

Although life-cycle assessment of buildings has become a basic model of regulatory mandates in many parts of the world (European Commission, 2018), this chapter is specifically dedicated to the influence of the relation between technology and behaviour on buildings performance, thus deliberately focusing on the buildings' operational phase.

The present chapter starts, in this first section, by providing the broad context of resource efficiency from the perspective of buildings' life cycle. It follows with the identification of the main behavioural issues affecting the energy performance of nondomestic buildings, which are inextricably connected to the comfort conditions provided to occupants. The third section deals with the role of those building automation and control systems that are installed in existing buildings, very frequently either not adapted to the buildings and the occupants' abilities, needs and expectations, or operating far from their design assumptions, and the corresponding consequences on building energy performance and internal comfort conditions. The fourth section deals with the requirements of human-technology interaction and the difficulties raised by the absence of user-centred approaches at the design phase of buildings and systems. Section 5 develops on the user-centred perspective of human-building interaction in several dimensions. The chapter closes with a section on some possible outcomes.

2 Main factors affecting the energy performance of nondomestic buildings: Focus on occupant behavioural issues

There is a considerable difference between the predicted energy consumption of a building at the design stage and the actual measured energy consumption when the building is at 'normal' operation and occupancy conditions. The occupant behaviour and operation control practices are dominant factors for the existence of this difference, which are difficult to predict at the design stage as they depend on several human factors (van Dronkelaar et al., 2016).

There are many situations when building occupants tend to passively accept technologies and building features of their surrounding environment, but some of their comfort needs are required to be met to keep them motivated (Haynes, 2008) to an energy-efficient behaviour. Otherwise, there is the risk of counterproductive behaviour that may lead to an increase in energy consumption (Xu et al., 2017).

Despite the aforementioned tolerant behaviour, people have a natural desire to have some degree of control over the surrounding environment, which may have consequences in building energy performance. Indoor environmental conditions may trigger occupants to interact with the building control systems, causing changes in energy demand. These adaptive actions undertaken by the occupants may generate a perturbation of the indoor environmental conditions (Hong et al., 2017). Adjusting the comfort temperature set points, switching lights, opening/closing windows, pulling up/down window blinds, and moving between spaces can have a significant impact on energy use and indoor environmental quality in buildings (IEA-ECBCS, 2017).

2.1 Paths to the diagnosis of behavioural influence on buildings energy performance

To guide the design and operation of low-energy buildings, either residential or nondomestic buildings, that integrate technological and human dimensions, it is crucial to understand occupant behaviour in a comprehensive way, integrating qualitative approaches with data- and model-driven quantitative approaches, and employing appropriate tools (Hong et al., 2017). There are several methods to collect occupant-related data for the purpose of characterising occupants' behaviour—each with its own strengths and weaknesses. Within the research conducted in the context of the Annex 66: 'Definition and Simulation of Occupant Behaviour in Buildings', from the International Energy Agency, Energy in Buildings and Communities Programme (EBC), the approaches for monitoring and data gathering occupant behaviour in buildings were summarised in four different categories (IEA-ECBCS, 2018):

- *In situ monitoring studies*, which involves monitoring occupant's actions, presence, and indoor environmental quality (IEQ) in operating existing buildings; data are normally acquired passively through building automation and control systems or through dedicated sensors installed for research purposes; typically, data collection dedicated to long-term studies (months or years).
- *Laboratory experiments*, requiring the construction of artificial environments similar to real ones but with high degree of control over the indoor conditions and with the possibility to select participants according to predefined criteria; it allows detailed monitoring of occupant´s actions and comfort perceptions under several controlled scenarios.
- *Surveys, interviews, focus groups, and diaries*, where occupants self-report actions, presence, IEQ perception, and other relevant information, either by filling out questionnaires or through interviews and focus groups; this method is used either alone or together with sensor-data gathering.
- *Virtual reality experiments*, using computer-designed environments to study the occupant's behaviour when exposed to certain type of stimuli; nowadays still limited to the visual and acoustic domains, they do not yet allow to perform thermal comfort and indoor air quality experiments.

Despite the first three methods being the most developed, the use of surveys overcomes key barriers to the adoption of state-of-the-art sensing technology, which include high costs for initial installation, operation, and maintenance and the difficulty of integrating the sensors with existing building automation and control systems. In addition, surveys are a valid alternative to behaviour sensing when direct monitoring techniques are not allowed or are insufficient in what concerns the scope of the research being conducted (Hong et al., 2017).

2.2 The role of management decisions and middle-out agents in building performance

Nondomestic buildings are used for a diverse range of activities, whether they are schools, hospitals, commercial spaces, etc. all with different management priorities and strategies, sometimes without considering energy cost. In general, there is a lack of awareness on energy performance issues amongst the organisations' top managers who are responsible for decision-making. Management decisions play a determinant role in building performance, from the point of view both of investment in technology and of the building operation.

The resulting operation and energy usage profiles depend on the building's operation and maintenance (O&M) practices and schedules, activity, shape, size, and age, amongst other factors, making it difficult to define a typical nondomestic building (Delay et al., 2009).

Janda and Parag (2013) proposed the use of an approach based on the action of middle-out agents for improving buildings' energy performance. This strategy is a normative approach that recognises those actors who are already performing various roles in society and are neither at the top nor at the bottom of an organisation hierarchy, examines their agency and capacity characteristics (or potential) with reference to the various aspects of change and/or barriers for change, explores the various directions in which they could act (upstream, downstream, and sideways), and examines ways to empower them to enable change to happen (Janda and Parag, 2013; Parag and Janda, 2014).

In the nondomestic sector, due to the extension of technical facilities and profusion of technical equipment, there are usually professionals responsible for building operation and FM. These professionals are key intermediaries between the users/occupants of the building and the control of the building's energy services, playing an important role on organisations' energy management strategy (Banks et al., 2016).

Considering the building operators/facilities managers as middle-out actors in the context of existing nondomestic buildings, the middle-out approach could be used as a strategy to initiate, motivate, support, and upscale change in the use of technology towards lower energy buildings (Fig. 1).

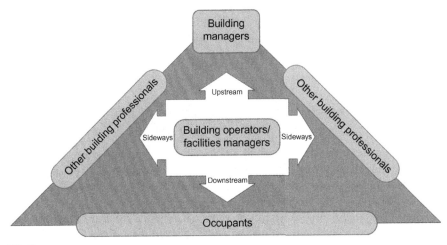

FIG. 1

Application of middle-out approach to existing nondomestic buildings.

Adapted from Parag, Y., Janda, K.B., 2014. More than filler: Middle actors and socio-technical change in the energy system from the "middle-out." Energy Res. Soc. Sci. 3, 102–112.

Building operators and facilities managers may have influence on different directions across the building management and operation structure:

- *Upstream*: they can influence the building management to the need of investment in technology, equipment upgrade through the procurement of new low-energy technical solutions and supporting the investment decisions; they also play an important role in performing energy accountability and communicating to managers the benefits of an effective energy management strategy.
- *Downstream*: they are crucial agents in meeting the occupants' demand for comfort conditions as they are the intermediaries between occupants and technology, often being the only ones that have permission to change parameters in the BACS, such as temperature set points and mechanical ventilation schedules; they have the responsibility of keeping facilities operation as efficiently as possible, mediating the occupants' comfort needs and management restrictions in terms of reducing the energy cost.
- *Sideways*: innovative FM and energy-efficient practices are transferred not only within building operators and FM professionals but also to other professionals working in the sector, such as consultants and engineers; this could be achieved through networks of formal contracts when they work in multidisciplinary teams, professional affiliation in professional associations, and also informal relations with other building professionals.

On one hand, facilities managers are in a unique position to understand the occupants demand for indoor comfort conditions. On the other hand, they know how systems and building technology work to contribute for the implementation of low-energy buildings (Min et al., 2016).

Due to the important role played by building operators and facilities managers, they need knowledge and tools to support them on managing and operating technical building systems, maximising occupant's satisfaction with adequate indoor environmental conditions and low energy usage.

3 The role of existing building automation and control systems

An important way of saving energy and reducing GHG emissions in nondomestic buildings relies on the use of ever more sophisticated automation systems to monitor and control the active and natural systems providing heating, cooling, lighting, etc. to the building.

Much of the energy used to heat, cool, ventilate, or light a building may be wasted in periods of low or even zero occupancy whilst also being insufficient in higher occupancy periods. An artificial lighting system may be at once excessive in daylight periods and insufficient without daylight. Some buildings have simultaneous heating and cooling needs (due to solar orientation, occupancy, etc.) that are provided by nonintegrated/interlocked HVAC systems, increasing the energy needs compared with an integrated system. These are just a few examples to illustrate the potential savings that can be obtained through building automation.

According to EN ISO 16484-2:2004, a building automation and control system (BACS) is a 'system, comprising all products, software and engineering services for automatic controls (including interlocks), monitoring, optimisation, for operation, human intervention, and management to achieve energy efficient, economical, and safe operation of building services'. Some systems known as building (energy) management (and control) systems (BMS, BMCS, or BEMCS) fit the BACS definition in EN ISO 16484 and should therefore be designated as BACS. These systems have gained a prominent role in the management of daily maintenance and energy-related operations with significant impact on the energy performance and indoor environmental quality of buildings (Oliveira and Bernardo, 2019).

A BACS can be described as a centralised, automated system that receives and monitors information from the various sensors installed across the building, allowing building facilities managers and operators to control actions based on schedules, inputs from sensors, and preferences expressed by occupants. It can be programmed to control all building energy-related systems, including heating, cooling, ventilation, domestic hot water production, lighting, on-site energy generation, mechanical systems for shading devices, window actuators, double façade elements, and also nonenergy functions, such as building security/intrusion detection, and fire alarms. Whilst some of these systems may be very limited (such as only performing system monitoring and data visualisation), others may integrate all the building's systems and include automated control, enabling the automation of various physical tasks that would otherwise have to be performed manually and in situ (Brambley et al., 2005).

An important effort to create a standard framework to characterise energy efficiency in these systems was introduced by European standard EN 15232-1:2017 (CEN, 2017). This standard presents an energy efficiency classification of BACS (into four efficiency classes, from A to D) based on a structured list of BACS and technical building management (TBM) functions that systems should be able to implement and on minimum requirements for these systems for different building complexities. It also provides standard, simplified methods to estimate and detailed methods to assess the impact on the energy performance of buildings when such systems are introduced, upgraded, or retrofitted.

---

4 Challenges of human-technology interaction

Nowadays, there is still a widespread belief that all energy-related concerns could be overcome by technology in spite of a progressive awareness of the importance of human behaviour, directly linked with technology usability. On one hand, there are the intentions of technology and control designers and manufacturers, and on the other hand, there are users' awareness and perceptions.

Some designers are more influenced by technology manufacturers than by users' needs, corresponding to an expectation that technology-based approaches will be enough to the provision of both comfort and energy efficiency (Bordass et al., 2007). Users are not interested in technology itself, but in the results of their interaction with the control systems to achieve comfort. Interactivity between users and technology is provided by the user interface, which desirably allows users to perform adjustments in environmental parameters as simply and quickly as possible—it may be a touchscreen, a thermostat knob, a voice command receiver, a mechanical lever, a switch, etc. If the user interface does not meet some elementary usability requirements, occupants will give up since they cannot achieve what they want quickly and easily (Bordass et al., 2007).

Frequently, design purposes do not match user needs and perception of the control functions, actually tampering the effective and efficient operation of the system. In most cases, this may be due to a lack of communication between technology designers and building owners, operators, and users, contributing to a lack of general awareness about the causes and consequences of poor control and user interface design. Some of the barriers and drawbacks identified are summarised in Table 1.

There are cases of buildings provided with a large range of control functions, which are operated using only few of them, mainly due a lack of knowledge and empowerment of the building's operators and facilities managers (Paone and Bacher, 2018). Fig. 2 provides some insight into the general range of available BACS functionality and the degree to which buildings' operators exploit the available functions. The information presented was collected through interviews to building operators conducted in eight school buildings in Portugal.

It is patent that most BACS are used to perform basic plant control functions, such as time scheduling and room temperature adjustments. However, many BACS have a certain number of more sophisticated functions, such as daylight control, free cooling (night cooling), or demand-control ventilation. The results of this survey clearly support the argument that a significant portion of the potential functionalities of BACS are not used in many cases. Other studies also state that building operators tend to use only a fraction of possible BACS functionalities, thus limiting the performance gains, and also that in many buildings the system does not operate correctly (Lowry, 2002).

During the interviews conducted in Portuguese schools, it was found that only one building operator had previous experience and knowledge on the use of BACS and only half of them had initial training for operating the system when starting their professional activity related to building O&M.

Another factor that usually affects the BACS performance is the nonexistence of a stage of system commissioning or a poorly conducted commissioning. This stage plays an important role in building performance since it is required to ensure that building equipment and systems are integrated in such a way that they perform together effectively and efficiently and meet the building operation management requirements and expectations (Aghemo et al., 2014).

Table 1 Summary of barriers and drawbacks in design, installation, and operation phases, affecting systems' usability on the perspectives of designers and users.

	Designer	User
Design	Design reality may not match practical reality – The lack of good specifications for usability – Poor feedforward of supplier experience	Poor understanding and discussion on user experience – Building owners do not realise the importance of user interfaces – Narrow engagement of building owners in setting the user interface requirements
Installation	Design intent becomes blurred during implementation – The lack of products with adequate usability requirements – Insights lost down the supply chain – Fine details (location and labelling) are poor	Actuality drifts away from client expectation during implementation – Detailed provisions not discussed with building owners and users – Management not involved in agreeing details – Handover procedures and data are often poor
Operation	– Management gets involved when it does not need to be – The lack of feedback on in-use performance	– Users experience both intended and unintended consequences – Design intent and system response often unclear to user
All phases	– The lack of feedback to controls manufacturers	

Adapted from Bordass, B., Leaman, A., Bunn, R., 2007. Controls for end users: a guide for good design and implementation. BSRIA Ltd, Reading.

Technology also faces some challenges to integrate future energy-related issues, such as shifting to a demand-control paradigm where the technical building services should have the capability of automatically adjusting energy demand to follow actual occupancy profiles, instead of operating at full load to meet standard predefined or peak occupancy profiles.

In the context of smart grids, the automation and control systems must have the capability to deal with and to optimise, in an integrated way, the use of renewable energy sources, electric vehicles, energy storage, and demand response actions (Carr et al., 2017).

5 Human-building interaction

Human-building interaction (HBI) deals with the ways through which occupants of a building experience their relation with the building, its components, and technical

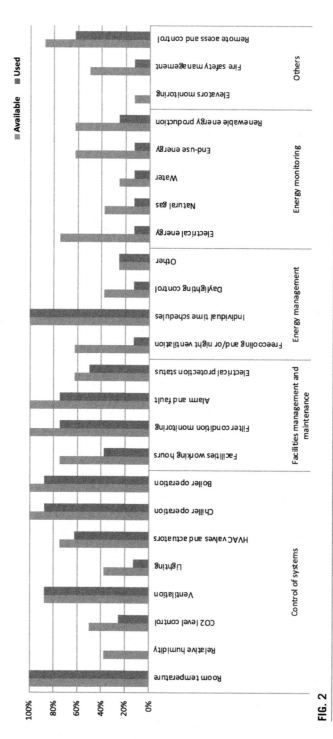

FIG. 2

Surveyed prevalence and usage rates of selected BACS functions (Y-axis represents percentage of occurrences).

systems and with the influence that occupants' behaviour have on the overall energy performance of the building. The aim of this study domain is not only to identify the forms assumed by that influence of behaviour but also, on a generalised perspective, to assess how to influence behaviour in such a way that energy performance of buildings can be maximised whilst preserving or improving occupants' comfort (Shen et al., 2016).

Occupants are therefore agents of adaptation of local internal comfort, in a setting where immersion into the interactive object itself is a singular characteristic of HBI (Nembrini and Lalanne, 2013). They act on their surroundings to influence the building's response, which causes changes in energy consumption and consequently affects the building's performance (Delzendeh et al., 2017).

The assumptions made about occupants' behaviour in the context of software tools for assessing the energy performance of buildings are usually simplified, based on rational decisions of occupants or simply neglecting the role of occupants on the use of the building or its technical systems. In many cases, assumptions are made only on the number of occupants for the sake of identifying the influence of biological internal gains on the building's thermal load. This inevitably leads to estimations on energy use that are far from being coincident with reality when comparing simulations with measured data. The difference between these values is usually designated energy gap (Paone and Bacher, 2018; Delzendeh et al., 2017) (IEA-ECBCS, 2018). Chapter 3.2 of this book, by Hong et al., deals extensively with current progress and research on this topic.

The energy gap is the target of many research initiatives that aim at improving the capacity of software tools to identify more accurately the actual performance of buildings. This is a key issue either when designing new buildings, or when renovating existing buildings or when performing postoccupancy evaluations of existing buildings, or even when energy audits are accompanied by energy performance simulation to test or fine-tune energy efficiency measures to be adopted (Robinson et al., 2015; Hong et al., 2017).

It has been shown that building occupants assign a very high value to the capacity of influencing their surrounding environment through acting, for example, on windows, shades, lights, or thermostats, towards their perceived comfort, not necessarily towards an efficient use of energy. Comfort affects productivity and, ultimately, occupants' health, which must be taken into account not only in business management decisions but also, in the first place, in the design of spaces and their functionality. Too many constraints, or too intense attempts of persuasion, to influence occupants' behaviour usually lead to high levels of rejection and are counterproductive (D'Oca et al., 2018).

5.1 Human-building interaction and highly efficient buildings

Buildings that have been conceived as potentially highly efficient usually require adequate instructions and training of occupants along with appropriate feedback mechanisms that ensure a smooth adaptation of users to the building's characteristics, devices, and technical systems (Hauge et al., 2011). This is due to the operational features of these types of buildings that are usually different, in one way or the other, from the commonly designated conventional buildings.

Highly efficient buildings are actually capable of providing high levels of comfort to occupants with a modest energy consumption. A high energy intensity does not necessarily mean a high comfort level. The main requisite for a convenient adaptation of occupants to an energy-efficient building lies at the design phase, when a user-centred approach should be adopted for the best results, that is, provide comfort and allow easy adaptation actions of occupants with minimum energy consumption. Postoccupancy adaptation of existing buildings should follow a similar path for the best results (Steemers and Manchanda, 2010).

The particular demands posed by energy-efficient buildings towards an optimum use of resources, including water, for example, are frequently causes of rejection by occupants, who develop a negative attitude towards these demands (Ornetzeder et al., 2016). In fact, in many cases, in the absence of training and meaningful feedback, together with overly complex operational requirements caused by a non-user-centred approach, occupants feel incapable of developing adaptation strategies that allow them to obtain the comfort conditions they aim (Hauge et al., 2011; Shen et al., 2016). These cases, although technologically efficient, hardly, if ever, lead to an energy-efficient operation of buildings and actually correspond to a poorly designed HBI (Leaman and Bordass, 1997).

5.2 User-centred human-building interaction

A user-centred design must address all levels of HBI requirements: the building itself, including all the nonstatic components, that may be influenced by user actions, technical systems—those that provide essential energy services towards visual and thermal comfort, as well as air quality, control systems, and O&M. The main features of a user-centred HBI design are as follows:

- Simplicity of use: interfaces should be intuitive, with a smooth and short learning path, under the principle of making things easy and avoiding any expendable complexity for the sought purpose, which should be to make occupants feel comfortable with minimum adaptation action; this, in turn, makes the building operate as near as possible to the expected operational conditions, that is, an energy-efficient operation.
- Ergonomics, which is closely related to simplicity, since ergonomics intrinsically ensures ease of use.
- Eco-feedback, to provide appropriately persuasive stimuli. Codesign and cocreation with users are advantageous whenever possible.

The temptation to avoid appropriate HBI requirements through extensive automation aiming at a total and unquestioned satisfaction of users is usually doomed to failure, leading to the standard user rejection effect (Paone and Bacher, 2018).

D'Oca et al. (2018) illustrates the influence of simplicity with the results of a comparative study that showed small office buildings to have a better energy performance than large office buildings, having in mind that small buildings were almost exclusively manually operated, with simple to use devices.

Plug loads are a specific type of load in the sense that they are not prone to automatic control. Therefore, they require innovative HBI solutions, which creatively integrate elements such as suggestive feedback, comparison, emulation, and awareness raising, to contribute to an efficient use of energy in office buildings, where space and labour organisation have a strong influence (Metzger et al., 2011).

The most common types of devices and the most common types of actions performed by occupants on those devices are the subjects of ongoing research to suitably integrate their influence on energy consumption of buildings in the context of simulation software packages. Sociology plays an important role in this research in the scope of interdisciplinary teams, to identify and interpret the influence of specific barriers, stimuli and incentives, the importance of contextual issues, or the influence of the variety of occupants' profiles, bearing in mind that the behaviour of an individual person may also vary with acquired experience of building use (D'Oca et al., 2018).

Shen et al. (2016) used the concept of Fogg behaviour model (FBM) (Fogg, 1992) to devise a strategy for influencing the occupants' behaviour at a commercial building, dealing with the need to reduce energy consumption of office plug loads. In FBM, there are three basic elements for persuasion: motivation, ability, and triggers. The ability of a person to perform a certain action depends on the simplicity of that action, which emphasises the need to ensure simplicity of HBI mechanisms, by design. Motivation can be stimulated, leading to an increase of the motivation level, an example of which may be the aforementioned eco-feedback. A trigger is a stimulus that defines the moment when the action is taken. It may have a double nature if, for example, it simultaneously contributes to raise motivation or if it simultaneously turns the simplicity of the action evident to the target person. Or it may just be a reminder when motivation and ability are already sufficient. Other chapters in this book address motivational factors to change energy behaviour, such as Abrahamse and Schuitema (Chapter 1.1) and Heiskanen et al. (Chapter 4.1).

5.3 Building automation and control system

Building automation and control system (BACS), where they exist, are an intrinsic part of HBI. From a rationality standpoint, BACS play a very important role in buildings in general and specifically in energy-efficient buildings. In principle, the existence of a BACS in a building designed to be passively efficient and equipped with highly efficient devices delivering the energy services needed for internal comfort allows a potentially optimal use of energy since the automation of energy service provision potentially eliminates the influence of human decisions based on imprecise reasoning or less rational behaviour. The same can be said about not so efficient buildings (Bordass et al., 2004). The overall result will be less interesting than in the case of highly efficient buildings, but in the context of the existing physical constraints, the results would also be optimal.

However, BACS are not a panacea for efficient use of energy. On a less bright perspective, they add complexity to the building systems, and they require maintenance and regular updates, as well as specialised personnel to operate with them (Leaman et al., 2010). Besides, BACS also play an important role in HBI, with all the consequences, positive or negative, depending on specifications and operation (Martins, 1988).

Plug loads are, by definition, not prone to automatic control. This is why they constitute a special category that has to be tackled carefully, taking into account all the eligible behavioural aspects. However, the possibility of automating the control of the other types of loads does not ensure *per se* an optimum operation of a building. Operational BACS problems have long been reported in the literature, both technical and scientific (Andrews, 1982; Martins, 1988). Some of these problems are directly connected to HBI aspects, namely, those related to interface and feedback channels to operators and users.

Poorly conceived control strategies, hard-to-understand interface devices, the absence of manual override options provided to users are some of the problems classically identified in BACS. These types of problems, although identified a long time ago, are presently being researched also from the point of view of the behavioural response of occupants, aiming at improving control strategies and interfaces towards a better acceptance by users and a consequent more efficient building operation. The natural trend of building occupants towards adapting their closest environment to obtain a more comfortable situation, when in the context of a rejected BACS, usually leads to an inefficient operation of the building, inverting the expected positive results of the use of a BACS. (Leaman and Bordass, 1997) align a systematic view of causes for BACS rejection by occupants, partially organised on an energy service basis, along with possible remedies.

5.4 Operation and maintenance, facilities management

The building as a physical system and the whole set of technical systems and the BACS all require regular O&M. FM personnel, when available, either part of the building staff or under contract, as stated before, can play a relevant role as middle-out agents, influencing both management decisions and occupants' attitudes towards an efficient use of energy. Occupants tend to appreciate positively the actions of FM personnel when these implement a systematic workplan that includes interaction with building occupants on good practices (Ornetzeder et al., 2016). It not only corresponds to the component of awareness raising of any coherent programme for influencing behaviour but also ensures an effective use of the available HBI mechanisms by the occupants.

In those cases where O&M is not effective, the inevitable degradation of the response of the building and its technical systems contributes to some degree of dissatisfaction, since occupants are no longer able to adapt the operating conditions of their surrounding spaces to their convenience (Leaman et al., 2010). In the worst case, some technical systems may eventually cease to provide energy services, reducing energy consumption at the cost of degrading productivity. In other cases, occupants' actions through a defective HBI usually tend to increase energy consumption without achieving adequate comfort (Paone and Bacher, 2018).

The rejection behaviour of uncomfortable occupants may be potentiated by the actions of a BACS that does not respond to comfort requirements. Improper settings of control parameters or unplanned modifications of some circuitry or software components are likely to happen even if O&M is apparently being performed (Stevenson and Williams, 2007), in the case of external personnel under contract who may not be familiar enough with the system design and operation.

5.5 A holistic view of human-building interaction

The increasing penetration of artificial intelligence (AI), mechatronics, and robotics in the building environment, together with innovations in the field of sensors and new materials, as well as the use of augmented reality techniques for architectural design support, have been contributing to the emergence of a new perspective on HBI (Alavi et al., 2017). This new perspective goes beyond the objective of increasing energy efficiency, acknowledging the growing importance of embedded interactive technologies in the built environment, to rethink the relation of buildings with their occupants by taking advantage of computing power and new interaction techniques, such as self-adaptive systems and devices that are able to automatically adjust the response of technical systems to comfort requirements. Although the energy efficiency perspective remains as an essential ingredient, the adaptation of ambient conditions by occupants, intrinsic to HBI, addresses all aspects of subjective well-being and comfort perception, even if not directly related to energy use. This emergent perspective requires a multidisciplinary approach gathering architects and interaction designers. It 'address[es] the physical, spatial, and social opportunities and challenges that emerge as built environments become increasingly interactive' as well as collateral issues such as privacy, cybersecurity, or technology obsolescence (Alavi et al., 2016).

The increasing sophistication of the built environment assumed in this perspective raises again the question of the legibility (simplicity) of HBI to both occupants and FM personnel, which is key to a seamless adaptation of these agents towards an efficient use of energy (Leaman and Bordass, 1997).

Besides, since a great number of installed BACS in existing buildings operate in conditions that are far from optimal, the question arises of the need of postoccupancy interventions aiming at a more effective HBI if a large-scale result is sought (Delzendeh et al., 2017).

It is not by chance that the most recent version of the European directive on the energy performance of buildings (European Commission, 2018) is oriented to the promotion of building automation, introducing the concept of 'smart readiness' as a means of quantifying in the future the ability to improve HBI through BACS dissemination. This policy trend towards profiting the most from technology progress has to be informed by accumulated knowledge on the best HBI practices in relation to human behaviour.

6 Conclusion

There is a clear evidence of the intimate relation between technology and occupants' behaviour on the energy performance of buildings. From a life-cycle perspective, an efficient building has the potential to have a much smaller impact on the environment than a building designed and built without a life-cycle perspective. Technology, though, is not enough to achieve that full potential, since the actions of occupants determine, to a great extent, how a building will perform during the operating phase of its life-cycle, which is the phase with the greatest impact.

Occupants of nondomestic buildings tend to try to adapt their local environment to obtain the highest possible perception of comfort, which turns out to be an essential precondition to labour productivity. Therefore, building design, including internal layout, building components, and technical systems, should be achieved using an occupant's centred approach.

This means, for example, that occupants should be able to easily apprehend the requirements of operating their closest environment, for which purpose the interfaces should have adequate readability and should provide flexibility of operation whilst ensuring minimum use of resources, namely, energy.

A user-centred approach is also a requirement for the operating phase of buildings, where FM personnel, as middle-out agents, may be the connecting link between occupants and their working environments. They keep systems operational, they guide or influence occupants' actions, and ultimately, they avoid revenge reactions of occupants towards what these could perceive as aggressive or not responding environments. Simultaneously, these agents may influence managers' decisions to overcome organisational barriers to efficient use of resources within a building.

The difficulty of quantifying the effects of occupants' behaviour usually leads to a gap between the energy consumption of buildings predicted through computer simulators and the actual values. There is active research on data collection in real environments, which aims at grasping behavioural effects towards the improvement of computer simulators.

Building automation and control systems are an essential part of efficient buildings in the sense that they leave to occupants only those roles that cannot or should not be automated. Again, a user-centred approach is required to provide comfort and avoid waste, simultaneously providing a friendly dialogue with occupants' requirements and actions.

Human-building interaction is evolving to more elaborate forms of relation between occupants and buildings. It is supported by emergent technologies, seeking to go beyond the rational compromise between comfort and energy efficiency. It aims at responding to new expectations and demands of humans in a developed society, also addressing collateral issues such as privacy or cybersecurity.

Acknowledgement

This work was partially supported by the European Regional Development Fund in the framework of COMPETE 2020 Programme through projects UID/MULTI/00308/2019, ESGRIDS

(POCI-01-0145-FEDER-016434) and Manager (POCI-01-0145-FEDER-028040), and the FCT—Portuguese Foundation for Science and Technology.

References

Aghemo, C., Blaso, L., Pellegrino, A., 2014. Building automation and control systems: A case study to evaluate the energy and environmental performances of a lighting control system in offices. Automat. Construct. 43, 10–22. https://doi.org/10.1016/j.autcon.2014.02.015.

Alavi, H.S., et al., 2016. Future of human-building interaction. In: Proceedings of the 2016 CHI Conference Extended Abstracts on Human Factors in Computing Systems - CHI EA'16, pp. 3408–3414. https://doi.org/10.1145/2851581.2856502.

Alavi, H.S., Churchill, E., Lalanne, D., 2017. The evolution of human-building interaction: an HCI perspective. Interact. Des. Arch. J. - IxD&A 32, 3–6.

Andrews, W., 1982. How to avoid the ten most frequent EMS pitfalls. Energy User News . (April 19).

Banks, N., Fawcett, T., Redgrove, Z., 2016. What are the factors influencing energy behaviours and decision-making in the non-domestic sector? A rapid evidence assessment.

Bordass, B., Cohen, R., Field, J., 2004. Energy performance of non-domestic buildings: closing the credibility gap. In: 8th International Conference on Improving Energy Efficiency in Commercial Buildings, pp. 1–10. August. Available at: http://www.usablebuildings.co.uk/Pages/Unprotected/EnPerfNDBuildings.pdf.

Bordass, B., Leaman, A., Bunn, R., 2007. Controls for end users: a guide for good design and implementation. BSRIA Ltd, Reading.

Brambley, M.R., Haves, P., Torcellini, P., Hansen, D., 2005. Advanced Sensors and Controls for Building Applications: Market Assessment and Potential R & D Pathways. Pacific Northwest Natl. Lab.

California Energy Commission 2007, 2007 Integrated Energy Policy Report, CEC-100-2007-008-CMF.

Carr, J., Brissette, A., Omati, L., 2017. Managing smart grids using price responsive smart buildings. In: Energy Procedia. Elsevier Ltd, pp. 21–28. https://doi.org/10.1016/j.egypro.2017.09.593.

CEN, 2017. EN 15232-1: Energy Performance of Buildings—Energy performance of buildings—Part 1: Impact of Building Automation, Controls and Building Management. Belgium.

D'Oca, S., et al., 2018. Human-building interaction at work: Findings from an interdisciplinary cross-country survey in Italy. Build. Environ. 132 (September 2017), 147–159. https://doi.org/10.1016/j.buildenv.2018.01.039. Elsevier.

Delay, T., Farmer, S., Jennings, T., 2009. Building the future, today: transforming the economic and carbon performance of the buildings we work in, Report CTC765. The Carbon Trust, London.

Delzendeh, E., et al., 2017. The impact of occupants' behaviours on building energy analysis: A research review. Renew. Sustain. Energy Rev. 80 (August), 1061–1071. https://doi.org/10.1016/j.rser.2017.05.264. Elsevier Ltd.

European Commission, 2018. Directive (EU) 2018/844. Off. J. Eur. Union 2018 (May), 75–91.

Fogg, B., 1992. Photomorphogenesis in Plants. Photochem. Photobiol. 56 (5), VII–VIII. https://doi.org/10.1111/j.1751-1097.1992.tb02203.x.

Hauge, Å.L., Thomsen, J., Berker, T., 2011. User evaluations of energy efficient buildings: Literature review and further research. Adv. Build. Energy Res. 5 (1), 109–127. https://doi.org/10.1080/17512549.2011.582350.

Haynes, B.P., 2008. The impact of office comfort on productivity. J. Facil. Manage. 6 (1), 37–51. https://doi.org/10.1108/14725960810847459.

Hong, T., et al., 2017. Ten questions concerning occupant behaviour in buildings: The big picture. Build. Environ. 114, 518–530. https://doi.org/10.1016/j.buildenv.2016.12.006. Elsevier Ltd.

Huang, Y., Niu, J., 2016. Optimal building envelope design based on simulated performance: history, current status and new potentials. Energ. Buildings 117, 387–398. https://doi.org/10.1016/J.ENBUILD.2015.09.025. Elsevier.

IEA-ECBCS, 2017. Technical Report: Studying Occupant Behaviour in Buildings—Methods and Challenges.

IEA-ECBCS, 2018. Annex 66: Definition and Simulation of Occupant Behaviour in Buildings—Final Report 1–155.

Janda, K.B., Parag, Y., 2013. A middle-out approach for improving energy performance in buildings. Build. Res. Inf. 41, 39–50.

Khasreen, M., Banfill, P.F., Menzies, G., 2009. Life-cycle assessment and the environmental impact of buildings: A review. Sustainability 1 (3), 674–701. Available at: https://doi.org/10.3390/su1030674.

Leaman, A., Bordass, B., 1997. Strategies for better occupant satisfaction. In: 5th Indoor Air Quality Conference, UK, (July), pp. 1–10. Available at: http://www.usablebuildings.co.uk/Pages/Unprotected/StrategiesForBetterOccupantSatisfaction.pdf.

Leaman, A., Stevenson, F., Bordass, B., 2010. Building evaluation: Practice and principles. Build. Res. Inf. 38 (5), 564–577. https://doi.org/10.1080/09613218.2010.495217.

Lowry, G., 2002. Factors affecting the success of building management system installations. Build. Serv. Eng. Res. Technol. 23, 57–66. ST-Factors affecting the success of build.

Martins, A.G., 1988. Energy management systems. In: De Almeida, A.T., Rosenfeld, A.H. (Eds.), Demand-Side Management and Electricity End-Use Efficiency. Springer, Netherlands, pp. 127–144. https://doi.org/10.1007/978-94-009-1403-2_7.

Metzger, I., Kandt, A., VanGeet, O., 2011. Plug load behavioural change demonstration project. . Contract.

Min, Z., Morgenstern, P., Marjanovic-Halburd, L., 2016. Facilities management added value in closing the energy performance gap. Int. J. Sustain. Built Environ. 5, 197–209.

Nembrini, J., Lalanne, D., 2013. Human-computer interaction INTERACT'97. In: Human-Computer Interaction INTERACT '97, pp. 348–369. https://doi.org/10.1007/978-0-387-35175-9.

Oliveira, F., Bernardo, H., 2019. Energy management tools for sustainability. In: Filho, W.L. (Ed.), Encyclopedia of Sustainability in Higher Education. Springer Nature.

Ornetzeder, M., Wicher, M., Suschek-Berger, J., 2016. User satisfaction and well-being in energy efficient office buildings: Evidence from cutting-edge projects in Austria. Energ. Buildings 118, 18–26. https://doi.org/10.1016/j.enbuild.2016.02.036. Elsevier B.V.

Paone, A., Bacher, J.-P., 2018. The impact of building occupant behaviour on energy efficiency and methods to influence it: A review of the state of the art. Energies 11 (4), 953. https://doi.org/10.3390/en11040953.

Parag, Y., Janda, K.B., 2014. More than filler: Middle actors and socio-technical change in the energy system from the "middle-out.". Energy Res. Soc. Sci. 3, 102–112.

Robinson, J.F., Foxon, T.J., Taylor, P.G., 2015. Performance gap analysis case study of a non-domestic building. Proceedings of the Institution of Civil Engineers—Engineering Sustainability 169 (1), 31–38. https://doi.org/10.1680/ensu.14.00055.

Shen, L., et al., 2016. Human-building interaction (HBI): A user-centered approach to energy efficiency innovations. In: ACEEE Summer Study on Energy Efficiency in Buildings, (Norman 2002), pp. 1–12. Available at: http://aceee.org/files/proceedings/2016/data/papers/8_714.pdf.

Soares, N., et al., 2017. A review on current advances in the energy and environmental performance of buildings towards a more sustainable built environment. Renew. Sustain. Energy Rev. 77, 845–860. https://doi.org/10.1016/J.RSER.2017.04.027. Pergamon.

Sørensen, Å.L., et al., 2017. Zero emission office building in Bergen: Experiences from first year of operation. Energy Procedia 132, 580–585. https://doi.org/10.1016/J.EGYPRO.2017.09.747. Elsevier.

Steemers, K., Manchanda, S., 2010. Energy efficient design and occupant well-being: Case studies in the UK and India. Build. Environ. 45 (2), 270–278. https://doi.org/10.1016/j.buildenv.2009.08.025. Elsevier Ltd.

Stevenson, F., Williams, N., 2007. Case study lessons in sustainable housing design guide. In: Sustainable Housing Design Guide for Scotland.

Thomas, W.D., Duffy, J.J., 2013. Energy performance of net-zero and near nct-zero energy homes in New England. Energ. Buildings 67, 551–558. https://doi.org/10.1016/J.ENBUILD.2013.08.047. Elsevier.

Tian, Z., et al., 2018. Towards adoption of building energy simulation and optimization for passive building design: A survey and a review. Energ. Buildings 158, 1306–1316. https://doi.org/10.1016/J.ENBUILD.2017.11.022. Elsevier.

van Dronkelaar, C., Dowson, M., Spataru, C., Mumovic, D., 2016. A review of the energy performance gap and its underlying causes in non-domestic buildings. Front. Mech. Eng. 1, 1–14.

Xu, X., Maki, A., Chen, C., Dong, B., Day, J.K., 2017. Investigating willingness to save energy and communication about energy use in the American workplace with the attitude-behaviour-context model. Energy Res. Soc. Sci. 32, 13–22. https://doi.org/10.1016/j.erss.2017.02.011.

The challenge of improving energy efficiency in the building sector: Taking an in-depth look at decision-making on investments in energy-efficient refurbishments

2.2

Katrin Arning[a], Elisabeth Dütschke[b], Joachim Globisch[b], Barbara Zaunbrecher[a]

[a]Chair of Communication Science/Human Computer Interaction Center, RWTH Aachen University, Aachen, Germany
[b]Fraunhofer Institute for Systems and Innovation Research ISI, Karlsruhe, Germany

1 Introduction

1.1 Investments in energy-efficient measures and technologies in the building sector in Germany

Next to the energy and transportation sector, the building sector harbours great potential for reducing CO_2 emissions. According to the estimations of the International Energy Agency (IEA), the building sector is responsible for about 30% of the energy consumption and over 50% of the electricity consumption worldwide (IEA, 2018). The IEA expects these figures to increase by 30% until 2060 (IEA, 2018) if no action is taken. To achieve the goal of a climate-neutral building stock for residential and nonresidential buildings, efficiency measures (e.g. thermal insulation or efficient/renewable heating technologies) represent a central lever to reduce primary energy demand and CO_2 emissions. However, the investment rate in energy-efficient (EE) refurbishment measures and technologies in the building sector is below the level needed to achieve these goals. At the same time, numerous policies to encourage such EE measures already exist in many countries. For example, since 2010

---

[a] https://eur-lex.europa.eu/eli/dir/2018/844/oj last access 13/03/19.

Energy and Behaviour. https://doi.org/10.1016/B978-0-12-818567-4.00002-8

in Europe, the Energy Performance of Buildings Directive (EPBD)[a] requires that EE measures must be implemented in case of major renovations. The directive also includes the obligation that EU Member States introduce cost-optimal energy performance requirements for new buildings. Since 2012, the Energy Efficiency Directive (EED)[b] requires EU member states to establish strategies for the renovation of their national building stocks and report on them every 3 years. Despite this, refurbishment rates are stagnating at a very low level across the EU and in Germany as well (ca. 1%) (dena, 2018). To achieve the climate protection goals, the refurbishment rate needs to increase, and more buildings must be refurbished to a higher energy efficiency standard in a shorter period of time (dena, 2018).

In this chapter, we take a closer look at refurbishment decisions in the building sector. We focus on the investor side of the decision-making process for EE refurbishments to identify the most promising levers for policies supporting a more rapid transition towards a sustainable building stock. Our analysis concentrates on refurbishments in the existing building stock as implementing EE standards in new buildings is more easily governed by legislation. We examine decision-making processes around measures related to the thermal conditioning of buildings, mainly heating and insulation, but also cooling and ventilation, as this is the main source of energy demand in buildings.

We give an overview of barriers to investments in EE refurbishment (Chapter 1.2), the state of research on EE refurbishment decision-making in the building sector (Chapter 1.3) and describe the empirical research approach of the study (Chapter 2). In Chapter 3, we present an EE refurbishment decision-making model for the residential and nonresidential building sectors, which integrates the role and impact of intermediaries like craftsmen and architects in the decision-making process, because they have been found to influence the refurbishment decisions of investors. The chapter closes with a discussion of the decision-making model and the actors involved and draws implications for promoting a low-carbon transition in the building sector.

1.2 Enablers and barriers to energy-efficient refurbishment investments in the building sector

EE refurbishment decisions have attracted increasing attention in recent years. The main influencing factors have been identified, which act as enablers or barriers to such decisions (e.g. Jafari and Valentin, 2017; Stieß and Dunkelberg, 2013; Wilson et al., 2015). The decision to implement an EE measure is regarded as based on the trade-off between the costs (usually not only investment costs but also nonmonetary costs such as restricted living comfort during construction) and benefits of such a measure (economic benefits from reducing the operating costs of a property, ecological benefits such as reducing CO_2 emissions, and further cobenefits such as improved comfort) (Jafari and Valentin, 2017).

Financial aspects are the most frequently identified driver of EE refurbishment decisions (energy cost savings, increase in the value of real estate due to modernisation

---

[b]   https://ec.europa.eu/energy/en/content/introduction-10 last access 03/01/19.

measures) (Wilson et al., 2015). The expectations of house owners concerning the positive effects on the living environment in terms of thermal comfort act as an additional important driver (Michelsen and Madlener, 2013). Aesthetic considerations, such as embellishing a house facade as part of an insulation measure, can also drive the decision to implement EE measures (Galvin and Sunikka-Blank, 2014). A recent stream of literature discusses the nonenergy benefits or cobenefits of EE measures in general (Cagno et al., 2019). Some of the topics discussed here include issues like improved indoor climate or improved workplaces. However, this new stream of literature focuses more on (potential) drivers after implementation and less on their relevance for the decision process.

A large body of literature in the field of energy efficiency explores barriers to implementing EE measures (e.g. Stieß and Dunkelberg, 2013; Wilson et al., 2015). The literature on barriers is more developed than that studying enablers or ways to overcome barriers (Cagno et al., 2019). Barriers are described as 'a mechanism that inhibits a decision or behaviour that appears to be both energy efficient and economically efficient' (Sorrell et al., 2004), a phenomenon that has also been described as the *energy efficiency gap*. Typical barriers to the adoption of EE measures discussed in the literature (e.g. Sorrell et al., 2004) include the following: (1) no monitoring of energy consumption or knowledge about the current energy demand or efficiency level; (2) the lack of awareness regarding EE issues; (3) the lack of knowledge about appropriate measures, for example, refurbishment; (4) split incentives, also discussed as the renter–tenant dilemma; (5) the investor's lack of time to engage in information searches and decision-making; (6) the lack of capital; and (7) unreliability of technology/measures. These barriers cover technological, economic, and behavioural-institutional aspects. While these barriers certainly describe issues that are relevant in this context, they are not able to identify internal processes in households and organisations. This shortcoming is also mirrored in recent publications. Cagno et al. (2013) developed a taxonomy, which explicitly distinguishes internal and external barriers. They also try to connect the knowledge on barriers with a process perspective and psychological concepts: 'It is apparent that every barrier is associated with the perception of the decision-maker and the value that he/she attributes to this perception' (p. 300).

The multitude of potential barriers affecting EE refurbishment decisions indicates that decision-making on EE measures, especially in the residential sector, is highly individual and characterised by a combination of factors. In order to locate and structure the effects of benefits and barriers as influencing factors, models have been developed that map decision-making on EE measures as a sequence of distinct process steps. The following section provides deeper insights into decision-making on EE refurbishment.

1.3 EE refurbishment decision-making in the building sector

While many studies have focused on influencing factors in general, that is, barriers and enablers of EE investment decisions, others have tried to conceptualise investors' decision-making behaviour in models that take a process perspective and/or capture more complex relationships between the influencing factors. While many different models have been developed for EE refurbishment decision-making of house owners as main actors in the residential sector, so far, there are hardly any model-like representations of the decision-making process for the nonresidential sector.

In the decision model of Stieß and Dunkelberg (2013), EE refurbishment decisions of house owners are regarded as strategic consumer decisions. The decision process is divided into the different steps of 'information procurement', 'planning', 'decision-making and implementation', and 'use'. The model comprises sociodemographic data, the life situation of investors and attitudes towards their own homes as well as lifestyle orientations. The authors assume that the respective life situation has a major influence on the decision to engage in EE refurbishment. Furthermore, the attitudes of the investors play an important role, both with regard to the refurbishment process and the outcome. These attitudes include, for example, the goals and causes of the refurbishment as well as resources, such as knowledge or access to knowledge, for example, via social networks, the financial capabilities of the investor, or their ability to carry out the refurbishment work on their own. The model also contains framework conditions of EE refurbishment decisions such as the house's structural condition and current legislation. Although the model subsumes all relevant influencing variables, it does not explicitly adopt a process perspective, and it remains unclear which factors affect the refurbishment decision at which point in the process. Nair et al. (2010) conceptualise the refurbishment decisions of house owners as a process and assume that a decision is reached via the following steps: first, homeowners must feel the need for refurbishment due to technical reasons (maintenance requirements or technical malfunctions) or attitudinal factors, for example, sustainable lifestyle orientations. In the next step, homeowners collect information from various sources. Mass media have a particular influence on innovators and early adopters, while the majority relies on interpersonal information sources (Nair et al., 2010). In this second stage of information acquisition, the authors introduce change agents as an important influencing factor. The decision to implement EE measures is finally made by the investor based on the perceived characteristics of possible alternatives.

Wilson et al. (2018) also illustrate the process character of the EE refurbishment decision in their model, which is based on Rogers' model of innovation diffusion (Rogers, 2010). In stage 0 of the model, homeowners do not yet think about renovation. The actual adoption process begins with stage 1 'thinking about renovations' in which homeowners consider, for the first time, the possibility of EE refurbishment due to 'domestic events' and are very receptive to information. This is followed by the second stage 'planning renovations' and 'finalising renovations' in the third stage, where intentional decision-making processes take place. Step 4 'experiencing renovations' refers to the situation after EE measures have been installed and how house owners get used to them. Up to the third stage, the process can be illustrated by a funnel shape: from a vague idea in stage 1, the process becomes more and more concrete, more focused, and associated with an increasing intention to install EE measures.

In contrast to the variety of models for the decision process in the residential sector, the findings for nonresidential buildings are still scarce, despite their relevance in the building sector. Relatively, little research has looked into decision-making processes for these kinds of buildings. Additionally, the academic research has so far been dominated by studies from the United States, many of them explorative in nature.

Kontokosta (2016) report the motivating factors for energy retrofits from a survey covering 393 nonresidential buildings in the United States. While reduced energy bills is the most frequently stated reason, repairs are the second most common, that is, implementing measures in combination with necessary modernisations. Other reasons include not only financial aspects like increased marketability or reduced operating costs but also broader issues like market recognition, increased tenant comfort, environmental benefits, and peer influence. Barriers mainly comprise financial constraints and risk issues. They also find that owner-occupied commercial buildings are more likely to undergo a retrofit. The study provides a more detailed overview of the factors influencing the probability that ambitious measures will be implemented, but does not go deeper into the conceptualisation of the actual decision-making process. Other studies take a similar approach. Based on findings from 178 participants in the United States, Gliedt and Hoicka (2015) state that financial gains are important, but are not the only motivation for retrofits as sustainability and compliance with regulations also play a role. Similarly, Andrews and Krogmann (2009) find some relationships of the implementation of EE measures with building characteristics and with local conditions like regional energy prices and climate. Additionally, their data from more than 5000 commercial buildings across the United States confirm the challenge of split incentives in the case of renting property. These findings were extended in a survey by Klotz et al. (2010), who analyse situational factors and are able to show the influence of an anchoring bias on retrofit preferences in an experimental design. More specifically, the questionnaire in this study differentiated several levels of goals, and higher goals lead to higher ambition levels for a retrofit.

An actor perspective is taken by Curtis et al. (2017), who explore the potential role that facility managers play in initiating and implementing retrofits. Based on 10 interviews, they conclude that targeting facility managers might be promising, but that success is subject to framework conditions. First, some facility managers do influence the owner's decision-making, but other groups are also important, for example, consultants or renters (Kontokosta, 2016). Second, the influence of facility managers depends on their specific role, for example, if this is limited to maintenance work, then their influence is likely to be low. Their influence also depends on their relationship and proximity to the building owner and their personal interest and expertise in energy efficiency issues.

A process perspective is developed by Birk, Jones, and Bogus (2010), who look at the different steps taken by a variety of organisations in order to decide on and implement an energy efficiency measure. Across the 12 organisations analysed, they find a large degree of heterogeneity and a low degree of systematic procedures when comparing the actual procedures with an ideal one.

To sum up, there has been less research on decision-making processes in organisations than on decision-making in households. Therefore, from a conceptual point of view, the findings are less developed. They not only underline the conclusions from the classic research on barriers to energy efficiency but also point out the variety of relevant actors (e.g. renters, cf. Kontokosta, 2016, and facility managers, cf. Curtis et al., 2017). Finally, the decision to refurbish is not yet understood as a process, that is, a development over time.

The dominant approach of focusing on the investor as a single actor is similar in the current research on refurbishments in both the residential and nonresidential sector. This is an important research gap as this single-actor approach ignores the interaction with other relevant actors in the decision process such as craftsmen, energy advisors, or architects. These actors, also referred to as *intermediaries*, which can be either a group of people or institutions, promote the diffusion of technologies through knowledge exchange, skill development, and mediator functions (Bush et al., 2017). Their role in supporting the low-carbon transition in the building sector, especially with regard to their specific impact on investment decision-making, has not yet been systematically researched.

Furthermore, the decision in favour of a refurbishment measure is not limited to a single point of time, but is a process that evolves across several steps over time. At each of these steps, there is the risk that the track towards refurbishment could be abandoned. Some authors have developed models to describe this evolvement over time for the residential sector. We build on this earlier work in the next chapter, combine it with a multiactor perspective, refine the current process models, and apply them to the nonresidential sector.

2 Research approach

We applied a qualitative research approach to obtain a deeper understanding of the EE refurbishment decision-making process of (non-)residential investors. We conducted detailed interviews to gain better insights into the decision-making process from the perspective of private house owners and decision-makers in nonresidential buildings. We placed special focus on the process character of decision-making and on identifying and uncovering the influence of intermediaries as further actors in different stages of decision-making. We conducted interviews with homeowners in the residential sector who had renovated their homes including energy efficiency measures or who had planned such measures and then rejected them (Section 2.1). In addition, we conducted further interviews focused on the nonresidential sector, which included expert interviews and interviews with decision-makers in organisations (Section 2.2).

2.1 Residential buildings

2.1.1 Research design and interview structure

The interviews were designed as semistructured or guideline interviews. The interviews with *residential house owners* began with a welcome section explaining the topic and procedure. After collecting demographic data and key data on the property and the refurbishment project (installed EE measures, cost, and duration), house owners were asked about their general attitude towards EE refurbishment, its advantages and disadvantages. The interview structure then followed the course of an EE refurbishment project, moving through the individual stages from the initial idea to implementation. First, house owners were asked to outline the occasion or starting point for refurbishment activities and to describe the subsequent steps from their perspective. The next part focused on the information search of house owners on the topic of EE refurbishment, that is, the sources of information about EE refurbishment, their credibility, and usefulness. Another focus was on the selection of intermediaries and an assessment of the relationship with them. House owners were also asked what information they required in the consulting stage and how the planning process led to selecting and planning concrete EE measures. Finally, house owners were asked about the final decision, about other persons influencing the decision process, and about the intermediary's influence on the type and scope of the EE refurbishment measures.

The interviews were audio recorded, transcribed, and analysed qualitatively. We applied the criteria of qualitative content analysis following Mayring (2003). In a detailed procedure, the interview content was systematically reduced and coded based on a category system.

2.1.2 Sample
The sample of house owners consisted of $n=20$, who owned the house they lived in. Two interviews were conducted with couples, the others were individual interviews. All the interviewees were volunteers. Their statements were anonymised, and quotations are identified by I(nvestor)—participant number (age and gender) in the text.

The 'persona technique' was applied to select interview partners in order to account for the heterogeneity of households and to cover a broad spectrum of EE renovation measures and house types. Selection criteria were as follows: (a) date of house purchase (long-term ownership vs new acquisition of an existing property), (b) house type (detached, semidetached, and multifamily), (c) age or life phase of the investor, and (d) type of EE measure implemented.

Most of the investors work in a nontechnical profession; only one interviewee was an expert in the field of EE refurbishment. The house owners lived in mostly (detached) single-family houses or terraced houses in cities. The houses were all existing properties built between 1800 and 1996, the majority after 1960. A wide range of thermal conditioning measures was applied during the refurbishment: insulation of facades and roofs, replacement of windows, and the replacement of heat generators. In most cases, more than one EE measure was implemented.

2.2 Nonresidential buildings
2.2.1 Research design and interview structure
The interviews focused on owner-occupied buildings because of the high degree of heterogeneity in the nonresidential sector regarding building types and the structures and responsibilities for building management and retrofits within organisations. Furthermore, to capture more of this heterogeneity, we interviewed not only

organisational decision-makers but also experts who support organisations in deciding on and implementing EE retrofits.

These interviews also followed a semistructured guideline. The *expert interviews* started with background questions on their expertise and experiences with the topic under study. Experts were then asked about their personal evaluation of developments in the field of thermal retrofits. The second block focused on office and commercial buildings as a target group for thermal retrofits and the heterogeneity versus homogeneity in this field (e.g. whether it is relevant to consider smaller and bigger organisations differently). Findings from this part of the interviews were not very consistent and will therefore not be discussed further in this chapter. The third block included detailed questions about how experts experience decision-making processes in the nonresidential sector during their research or work. This part of the interview also contained questions about the occasions that lead to decisions on retrofits as well as the internal and external actors involved.

The interviews with *decision-makers* also started with background questions about the interviewee. Following this part of the interview, questions were asked about thermal retrofit measures and procedures that had been implemented or were ongoing in the respective organisation. All the interviewees had experience in this field. Corresponding to the expert interviews, the next section asked detailed questions about the occasions that did or did not lead to retrofits, who was involved in the process, and how it was implemented. Originally, we planned to focus on one or two retrofit measures and describe the processes around them in detail. However, in practice, this proved very difficult, as these measures were not implemented in isolation from the respondent's perspective but were part of a fabric of decisions and developments where one thing led to another. Thus, respondents often talked more generally about the developments that had taken place in their respective organisations in the field of EE measures and thermal retrofits.

Interviews were audio recorded and fully transcribed. A code system was developed based on the interview guideline and modified according to the content of the interviews. Expert and decision-maker interviews were analysed simultaneously using an identical coding system.

2.2.2 Sample

Five experts were interviewed, two of them together in the same interview. The experts included two representatives from relevant NGOs: one represented an energy efficiency NGO from German industry and the second an NGO focusing on reducing CO_2 emissions from buildings. Both NGOs had recently completed projects in the field under study. Another expert was an engineer from a technical consultancy with a strong focus on EE. The final two experts were architects from a big company specialised in commercial and office buildings.

The decision-makers interviewed consisted of representatives from five SMEs, three of them with larger sale areas, all of them with offices. There was one large company, that is, a supermarket chain, one company from the public sector, and one landlord of nonresidential buildings. Overall, the interviews focused as much as possible on retail buildings and/or offices and only to a lower extent on production sites or warehouses.

In the description of results, interview partners were anonymised. Quotes are identified by 'Ex plus interview number' for experts and by 'Dec plus number' for decision-makers.

3 Results

This chapter presents the decision-making process model on EE refurbishment decisions (Section 3.1) for the residential (Section 3.1.1) and nonresidential sector (Section 3.1.2). The decision-making process model also integrates the impact of a highly relevant but so far understudied actor group: the group of intermediaries (Section 3.2), highlighting their specific roles for renovation decisions in residential (Section 3.2.1) and nonresidential buildings (Section 3.2.2).

3.1 Decision-making process model

3.1.1 Residential buildings

The EE refurbishment decision-making process is represented as a sequence of different phases from the first idea to the implementation of the EE measure (Fig. 1). The model not only builds on the models of Stieß and Dunkelberg (2013), Nair et al. (2010), and Wilson et al. (2018) (see Section 1.3) but also considers other influencing factors that were mentioned in the interviews. Please note that the model does not cover the 'implementation' stage or a final evaluation in detail, since these two stages happen *after* the EE refurbishment decision. Further, the sequence or length of the stages in individual decision processes vary as shown by the comparison of reported EE refurbishment projects in private households.

3.1.1.1 Refurbishment need

The first stage of the EE decision-making model refers to the initial perception of a need to pursue an energy-efficient refurbishment. Based on the interviews with house owners, we identified both external factors and internal motivational factors as 'triggers', which lead to a perceived refurbishment need. *Internal motivational factors* included a positive attitude towards energy efficiency, the desire for energy savings (not only in an economic but also in an ecological sense), an increase in property

FIG. 1

Decision-making model of EE-refurbishments in residential buildings.

value, the recommendation or advice by others, the idea of a self-sufficient energy supply, and increased thermal/living comfort. As *external factors*, we identified technical defects of appliances, other renovation activities in the house, house purchase, legal regulations, or adaptations to a changed life situation (e.g. having children, children leaving the house, or adapting the house to age-specific requirements).

Apart from these internal and external factors promoting the idea of EE refurbishment, house owners also addressed barriers, such as the expected costs or inconveniences during construction work (*'We were really afraid of all the dirt and also of all these people running around in our house' [I-5, 56, w]*). Another important barrier was the high technical complexity of EE refurbishment and the subjectively perceived low knowledge level that either deterred homeowners from continuing with the EE refurbishment process or prompted them to engage more intensively with the issue in the following 'information search stage'.

3.1.1.2 Information search

The 'information search stage' describes the information search activities of the investor prior to the actual planning and decision on EE refurbishment. Typically, the information search of private house owners was directed at EE measures in general, as well as at specific technical measures. House owners also gathered information about intermediaries, who are specialised in EE refurbishment projects.

Most homeowners reported having relatively low prior knowledge of EE refurbishment and feeling overwhelmed by the complexity of this issue. Accordingly, at this stage, they reported having only a few precise ideas about EE refurbishment options. Their information search typically started with an Internet search for EE measures and experts, where they collected basic information on the topic. One frequently mentioned problem in the context of Internet searches was the credibility of information, that is, doubts about the independence and objectivity of the sources. This basic information was supplemented by discussions with architects, craftsmen, and persons from the interviewee's personal network who had already retrofitted their house with EE measures in the past: *'I talked a lot with my brother, who had built his house a few years before and had already lived through the whole thing'. [I-3, 51, m]*. More technically experienced house owners consulted technical literature; in some cases, local EE advisory centres were approached. Personal information sources and professional advice by advisory centres were evaluated as more reliable, competent, and constructive with regard to the individual EE refurbishment project.

It is difficult to differentiate the 'information search stage' from the following stage 'consultation/planning', because homeowners behave differently: some investors searched intensively for information regarding EE measures and intermediaries on their own, some did not look for information themselves but relied on the advice of a specific intermediary, and some consulted an energy advisor.

One important characteristic of the 'information search stage' is that the influence of intermediaries becomes apparent for the first time in the decision-making process. Even if the information search is initiated mainly by the house owner, the intermediary becomes part of the process and potentially influences it with the information offered, particularly when homeowners exclusively consulted one intermediary.

3.1.1.3 Consultation/planning

In the consultation and planning stage, most house owners seek advice from intermediaries and, starting from the initial EE refurbishment idea, plan and refine the EE refurbishment project. The planning stage was traversed without consulting an intermediary only in a few exceptional cases, for example, when the homeowner was also an expert in the field of EE refurbishment or when house owners had already selected single EE measures.

The interviews showed that the majority of house owners had already made the fundamental decision to invest in EE refurbishment before they consulted intermediaries. However, intermediaries strongly influenced the scope of the refurbishment project and the type of EE measure chosen during the consultation and planning stage.

Even though house owners had already gathered information on EE measures and potential intermediaries in the preceding information search stage, most of them still perceived the topic of EE refurbishment as too complex. Most house owners were not able to assess the effect and suitability of specific EE measures (e.g. only replacing windows or also improving insulation) and were overwhelmed by the large variety of technical options (e.g. choosing a heating technology based on renewable or conventional energy). In addition, many house owners were unable to assess the consequences of single EE measures—which then required adjustments in other areas (e.g. changing the roof overhang after improving the facade insulation). Thus, more concrete ideas for potential EE refurbishment measures addressed aesthetic aspects or personal preferences (e.g. the type of house facade).

To sum up, the high complexity of EE refurbishment and the vast amount of information needed by house owners were the main reasons given for consulting EE expert intermediaries. Legal requirements were another reason for consulting EE experts but mentioned less frequently, for example, to obtain planning permission.

Interestingly, the consultation stage was perceived and evaluated differently by the interviewed house owners. On the one hand, they acknowledged the knowledge gained on the technical background of EE measures, time, and cost savings, as well as reduced workload with regard to refurbishment process management (*'If we hadn't had the architect's network, that refurbishment wouldn't have happened. So, if I had to look for every craftsman on my own...'* [A-2, 66, m]). Most homeowners expressed the expectation that the intermediary should act as a 'complexity reducer' and develop a 'retrofit roadmap' for the individual property. In addition, facing the multitude of different technology and material options, the intermediary was expected to propose and evaluate a reduced number of decision alternatives.

On the other hand, house owners criticised the obligation to seek expert advice to get planning permission and the unrealistic promises of energy savings and retrofit costs made by experts. The role and function of energy advisors were questioned, especially by the majority of house owners who did not consult an energy advisor (*'If I asked an energy consultant... I know what he would tell me. I do not need one'.* [*I-14, 41, m*]).

3.1.1.4 Decision

In the decision stage, homeowners select a specific EE measure, as well as the contracted intermediary. Typically, the fundamental decision to carry out EE refurbishment has already been made at an earlier stage. (*'So he didn't influence me in the basic decision, he influenced me in the implementation of the details, because in some places, he slowed me down and said don't do it like that, don't do it like that, it's enough like that and we're no longer efficient here'.* [*I-11, 51, m*]).

The final EE refurbishment decision among house owners was influenced by a *bundle of motives*. Most house owners reported that the intermediaries involved strongly influenced the measures implemented (*'We did it the way the plumber suggested it'.* (*I-7, 48, m*). This does not mean, however, that the planned EE measures always became more extensive as a result of the intermediary's intervention; intermediaries also advised against implementing certain measures. Other decision criteria included the financial budget and the amortisation period of the EE refurbishment project, structural conditions or constraints of the building, legal requirements concerning EE standards, desired living comfort, and anticipated inconveniences during the construction work. The available financial budget set the framework within which possible EE alternatives were discussed and selected. Only a few very environmentally conscious house owners with higher financial capabilities used maximum energy efficiency as the sole guiding criterion for their decisions.

3.1.2 Nonresidential buildings

Overall, the decision-making process in organisations features similar phases to the one in households. However, in the interviews, we observed fewer distinct steps, and—more importantly—isolated EE measures were discussed, planned, and implemented less often. Instead, EE measures were more likely to be closely linked to further measures in the field of refurbishment and further business activities. This meant that interviewees rarely reported one or two specific measures that were implemented as most organisations are involved in a constant process of keeping up standards and improving their buildings and adapting them to their business development. This also implied a close connection with the organisation's core business. For example, for retailers, measures were always seen in connection with sales and how they would affect customers—during their implementation and afterwards. Thus, measures that make the rooms more attractive to customers were viewed quite differently to measures that do not. Vice versa, if construction work needed to be done to make the shop more attractive to customers, then this (sometimes) allowed the implementation of EE measures at the same time.

Emerging from these dimensions of interconnectedness were two additional layers that we integrated into the phase model of decision-making developed for households (see Fig. 2). First, in the earlier phases and closely related to the identification

FIG. 2

Decision-making process model for EE investments in nonresidential buildings.

of refurbishment needs, is the process of agenda setting, that is, whether the topic of energy-efficient refurbishment arises at all. This is closely related to certain occasions ('occasions and agenda setting'). Second, in parallel to the consultation and planning and the final decision-making, the outcome of the decision-making process is shaped by a variety of motivations, often connected to the earlier occasions that triggered the process and by a greater or smaller number of actors within and outside the organisations ('convincing motivations and actors').

3.1.2.1 Occasions and agenda setting

Occasions that lead to the identification of refurbishment needs and to the search for information include internal and external factors, some of which are related to the field of energy use and sustainability, while others are not (see Table 1). Energy-related issues comprise the influence of broader energy initiatives, like the implementation of energy management systems, participation in certain initiatives like EE networks, or, in some cases, (obligatory) energy audits. Another energy-related issue was the aim to reduce energy costs. Furthermore, internal goals and strategies play a role, for example, not only concerning energy but also broader sustainability issues or societal benefits, and may initiate processes that lead to the identification of a refurbishment need. A very important area are disturbances to work processes or changes that alter how the buildings are used, for example, an increase or decrease in the number of employees or certain products, and, of course, the failure of certain installations, for example, when they reach the end of their lifetime.

Legal requirements or rising standards like the energy saving regulation, fire protection issues, or accessibility also play a role. Some interviews also referred to the perceived pressure from customers and the public. This was partly derived from a general perception of sustainability or energy debates in society, but in some cases, the perception was more specific, for example, in organisations who sell to other companies (b2b, business to business), where customer standards were also sometimes requested of companies along the supply chain. In addition, needs were sometimes detected by accident, for example, following external events like trade fairs, or through new or regular contacts to intermediaries, for example, consultants and craftsmen.

There were broad categories of actors who introduced a refurbishment need onto the organisation's agenda. Sometimes, the party responsible for building management identified the refurbishment need, that is, facility manager if existent or the

Table 1 Occasions that put EE refurbishments on the agenda as mentioned by the interviewees

	Energy related	Nonenergy related
Internal	• Comprehensive activities as an occasion for individual measures, for example, energy audit and energy/environmental management system • Energy costs (absolute and relative compared with benchmark) • Company policy/guidelines	• Disturbances to work processes • Conversion/renewal/upgrading/refurbishment of buildings • Proximity to core business, for example, exemplary implementation of own products/services • Damage or cyclical renewal of installations • Broader goals on sustainability and societal benefits
External	• Legal obligations, for example, energy saving directive • External pressure (customers) • Customer requirements • Suggestions from third parties, for example, intermediaries, participation in research project, and trade fair	• Legal obligations, for example, fire protection • Increasing attractiveness for customers

manager/managing board. Some companies also had energy managers. However, sometimes the initiative was taken by a single employee or by a group of employees and even by external actors like technical consultants or planners who were in a position to influence agenda setting.

3.1.2.2 Convincing motivations and actors

Economic aspects were very important in the final decision to actually implement an EE measure. Some organisations state to have very fixed rules ('We implement it if it has a return on investment within 5 years' Dec-10). In other cases, this was vaguer and decisions just needed to be perceived as 'sensible' in economic terms. Interviewees also mentioned that besides the economic evaluation, the final decision also depended on parallel, ongoing processes. One interviewee from a family-owned business explained: 'It does not always go exactly according to plan, but it also [depends on] (…) how does the family [currently] feel? (…) It's not always strictly according to that [return on investment] (…), but then it's a prestige project for them and then the money is taken for it. Meanwhile for other things, which would be much more worthwhile, stand in the back and [they] say, "No, we don't have any money for that right now"'. *(Dec-5).*

Furthermore, the alignment with the triggering occasion plays a role. For example, if the main motivation was to increase the attractiveness of the building for customers or employees, then the decision to implement the efficiency measure was strongly influenced by its fit with this overarching goal, for example, if a new heating system is able to improve the indoor climate. Aspects of visibility or a contribution to promotional activities or the broader vision of the company were also important. Additional arguments that led to a decision for or against a measure were related to the (additional) disturbances resulting from the implementation, for example, impact on the retail area of a building. A personal motivation or interest of the final decision-maker, often the CEO, was important to stabilise the decision. Measures were reported as more likely to receive a positive evaluation if they were likely to have quick results.

Besides the final decision-maker, who was either on the managing board or the CEO (in smaller companies or for bigger measures) or the facility or energy manager if existent (in bigger companies or for smaller measures), other parties also had a say in the decision-making process. These include the sales department, workers' council, affected employees, and supervisors. External actors were not mentioned as influential for the final decision.

3.2 Types and impact of intermediaries on the EE refurbishment decision-making process

3.2.1 Residential buildings

Different *types of intermediaries* with specific functions were identified over the course of the refurbishment process. We differentiate between 'advisory intermediaries' such as architects, civil engineers and energy consultants, whose main focus is on consulting and planning EE refurbishments, and 'implementing intermediaries', mainly craftsmen (e.g. builders, roofers, and heating installers), who actually conduct the refurbishment work and implement the EE measures in the buildings.

In the majority of renovation projects in private households, after an initial information search phase, a single intermediary was selected to plan and carry out the entire renovation process. Especially for larger refurbishment projects, house owners used the services of an advisory intermediary, for example, an architect or an energy consultant. Interestingly, *architects* were initially contacted for planning other (non-EE) renovation measures, especially when a building permit was required. In these cases, a construction measure turns into an EE refurbishment, even if the homeowner did not intend this originally. *Energy advisors* are rarely consulted by house owners, and their influence is usually limited to the information search and early planning stage. However, they can provide the initial trigger for EE refurbishment, influence the information-searching phase by informing the homeowner about EE standards and measures, and give detailed advice on which measures should be implemented. The planning and installation of EE measures are then usually discussed with implementing intermediaries, that is, craftsmen, who are recommended by energy advisors.

For private households, *craftsmen* are the most important intermediaries. As described before, in the majority of EE refurbishments, only one craftsman is chosen by house owners as the main actor in the information search, consultation and planning, and implementation stages. If the refurbishment project requires other competencies beyond the specific expertise of the craftsman (e.g. final 'cosmetic' work such as laying tiles and painting work), it is often also carried out by this craftsman or he chooses and coordinates subcontractors. This mostly corresponds to the house owner's expectation and desire to get 'everything from one source' during the renovation project.

Apart from the different roles and functions of intermediaries, different points in time were identified at which intermediaries *influence the decision-making process of private house owners*. Accordingly, we suggest that implementing EE refurbishments should not be regarded as the exclusive decision of the investor, but rather—from the moment the intermediary enters the process—as a joint negotiation process between the investor and the intermediary to which both sides contribute their ideas, interests, wishes, and knowledge (see Fig. 3).

In the first stage of the decision-making process, the perception of a *refurbishment need* and the influence of intermediaries are rather low, but they can stimulate the engagement with EE refurbishment ideas through acquisition activities or local advisory campaigns.

In the *information search* phase in which homeowners also search specifically for experts who can plan or carry out the measure, intermediaries become part of the decision-making process. Typically, the investor contacts the intermediary either for general information or with specific requirements, but as soon as the intermediary gives expert advice, he/she frames the course of interaction, depending on his/her expertise, technical portfolio, and interests.

Since, as already described, the intermediary influences the scope of the EE refurbishment project and the choice of EE measures, the *selection of the intermediary* is of central importance to the decision-making process. Overall, personal reasons were the key factor in private households when selecting intermediaries. In addition to personal acquaintanceship or the recommendation by others, liking the intermediary was a relevant selection criterion. The availability of an intermediary in terms of time, local residence, and specific professional expertise for the refurbishment project and qualification were additional factors that determined which intermediary was chosen by house owners. Especially in the case of personal acquaintanceship or

FIG. 3

Impact of intermediaries on EE refurbishment decisions in residential buildings.

recommendation, house owners stated that they trusted their intermediaries as competent 'navigators' and decision supporters.

In the *consultation and planning* stage, house owners often ask for general advice on EE measures and expect expert input not only on the general value of EE measures but also on details such as the efficiency, technical background, and functionality of specific EE technologies. For many homeowners, inadequate mental models play a role, such as the fear of damp walls or mould after facade insulation and the replacement of windows. Personal preferences or an aversion to certain materials (e.g. plastic for window renovations or polystyrene for façade insulation) as well as construction costs or expected energy cost savings are also addressed in the consultation talks. Further topics of the consultation meetings with intermediaries concern the maintainability of technology and legal issues, for example, the obligation to obtain permission for renovation measures.

Regarding the final *decision* on the EE measure, homeowners consistently stated that they made the decision together with their spouses and the involved intermediary. Interestingly, it became apparent in the interviews that homeowners are not looking for the (ecologically or economically) optimal solution by examining all the alternatives intensively, but based their decision mainly on the intermediary's assessment and recommendation.

3.2.2 *Nonresidential buildings*

Intermediaries also play a relevant role in the nonresidential sector. Especially in cases where the main motivation of the company comes from outside the energy field, intermediaries often have decisive influence on whether EE issues become part of the refurbishment. Furthermore, only a few organisations are able to consider all the relevant options, so it is again the role of the intermediaries to introduce further options to the decision-making process, for example, regarding heating systems. Here, it is also important to evaluate which option fits best with the individual needs of the organisation.

If either an energy consultant or a specialised architect or engineer is involved in the process, usually as a consulting intermediary, they tend to have broader influence, for example, on which types of measures are discussed and finally implemented. The role of craftsmen as implementing intermediaries is geared more towards the actual implementation and providing options for it. However, this is then usually limited to the respective field, for example, the decision has already been narrowed down to the heating system, and the craftsmen then propose different heating systems.

The interview partners reported limitations for all types of intermediaries. Some found it difficult to find competent consultants, while emphasising that their role could be crucial for a systemic view of the refurbishment process. Craftsmen are sometimes recruited by standardised and formal procedures, for example, tendering. In other cases, the organisations draw on their private or professional networks or have long-term relationships with certain craftsmen. Many of the interviewees reported experiences with craftsmen who were not qualified enough to implement innovative technologies, which led either to additional efforts, costs, and disturbances during installation or, even worse, during the usage phase.

4 Discussion and implications

In order to achieve a climate-neutral building stock, the diffusion of EE refurbishment measures and technologies in the building sector has to increase. We took an in-depth look at decision-making on investments in EE refurbishment among private homeowners and decision-makers in nonresidential buildings in order to improve the understanding of the process and investigate the role and impact of intermediaries. In the following, we discuss the insights gained and derive recommendations on how to increase the diffusion of EE refurbishments in the building sector.

4.1 The decision-making process for investments in EE refurbishment

The decision-making process for investments in EE refurbishment is a longer one comprising several separate stages. Since decision processes in this field are highly individual due to the different structural conditions of the building involved, different technology options, and varying financial capabilities of the investor, the sequence or length of stages can vary. The decision-making model presented here confirms some already existing models on decision-making processes (e.g. Wilson et al., 2018) but expands and differentiates some stages.

In our decision-making model, we introduced the concept of the 'first idea' or perception of a refurbishment need to the first stage. In contrast to other EE decision-making models, which start from the information search (e.g. Stieß and Dunkelberg, 2013), we found that the decision-making process of house owners has already started with the '*first idea*', the phase in which homeowners first think about EE refurbishments, which is strongly characterised by personal motivation. The internal factors identified here can be regarded as *positive preconditions*, which do not necessarily lead to an EE refurbishment decision. However, this 'first idea' can also be triggered externally by local advice and acquisition campaigns by energy advisors or craftsmen (e.g., Scott et al., 2016). Starting from the 'first idea' in the decision-making model, it is important to integrate every point of possible intervention to increase EE refurbishment rates. More technically oriented approaches (e.g. Achtnicht and Madlener, 2014) assume that the refurbishment need stage is determined by technical life cycles (malfunctions or end of life of technical devices), but the multitude of internal motivational factors identified here indicates that the decision-making process for EE refurbishment in private households is also affected by nontechnical triggers.

For the nonresidential sector, this phase of the first idea is even broader, and we describe it using the term agenda setting. It encompasses the idea of EE refurbishment and placing this idea on the organisation's agenda. Therefore, this phase also merges with the information search phase, because the individual(s) putting the idea forwards sometimes need to provide more information at a very early stage in order to convince others and actually trigger the start of the decision-making process.

The subsequent stages of our decision-making model also feature further refinements compared with existing models. In the model of Wilson et al. (2018), the stages '*planning and finalising*' describe how planning the EE refurbishment project becomes more and more concrete and ultimately leads to a decision. This rather broad conceptualisation leaves space for individual decision-making processes but is too generic to pinpoint the influence of intermediaries and investors. This is why we introduce 'information search' and 'consulting' to our decision process model as separate stages, although these do not necessarily take place in the order given but might also follow different courses. In the information search stage, the investor is the main actor gathering information and initiating the contact to intermediaries. However, from the 'consultation and planning stage' onwards, our decision-making model abandons the single-actor perspective and suggests integrating the intermediary as an additional actor into the process.

4.2 Impact of intermediaries on EE refurbishment decision-making

The decision to install EE measures in private households and organisations is influenced not only by the investor in a single-actor process but also by intermediaries such as craftsmen, energy advisors, or architects, which take different roles and functions in the EE refurbishment decision-making process in the residential and nonresidential sector (Table 2). For the residential sector, the decision-making of investors can be seen as a joint interaction process to which both actor groups contribute their interests, requirements, and needs. In previous models on EE decision-making, the intermediaries' advice was either disregarded or conceptualised as one possible source of information among others. In the model of Stieß and Dunkelberg (2013), for example, the consulting and information input of intermediaries are grouped under 'resources', which underestimates and does not sufficiently differentiate the influence of intermediaries on EE investment decisions. The interviews with house owners clearly indicated that intermediaries have considerable influence on the decision-making process at several stages and that their influence increases over the course of the process. Most house owners who retrofitted their homes energy efficiently took advice from only one or a few intermediaries (mainly craftsmen) and chose this intermediary at a very early stage of the decision-making process. Very few homeowners relied on professional advice from energy advisors. This shows that the selection of the intermediary by house owners is the most significant decision point in the decision-making process model, since it determines the outcome and quality of the EE refurbishment because of the high impact of the intermediary on the chosen EE measure. The selection of the intermediary by house owners was guided primarily by personal acquaintanceship or by personal recommendation and not by professional qualification criteria. Accordingly, it is important to increase homeowners' awareness of the qualifications (and certifications) in the field of EE refurbishment.

In organisations (cp. Table 2) for an overview, advising intermediaries are not always involved, and if they are, the motivation for doing so can come from not only the field of energy but also outside it. In both cases, they have a broad influence on where EE refurbishment is considered and thereby which options actually enter the

Table 2 Role and functions of intermediaries in the EE refurbishment decision-making process in the residential and nonresidential sector

	Role and function of intermediaries			
	Residential buildings		**Nonresidential buildings**	
Refurbishment need	Acquisition activities by advisory or implementing intermediaries or local community actions as potential triggers of a refurbishment need		Occasions and agenda setting	Advisory and implementing intermediaries sometimes put EE on the agenda
Information search	Intermediaries are contacted by homeowners			
Consultation/ planning	Consultation and EE refurbishment planning activities: • by advisory intermediaries (e.g., architects and energy advisors) in larger or ambitioned EE refurbishments • by implementing intermediaries (craftsmen) in the majority of EE refurbishments		Convincing motivations and actors	Advisory and implementing intermediaries add information on available options
Decision	• If advisory intermediaries are included, they develop the EE refurbishment concept and implementation plans, suggest implementing intermediaries • If exclusively implementing intermediaries are consulted, craftsmen strongly affect the dimension and details of the EE refurbishment			
Implementation	Outcome and quality of the EE refurbishment implementation is influenced by the competencies of the contracted intermediary			Implementing intermediaries are sometimes perceived limiting factors due to the lack of knowledge and expertise

decision-making process. Sometimes, this role is internalised if an organisation has an energy or sustainability manager. The influence of implementing intermediaries is usually limited to the different options considered once the type of measure has already been decided, so their role is smaller. In contrast to households, at least some organisations do not rely strongly on craftsmen to provide decision options and are also more ready to search for different craftsmen who are able or willing to implement the preferred option. This is closely connected with the experience of some organisations that some craftsmen lack the motivation, competence, or ambition to implement innovative solutions successfully, especially if this involves going beyond standard approaches.

5 Policy implications and future research

We took a closer look at the decision-making processes for refurbishments in the residential and nonresidential sectors. The interview study supported the conceptualisation of the decision as a process that develops over time and where different factors and actors are influential at different points in this process. Since the overall objective is to increase the diffusion of EE measures and technologies in buildings, it seems sensible to consider this decision-making process in future research and when designing policy measures.

Advisory and implementing intermediaries represent an important intervention point to increase the renovation rate and the quality of EE refurbishments. So far, they have been given too little consideration. Intermediaries do not affect the fundamental decision of house owners to engage in EE refurbishment, but they do influence the quality and scope of the installed EE measures. Policy measures should therefore involve craftsmen, in particular, as change agents. This can be achieved by the following: (a) offering training courses to inform and convince craftsmen of the benefits of EE technologies; (b) providing financial incentives and solving liability issues, which deter craftsmen from installing innovative EE technologies; and (c) introducing legal provisions obliging craftsmen to comply with existing energy standards.

Based on the homeowners' needs and perceived barriers identified during the decision-making process, approaches to increase the diffusion of EE measures and technologies in buildings should focus on reducing the perceived complexity of EE refurbishment and on increasing its quality and depth. Comprehensive and credible information about EE refurbishment needs to be provided and (financial) access barriers to energy consultants should be lowered (e.g. through local advisory campaigns). Refurbishment projects in the residential sector should be managed and carried out by 'a single source', which requires more comprehensive training or merging different enterprises into refurbishment centres. In order to achieve not only a higher number but also a higher quality of renovations, EE refurbishment strategies are also advisable for the residential sector.

For the nonresidential sector, it turned out that cobenefits and caveats are highly relevant. They create windows of opportunity for refurbishments, for example, if the motivation to renovate originates from a desire to not only improve the building standard but also set important limitations, for example, preference for visible measures and low level of disturbance during implementation. Thus, policy measures could try to target these windows of opportunity and use them to achieve higher energy standards.

Due to the high relevance of intermediaries for the diffusion of EE measures in the building sector, further research should focus on exploring the interaction between intermediaries and investors in more detail. In particular, this involves the selection of intermediaries by investors, the formation of intermediaries' technology portfolios, the differing perspectives of intermediaries and investors on the decision-making process, and the role of trust and confidence in the relationship between investor and intermediary. In addition, a quantitative validation of the decision-making process model should be conducted based on sufficiently large samples. A further avenue of research involves extensions to energy system models, which usually model refurbishments based on technology life cycles. Model refinements could try to represent additional windows of opportunity for EE refurbishments, for example, in relation to the household life events or organisational changes identified in the interviews. These extensions could be especially promising if combined with modelling the influence of special policy measures that either target these additional opportunities or the influence of intermediaries.

Acknowledgements

The authors would like to thank the German Federal Ministry for Economic Affairs and Energy (BMWi) for supporting and funding the project Diffusion EE. We also thank Julian Halbey and Julius Wesche for their valuable research support in conducting the interviews in the residential and nonresidential sector.

References

Achtnicht, M., Madlener, R., 2014. Factors influencing German house owners' preferences on energy retrofits. Energy Policy 68, 254–263.

Andrews, C.J., Krogmann, U., 2009. Explaining the adoption of energy-efficient technologies in U.S. commercial buildings. Energ. Buildings 41 (3), 287–294. https://doi.org/10.1016/j.enbuild.2008.09.009.

Bush, R.E., Bale, C.S.E., Powell, M., Gouldson, A., Taylor, P.G., Gale, W.F., 2017. The role of intermediaries in low carbon transitions—Empowering innovations to unlock district heating in the UK. J. Clean. Prod. 148, 137–147. https://doi.org/10.1016/j.jclepro.2017.01.129.

Cagno, E., Moschetta, D., Trianni, A., 2019. Only non-energy benefits from the adoption of energy efficiency measures? A novel framework. J. Clean. Prod. 212, 1319–1333. https://doi.org/10.1016/j.jclepro.2018.12.049.

Cagno, E., Worrell, E., Trianni, A., Pugliese, G., 2013. A novel approach for barriers to industrial energy efficiency. Renew. Sustain. Energy Rev. 19, 290–308. https://doi.org/10.1016/j.rser.2012.11.007.

Curtis, J., Walton, A., Dodd, M., 2017. Understanding the potential of facilities managers to be advocates for energy efficiency retrofits in mid-tier commercial office buildings. Energy Policy 103, 98–104. https://doi.org/10.1016/j.enpol.2017.01.016.

dena, 2018. DENA GEBÄUDEREPORT KOMPAKT 2018. In: Statistiken und Analysen zur Energieeffizienz im Gebäudebestand. Available at https://www.dena.de/fileadmin/dena/Dokumente/Pdf/9254_Gebaeudereport_dena_kompakt_2018.pdf. (retrieved 14.2.19).

Galvin, R., Sunikka-Blank, M., 2014. The UK homeowner-retrofitter as an innovator in a socio-technical system. Energy Policy 74, 655–662. https://doi.org/10.1016/j.enpol.2014.08.013.

Gliedt, T., Hoicka, C.E., 2015. Energy upgrades as financial or strategic investment? Energy Star property owners and managers improving building energy performance. Appl. Energy 147, 430–443. https://doi.org/10.1016/j.apenergy.2015.02.028.

IEA, 2018. Buildings. Retrieved 14 February 2019, from, https://www.iea.org/topics/energyefficiency/buildings/.

Jafari, A., Valentin, V., 2017. An optimization framework for building energy retrofits decision-making. Build. Environ. 115, 118–129.

Klotz, L., Mack, D., Klapthor, B., Tunstall, C., Harrison, J., 2010. Unintended anchors. Building rating systems and energy performance goals for U.S. buildings. Energy Policy 38 (7), 3557–3566. https://doi.org/10.1016/j.enpol.2010.02.033.

Kontokosta, C.E., 2016. Modeling the energy retrofit decision in commercial office buildings. Energ. Buildings 131, 1–20. https://doi.org/10.1016/j.enbuild.2016.08.062.

Mayring, P., 2003. Qualitative Inhaltsanalyse. (Qualitative content analysis). Beltz (Weinheim Basel) 8, 5–135.

Michelsen, C.C., Madlener, R., 2013. Motivational factors influencing the homeowners' decisions between residential heating systems: an empirical analysis for Germany. Energy Policy 57, 221–233.

Nair, G., Gustavsson, L., Mahapatra, K., 2010. Factors influencing energy efficiency investments in existing Swedish residential buildings. Energy Policy 38, 2956–2963. https://doi.org/10.1016/j.enpol.2010.01.033.

Rogers, E.M., 2010. Diffusion of Innovations, fourth ed. The Free Press, New York.

Scott, M.G., McCarthy, A., Ford, R., Stephenson, J., Gorrie, S., 2016. Evaluating the impact of energy interventions: home audits vs. community events. Energ. Effic. 9, 1221–1240. https://doi.org/10.1007/s12053-015-9420-9.

Sorrell, S., Schleich, J., O'Malley, E., Scott, S., 2004. The Economics of Energy Efficiency: Barriers to Cost-Effective Investment. Edward Elgar, Cheltenham.

Stieß, I., Dunkelberg, E., 2013. Objectives, barriers and occasions for energy efficient refurbishment by private homeowners. J. Clean. Prod., Environmental Management for Sustainable Universities (EMSU) 2010. European Roundtable of Sustainable Consumption and Production (ERSCP). 48, 250–259. https://doi.org/10.1016/j.jclepro.2012.09.041.

Wilson, C., L. Crane, G., Chryssochoidis, G., 2015. Why do homeowners renovate energy efficiently? Contrasting perspectives and implications for policy. Energy Res. Soc. Sci. 7, 12–22.

Wilson, C., Pettifor, H., Chryssochoidis, G., 2018. Quantitative modelling of why and how homeowners decide to renovate energy efficiently. Appl. Energy 212, 1333–1344. https://doi.org/10.1016/j.apenergy.2017.11.099.

Further reading

Birk Jones, E.I.T., Bogus, S.M., 2010. Decision process for energy efficient building retrofits: The owner's perspective. J. Green Build. 5 (3), 131–146.

Reframing energy efficiency in industry: A discussion of definitions, rationales, and management practices

2.3

Jenny Palm[a], Patrik Thollander[b,c]

[a]*IIIEE, International Institute for Industrial Environmental Economics, Lund University, Lund, Sweden*
[b]*Department of Management and Engineering, Division of Energy Systems, Linköping University, Linköping, Sweden*
[c]*Department of Building, Energy and Environment Engineering, University of Gävle, Gävle, Sweden*

1 Introduction

The United Nation's 2030 Agenda for Sustainable Development emphasises that present needs are to be met without affecting future needs in a negative way. An important area is the energy sector and how energy not only is produced but also consumed.[a] Industry is, in this perspective, of vital importance since it is the highest energy-using sector in the world[b] (Dudley, 2018). The process of improving industrial energy systems is a cornerstone in climate change mitigation.

Within the European Union (EU), the European Commission (EC) has emphasised the need to transform the economy to become more sustainable in a cost-efficient way. The EC's low-carbon economy roadmap builds on the 2020 goals (EC, 2009) and sets the goal of at least a 27% improvement in energy efficiency

---

[a] We are aware that the terms "consumed" and "produced" are not in line with the laws of thermodynamics, but we use these terms in this chapter as they are commonly used.

[b] Depending on how the term industry is defined in different reports, the industry accounts for between 25% and 50% of global energy end use. In 2014, the industrial sector used 43,300 TWh/year, which represented 36% of the world's total energy end use (IEA, 2017). The world's most energy-intensive sectors are, according to the IEA (2017), chemicals and petrochemicals, iron and steel, cement, pulp and paper, and aluminium representing about 69% of the global energy end use (IEA, 2017).

Energy and Behaviour. https://doi.org/10.1016/B978-0-12-818567-4.00007-7

by 2030. All sectors must contribute, and the industry sector is, according to the EC, capable of cutting their emissions by more than 80% by 2050 (European Commission, 2018). Energy efficiency in industry is thus emphasised as a priority goal by the EU.

Energy efficiency also remains an important competitive edge amongst individual industrial companies, since it can cut costs and increase revenue. The dominating means of supporting energy efficiency has been via information provision of potential energy-efficient measures, for example, using energy audits and energy audit programmes and through in-house energy management. There is, however, an obvious risk that these measures alone will not be sufficient to contribute to the transformation of the system. A reduction of energy use in industry will not only require these organisations to improve energy efficiency through investments in more energy-efficient equipment but also require them to 'transform' their behaviour, values, and routines towards improved energy efficiency. However, these aspects are too often not considered in research on how to improve energy efficiency in industry.

The major model used to explain the discrepancy between optimal level of energy efficiency and the current level is the barrier model, that is, different barriers to energy efficiency inhibit the adoption of cost-effective measures (Cagno and Trianni, 2013; Decanio, 1998; Engelken et al., 2016; Sorrell, 2004; Thollander et al., 2010; Trianni et al., 2016; Trianni and Cagno, 2012). These barriers are technoeconomic oriented, and most research in the field is grounded in engineering science (Andrews and Johnson, 2016; Thollander and Ottosson, 2010; Thollander and Palm, 2015). Most measures outlined in research and in policy action plans are also dominated by a technological perspective. Lutzenhiser (2014) argues, for example, that the problematisation of organisational energy consumption and environmental performance has been dominated by the physical-technical-economic model (PTEM).

The models or underlying principles for improving energy efficiency have so far remained quite unchallenged. In this chapter, we aim, firstly, to outline the major underlying principles for energy efficiency and energy management in industry and, secondly, outline some important methodological implications that need to be regarded in the formulation of future models for promoting energy management and energy efficiency in industry.

We will explore how to develop an energy management perspective with theories emphasising social contexts and how local values and norms can influence decision-making. We will elaborate how a cross-pollination of these fields may contribute to furthering the existing knowledge and at the same time be usable for practitioners. We will discuss an alternative model for energy management including an understanding not only on how to lead an organisation in energy efficiency but also on how to implement sustainable values, routines, and behaviours in a way that energy efficiency becomes embedded in the industrial organisation. We will start by briefly discussing the need for interdisciplinarity, that is, the need to combine several perspectives when studying energy efficiency in industry.

2 Improved efficiency in industry requires an interdisciplinary approach

To be able to discuss energy efficiency in industry in a new way calls for the application of a range of perspectives and methods and the cross-pollination of different fields. This interdisciplinary approach requires some clarification.

Interdisciplinarity is the joint use and interaction of different knowledge areas to solve a problem. In interdisciplinary research, researchers cross boundaries to different disciplines, also often including the formulation of new problems and the development of new research fields. The researchers go beyond the questions and answers established within their discipline and form something new (Klein, 1990). There are different levels of interdisciplinarity. A lower degree is often called multidisciplinarity, where researchers from different disciplines work together without merging or altering their methods or assumption of the world (Coast et al., 2007). Transdisciplinarity is a common name for the next step, when there is a high integration of theories, models, and methods. This is not common and is probably more difficult because researchers need to understand more than their own paradigm.

We have, in our earlier research, aimed at transdisciplinary research combining perspectives from social science and engineering to consider, both empirically and theoretically, how an energy transformation of industry can be fostered (e.g. Palm and Thollander, 2010; Thollander and Palm, 2013, 2015). In our research, we have moved within and across the boundaries of various disciplines such as energy engineering, political science, and science and technology studies and assessed what we can learn from each of them in our discussions (compare also Axon et al., 2012; Moezzi, 1998; Lutzenhiser, 2010, 2014; Sanquist et al., 2010; Sovacool et al., 2015). This is done with the purpose to develop research on energy efficiency in industry in a way to complement existing theories by asking different questions and discovering new aspects of the problem. The purpose is not to criticise other research in the area, but to give another perspective and to ask different questions. We cannot cover all relevant theories or earlier literature in all fields, but we will focus on a few and discuss how combinations of them can benefit the advancement of knowledge on improved energy efficiency in industry.

3 Energy efficiency in industry in a technoeconomic framing

The understanding of energy efficiency in industry has traditionally been based on mainstream economic theory and Adam Smith's idea that governments should only intervene in a market when the so-called market failures exist. In relation to energy efficiency, two market failures are in the forefront: information asymmetries and imperfections. Lack of information (Sanstad and Howarth, 1994) and information asymmetries are examples of imperfect information. This means that there are imperfections and asymmetries associated with searching and acquiring information about energy-efficient solutions (Sorrell, 2004). An example of asymmetric information is adverse selection, that is, if, for example, a supplier knows more about the energy performance of machines than a purchaser, the purchaser may select the machine only in relation to some features such as the price (Jaffe and Stavins, 1994). A common policy recommendation to overcome these barriers is to implement an energy information programme such as an energy audit programme providing tailor-made specific information to companies (Thollander et al., 2010).

Another assumption dominating the energy efficiency discourse is the fully rational market actor. A fully rational actor will implement all energy-efficient measures that are technically feasible and financially profitable. This assumption, together with the model that policy should only be launched if a market failure exists, has resulted in an energy policy gap, that is, too few energy efficiency programmes in action (Backlund et al., 2012; Palm and Thollander, 2010; Thollander and Palm, 2013). A consequence of the rational actor assumption is that the nonuptake of energy-efficient measures will be interpreted as these measures are not perceived as cost-effective. Other possible explanations, such as the actors simply lacking information about energy-efficient measures or other issues occupying the mind of the company's staff, are not often considered. This rationality paradox implies that it becomes too easy to dismiss energy-efficient measures from governments based on applied scientific perspectives dominated by economic reasoning (Revell, 2007; Revell and Blackburn, 2007).

Following this technoeconomic paradigm, industrial companies are perceived to pay for energy efficiency in two occasions. One is when the energy prices are high, and the company sees a need to reduce their energy expenses to stay competitive. Higher energy prices will, however, affect industrial companies differently. Energy-intensive industries, like pulp and paper mills, are much more exposed to higher prices than nonintensive industries. But for all companies, increased energy costs will have a negative impact, while increased energy efficiency will positively directly affect a company's overall costs (Cagno et al., 2013; Worrell et al., 2003; Worrell and Price, 2001). The other occasion when a company should be prepared to pay for energy efficiency is in a global environment crisis, when energy efficiency may be the only mean left.

Energy efficiency has a risk of becoming neglected or, at least, not a priority. Decisions on energy efficiency are often based on costs and not related to, for example, environmental concerns or security of supply perspectives. But not even when industries benefit financially from improved energy efficiency, these measures are always implemented, as will be discussed later.

Before going into energy management practices, we will also say something about how a company can reason about energy efficiency, for example, taking the perspective of the company. In manufacturing industries, we have earlier discussed two basic means for them to improve energy efficiency (Thollander and Palm, 2013):

1. Focus on cost-efficient supply of energy through negotiating with energy suppliers or invest in own or external supply.
2. Adopt more efficient technologies, management practices, and behavioural measures.

Table 1 Four basic means of reducing industrial energy-related costs.

Means for energy efficiency	Comment
Adoption of energy-efficient technologies	Improved efficiency amongst technologies using energy
Load management	Reducing power costs by performing load shifting and load curtailment
Change of energy carriers	Changing energy carriers, for example, switching from electricity heating to heating performed by other energy vectors
Adoption of energy-efficient behaviour or energy conservation practices	Energy-efficient behaviour involving changing routines and more reflexive use of energy amongst staff at the industry

Based on Thollander, P., Palm, J., 2013. Improving Energy Efficiency in Industrial Energy Systems: An Interdisciplinary Perspective on Barriers, Energy Audits, Energy Management, Policies, and Programs, Springer, London.

In practice, we can see four basic means to reduce industrial energy costs, presented in Table 1.

Of the four means presented in Table 1, research shows that investment in energy-efficient technologies has been one of the most common measures used by industrial companies. In nonenergy-intensive industries, major investments have been made in support processes such as ventilation, heating, lighting, and compressed air (Fleiter et al., 2012; Thollander et al., 2007). For energy-intensive industries, the major technologies implemented are related to production processes (Paramonova and Thollander, 2016). In a novel attempt to classify energy-efficient measures as purely management or purely technology related, a study concluded that roughly 40% of implemented measures were management related while the remaining were technology related (Paramonova et al., 2015).

Conducting an energy audit (which will be further discussed later) is usually the first step when an industry wants to improve its energy efficiency. However, it also contributes to conserving a technological focus. Energy audits need to be complemented with energy management and other strategies to also cover behavioural issues. This will be discussed further in the succeeding texts.

4 Energy management in Small and Medium Enterprises (SMEs)

Energy management is strategic in its nature, and having an internal energy management system is one of the most important means an industry can have to overcome barriers to improve energy efficiency (Gordić et al., 2010; Thollander and Palm, 2013). In 2011, the new international energy management standard ISO 50001 (International Organisation for Standardization) came out. This standard was designed according to the plan-do-check-act (PDCA) cycle (Deming, 2000) and is similar to quality and environmental management standards (ISO 9001 and 14001). The relationship between the elements can be described as follows:

- **Plan**: Conducting an energy review; establishing the baseline; calculating energy performance indicators; and deciding on objectives, targets, and action plans
- **Do**: Implementation of the energy management action plans
- **Check**: Monitoring and measuring of processes and key characteristics of operations to determine energy performance in relation to objectives and report of results
- **Act**: Taking actions for continual improvement of energy performance

The ongoing adoption of ISO 50001 indicates, however, that a strategic view on these issues has become of increased interest for companies.

4.1 Energy audits: A first step to introduce an energy efficiency strategy

An energy audit is a typical first step when initiating energy management practices or related policy programmes. Traditionally, energy audits have had a technical approach performed by professionals with a background in engineering (Fleiter et al., 2011). In general, an energy audit investigates the energy used by a company's support and production processes. Support processes relate to the company's manufacturing of products. Examples of support processes are ventilation, heating, and compressed air. Production processes are linked to the actual manufacturing of products, for example, melting and heating. For capital- and energy-intensive industries, such as the paper industry, where a paper machine costs hundreds of million euros, a change in the production process is complicated. In these industries, a change in the support system, such as upgrading lighting, is a much more appealing alternative.

In general, an energy audit covers both a company's support and production processes. An energy audit can last from a single day to months to accomplish. Two international standards are now in place for energy audits, ISO 50002 and the European standard for energy audits EN 16247. Energy audits according to EN 16247 generally include the following main steps:

1. Preliminary contact
2. Start-up meeting
3. Collecting data
4. Field work
5. Analysis
6. Report
7. Final meeting (EN 16247)

One of the largest energy programmes for industries is the American Information Assessment Center's (IAC) programme. Since 1976, more than 10,000 SMEs have participated. These companies have been offered energy audits, and the results show that half of the suggested energy-efficient measures were implemented.

The reason for not implementing the suggested measures was that they were perceived as economically undesirable. Evaluation of energy audit programmes reveals that 60%–90% of the implemented measures from industrial SMEs were allocated to the support processes (Anderson and Newell, 2004; Harris et al., 2000; Thollander et al., 2007).

Norms are clear and state that reports need to be presented to the top management. The actors most often in charge of energy issues and the receivers of an audit report are the energy managers or environmental managers. In industrial SMEs, the responsible person for the energy issues might well also be the CEO or the production manager. The CEOs usually have a general knowledge of the industrial processes of the company, and the production managers have detailed knowledge of the production process. Financial managers are responsible for providing financial guidance, so they can make sound business decisions. The energy manager and the financial manager can both assess potential cost-effective energy efficiency investments. Further, the production manager is normally the one closest to the actual implementation of a measure. The production manager has control over energy efficiency through the production processes and s/he decides *how* a measure will be implemented. The same holds for the maintenance manager, even though the latter often may have a more risk-averse approach than the production manager. For the maintenance manager, no production disruption is the highest priority, since that lead to the loss of revenue. Further, the owner and the CEO are in charge of the company's overall strategy, which imply they have a word to say in relation to the energy efficiency issues. Then, we have the staff, for example, process and technology operators, that work in the production, whose role in implementing energy-efficient measures is a vital process, be them related to technology. When energy efficiency is not the result of technology improvement, it means that it is the result of a change in the process, the infrastructure, or the individuals' behaviours. All these roles are example of actors that can be involved in decision-making on energy efficiency in a company.

The most common outcome of an energy audit is a technical report including possible energy-efficient measures. The result from this practice is that aspects such as behaviour of the staff, routines, culture, and knowledge formation and decision-making processes become less focussed even if they can be of immense importance for improved energy efficiency in a company. Another problem discussed by Backman (2018a,b) is that most energy audits are made in SMEs where the receiver of the programme lacks the same background and knowledge as the expert conducting the energy audit. Energy and environmental literacy are low amongst SMEs (Coles et al., 2016), and many fail to take up cost-effective opportunities for energy savings (Revell and Blackburn, 2007). It is difficult for the SMEs to interpret and understand the energy audit, with the most probable outcome that the energy-efficient measures presented in the report will not be implemented in practice. The well-known energy efficiency gap (Backlund et al., 2012; Blumstein et al., 1980) is substantiated by this communication deficit of the energy audit practice. Conducting an energy audit is one practice that needs to be supported by another practice focussing on how to communicate the results to actors with a totally different composition of knowledge, experiences, and interests (Carrico and Riemer, 2011). However, so far, this has not been discussed to any larger extent in the energy audit discourse.

Asymmetric and imperfect information are often discussed as major barriers when trying to process complicated and (for the receiver) often unfamiliar data (Golove and Eto, 1996; Henriques and Catarino, 2016). *Communication* as a barrier is seldom discussed in this context. The companies are informed about the results from energy audits, but it is often a one-way communication (Backman, 2018a,b), rather than a process characterised by dialogue and a discussion around the results.

The companies are informed about the results, and then, it is up to them to interpret the results and find out how to implement different measures. Further, the underlying model for understanding energy efficiency, as stated earlier, lies in the fact that information of available cost-effective energy-efficient measures is needed, while knowledge on how to understand this information is overseen, that is, not included in the underlying model.

It is, however, difficult to capture people's attention and exert any impact by just providing information. Information needs to be adapted by peoples' experiences, values, and knowledge. Information is applied in the minds of knowers, as Davenport and Prusak (1998) put it. Knowledge is not just flowing unchanged between a sender and the receiver. That is also why social interaction benefits from information exchange. Through a dialogue, it is possible to negotiate and interpret the information in an energy audit (Noorderhaven and Harzing, 2009). This is less discussed within research on energy audits.

Research by, for example, Cooremans and Schönenberger (2017) has tried to open up the black box of energy efficiency decision-making showing that it is a process, starting with an initial idea. Further, for process industry, measures may only be undertaken if one has extensive knowledge about the process, meaning that an energy audit programme where an external expert is brought in to propose (mostly technical) measures has a lower effect compared with when in-house experts are used. One challenge to overcome is then that a process industry might need both a person with competences on how the process is run and a person with knowledge on technical dimension of energy efficiency and in the decision-making; otherwise, this energy efficiency potential may remain hidden to the company.

How information is perceived and understood is also dependent on the social context.[c] When an energy audit is presented, for example, a management team, it opens up for different interpretations. We have what Star and Griesemer (1989) call an interpretive flexibility. Information is not simply reextracted in its new context, and most likely the information will be interpreted by the ones taking part of the information, which includes a process of transformation of the original meaning. The information will be embedded in the receivers existing knowledge, experiences, and values (Hildreth and Kimble, 2002). That is also why an energy audit needs to be communicated and accompanied with a dialogue. Another aspect deserving more attention in this research field is that information translated into personal knowledge not necessarily will be transformed into behaviour (Aronson and Stern, 1984).

---

[c] Another word for social context is milieu, which is someone's immediate social or physical environment.

There are examples where governments' funded energy programmes have included educational and training programmes for SMEs, which contribute to translate information to personal knowledge (Anderson and Newell, 2004; Backman, 2018a,b; Väisänen et al., 2002). Thollander et al. (2007) found that local energy programmes, which start with the needs of individual SME, were more successful than those trying to approach SMEs with general information. There are also several studies indicating the implementation of suggested measures benefits from targeted information and the existence of practice-oriented networks empowered by the management to improve energy efficiency within the company (Backman, 2018a,b; Koewener et al., 2011, 2014; Thollander and Palm, 2015).

Viktorelius (2017) presented an alternative approach to energy efficiency in shipping, experimenting with combining energy efficiency with social practice theory and cultural–historical activity theory. The theoretical approach contributed to uncover how energy efficiency incorporates various forms of knowledge and how energy efficiency is interwoven in processes by which organisations' identity was constructed. This kind of experimentation is an interesting way forward to enhance energy efficiency in SMEs.

4.2 Energy management is important for improving energy efficiency in SMEs

If a company has someone committed to work on energy efficiency and a long-term energy strategy is in place, the company's possibilities to achieve improved energy efficiency increase (Bull and Janda, 2018; Thollander et al., 2009). An early study by Caffal (1995) found that companies strategically working with energy management practices reduced their total energy use by 40%. These companies usually had a strategy including energy audit, senior management support, and energy use monitoring and a clear commitment from the company's management team.

Johansson and Thollander (2018) conducted a review of studies on barriers, drivers, and energy management and identified the following success factors:

1. Full top-management support of the energy management activities
2. The existence of a long-term energy strategy, preferably with quantified goals
3. A two-step energy plan: one covering one-year and one covering multi-year periods
4. A clear energy manager position, but not necessarily a full-time position
5. Real energy cost allocation based on submetering, not just energy costs allocated per square metre or number of employees
6. Clear key performance indicators (KPIs) that enable follow-up of results
7. Energy controllers at floor-level position, that is, one person per shift responsible for energy-efficient operation
8. Continuous energy efficiency education for employees
9. Visualisation of energy efficiency progress at company level and at division level
10. Energy competition between divisions within the same company, encouraging improved energy efficiency

In earlier research, the main focus has been to find concrete investments that would contribute to improved energy efficiency. Technological and economic factors are well analysed. Factors related to staff education and motivation are dealt with, but these factors are less elaborated upon. Organisational routines and practices are usually dealt with in the context of energy management systems. As mentioned earlier, energy management systems are based on the PDCA cycle (Deming, 2000), and earlier research is often referring back to Deming's cycle (Thollander and Palm, 2013). But it is possible to discuss these results also in a broader context. Later, we will relate these earlier findings on organisational roles, motivations, etc. and the definition of knowledge, comprising the three parts, episteme, techne, and phronesis (Flyvbjerg, 1992, 2001).

5 Broadening the view of energy efficiency in SMEs by Aristotle's three forms of knowledge

Aristotle defined various forms of knowledge. Three of the more common definitions are episteme, techne, and phronesis. Episteme knowledge means 'to know' and relates most often to scientific knowledge. Techne is defined as a form of knowledge related to craftsmanship. Phronesis is defined as a form of wisdom (Flyvbjerg, 2004; ISO 19011:2011).

While general information can be transferred, for example, via one-way communication in a big auditorium to create episteme knowledge, techne knowledge is rather the opposite. This form of knowledge, when translated as craftsmanship, demands a mentor who, at least initially, provides feedback on the welds or other technical installations attempted to be created. This requires a close relationship between the mentor and the apprentice and a learning process extended over time. Techne knowledge improves over time and is gained through years of practice, where mistakes and feedback from more skilled persons help the individual to improve. Phronesis, the third form of knowledge, often translated as 'wisdom' might come with practically no or only minor episteme and techne, but with years of experience. Phronesis as knowledge is good for a leader to have, and this knowledge tends to come with age. Phronesis knowledge can, however, be acquired also by young leaders with years of practice in social environments such as school, sports, and family life. As with the other forms of knowledge, phronesis is also improved via training or practice.

Managers usually have excellent episteme knowledge (general know-how of the company and its divisions, functions, etc. and often a master degree in business administration or equivalent). However, they often have lower techne knowledge (practical technical know-how on how the production processes work and are operated). The manager therefore needs staff in the team with low episteme and high techne knowledge to embrace an energy efficiency programme.[d] Launching an

---

[d] Energy efficiency programs usually include identification of opportunities for technology improvement, generally through energy audits or other technical assistance, and direct financing or other implementation facilitation of identified opportunities.

energy management programme could be seen as equivalent to an entrepreneur or a company launching a new product to the market. It involves a certain degree of uncertainty where the lack of control from the initiator (the manager) is apparent and s/he needs to have trust in the staff, with the capacity (techne knowledge) to actually carry out improvements. Research on energy efficiency in industry clearly shows that when aiming to achieve system and operational improvements, deep techne knowledge of the production processes is needed. Earlier research has, however, not analysed the need for these different kinds of knowledge when improving energy efficiency but has used a black box view on education as one unanimous factor. Just acknowledging that improvements in energy efficiency relate to all three types of knowledge defined by Aristotle (techne, episteme, and phronesis) could contribute to new questions and insights.

There is also a lean in earlier studies towards finding standardised solutions suitable for all companies. There are some authors discussing the problems with finding a solution that fit all companies (Christoffersen et al., 2006; Russell, 2005). But there are not many studies moving beyond a technoeconomic perspective towards a less mechanistic view of decision-makers (e.g. engineers and economists, as fully rational). Later, we will discuss some more complementary perspectives that would enrich the energy management field simply by bringing in some other rationale for the analysis.

6 'Lifestyle' categorisation of companies

Earlier studies of energy efficiency in industrial enterprises most often identify and discuss barriers. In these studies, the focus is categorising barriers rather than companies. Another research field is energy efficiency in households. A common theme in this field is to categorise households in different lifestyle categories and discuss energy consumption and energy-saving potential for each of these lifestyle categories (Hallin and Petersson, 1986; Aune, 1998, 2007; Palm and Eriksson, 2018). In a study, Palm (2009) explored what would happen if the analysed model usually used for households was used in relation to companies. Based on in-depth studies of 10 companies in Borås, Sweden, four lifestyle categories were developed. In the research project, the 10 studied companies received an energy audit, and after around 6–12 months, these audits were followed up by a visit and semistructured interviews. From this material, the four lifestyle categories were developed: the *ignorant* company, the *implementer of easy measures*, the *economically interested* company, and the *innovative environmentalist*.[e] These categories are further described in the succeeding texts.

6.1 The ignorant company

The ignorant company had no focus on energy efficiency and lacked anyone working on these issues. A typical statement was

---

[e] This section is based on Palm (2009).

Energy is something that I handle in addition to my main tasks, but I am trying to integrate it into other issues. It is not a prioritised issue, however.

In general, the company lacks incentives to focus on energy issues, and the company's customers do not impose any demands in relation to energy, which one respondent meant was the main reason for not engaging in energy efficiency:

We cannot really earn anything from it [energy efficiency]. Because we deliver to retailers and not directly to consumers, we don't have the same pressure as other industrial companies regarding these issues. It is up to the next level in the production line to make environmental demands.

Some companies meant that they lacked enough knowledge, which they felt was a barrier to be more active in relation to energy efficiency. Another company was in a process of establishing on the market and felt they needed to focus on their core activity. In summary, ignorant companies lack forces driving them to focus on energy issues.

The ignorant company has no special focus on energy efficiency and lacks someone responsible for energy. Energy is seen as an unimportant issue, energy costs are low, and their customers do not demand any action in the area.

6.2 The implementer of easy measures

This company has knowledge about the cheap and easy measures to reduce energy. They had engaged in picking the low-hanging fruits. As one company explained, 'We have looked at ventilation and lighting ourselves, so that parts are efficient.… It is hard to move on'. These companies focus on measures with a quick payback, and they are reluctant to take a longer planning perspective on energy issues:

We will absolutely, as fast as possible, implement the easily captured things. The more stubborn and harder ones we will do if we have time and the strength /…/ but they are not priority for the moment.

The implementer of easy measures: This company is quite aware of existing cheap and easy measures and has already implemented those. It is fairly satisfied with its energy-efficient activities and is convinced that only expensive and complicated measures remain to be implemented in the organisation. It will not take any action without incentives such as subsidies.

6.3 The economically interested company

Similar to the companies that invest in easy measures, these companies have strong focus on payback time. All investment made need to give economic benefits, and this is expressed in different ways as exemplified in the succeeding text:

The investments must bear their own costs. A suitable pay-back time is two years.

We see it [energy efficiency] simply as the potential to save money. That is the driving force of the company, to become cost efficient.

Measures with longer payback time like 5 years did not interest these companies. The only other reason for those companies to engage in energy-efficient activities is if the law requires it.

The economically interested company: This company views reduced energy consumption and improved energy efficiency as means to cut costs. It has strong focus on payback time, and investment is seen as a means to give direct and calculable economic benefits for the company.

6.4 The innovative environmentalist

This group is well aware of both energy and environmental issues in general and had for long worked successfully on these issues. These companies have set goals for reducing their energy consumption and formulated action plans for different parts of the process. These companies often experienced a lack of good solutions on the market and felt a need to innovative new ways to reduce energy:

I have found a special solution for the air compressors, so that they turn off automatically after 10 minutes when not in use. Previously, they were running 365 days a year, so I needed to make this improvement.
And there seems to be no end of new possibilities to innovate:

We can do many more things, I believe. For example, we do not know how to use the heat that results from our processes. It would be great to be able to store the heat and redistribute it in winter.

To summarise, innovative environmentalist companies have great environmental concern and have implemented both easy and more complicated and costly measures.

The innovative environmentalist: This group is well aware of both energy and environmental issues, in general, and has worked successfully on these issues for some time. Often, it has an enthusiast who is a driving force for these issues. Energy efficiency is not seen as a problem, but a challenge. It faces customers who require the company to take into account environmental concerns, and its managers are supportive of all kind of environmental activities.

The contribution from bringing in a lifestyle thinking to the energy management field is that it adds a shift in perspective and interpretation of how energy efficiency can be understood. In studies using lifestyle categories as an analytical tool, the emphasis is more on contextual factors. How technology is used depends on culture and routines and how a practice is performed. This gives a slightly different perspective to energy efficiency and starts a reflection on how everyday activities are performed in mundane industrial processes. This gives way for new interpretation of energy use in industry and how to approach the improvement of energy efficiency. Energy policies targeting the ignorant company should, for example, differ from policies targeting the innovative environmentalist. Changing theoretical glasses to lifestyle glasses also highlights that other issues than energy are involved and it is important to study energy use in its everyday context. This takes us to another perspective often discussed in energy management literature—decision-making.

7 Embedded decision-making and situated action

Decision-making on energy efficiency in companies is most often not analysed per se, but it is assumed that decisions are made by more or less fully rational market actors. When decision-making is not rational in the sense that it was not the most energy-efficient option that was chosen even if it was cost-effective, the explanations used are related to imperfect markets and barriers.

There are, however, alternative ways to understand why a certain decision is made. Janda (2014) emphasises that firms and organisations do not behave like individuals (compare also with March, 1988). Instead, companies exhibit their own dynamics that can explain the lack of investments in energy efficiency (Janda, 2014). To not choose the most energy and cost-efficient technology can be very rational if one just takes another perspective into account. March and Olsen (1989) discuss that people's action are governed by experience-based decision rules. Institutional identity and standardised procedures decide what is seen as acceptable behaviour in a situation and influence the choices that are made. Institutions (formal rules and laws and informal such as conventions, routines, and roles) influence individuals' ways of thinking (Bedwell et al., 2014). An actor decides in relation to what situation s/he is in, what role s/he is being fulfilled, and what obligations come with this specific role. All decisions are embedded in that exact moment and in that specific context where an actor is situated when deciding.

Yet, another way to understand why a company works towards improving energy efficiency in a certain way is to analyse the informal networks existing in all organisations. Actors act and decide in a certain context wherein they consciously or unconsciously are influenced by the values and norms established in a group. This is analysed in literature using a community of practice (CoP) perspective (Wenger et al., 2002). Wenger (1998) states that an organisation is a social construct consisting of various constellations, or communities, of practice. Wenger et al. (2002) define a CoP as a set of people who 'share a concern, a set of problems, or a passion about a topic, who deepen their knowledge and expertise in this area by interacting on an ongoing basis' (Wenger et al., 2002, p. 4). It is an informal grouping that is defined by its members. A CoP includes features such as well-defined roles, a common language, tools, informal rules, symbols, embedded understandings, and tacit agreements (Palm and Törnqvist, 2008). For most people, learning takes place amongst and through other people, in informal networks (Argote and Ingram, 2000; Retna and Ng, 2011; Tennant, 1997). To become a member of a CoP, it is necessary to absorb the communities' unspoken knowledge and learn the informal procedures that help the members to know how to act in a certain situation. It is within the CoP that useful tips and anecdotes, not found in any formal documents, are exchanged (Retna and Ng, 2011), also those related to energy efficiency.

Suchman (1987) has discussed decision-making in organisations in a situated action perspective, which has many similarities to the CoP perspective. If taking Suchman's perspective, decisions on energy efficiency are made locally, in the practices where people meet, act, and perform. Taking this perspective to its extreme, all decisions would be constructed in the exact moment they are made, and therefore, the outcome would be impossible to predict. Suchman does not go so far, but we consider it an interesting thinking experiment in this context, and it could be a useful way to contrast mainstream analysis within energy management. It would lay the ground for new questions and alternative explanations to the existence of an energy efficiency gap.

Consider a situation when a decision will be made concerning energy efficiency in a company; from an energy management perspective, the outcome, to a high degree, will depend on existing policies, decided energy efficiency goals, and established procedures. From a situated action perspective, the outcome of the same meeting would be much more of an open issue. Rather than depending on existing policies and goals, it will depend on the roles and individuals that were at the meeting. The actors attending the meeting will most likely not have memorised the company's policies, standards, and goals. The actors are more likely to contribute to the meeting in relation to the role they have in the organisation, their culturally embedded understanding of how to act, and what decision seems to be suitable in this specific setting. In a situated action perspective, the outcome of the meeting will be dependent on which actors participated on that occasion, while in an energy management perspective, the outcome will depend on existing policies, standards, and procedures. The 'truth' probably lies somewhere in between our two extreme interpretations, but what we want to highlight is that what theoretical lenses we have when studying energy efficiency in industry will very much decide what problems we detect and how we interpret the decisions. To use different perspectives can therefore add to a broader understanding of decisions on energy efficiency, covering more aspects than usually identified as important.

Another aspect that is less recognised in energy management literature is that decisions are made in the context of group dynamics in a meeting and the participating actors' mutual relationships. Individuals have or take on different roles in a group. The group context influences how a person acts and what is said (Deline, 2015). For this reason, it is not at all odd that an actor can appear to have different and sometimes opposing opinions when it comes to energy efficiency. The opinion can be related to a role, and when the individual change roles also, his or her opinion might need to change. Therefore, an opinion or a statement needs to be contextualised and related to the situation within which it is expressed. This is also why an energy-efficient measure can be interpreted as beneficial and valid in one situation while in another situation the same measure can be dismissed as inappropriate (Ryghaug and Sørensen, 2009).

Over time, certain ideas, ways of thinking, and ways of acting tend to converge into a harmonised social construction (Palm and Thollander, 2010). This is often criticised from a social science point of view. At the same time, convergence and harmonisation of norms and actions are not only negative, rather the opposite. Organisations need harmonised understandings of how to behave to function. It is not possible to negotiate and come to an agreement in all thousands of mundane decisions that are made every day.

A critique that is possible to raise in relation to social science studies is that they too often advocate changes in social norms and social constructs without delivering an alternative way of thinking. The problematisation is the goal, which leaves the stakeholders with no real clue how to continue a process.

One important thing to bear in mind is that the 'transformer' of the phronesis knowledge and culture in general and the common notions of perceived 'truths', social constructions, are quickly released by knitting close relationships with the other individuals of the group. One example of such initiative was when Volvo CE in 2013 took a rather advanced approach towards improving energy efficiency by adopting principles of sound Lean energy management practices (Wollin et al., 2016). The programme was launched with an emphasis towards energy efficiency with no investments. The action implemented led to cuts in idle loads for Volvo CE globally of more than 40% (Wollin et al., 2016). Based on embracing this concept or social construct, a target on idle load for the whole company group was set.

8 Discussion: Combining different perspectives for successful energy management

When discussing energy efficiency in industry, it is important to have a holistic view and include more aspects than just finance and technology in the discussions. The user phase including actors, various forms of knowledge, routines, and behaviours are also important. The management level is important, and the support from a company's management team is essential when promoting energy efficiency. All decisions are also made in a context and embedded in structures. We mean that in earlier research on energy efficiency, there has been a domination of a top-down perspective where a restricted number of factors have been included in the analysis. As Lutzenhiser (2014) emphasised, the PTEM paradigm is epistemologically narrow, downplaying questions relating to how and why energy is used by organisations. If widening the perspectives, more factors will be included, and the number of possible explanations will increase. No perspective is necessarily more correct than another; it simply reveals different dimensions and possible explanations to an issue (Björkman et al., 2016; Cooremans and Schönenberger, 2017; Bull and Janda, 2018). Continuing further along this line of argument, a simple, yet enhanced, model for understanding energy efficiency in general and energy management practices in particular may be launched.

Fig. 1 shows an alternative model on how to initiate transformation in an organisation. In this model, the legislation and the energy management system are included but not in the centre, but knowledge, communication, and activities/behaviours are focussed. Legislation and energy management systems frame the individual's behaviour and incentives and restrict actions. CoPs influence individual's values and norms, which also influence how the employees will behave. If energy efficiency is demanded by legislation and/or the management system and captured by a CoP, a transition process has potential to start. Pressure will be put on energy managers and process operators. Energy managers with phronesis knowledge (wisdom and leadership) might, by communicating fundamental principles of group dynamics and cultural change, change a group's perception to improve energy efficiency. Techne knowledge of process operators and staff are also critical in making improvement happen; moving from information to communication and from information to focussing knowledge in the model emphasises that successful energy management relies

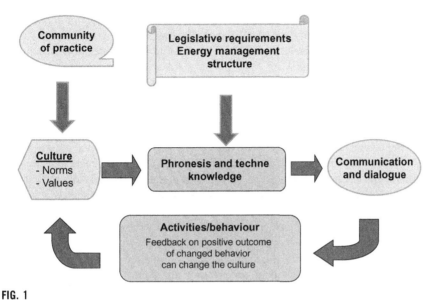

FIG. 1

A model for how energy efficiency processes in SMEs can be understood.

on phronesis and techne knowledge as well as communication skills. Good leadership becomes essential, which relates to earlier research showing that employees with real ambition is a major driver for improved energy efficiency in industry (compare Bull and Janda, 2018). Sometimes, this driver is even ranked in parity or higher than the cost reductions resulting from reducing the energy use driver (Thollander and Palm, 2013).

9 Conclusions

Energy efficiency problems are multifaceted and should be approached accordingly. Taking the starting point that there are technical and social reasons why optimal energy-efficient measures are not being implemented in industry should prompt us to formulate new questions, be open-minded to new and sometimes opposing perspectives and theories, and try out different and sometimes also provocative ideas to reach effective solutions.

This chapter discussed the need to further develop the research field of energy efficiency in industry by cross-pollinating existing research with perspectives from social sciences. We think that energy management in industry is one very important factor to improve energy efficiency, but that is underdeveloped in both research and practice. We mean that the energy management approach would benefit from a more interdisciplinary approach allowing for alternative ontological approaches to complement the research. We have discussed that the field could benefit from including other research perspectives and include theories focussing on decision-making in situated action and in social networks. By incorporating, for example, situated action and CoP in energy management models, it is possible to give the models another framing, as discussed earlier.

We think it is not possible to find one single model for energy management suitable for all companies, as also argued by Christoffersen et al. (2006). An alternative approach is to find a reflexive model, that is, a model that can be adjusted in relation to the local context will be used. We have suggested one model where successful energy management practices include a leader with phronesis knowledge (wisdom), who is able to communicate values and changes needed to improve energy efficiency. The idea is that a change in the factors emphasised in the model leads to a shift in the focus, which gives other connotations. These connotations might invite new actors to reflect on energy efficiency, which can bring in new ideas and solutions.

Recent findings often indicate a quite high potential for energy management systems to contribute to the improvement of energy efficiency in the industry. So far, less attention has been given to the theoretical background of those systems. Despite the potential of energy management systems, it is also stated in earlier research that the systems are sensitive with respect to how individuals act when implementing these systems and how the implementation processes fold out. Energy management systems will not alone be able to fully explain the outcome. The energy management system perspective thus benefits from adding theories focussing communication, behaviour, culture and social interaction, etc.

A broadened perspective on energy efficiency industry is required including more factors than just technical and economical ones. It calls for the application of a range of perspectives, theories, and methods to be able to capture, understand, and develop how existing formalised energy management systems, such as the ISO standards and tools, are translated and changed in industrial organisations. We have touched upon some possible perspectives to be integrated, but there are of course more theories, models, and aspects to consider in future research. Improved energy efficiency is done in situated practices. It is by integrating perspectives and applying a system perspective on energy efficiency that we will be able to achieve the UN 2030 Agenda targets discussed in the introduction to this chapter.

References

Anderson, S.T., Newell, R.G., 2004. Information programs for technology adoption: the case of energy-efficiency audits. Resour. Energy Econ. 26, 27–50. https://doi.org/10.1016/j.reseneeco.2003.07.001.

Andrews, R.N.L., Johnson, E., 2016. Energy use, behavioral change, and business organisations: reviewing recent findings and proposing a future research agenda. Energy Res. Soc. Sci. 11, 195–208. https://doi.org/10.1016/j.erss.2015.09.001.

Argote, L., Ingram, P., 2000. Knowledge transfer: a basis for competitive advantage in firms. Organ. Behav. Hum. Decis. Process. 82, 150–169. https://doi.org/10.1006/obhd.2000.2893.

Aronson, E., Stern, P.C., 1984. Energy Use: The Human Dimension. W. H. Freeman and Company, New York.

Aune, M., 1998. Nøktern eller nyttende: Energiforbruk og hverdagsliv i norske husholdninger. Senter for teknologi og samfund. Norges teknisk-naturvidenskapelige universitet, Trondheim, Norway.

Aune, M., 2007. Energy comes home. Energy Policy 35, 5457–5465. https://doi.org/10.1016/j.enpol.2007.05.007.

Axon, C.J., Bright, S.J., Dixon, T.J., Janda, K.B., Kolokotroni, M., 2012. Building communities: reducing energy use in tenanted commercial property. Build. Res. Inf. 40, 461–472.

Backlund, S., Thollander, P., Palm, J., Ottosson, M., 2012. Extending the energy efficiency gap. Energy Policy 51, 392–396. https://doi.org/10.1016/j.enpol.2012.08.042.

Backman, F., 2018a. Energy efficiency in Swedish SMEs: exploring barriers, knowledge creation and the role of municipal energy efficiency programs. Linköping University, Department of Thematic Studies, Linköping.

Backman, F., 2018b. Local knowledge creation with the use of industrial energy efficiency networks (IEENs): a Swedish case study. Energy Res. Soc. Sci. 42, 147–154. https://doi.org/10.1016/j.erss.2018.03.027.

Bedwell, B., Leygue, C., Goulden, M., McAuley, D., Colley, J., Ferguson, E., Banks, N., Spence, A., 2014. Apportioning energy consumption in the workplace: a review of issues in using metering data to motivate staff to save energy. Technol. Anal. Strateg. Manag. 26, 1196–1211. https://doi.org/10.1080/09537325.2014.978276.

Björkman, T., et al., 2016. Energy management: a driver to sustainable behavioural change in companies. In: ECEEE Industrial Summer Study Proceedings. pp. 379–387.

Blumstein, C., Krieg, B., Schipper, L., York, C., 1980. Overcoming social and institutional barriers to energy conservation. Energy 5, 355–371. https://doi.org/10.1016/0360-5442(80)90036-5.

Bull, R., Janda, K.B., 2018. Beyond feedback: introducing the "engagement gap" in organisational energy management. Build. Res. Inf. 46 (3), 300–315. https://doi.org/10.1080/09613218.2017.1366748.

Caffal, C., 1995. Energy management in industry. Centre for the Analysis and Dissemination of Demonstrated Energy Technologies (CADDET). Anal. Ser. 17.

Cagno, E., Trianni, A., 2013. Exploring drivers for energy efficiency within small-and medium-sized enterprises: first evidences from Italian manufacturing enterprises. Appl. Energy 104, 276–285.

Cagno, E., Worrell, E., Trianni, A., Pugliese, G., 2013. A novel approach for barriers to industrial energy efficiency. Renew. Sustain. Energy Rev. 19, 290–308. https://doi.org/10.1016/j.rser.2012.11.007.

Carrico, A.R., Riemer, M., 2011. Motivating energy conservation in the workplace: an evaluation of the use of group-level feedback and peer education. J. Environ. Psychol. 31, 1–13. https://doi.org/10.1016/j.jenvp.2010.11.004.

Christoffersen, L.B., Larsen, A., Togeby, M., 2006. Empirical analysis of energy management in Danish industry. J. Clean. Prod. 14, 516–526.

Coast, E.E., Hampshire, K.R., Randall, S.C., 2007. Disciplining anthropological demography. Demogr. Res. 16, 493–518. https://doi.org/10.4054/DemRes.2007.16.16.

Coles, T., Dinan, C., Warren, N., 2016. Energy practices among small- and medium-sized tourism enterprises: a case of misdirected effort? J. Clean. Prod. 111, 399–408. https://doi.org/10.1016/j.jclepro.2014.09.028.

Cooremans, C., Schönenberger, A., 2017. Energy management: a key driver of energy-efficiency investment?, in ECEEE Summer Study Proceedings, 221231.

Davenport, T.H., Prusak, L., 1998. Working knowledge: How organisations manage what they know. Harvard Business School, Boston, MA.

Decanio, S.J., 1998. The efficiency paradox: bureaucratic and organisational barriers to profitable energy-saving investments. Energy Policy 26, 441–454. https://doi.org/10.1016/S0301-4215(97)00152-3.

Deline, M.B., 2015. Energizing organisational research: advancing the energy field with group concepts and theories. Energy Res. Soc. Sci. 8, 207–221. https://doi.org/10.1016/j.erss.2015.06.003.

Deming, W.E., 2000. Out of the Crisis. MIT Press, Cambridge, MA.

Dudley, B., 2018. BP Energy Outlook 2018.

EC, 2009. Directive 2009/28/EC of the European Parliament and of the Council of 23 April 2009 on the promotion of the use of energy from renewable sources and amending and subsequently repealing Directives 2001/77/EC and 2003/30/EC.

Engelken, M., Römer, B., Drescher, M., Welpe, I.M., Picot, A., 2016. Comparing drivers, barriers, and opportunities of business models for renewable energies: a review. Renew. Sustain. Energy Rev. 60, 795–809. https://doi.org/10.1016/j.rser.2015.12.163.

European Commission, 2018. 2050 Low carbon economy.

Fleiter, T., Gruber, E., Eichhammer, W., Worrell, E., 2012. The German energy audit program for firms—a cost-effective way to improve energy efficiency? Energ. Effic. 5, 447–469.

Fleiter, T., Worrell, E., Eichhammer, W., 2011. Barriers to energy efficiency in industrial bottom-up energy demand models—a review. Renew. Sustain. Energy Rev. 15, 3099–3111. https://doi.org/10.1016/j.rser.2011.03.025.

Flyvbjerg, B., 1992. Aristotle, foucault, and progressive phronesis: outline of an applied ethics for sustainable development. Plan. Theory 7–8, 65–83.

Flyvbjerg, B., 2001. Making social science matter: Why social inquiry fails and how it can succeed again. Cambridge University Press.

Flyvbjerg, B., 2004. Phronetic planning research: theoretical and methodological reflections. Plan. Theory Pract. 5 (3), 283–306.

Golove, W.H., Eto, J.H., 1996. Market barriers to energy efficiency: a critical reappraisal of the rationale for public policies to promote energy efficiency (No. LBL-38059). Berkeley, CA.

Gordić, D., Babić, M., Jovičić, N., Šušteršič, V., Končalović, D., Jelić, D., 2010. Development of energy management system—case study of Serbian car manufacturer. Energ. Conver. Manage. 51, 2783–2790. https://doi.org/10.1016/j.enconman.2010.06.014.

Hallin, P.O., Petersson, B.A., 1986. De glömda aktörerna. Energiforskningsnämnden (Efn), Stockholm.

Harris, J., Anderson, J., Shafron, W., 2000. Investment in energy efficiency: a survey of Australian firms. Energy Policy 28, 867–876. https://doi.org/10.1016/S0301-4215(00)00075-6.

Henriques, J., Catarino, J., 2016. Motivating towards energy efficiency in small and medium enterprises. J. Clean. Prod. 139, 42–50. https://doi.org/10.1016/j.jclepro.2016.08.026.

Hildreth, P.M., Kimble, C., 2002. The duality of knowledge. Inflamm. Res. 8, 1–18.

IEA, 2017. Tracking Clean Energy Progress 2017 Energy Technology Perspectives 2017 Excerpt Informing Energy Sector Transformations.

ISO 19011:2011, 2011. Guidelines for auditing management systems (ISO 19011:2011)

Jaffe, A.B., Stavins, R.N., 1994. The energy-efficiency gap: What does it mean? Energy Policy 22, 804–810. https://doi.org/10.1016/0301-4215(94)90138-4.

Janda, K.B., 2014. Building communities and social potential: between and beyond organizations and individuals in commercial properties. Energy Policy 67, 48–55. https://doi.org/10.1016/j.enpol.2013.08.058.

Johansson, M.T., Thollander, P., 2018. A review of barriers to and driving forces for improved energy efficiency in Swedish industry—recommendations for successful in-house energy management. Renew. Sustain. Energy Rev. 82, 618–628. https://doi.org/10.1016/j.rser.2017.09.052.

Klein, J.T., 1990. Interdisciplinarity: History, Theory, and Practice. Wayne State University Press.

Koewener, D., Jochem, E., Mielicke, U., 2011. Energy efficiency networks for companies—concept, achievements and prospects. In: ECEEE Summer Study Proceedings, Energy Efficiency First: The Foundation of a Low-Carbon Society. pp. 725–733.

Koewener, D., Nabitz, L., Mielicke, U., Idrissova, F., 2014. Learning energy efficiency networks for companies saving potentials, realization and dissemination. In: ECEEE Industrial Summer Study Proceedings. pp. 91–100.

Lutzenhiser, L., 2010. The evolution of electricity efficiency policy, the importance of behaviour, and implications for climate change intervention. In: Reeve, D., Dewees, D.N., Karney, B.W. (Eds.), Current Affairs: Perspectives on Electricity Policy for Ontario. University of Toronto Press, Toronto, pp. 158–193.

Lutzenhiser, L., 2014. Through the energy efficiency looking glass. Energy Res. Soc. Sci. 1, 141–151. https://doi.org/10.1016/j.erss.2014.03.011.

March, J.G., 1988. Decisions and Organisations. Basil Blackwell, Oxford.

March, J.G., Olsen, J.P., 1989. Rediscovering Institutions: The Organisational Basis of Politics. Free Press, New York.

Moezzi, M., 1998. The predicament of efficiency. In: Proceedings of the 1998 Summer Study on Energy Efficiency in Buildings. Citeseer.

Noorderhaven, N., Harzing, A.-W., 2009. Knowledge-sharing and social interaction within MNEs. J. Int. Bus. Stud. 40, 719–741. https://doi.org/10.1057/jibs.2008.106.

Palm, J., 2009. Placing barriers to industrial energy efficiency in a social context: a discussion of lifestyle categorisation. Energ. Effic. 2, 263–270. https://doi.org/10.1007/s12053-009-9042-1.

Palm, J., Eriksson, E., 2018. Residential solar electricity adoption: how households in Sweden search for and use information. Energy Sust. Soc. 8, https://doi.org/10.1186/s13705-018-0156-1. Article-ID: 14.

Palm, J., Thollander, P., 2010. An interdisciplinary perspective on industrial energy efficiency. Appl. Energy 87, https://doi.org/10.1016/j.apenergy.2010.04.019.

Palm, J., Törnqvist, E., 2008. Governing the sea rescue service in Sweden: communicating in networks. J. Risk Res. 11, 269–280. https://doi.org/10.1080/13669870801939449.

Paramonova, S., Thollander, P., 2016. Ex-post impact and process evaluation of the Swedish energy audit policy programme for small and medium-sized enterprises. J. Clean. Prod. 135, 932–949. https://doi.org/10.1016/j.jclepro.2016.06.139.

Paramonova, S., Thollander, P., Ottosson, M., 2015. Quantifying the extended energy efficiency gap-evidence from Swedish electricity-intensive industries. Renew. Sustain. Energy Rev. 51, 472–483. https://doi.org/10.1016/j.rser.2015.06.012.

Retna, K.S., Ng, P.T., 2011. Communities of practice: dynamics and success factors. Leadersh. Organ. Dev. J. 32, 41–59. https://doi.org/10.1108/01437731111099274.

Revell, A., 2007. The ecological modernisation of SMEs in the UK's construction industry. Geoforum 38 (1), 114–126. https://doi.org/10.1016/j.geoforum.2006.07.006.

Revell, A., Blackburn, R., 2007. The business case for sustainability? An examination of small firms in the UK's construction and restaurant sectors, Business strategy and the environment. Wiley Online Library 16 (6), 404–420.

Russell, C., 2005. Energy management pathfinding: understanding manufacturers' ability and desire to implement energy efficiency. Strateg. Plan. Energy Environ. 25, 20–54. https://doi.org/10.1080/10485230509509690.

Ryghaug, M., Sørensen, K.H., 2009. How energy efficiency fails in the building industry. Energy Policy 37, 984–991. https://doi.org/10.1016/j.enpol.2008.11.001.

Sanquist, T., et al., 2010. Transforming the energy economy—the role of behavioral and social science. In: Proceedings of the Human Factors and Ergonomics Society. pp. 763–765. https://doi.org/10.1518/107118110X12829369605568.

Sanstad, A.H., Howarth, R.B., 1994. 'Normal' markets, market imperfections and energy efficiency. Energy Policy 22, 811–818.

Sorrell, S., 2004. The economics of energy efficiency: barriers to cost-effective investment. Edward Elgar, Cheltenham.

Sovacool, B.K., et al., 2015. Integrating social science in energy research. Energy Res. Soc. Sci. 6, 95–99. https://doi.org/10.1016/j.erss.2014.12.005.

Star, S.L., Griesemer, J.R., 1989. Institutional ecology, 'Translations' and boundary objects: amateurs and professionals in Berkeley's Museum of Vertebrate Zoology, 1907–39. Soc. Stud. Sci. 19, 387–420. https://doi.org/10.1177/030631289019003001.

Suchman, L.A., 1987. Plans and Situated Actions: The Problem of Human-Machine Communication. Cambridge Univ. Press, Cambridge.

Tennant, M., 1997. Psychology and Adult Learning, 2nd ed. Routledge, London.

Thollander, P., Danestig, M., Rohdin, P., 2007. Energy policies for increased industrial energy efficiency: evaluation of a local energy programme for manufacturing SMEs. Energy Policy 35, 5774–5783. https://doi.org/10.1016/j.enpol.2007.06.013.

Thollander, P., Mardan, N., Karlsson, M., 2009. Optimization as investment decision support in a Swedish medium-sized iron foundry—a move beyond traditional energy auditing. Appl. Energy 86, 433–440. https://doi.org/10.1016/j.apenergy.2008.08.012.

Thollander, P., Ottosson, M., 2010. Energy management practices in Swedish energy-intensive industries. J. Clean. Prod. 18, 1125–1133. https://doi.org/10.1016/j.jclepro.2010.04.011.

Thollander, P., Palm, J., 2013. Improving Energy Efficiency in Industrial Energy Systems: An Interdisciplinary Perspective on Barriers, Energy Audits, Energy Management, Policies, and Programs. Springer, London.

Thollander, P., Palm, J., 2015. Industrial energy management decision making for improved energy efficiency-strategic system perspectives and situated action in combination. Energies 8, 5694–5703. https://doi.org/10.3390/en8065694.

Thollander, P., Palm, J., Rohdin, P., 2010. Categorizing barriers to energy efficiency—an interdisciplinary perspective. In: Palm, J. (Ed.), Energy Efficiency. INTECH Open Access Publisher, Croatia, pp. 49–62.

Trianni, A., Cagno, E., 2012. Dealing with barriers to energy efficiency and SMEs: some empirical evidences. Energy 37, 494–504. https://doi.org/10.1016/j.energy.2011.11.005.

Trianni, A., Cagno, E., Marchesani, F., Spallina, G., 2016. Classification of drivers for industrial energy efficiency and their effect on the barriers affecting the investment decision-making process. Energ. Effic. 1–17. https://doi.org/10.1007/s12053-016-9455-6.

Väisänen, H., Christensen, W., Despretz, H., Aamodt Espegren, K., Gaspar, K., Lytras, K., Meyer, B., Reinikainen, E., Sattler, M., Starzer, O., 2002. Guidebook for Energy Audit Programme Developers.

Viktorelius, M., 2017. The Social Organisation of Energy Efficiency in Shipping: A Practice-Based Study. Department of Mechanics and Maritime Sciences, Chalmers University of Technology, Göteborg.

Wenger, E., 1998. Communities of Practice: Learning, Meaning, and Identity. Cambridge University Press, Cambridge.

Wenger, E., McDermott, R., Snyder, W.M., 2002. Cultivating communities of practice: a guide to managing knowledge. [Elektronic resource]

Wollin, J., Nehler, T., Rasmussen, J., Johansson, P.-E., Thollander, P., 2016. Idle electricity as energy conservation within Volvo Construction Equipment. In: ECEEE Industrial Efficiency Summer Study.

Worrell, E., Laitner, J.A., Ruth, M., Finman, H., 2003. Productivity benefits of industrial energy efficiency measures. Energy 28, 1081–1098. https://doi.org/10.1016/S0360-5442(03)00091-4.

Worrell, E., Price, L., 2001. Policy scenarios for energy efficiency improvement in industry. Energy Policy 29, 1223–1241.

Further reading

Churchman, C.W., 1968. The Systems Approach. Dell, New York.

Lave, J., Wenger, E., 1991. Situated Learning: Legitimate Peripheral Participation. Cambridge Univ. Press, Cambridge.

What do we know about the role the human dimension plays in shaping a sustainable low-carbon transport transition?

2.4

Maria J. Figueroa[a], Oliver Lah[b],

[a]Department of Management Society and Communication, Copenhagen Business School, Frederiksberg, Denmark
[b]Mobility and International Cooperation, Wuppertal Institute for Climate, Environment and Energy, Berlin, Germany

1 Introduction

A transformative systemic change is required to limit global warming to 1.5°C above preindustrial levels in an integrated manner with sustainable development (de Coninck et al., 2018). Sustainable mobility aims at finding a balanced way of advancing social, environmental, and economic transport-related goals[a] whilst reducing negative impacts. Key objectives for sustainable mobility are accessibility; affordability; functionality; efficiency; public participation; traffic safety and energy security; and, concurrently, reductions of climate emissions, use of fossil fuels, air pollution, noise, congestion, and traffic accidents (Banister, 2008; Kahn Ribeiro and Figueroa, 2012). In addressing these goals, the interplay between mobility patterns, the availability of transport modes and vehicle technologies, and the associated individual and collective choices makes sustainable mobility and transport decarbonisation highly dependent on the human dimension.

The question that motivates this chapter is as follows: what role can the human dimension play in shaping a low-carbon energy transition of the transport sector? A great emphasis of the literature concerning energy transition and energy demand for transport is placed on the technical findings. However, the feasibility of

---

[a] The chapter will refer indistinctly about these goals as part of sustainable mobility or sustainable transport mobility. Goals for low-carbon transport or low-carbon mobility targets relate to increasing transport energy use efficiency, reducing or avoiding energy use for transport, and changing the use of fossil fuels to low-carbon fuels.

Energy and Behaviour. https://doi.org/10.1016/B978-0-12-818567-4.00008-9

deployment of new energy technologies and the implementation of policies and actions on low-carbon transport solutions requires understanding and balancing the human dimensions in context (de Coninck et al., 2018). A departing premise for this chapter is that beyond technical advances of fuels and vehicles, systemic changes will be needed to advance transport decarbonisation. Therefore, the essential question addressed in this chapter is as follows: what policies and actions can facilitate a collective transition and supportive human response to the adoption of fast systemic changes towards a decarbonised transport sector?

Knowledge about how the human dimensions impact energy demand in transport is found at the intersection of several disciplines. We take a brief overview to this variety of disciplinary approaches from sources available in scientific journals and in recent official country and international organisations' reports. Our initial approach is broad, but we will seek to narrow it down to discuss concrete principles, policies, and practices directly influencing the capacities of individual users and decision-makers in charge of facilitating the scaling up of actions towards a low-carbon transport transition.

The literature covering human dimensions of transport is vast and spanning a wide variety of disciplines. Seminal insights from urban and regional planning indicate that a strong relationship exists between the built environment, the type of vehicle and the amount of fuel used for personal travel (Newman and Kenworthy, 1989; Ewing and Cervero, 2010). From comprehensive reviews on the scope of transport policy and better urban planning, we learn the variety of interventions targeting modal choice in personal travel (Santos et al., 2010; Lah, 2017a). From transport environmental economics, we gain insights on the effects that pricing policies have on transport behaviour (Small and Verhoef, 2007). From social scientist focusing on transport behaviour, we understand how symbolic motives and lifestyles (Steg, 2005; Steg and Gifford, 2005), motivations and contextual factors (Steg et al., 2015), and peer-effects motivate transport human behaviour (Bamberg et al., 2011). Engineering disciplinary approaches relate to the way in which new technologies can accommodate service or vehicle improvements (Short and Kopp, 2005). Innovation scholars seek to explain the effects of technological change and impacts of evolution of transport innovation in society (Zhang and Gallagher, 2016). The more recent literature on technology innovation is discussing, the potential impacts of disruptive forms of transport service innovation on aggregated transport passenger demand (Wilson, 2018). The environmental research literature links human behaviour and energy use in transport (Anable, 2005; Banister et al., 2011; Figueroa et al., 2014; Newman and Kenworthy, 2011) and discusses other environmental impacts of personal mobility choices (Hensher, 2008; Noy and Givoni, 2018). Following further traits of literature to match our interest here in understanding the role that human dimension plays to scaling up low-carbon sustainable transport solutions, we found that social science and transport literature offers examples of new organisational models for collective solutions like sharing of transport means in the city (Shaheen and Cohen, 2013; International Transport Forum, 2017; McLaren and Agyeman, 2017). The expansion of literature on transport and human behaviour covers the impact that humans, through collective actions and forms of social innovation, can have in the promotion of low-carbon mobility options (Figueroa et al., 2019). The planning literature addresses how diverse forms of participatory planning processes with multistakeholder partnerships can contribute to sustainability in transport (Koppenjan and Enserink, 2009). We review key points made in the literature and, when possible, present practical examples.

The chapter follows a four-part structure: Section 1 is this introduction; Section 2 discusses the increasing energy demand for transport as one of the major challenges of the human dimension in transport and reviews the opportunities and cobenefits that can facilitate transport decarbonisation; Section 3 outlines changing conditions affecting human dimensions and energy for transport, paying attention to policies, and technologies delimiting the solution space for decision-makers; Section 4 discusses synergies between actors, collective community approaches, and the role these may play to facilitate scaling up policies and solutions; and Section 5 summarises the answer to the leading chapter question and concludes.

2 Energy use for transport and sustainability

Transport is relatively unique among the energy end-use sectors in its near complete reliance on liquid fossil fuels, with natural gas, biofuels, and electricity still making up only a small fraction of the global transport energy demand (Sims et al., 2014). Transport CO_2 emissions have more than doubled globally since 1970, increasing at a faster rate than any other end-use sector to reach 6.7 $GtCO_2$, representing 23% of total energy-related CO_2 emissions in 2016 (IEA, 2018b). Final energy consumption for transport reached 27% of total end-use energy in 2017, and over 53% of global primary oil consumption was used to meet 94% of the total transport energy demand (Sims et al., 2014; International Transport Forum, 2017; Roy et al., 2018).

Trends are rapidly changing in some cities and market segments, for example, concerning fleets of electric buses, bicycles, and scooters, which are being adopted at a very fast pace in several cities and regions around the world (IEA, 2018a; SLoCaT, 2018). These new trends are significant to reducing transport energy demand because the modal choices with which people, economic goods, and services move around vary greatly with regard to their energy intensity and sustainability. Vital aspects of a low-carbon transition pathways for the transport sector include the following: *avoiding* the need to travel by managing overall travel demand; *improving* efficiency by increasing the share of efficient transport modes such as public transport, walking, and cycling; and *shifting* to low-carbon energy carriers for vehicles, for example, through electrification. Beyond the decarbonisation of the sector, there are other key sustainability challenges that need to be addressed within transport, such as ensuring safety and security, improving air quality and physical health, and ensuring accessibility for all to jobs, economic opportunities, and social activities. With growing population and increasing living standards, decision-makers are finding that meeting an increasing mobility demand for both passenger and freight transport, whilst simultaneously meeting strong transport decarbonisation and sustainability goals, is a highly complex challenge. Effective solutions require a multi-objective strategy that goes beyond a pure technological shift into finding synergies between actors and the promotion of solutions spanning collective and individual mobility choices (Figueroa et al., 2013).

2.1 Human dimensions and their significance to energy for transport

Socio-economic aspects such as increasing population; the age structure of population; urbanisation and economic growth; and sociology of consumption dimensions like practices (Shove et al., 2012), consumption, and status seeking (Schor, 1998) all induce rapid travel demand. Generally, providing affordable and safe mobility for people and goods is a necessary condition to reach goals of poverty alleviation, access to health services, food, jobs, education, gender equality, and other social and economic sustainable development goals (UN-Habitat, 2013). The recognition of the urgency of these goals is manifested in the adoption of 17 Sustainable Development Goals with Agenda 2030 in 2015 (UN General Assembly, 2015).

Accelerated urbanisation growth increases passenger and good transport demand. Rapid rates of passenger transport demand are present in many urban areas in Africa and Asia, whilst shrinking cities in Europe, Russia, and the United States are experiencing low-growth population and decreasing travel demand (Dahiya, 2012; Haase, 2013). Over the next decades, the United Nations (UN) estimates that urbanisation will continue to increase, particularly in Asia and Africa, and by 2050, two out of three inhabitants will live in urban areas, meaning around 65% of population in developing countries and nearly 90% of population in developed countries will live in urban areas (UN, 2014). These projections justify the dedication of 1 of the 17 Sustainable Development Goals (SDGs) to attaining sustainable infrastructure in urban areas. This goal recognises the human dimension challenges that urban areas will increasingly be facing in terms of access to basic services to secure human life, such as drinking water, food, clean air, healthcare (including basic sanitation requirements), basic transportation, and resilience to disasters and extreme weather events resulting from climate change (Dahiya, 2016). Attending to the scale of the human dimension challenge compounds the tension that decision-makers confront to meet competing goals within economic resource and environmental constraints.

The transport human dimension pressure in terms of scale and quality of transport service expectation will be consequential to framing the political opportunities for energy transition and transport decarbonisation. The most intense socio-political pressure will be in fast-growing developing nations, particularly in countries with high fuel for transport import dependency. An example of the type of complex political domestic interaction concerning access to affordable fuels for transport is the case of reducing transport fuel subsidies to the final consumer. There are many reasons why countries may offer fuel subsidies, but, often, subsidies are set in place with impetus from political leaders intending to buy political support or in the name of advancing redistributive policies. Irrespective of the motivation, fuel subsidies maintain low fuel prices that increase transport energy use. Once this type of subsidies is set in place, they have proven difficult to revert. Governments can risk facing strong opposition forces as a result of the increase in fuel prices. Protest can risk escalating or mobilising large and potentially violent demonstrations (Lockwood, 2015). This is not only the case applicable to developing countries context. At the time of writing this chapter, recent events demonstrated the emergence of the movement '*des gilets jaunes*', the yellow vest in France.[b] The initial motivation for the yellow vest protest was the opposition to the governmental political decision to increase fuel prices (Skovgaard and van Asselt, 2019). The social and economic need to secure transport accessibility at affordable prices puts pressure on policy and decision-makers for resolving how to decarbonise whilst increasing and improving passenger transport services that are essential for development.

2.2 Attending the human dimensions in transport transition managing cobenefits

Decision-makers have opportunities to tailor solutions to fit their city or country's own political contextual priorities by considering potential cobenefits and trade-offs of interventions oriented to reach simultaneously social, economic, and environmental transport goals. For example, policies and actions that relieve congestion can generate a number of cobenefits (e.g. health, clean air, reduced noise and energy use, improved traffic safety, and better quality of life), provided that the method for reducing congestion in one area does not induce additional traffic in another (Hymel et al., 2010). Policies that have an exclusive focus on technology and fuel-switching measures are unlikely to reduce congestion levels and traffic flows, indicating that these actions should also be part of a wider and more comprehensive strategy that measures cobenefits and trade-offs.

Another area of intervention that can offer multiple cobenefits and help reduce energy use for transport relates to reaching health benefit objectives connected to the various health impacts of transport (Anenberg et al., 2012). A number of cobenefits and positive synergies exist between activities tackling air pollution, noise, physical health, transport heavy weight producing vibration, road safety, and policies managing transport demand and energy use in transport (Maione et al., 2016; SLoCaT, 2018). China, for example, is investing in electrified transportation with the objective of improving air quality in cities, reducing oil imports, and fostering a pioneering electric transport national industrial production (Zhang and Gallagher, 2016; IEA, 2018a). Road safety is a key policy objective for addressing human dimensions of transport since well over one million people are killed in road accidents worldwide each year, 93% of those occurring in low- and middle-income countries (WHO, 2018). Policies and solutions that affect transport demand through individual choices for travel modes and types of vehicles can have an immediate effect on road

---

[b] https://en.wikipedia.org/wiki/Yellow_vests_movement Accessed 20-05-2019.

safety and affect transport demand, in both positive and negative ways. For example, encouraging people to walk or cycle may increase their health but may also expose them to air pollution and, also, to road safety risks. Hence, a combined approach for a safe system for walking and bicycling, high capacity public transport systems, along with low-carbon and nonpolluting vehicle technologies for individual mobility may yield the greatest potential for cobenefits in terms of reducing car travel and managing travel demand whilst improving health and quality of life in the city (Pucher and Buehler, 2010).

With the goal of improving air quality, decision-makers can directly impact energy use in transport, if not the total demand by, for example, replacing or shifting liquid fossil fuels with biofuels, or with renewable energy-produced electricity. A unique example is the case of replacing transport petroleum products with liquid biofuels, a policy that has been successfully implemented, almost exclusively, in Brazil (Schaeffer et al., 2015). However, from the point of view of managing transport energy demand, the shift to using biofuels has been associated with several uncertainties. The literature challenges whether biofuels can effectively contribute to GHG emission reductions or have limited contributions to reduce air quality and health impacts (Fargione et al., 2010). A major challenge is the land use competition between crops for food versus crops for fuel production (Tenenbaum, 2008). In addition, concerning impacts on air pollution, replacing gasoline or diesel fuels with biofuels, may reduce carbon monoxide and hydrocarbon emissions but increase nitrogen oxide (NO_x) emissions (Hill et al., 2009). This is likely to change with the production of more advanced biofuels. Indeed, a lot of research attention is required as shown by the experience of increased biofuel market share in Brazil or the United States, but the valid point is still here that clear cobenefits to local air quality improvements can be advanced from fuel switching. The air quality improvement cobenefits from switching to electricity in transport depend on the source of electricity generation and, when applicable, the location of coal-fired power plants (Hawkins et al., 2013). Cobenefits and trade-offs offer a framework to assessing interventions that can indirectly impact transport energy demand. The following sections consider strategies for changing conditions that directly influence human behaviour and the demand for transport.

3 Changing conditions affecting the energy demand for transport

Transport is the energy end-use sector with strongest dependency on liquid fossil fuels. Therefore, policies to reduce energy use for transport need to target directly vehicles, fuels, and behaviour. A practical framework that helps summarise what actions needs to be advanced for transport decarbonisation is the 'avoid, improve, shift' framework coined by Bongardt et al. (2011). In short, energy use in transport can be minimised by *'avoiding'* unnecessary trips reducing transport demand, *'shifting'* fuels to low-carbon or renewable produced fuels, and by *'improving'* the energy efficiency of vehicles. With the internal combustion engine being the most dominant vehicle technology across all transport modes (except for rail), changes already underway like the emergence of new fuels; vehicles and travel options may begin to challenge the strong association between fossil fuels and motorised transportation.

As an example of these emerging changes, electric vehicles (EVs)—hybrids and plug-in—are still representing a small market globally but, together with buses, e-electric two- and three-wheelers, or scooters and bicycles, EVs are beginning to make inroads into the share of energy use for passenger transport demand (IEA, 2018b). Over three million EVs were on the road in 2018, with China being the largest market of electric vehicles. According to the China Passenger Car Association,[c] electric passenger vehicle sales in China totalled 1,016,002 units in 2018, up 83% year over year. Including commercial vehicles, this total jumps to 1,256,000 units, with the vast majority of these vehicles made and used within China. Norway is the country with the largest share of EVs in the world, with nearly 40% of the new vehicle stock being electric in 2017 (IEA, 2018a). Aviation and maritime transport that combine passenger and good transportation are not discussed in this chapter, but are highly relevant and pose particular challenges, from the behavioural change and policy perspective. The following sections discuss in more detail policies that can help influence individual and collective mobility behaviour and choices for vehicles and technologies for land transport with consequences over total energy demand for transport.

3.1 Changing fuels and vehicles

A shift to electric mobility is expected to play the most significant role in the decarbonisation of the land transport sector (IEA, 2018b). As previously noted, many e-mobility solutions are readily available, several of which are already cost-effective and can deliver wider socio-economic benefits, which makes the prioritisation of these solutions pivotal for policy decisions in this area (Lah, 2017b; SLoCaT, 2018). From a societal scaled up perspective, the electrification of public and shared vehicles fleets is the most cost-effective option, since these vehicles tend to drive longer distances and these sectors should become priorities for policy intervention. The cost per passenger kilometre[d] for different vehicle types also changes over time as e-mobility technologies become more cost competitive in comparison with fossil fuelled ones (Nykvist and Nilsson, 2015). By 2030, the life-cycle costs of electric cars are likely to be lower than those of an internal combustion engine car (Lah, 2017a). By then, also the shared modes, discussed later in this chapter, will be cost-effective, in particular if automation for public transport becomes a viable option.

Many of the policies and key contributing factors to the growth of electric mobility like oil independence goals, energy efficiency goals, industrial policy, and innovations to reduce the cost of batteries used by EVs are decisions or policies taken at

---

[c] http://autonews.gasgoo.com/new_energy/70015566.html Accessed 03-03-2019.
[d] Cost per passenger kilometre represents the cost of transporting one person by type of vehicle over a distance of 1 km.

the national or state level. For example, the state of California in the United States, introduced standards for automakers, fuel economy, and zero emissions mandates, and several countries have introduced tax incentives and exemptions for the use of EVs in public transport services (SLoCaT, 2018). At the local level, municipal authorities also have a number of options to encourage the purchase and use of electric vehicles. To guide these decisions, focusing on the most (cost) effective electric mobility options is an important aspect. City authorities have the right and play the role of regulating and, sometimes, even directly operating local fleets of public transport services. Cities have the opportunity for establishing priorities and using policy levers as they make decisions about investment and procurement of transport services. As a regulator for taxi and car sharing fleets, there can be a similar role for cities to find ways of incentivising and encouraging the shift of entire fleets to electric mobility. From a societal scaling up perspective, subsidies to support electric public transport, car, and bike sharing schemes are cost-effective and more inclusive as opposed to the subsidies for private cars (Estupinan et al., 2007). Improving and facilitating accessible low-carbon forms of transport can induce more people out of their private cars to make their daily journeys connecting to the major public transport network.

3.2 Changing behaviour: Carrots, sticks, and rebounds

Encouraging production of more efficient vehicles (improving) is not the same as having people making a more efficient use of these vehicles. Changing behaviour towards efficiency or even towards discouraging the use of private vehicles (avoiding) can take the form of hard, direct economic and fiscal measures targeting vehicle use and soft, indirect regulatory policy measures that target vehicle fleet and fuel use, examples of which include the instruments detailed in the succeeding text.

3.2.1 Fiscal instruments

Fiscal instruments can guide individual purchasing decisions, for example, vehicle and fuel taxes that can have a significant impact on the efficiency of vehicles introduced into the fleet. Differentiated taxes reward energy efficient cars with considerably lower taxes and impose significant taxes on cars with poor fuel efficiency. Vehicle taxes based on CO_2 emissions with sufficient differentiation may result in significant improvements of the energy efficiency of the vehicle fleet (Ajanovic et al., 2016). Under a system where efficiency is rewarded, called a feebate system, the level of progression is being increased over time and thus results in even greater CO_2 emissions reduction, but considerable savings could also be attained by increasing the differentiation of existing taxes as will be reviewed in the next section.

3.2.2 Fuel taxes

Fuel taxes and excise duty rates ought to be set at a level internalising external costs (e.g. from carbon emissions) (Sterner, 2007). This would directly impact both travel demand and the vehicle technologies used and, in turn, fleet fuel consumption and CO_2 emissions. Also, fuel prices potentially have a considerable impact on the rate of vehicle ownership. The influence of fuel price changes on consumption is defined as its price elasticity. Fuel taxes can deliver a longer-term policy signal, provided they are substantive enough (Lah, 2017a).

3.2.3 Differentiated vehicle tax

Differentiated vehicle registration, purchase taxes, and/or feebate schemes can guide consumer demand. This in turn can help reducing split incentives between individuals and society. These schemes need to be responsive to developments in the vehicle fleet to guarantee sufficient demand for more efficient vehicles and to increase cost-effectiveness. Circulation and ownership taxes are a reoccurring charge (typically yearly), which can promote purchases of more efficient cars by calculating the charge according to cars' fuel economy, either directly or by using a proxy (CO_2 emissions, engine size, or power-to-weight ratio). Directly taxing greenhouse gases and harmful emissions is a well-established and researched policy measure, which has proven to be more cost-effective than enforcing direct controls (Newbery, 2005; Ajanovic et al., 2016).

3.2.4 Registration taxes

Through placing higher taxes on the purchase of less-efficient vehicles, registration taxes directly influence consumer behaviour at the point of vehicle sale (Fu and Andrew Kelly, 2012). Purchase or registration taxes are very visible, which is especially helpful in guiding buyers' decisions towards more efficient vehicles and may also give way to a reduction in car ownership rates, which would result in lower overall car use and a greater share of more efficient modes in urban areas (Fridstrøm and Østli, 2017). However, this may result in negative welfare or equity implications if not matched with modal alternatives at the local level (Lucas and Pangbourne, 2014). Taxes imposed at the time of the first registration could lead to the delay of vehicle fleet renewal, as car owners may keep their vehicles longer and may prefer to replace their current vehicle with used rather than new ones.

3.2.5 Parking management

Parking management can be a powerful tool for local authorities to manage car use and to raise revenue (Litman, 2006). Similar to road user charging, parking management and pricing can help discouraging the use of a privately owned car and raise revenue to fund public transport, walking, and cycling infrastructures and improve public spaces (Pitsiava-Latinopoulou et al., 2012). Parking management schemes reduce the number of cars entering the city, which can reduce congestion and encourage the use of public and nonmotorised transports (Pitsiava-Latinopoulou et al., 2012). The parking pricing structure and the level of enforcement are important aspects to consider (Shoup, 2011). A structured fee that differentiates between different zones of a city or times depending on the demand is one aspect that needs strong enforcement to be meaningful. Coordination of parking pricing and zoning among relevant municipal authorities is another vital aspect. Parking management also includes time restrictions and a control of the number of available parking spaces (Litman, 2006).

Parking time restriction for nonresidents, for example, to 2h, is a proven tool to reduce commuting by car without affecting accessibility to urban shops. In fact, in many cases, shops and other local businesses become more accessible when public space is freed up by a reduced number of parking spots.

3.2.6 Registration management

Managing vehicle registrations can help limiting the number of cars in a city by differentiating registration fees for vehicles, for example, according to their CO_2 emissions or engine size (Ajanovic et al., 2016). This can guide purchasing behaviour and can encourage people to opt for cleaner vehicles or more sustainable modes such as public transport and nonmotorised modes (Lah, 2015). This approach is not common in European cities but has been implemented very successfully in several cities in Asia, for example, Singapore and Shanghai. Vital to this approach is that the authority to register vehicles lies with the municipality (SOLUTIONS, 2016).

3.2.7 Rebound effects and strategies to address them

The rebound effect matters greatly to scaling up solutions and policies to manage energy demand. The rebound effect refers to the tendency for total demand for energy to decrease less than expected after efficiency improvements are introduced, due to the resultant decrease in the cost of energy services (Sorrell, 2010; Gillingham et al., 2013). A key distinction is made between direct rebound, for example, behaviour of increased utilisation of a more efficient device; indirect rebound effect, for example, behaviour of engaging in use of other services/energy devices after redirecting the savings obtained with purchase of an efficient unit; and economy-wide effect, for example, when an overall energy efficiency policy that is pursued by the nation makes the energy services cheaper and leads to overall readjustments of the supply and demand of both that sector and other sectors in the economy (Azevedo, 2014). As an example, the introduction of fuel efficiency standards for light duty vehicles may improve the efficiency of the overall fleet but may also induce additional travel as fuel costs decrease for the individual users (Yang et al., 2016). Research shows that the direct rebound effect is not significantly large: around 0%–12% for household appliances such as fridges and washing machines and lighting, whilst it is up to 20% in industrial processes and 12%–32% for road transport (IEA, 2013). The higher the potential rebound effect and also the wider the range of possible take back, the greater the uncertainty of a policy's cost-effectiveness and its effect upon energy efficiency (Ruzzenenti and Basosi, 2008).

Ignoring or underestimating the rebound effect whilst planning policies may lead to inaccurate forecasts and unrealistic expectations of the outcomes, which, in turn, lead to significant errors in the calculations of policies' payback periods (IPCC, 2014). Rebound exemplifies why decision-making on transport policy and infrastructure investments are as complex as the sector itself. It also illustrates why rarely a single measure achieves comprehensive climate change impacts and also generate economic, social, and environmental benefits (Lah, 2014; Creutzig, 2016). Countering this complexity, however, is the fact that many policy and planning decisions have synergistic effects, meaning that their impacts are larger if implemented together or in combination. The evidence of the existence of rebound effects supports the idea that it is better to implement and evaluate integrated programs rather than individual strategies (Hüging et al., 2014). For example, by itself, a public transit improvement may cause minimal reductions in individual motorised travel and associated benefits such as congestion reductions, consumer savings, and reduced pollution emissions. However, the same measure may prove to be more effective and beneficial if implemented with complementary incentives, such as efficient road and parking pricing, so that passengers have an incentive to shift away from individual car travel (Cuenot et al., 2012; den Boer et al., 2011). A vital benefit of combining measures is the potential for integrated packages to deliver synergies and contribute to minimise rebound effects. In general, the most effective programs tend to include a combination of qualitative improvements to alternative modes (e.g. walking, cycling, and public transit services), incentives to discourage carbon-intensive modes (e.g. fuel pricing, vehicle fuel efficiency regulation, and taxation), and integrated transport and land use planning, which creates more compact, mixed, and better connected communities with less need to travel (Figueroa et al., 2014; Sims et al., 2014).

3.3 Changing behaviour: From solo driving to sharing

Car sharing is arguably the pioneer mode of the sharing economy—a trend that follows the concept of collaborative consumption (McLaren and Agyeman, 2017). The sharing economy aims to pool products and services to gain access to certain resources in a more sustainable way (Bardhi and Eckhardt, 2012). Besides other sharing services such as bike and scooter sharing, the following analysis will solely focus on cars.

Car sharing is defined as a membership-based service that offers the user short-term vehicle access without full ownership, when other modes of transport are not available or convenient (International Transport Forum, 2017). Given that, on average, a car is parked and not in use most of the time, the overarching objective is to have multiple people using the same car at different times and thus increase utilisation. In that sense, shared mobility offers a scalable alternative to traditional car use, maximising use efficiency and partly eliminating the need for individual ownership. Despite great popularity, car sharing still accounts for a small portion of trips in most urban areas. Taking the San Francisco Bay Area as a representing example, only 1% of the overall distances travelled are conducted by car sharing. Public transport (24%), walking (25%), and driving alone (30%) still hold the major shares (International Transport Forum, 2017). Experts expect it to be a significant mode of transportation, meaning to contribute at least 10% to the transportation mix by 2030 (Shaheen and Cohen, 2013; International Transport Forum, 2017). However, car sharing requires the diffusion of smartphones and cellular connectivity and big data usage in cities to become a suitable service. A full environmental impact assessment of car sharing and the evaluation of whether car sharing represents a contributing or restrictive factor in achieving sustainable transportation in cities are still needed. Considering the initial costs of buying a vehicle, insurance, registration, vehicle tax, parking space, and maintenance, owning a car becomes increasingly uneconomical, and the alternative of considering car sharing schemes is becoming increasingly popular in particular in urban areas (Fellows and Pitfield, 2000; Shaheen and Cohen, 2018).

Car sharing business accommodates both business-to-consumer (B2C) and peer-to-peer (P2P) car sharing. Business-to-business (B2B) vehicles are owned by companies such as Zipcar[e] (Avis), DriveNow[f] (BMW), Car2go[g] (Daimler), or We Share[h] (Volkswagen). P2P sharing builds on private individuals offering their cars by using a centralised platform (e.g. BlaBla[i] or GoMore[j]). Among the various providers of car sharing schemes, there are free-floating and stationary systems. Free-floating schemes such as Car2go and DriveNow allow for their users to park the car anywhere within a designated zone and facilitate the pickup through GPS-assisted smartphone apps. Users of stationary car sharing systems such as Cambio and Zipcar return the vehicle to designated parking areas; reservation and payments are normally also handled through apps. Charges can be time based and/or be based on kilometres driven and usually cover all costs including fuel. Most of the free-floating car sharing providers focus on larger cities, and even there, they are focusing on the most densely populated areas, which can lead to more competition than complementarity with public transport. Many midsized cities work with providers of stationary sharing schemes, which often have a higher rate of replacing privately owned vehicles. Increasing the usage of shared vehicles has the potential to minimise the need to own a car and therefore of reducing car ownership.

The environmental benefits of car sharing include decreased land use (for parking) and lower overall usage of cars as opposed to use of privately owned vehicles (Loose, 2010; Glotz-Richter, 2012). A reduction on parking searching circulation pollution could also be the result of sharing, since cities increasingly allocate preferential parking spots to car sharing providers, therefore reducing the need for searching for one (Amini et al., 2017). This effect is particularly strong if smart parking is integrated into the intelligent transportation systems of a city. Gains in overall car fleet fuel efficiency can be estimated from contrasting the average age of the fleet of car sharing companies with the average of total vehicle fleet. Shared vehicles are on average newer than the average private vehicle in the fleet and therefore more fuel efficient (or likely EVs). This is because car sharing company providers aim at differentiating their car brand by promoting attractive new car fleet models. Considering total cost of ownership, car sharing providers favour attractive, innovative, and fuel-efficient vehicles over older ones (Shaheen and Cohen, 2013). The opportunity for expansion and diffusion of EVs in cities can be boosted with a number of car sharing

---

[e] https://www.zipcar.com/cities Accessed 07-03-2019.
[f] https://www.your-now.com/ Drive now is part of share now. Accessed 07-03-2019.
[g] https://www.car2go.com/US/en/join/?cid=c2g_ppc_us_all_cpc_performance_google_prsitelink_mainpage_none_none_none_none_none_none_none&gclid=EAIaIQobChMI1uaC5LHx4AIV0eWaCh3shgS9EAAYASABEgL8RPD_BwE Accessed 07-03-2019.
[h] https://www.volkswagenag.com/en/news/2018/08/VW_Brand_We_Share.html# Accessed 07-03-2019.
[i] https://www.blablacar.com/ Accessed 07-03-2019.
[j] https://gomore.dk/rental/how/owners Accessed 07-03-2019.

company providers operating electric fleets as such companies may help build trust among customers experiencing the new technology. One of the major expected potential impacts of car sharing is a reduction of the total vehicle fleet that could help mitigate congestion and air pollution and reduce total carbon emissions. However, at the time of writing this contribution, the authors found no substantive empirical evidence that quantitatively demonstrate how car sharing in practice has achieved this potential. With fewer cars needed on the fleet to provide the same services, another expectation is that the total number of cars produced and sold in the market could likely decrease. This could lead to reducing carbon emissions from vehicle manufacturing. Considering the total car life-cycle emissions, manufacturing makes up an estimate of approximately 30% of the vehicle life-cycle emissions (EEA, 2011). A recent study in the Netherlands on the effects of car sharing on car ownership, car use, and CO_2 emissions found that individuals that use car sharing own 30% less cars than prior to the availability of car sharing. Thus, the study concludes for a positive environmental impact for car sharing (Nijland and van Meerkerk, 2017). Yet, a caveat in the study is that in the majority of these incidences, the shared car was mainly used to replace a second or third car and did not encourage people to give up cars entirely. Based on the assumptions of reduced car ownership and car use, the study concludes that car sharing emit between 13% and 18% less CO_2 related to car ownership (Nijland and van Meerkerk, 2017), but further studies are necessary in this area.

Despite the described potentially positive environmental effects of car sharing, there are also concerns. In particular, there are doubts whether car sharing leads to a rebound effect, or whether it cannibalises on public transport demand.[k] Car sharing essentially increases car kilometres that would have otherwise not been travelled; or take on trips that may have been otherwise completed by using more environmentally friendly modes, such as public transport or walking (Martin and Shaheen, 2011). The controversial environmental question that needs further research is that for certain trips, public transport, and car sharing compete rather than complement each other. With the foreseeable development of autonomous cars, experts are considering that the overall effect may be for car use, travel demand, and energy use for transport to increase even further (International Transport Forum, 2017).

3.4 Changing the system: Intelligent transport approaches

The integrated application of computer, electronics, communication technologies and management strategies to improve road safety and increase transportation efficiency is named as intelligent transport systems (ITS) (Ali et al., 2018). The main applications in this system aim at minimising average travelled distance, travel time, and traffic density to improve citizens' driving behaviour and to minimise pollution. These potential reductions could also minimise travel energy use.

---

[k] https://onezero.medium.com/the-sharing-economy-was-always-a-scam-68a9b36f3e4b Accessed 07-03-2019.

The concept of smart or intelligent transport translates in multiple service applications such as traffic management, public transportation, electronic payment, emergency management, and maintenance management (Ali et al., 2018). In terms of ITS contributions to a sustainable energy transition, the expectation is that intelligent solutions will in turn be consolidated into a complete traffic management systems (TMS) (de Souza et al., 2017). A TMS will use wireless systems to maintain an efficient and open communication between the core of the built transport infrastructure and each vehicle. Up to the present time of writing this chapter, the overall target of existing TMS is to reduce the congestion queue time in traffic intersections (Woetzel et al., 2018). Keeping traffic floating smoothly at a moderate speed can reduce fuel consumption and therefore carbon-related emissions. In transport, fuel consumption can be reduced either by reducing the fuel use per person-kilometre travelled or by minimising the travel time. A TMS aims at influencing both, since it can reduce travel distance by estimating traffic flows and guiding the vehicle on the fastest route, for example, preventing the car to be stuck in congestion. To reduce the fuel use, on the other hand, it is necessary to influence directly the driver's behaviour. Therefore, the TMS will need to guide vehicles to reduce their CO_2 emissions by minimising situations of stop and go due to red lights and accelerations needed to catch the green light (de Souza et al., 2017).

In spite of widely accepted positive effects of TMS on reducing transportation-related environmental impacts, research lacks profound quantitative assessments on GHG reductions. Whereas the McKinsey Global Institute (Woetzel et al., 2018) predicts emissions to be cut by up to 5%, a study conducted by Barth et al. (2015) shows CO_2 emission reductions of 5%–15%. A plausible reason for the low results is that individual technologies in the TMS are closed-off from other technologies, and they are unable to operate simultaneously in the described intelligent architecture, aiming at a seamless form of information exchange.

3.4.1 Mobility as a service: Improving mass mobility

Mobility as a service (MaaS) takes a different perspective away from the vehicle as the main transport product towards multimodal integrated mobility services (Ho et al., 2018). The service created is not as a physical asset to purchase (e.g. a car) but a mobility service that is easily available well integrated across different transport services from cars to buses to rail and on-demand services (Mulley et al., 2018). The expectation is that MaaS may disrupt the current role and organisation of public transport (Hensher, 2017). Several demonstrations of MaaS (or elements of MaaS) have been first applied in the context of a Finish project called Whim.[l] In Gothenburg, for example, they have tried out UbiGo,[m] in Vienna, SMILE[n] and Switch[o] in Hannover,

---

[l] https://whimapp.com/ Accessed 07-03-2019.
[m] https://ubigo.me/ Accessed 07-03-2019.
[n] https://smartcity.wien.gv.at/site/en/smile-2/ Accessed 07-03-2019.
[o] https://www.moia.io/en/press/Go---ahead-for-ridesharing-in-Hannover-MOIA-switches-from-service-test-to-official-operations Accessed 07-03-2019.

the project is called MinRejseplanen[p] in Copenhagen, Entur[q] in Norway, and with other concepts on the way in Scotland, the United Kingdom, the United States, and Australia (Kamargianni and Matyas, 2017; Smith et al., 2018; Ho et al., 2018). These new alternative modes of transport provide extra resources to fulfil personal 'door-to-door' journeys, without need to resort to the use of private cars, reducing the need and potentially supporting a shifting in the paradigm of traditional car ownership (Ho et al., 2018).

3.4.2 Taking humans out with automatisation

Multiple forms of utopias from notable transport actors like automotive manufacturer; major tech players like Apple, Google, Alphabet, and Uber; and other companies that currently have a high stake in the ascendant autonomous systems are envisioning scenarios where humans are partially or fully removed from the driver's seat with transport automation.[r] The reason that justifies why societies need such a transition is the promise that driving will be safer for everyone involved and, in addition, people will be enjoying the freedom of time and space in a car that can potentially be transformed into the new commercial space of pleasure and entertainment and in the meantime finding some extra time to do some work.[s] Interestingly, in these utopian visions, environmental problems such as congestion or climate change are rarely, if at all, addressed. Machine-readable information, algorithms, sensors, and cameras that transform its surrounding environment into data that a computer can process and act upon are providing radically new ways to innovate, including substituting the human behind the steering wheel. Concerns over how the process of accelerated automation oversteps individual rights to privacy are increasingly voiced, prompting states to consider what kind of regulatory framework can be put into place and to what effect.

At the time of producing this chapter, the status is that, whilst a number of automakers have engineered vehicles that can pilot themselves with an ability unfathomable even a decade ago, research and practical evidence so far indicate that an autonomous future is not in any real sense here now, nor around the corner, but likely decades away.[t,u] There are many advances, and a profusion of data sources created and recognised to need standardisation of formats and technical integration to maximise the value autonomous vehicles contain. Public authorities in cities worldwide are increasingly using mobility platforms and big data to better predict current and future vehicle and passenger flows, to improve accessibility, liveability, and

[p] https://play.google.com/store/apps/details?id=de.hafas.android.nt&hl=da Accessed 07-03-2019.

[q] https://en-tur.no/?gclid=EAIaIQobChMIrPvu5vz14AIVQamaCh2NEgfEEAAYASAAEgJRq_D_BwE Accessed 07-03-2019.

[r] https://tech.co/news/self-driving-cars-ready-2016-12 Accessed 07-03-2019.

[s] https://www.wired.com/2012/01/ff_autonomouscars/ Accessed 07-03-2019.

[t] See a good overview at: http://fortune.com/2017/03/08/the-way-we-talk-about-autonomy-is-a-lie-and-thats-dangerous/ Accessed 20-05-2019.

[u] https://www.nytimes.com/2019/07/16/opinion/smart-cities.html?action=click&module=Opinion&pgtype=Homepage Accessed 17-07-2019.

sustainability (Yap and Munizaga, 2018). However, similar to the challenges companies face with MaaS, the difficulty is that external data sets still remain proprietary and not openly accessible. In sum, there is no rapid and easy path from automation utopias to effectively managing transport demand, reducing energy use and, particularly, for solving the problems at the heart of transportation.

The challenges of the transport sector vary regionally and by countries. Highly industrialised countries need to decarbonise their mature and vehicle-saturated transport systems. Conversely, fast-growing cities in developing countries face the problem of how to offer safe and reliable mobility services and how to expand the built infrastructure giving better access to services to the growing urban population. Simultaneously, they are in need of growing their transport infrastructure whilst keeping a carbon constrained approach to providing services within this sector. The following section considers the variety of changes and interventions advanced at different scales to address these transport challenges.

3.5 Changing the city: Moving people safely on a low-carbon budget

More than half of the world's population lives in cities, and this proportion is expected to become two-thirds of the total population by 2050 (UN, 2014). Most of this population growth is taking place in medium-sized cities in developing countries with great needs of investment and planning on infrastructure development. Public services together with safe nonmotorised form of mobility solutions can offer safe mobility and accommodate increasing volumes of passengers (Cuenot et al., 2012). A properly planned and implemented urban mobility system can reduce the dependence on fossil fuels, increase air quality, and improve access to jobs and services (Tiwari, 2016). Yet, the cross-sectoral nature of urban mobility also makes tackling transport a complex policy area.

Addressing the complexity of urban transportation and attaining sustainable mobility requires a strong policy framework that supports integrated urban design, enables the shift to low-carbon transport options, and accelerates the uptake of low-carbon vehicle technologies (Figueroa et al., 2013). Planning efforts and elaboration of a strong national level urban mobility policy framework can help decision-makers carry out good practices and provide access to methodologies that can be transferred to the local context (D'Cruz et al., 2014). At the local level, there are a large number of possible interventions that authorities can initiate to impact travel demand, influence travel behaviour, and vehicle choice and use. City governments have a key role to play in shaping urban form and planning transport infrastructure (Ewing and Cervero, 2010). With integrated urban planning and the ability to regulate, fund, and often even operate public transport services (small, medium, and large), cities can substantially shape the modal structure of its transport system. Even though many policies that can influence the energy efficiency of the transport sector are driven by the national (e.g. vehicle and fuel tax) or regional level (e.g. European fuel efficiency regulation), there are several local policies that can have a similar effect on the choice and use of vehicles, such as road and parking pricing or access limitations (Lah, 2017a).

Within the framework of sustainable transport planning, specific measures are required to actively manage travel demand and to improve transport energy efficiency (Bongardt et al., 2011). This includes improvements of the public transport system as reliable and affordable alternative to the private car and measures targeting the efficiency of the vehicle fleet (Sims et al., 2014). Vital to the success of sustainable urban transport concepts is a mix of measures that improves the efficiency of the vehicle fleet, reduces travel distances via integrated land use planning, and provides modal alternatives to the private vehicle (Santos et al., 2010). Whereas the vehicles fleet policies fall only partially in the jurisdiction of city councils, the land use planning and modal efficiency are key areas of responsibility for local governments. The following strategies are specifically available to cities.

3.5.1 Sustainable urban mobility planning

Sustainable Urban Mobility Plans (SUMP) are strategic planning documents that guide the integration of all transport modes and work towards a sustainable transport system within a city. The development of a SUMP includes a number of steps from the identification of the main transport issues in a city to the development of a joint vision and the identification of specific measures and processes to implement actions (SOLUTIONS, 2016). The European Commission provides detailed guidelines,[v] which are aimed at cities, in the development of SUMPs. A vital component of the process of developing a SUMP is the involvement of stakeholders and the active participation of the public, through dialogue to identify mobility problems and find common objectives and selecting solutions (SOLUTIONS, 2016). An example of initiatives and of the work of international partnerships is represented by the Urban Electric Mobility Initiative (UEMI), which is working with cities on the implementation of sustainable urban mobility measures in the context of the UN Habitat New Urban Agenda. As part of urban implementation actions, the UEMI is working with cities to assess the opportunities for electric mobility concepts in their wider sustainable transport strategy.[w]

3.5.2 Integrated urban planning

Integrated land use planning focuses on higher densities, mixed use, and the integration of public transport and nonmotorised transport infrastructure (Hymel et al., 2010). Integrated land use planning only takes effect over longer time scales, but impacts are lasting (Caspersen et al., 2006). Local authorities can largely influence future travel patterns (Figueroa et al., 2014). Thereby, land use planning decisions of today can ease the traffic demand management task in the future. Cities can limit the increase in car use and dense urban structures with mixed use, for economic and social activities playing an important role in achieving an efficient transport system (Aljoufie, 2016). Integrated urban and transport planning improves a city connectivity, providing a

---

[v] http://eltis.org/content/sump-process Accessed 07-03-2019.
[w] http://www.uemi.net/city-actions.html Accessed 07-03-2019.

better mobility service and shorter trips for inhabitants, and bringing people and places closer together (Newman and Kenworthy, 2006). As part of this planning, cities may want to consider integrating fares, infrastructure, and operations for public transport and create easy connections with nonmotorised transport (SOLUTIONS, 2016). Combined, these factors can reduce travel distances, enhance the role of nonmotorised modes, and improve accessibility and efficiency of public transport. Cities that have applied integrated urban strategies successfully, such as Copenhagen and Freiburg, show that, the integration of public transport with wider urban planning and nonmotorised infrastructure makes urban mobility more convenient and efficient and provides better access to services, jobs, education, and social activities (Colville-Andersen, 2018). Urban scale matters with mega cities are prominently receiving large investment in public infrastructure and attracting more attention from national governments, whereas medium-sized cities that may need even more direct support to develop tailored mobility solutions often lack planning and policy capacities.

3.5.3 Low emission zones

Access restriction schemes are applied in many cities in Europe in different forms and generally aim to restrict access to city districts or specific traffic hotspots in the city (Cervero, 2004). Basic access restriction schemes are easy to adopt but require enforcement efforts to operate in the intended way (Santos, 2008). There are different types of access restriction schemes, including those that control access at specific points (e.g. when crossing a bridge), cordons, or areas (e.g. around a specific location), which may differentiate further between different types of vehicles or times of the day. Restricting access to certain areas of a city, normally the city centre, can have a direct effect on local air quality and noise pollution, reduce pressure for parking spaces, and improve traffic safety in the area. The effect on GHG depends on the design and complexity of the scheme and the provision and integration of modal alternatives. Whilst the schemes can be very effective in managing congestion and noise and air pollution, they may have also unintended consequences, for example, by banning higher polluting but potentially more fuel-efficient diesel cars from entering the city, this can induce travel by redirecting to longer routes or encourage the shift to a less-efficient petrol-powered car. Hence, access restrictions should be implemented in combination with other measures that maximise synergies.

3.5.4 Road user charging

One very effective option to improve traffic flows and reduce overall travel demand by avoiding and shifting traffic to more sustainable transport modes is congestion charging, which is an urban road pricing scheme for peak hours (Börjesson et al., 2012; Liu et al., 2009). Congestion charging lies at the intersection of traffic management and travel demand management, as information gained from real-time traffic information systems could be used to improve the pricing mechanisms of congestion charging by introducing real-time variable pricing systems, which can encourage more efficient travel behaviour. Congestion charging systems have been operating in Singapore for several decades and have been implemented more recently in London and Stockholm.

As early as 1975, the road pricing was implemented in Singapore to manage the choked streets of the rapidly growing city. First an area license system was established, which required a permit to enter Singapore's central area (Ang, 1990). The city entry charge boosted public transport patronage almost immediately after its introduction and led to a 45% reduction in traffic, road site accidents decreased by 25%, and average travel speeds increased from about 20 to over 30 km/h (Lah, 2015). The system resulted in a public transport share of over 60% in daily traffic, an increase of nearly 20% (SOLUTIONS, 2016). The success of the system in improving infrastructure capacity, safety, and air quality and reducing travel demand, fuel use, and greenhouse gas emissions inspired the congestion charge systems in London and Stockholm and provided the basis for several feasibility studies for similar schemes for cities around the world (Prud'homme and Bocarejo, 2005).

3.5.5 Moving people supporting walking and cycling

Nonmotorised modes, that is, cycling and walking, can take a substantial share of the urban transport task, in particular on short distances, and help reduce congestion, emissions, and energy consumption. Walking and cycling is especially suitable for urban transport as in cities, the majority of trips are short distance trips (below 5 km) (Moudon and Lee, 2003). One fundamental advantage of nonmotorised modes is that they are low-cost modes compared with other transport options, not only for the individual but also for public authorities (Tirachini and Hensher, 2012). Wider benefits of cycling, electric scooters, and walking are health benefits for the cyclists or pedestrians and environmental and economic benefits resulting from zero emissions and energy consumption from use of low-carbon modes (Santos et al., 2010). Walking and cycling infrastructure consumes less space compared with roads, yet this infrastructure is often neglected in transport planning. The provision and maintenance of infrastructure for pedestrians and cyclists is crucial to make these modes more attractive. Separate crossing signals, cycle lanes, and buffers between road and lane can improve cycling and walking safety (Santos et al., 2010). Alongside public investments to build up a proper infrastructure for nonmotorised travel, the promotion of a shift from personal vehicles to the use of nonmotorised modes can be supported by efforts from civil society organisation initiatives through forms of social innovation as discussed in what follows.

3.5.6 Social innovation to promote sustainable mobility: How does it work?

When innovation is not about cutting edge technology but solving a social problem, the engagement of citizens and organisations has the potential to alter the structure of innovation systems and public and private governance (van der Have and Rubalcaba, 2016). Work of civil society organisations in social innovation practices[x]

---

[x] Social innovations are new solutions (products, services, models, markets, processes etc.) that simultaneously meet a social need (more effectively than existing solutions) and lead to new or improved capabilities and relationships and better use of assets and resources. In other words, social innovations are both good for society and enhance society's capacity to act (Caulier-Grice et al., 2012).

can potentially contribute to improving mutual understanding among individuals and communities and to facilitate advancing coordinated actions (Moulaert et al., 2013). Civil society's current engagement in providing sustainability transport practices and services has been expanding, and in some cases, it works towards changing unsustainable social, ecological, economic, and cultural patterns (Frantzeskaki et al., 2018). Civil society's collective social innovative practices are necessary because they can facilitate achieving a level of social coordination that is based on a common interpretation of the social context (Habermas, 1984; Cajaiba-Santana, 2014). Social innovative actions can offer space for public debate and support societal participatory processes of communication and action. Public acceptance of new mobility practices and better opportunities to attain a legitimate support to the adoption of new mobility patterns and services, can be aided by processes of social innovation that favor human communication and forms of collective action. An exemplary case is the introduction and use of bicycle infrastructure and sharing space for bicycling in cities. Collective practices can potentially create a positive cycle of reproduction of safe bicycle use and acceptance of that practice. A recent study evaluated the contribution of civil society to configuration of probicycling values among the residents of several European urban areas (Figueroa et al., 2019). The study reflected how even, when social innovation practices did not always work to support sustainability outcomes, they contributed to bringing forward a shared sense of normalisation and acceptance of the use of less carbon-intensive mobility transport modes and patterns. The study reviewed the role of voluntary civil society associations in contributing innovative practices to promote the use of bicycles. These groups contributed by helping showcase different safe bicycling practices that the rest of the community accepted as worth imitating, supporting, and sustaining (Cajaiba-Santana, 2014; Figueroa et al., 2019).

Deep decarbonisation of personal mobility requires profound changes in travel patterns and modes of passenger travelling. There is a need to further exploring how collective actions advanced by organised societal groups can create conditions of acceptability towards a fast transport sustainability transition (Frantzeskaki et al., 2018; Figueroa et al., 2019).

3.5.7 Public transport infrastructure, operation and vehicles

A reliable and affordable public transport system is a key element of a sustainable urban transport concept. Whilst providing a similar level of mobility, public transport with full passengers on board only requires a fraction of energy per passenger and space compared with the private car (Dirgahayani, 2013). Public transport not only contributes to lower energy consumption and emissions but also reduces congestion, which improves traffic flows and reduces travel times. Enhancing the share of public transport in urban passenger transport yields the potential to mitigate rising energy consumption and emissions because public transport is typically more than twice as energy efficient per passenger kilometres as individual motorised transport (Sims et al., 2014). Investments in capacity and reliability and physical integration of public transport with walking and cycling are essential to influence individuals to shift modes towards public transport from individual motorised transport

(Pucher and Buehler, 2010). Reliability is an important factor for modal choice. The predictability of travel times with metro, light rail, and/or bus rapid transit compared with a journey in the private car may provide enough incentive to shift from private to public transport (Brownstone and Small, 2005; SOLUTIONS, 2016). Public transport systems generally require substantial public investments and their operation often requires continued public subsidies. Linking public transport investments with road user charging and parking pricing schemes can help to reduce the pressure on public funds and at the same time create disincentives to use the private car and encourage the use of public transport (Johnstone and Karousakis, 1999).

4 Harnessing stakeholder's synergies

Harnessing human dimensions to achieve a transition for sustainable energy for transport is challenging and necessary. This chapter has reviewed the multifaceted complex character of the human dimensions of mobility. This complexity is observed in the form of the accelerating passenger mobility demand, which is reinforced in emerging fast-growing cities. Human complexity is difficult to address at a large societal level. Therefore, there is an emphasis in understanding individual mobility decisions, which may be in conflict with societal objectives. In practice, as we have seen throughout the chapter, harnessing human dimensions for the decarbonisation of the transport sector will vary greatly by context, organisational, and institutional capacities of individual actors and decision-makers. Sustainable energy for transport is an area where the impacts of organised interest groups of actors and businesses can be substantial. A recent example is with introduction of electric mobility in cities, where many actors, besides local authorities, are playing roles. We have discussed how policy makers at state and local levels have an important role to play, setting planning conditions and legal frameworks for private actors to operate.

Civil society encompassing actors in grassroot organisations, community-based organisations, advocacy groups such as nongovernmental organisations (NGOs), coalitions, professional associations, and other organisations that operate between the state, individuals, and the market have also capacities to generate innovative concepts and address public and social problems of many kinds. The important contributions from civil society to sustain transformative ideas in society were earlier discussed as forms of social innovation. The role of business is key in the transition to sustainable energy for transport. Large automotive companies drive innovation on vehicle efficiency and new car models. Oil companies produce and refine the fuels used in transportation. Large information technology companies are investing heavily in the digital transformation of personal mobility. Numerous synergies exist between these different large private companies and public authorities. The public sector still plays a central role in planning and implementing policies in practice, they can help in facilitating the engagement of relevant private stakeholders and transport users.

Table 1 summarises how the three key approaches to manage energy demand for decarbonisation in transport highlighted from the literature through this chapter

Table 1 Examples of how approaches and objectives for managing transport energy demand for climate mitigation link to other sustainable development policy objectives and actors.

Managing energy for transport carbon mitigation approach and objective	Economic implications and actors	Social implications and actors	Environmental implications and actors
Activity: Avoid vehicle travel by reduced trip distances, for example, by developing more compact, mixed communities, and telework (avoid)	Reduced congestion: *Local authorities (v) ↑* More efficient freight distribution: *Businesses and associations ↑ Economic development ministry (v) ↑*	Improved access and mobility *Social development ministry ↑* Accident reductions *Health ministry ↑*	Reduced land consumption *Local planning authority (v) →*
Structure: Shift to low-carbon transport modes, such as public transport, walking, and cycling (shift)	Improved productivity due to reduced urban congestion and travel times across all modes *Local authorities (v) ↑*	Reduced exposure to air pollution Health benefits from shifts to active transport modes *Local authorities (v) ↑*	Ecosystem benefits due to reduced local air pollution *Local environmental department and national ministry ↑*
Intensity: Improve the efficiency of the vehicle fleet and use (improve)	Reduced transport costs for businesses and individuals *Local authorities (v) and economic and social development ministries ↑*	Health benefits due to reduced urban air pollution *Health ministry ↑*	Ecosystem and biodiversity benefits due to reduced urban air pollution *Local authorities (v) ↑*
Fuels: Reduce the carbon content of fuels and energy carriers (improve)	Improved energy security *Economic development ministry ↑* Reduce trade imbalance for oil-importing countries *Finance ministry (v) ↑*	A shift to diesel can improve efficiency but tends to increase air pollution *Health and environment ministries (v) ↓*	Potential adverse effects of biofuels on biodiversity and land use *Environment and agriculture ministries (v) ↓*

Note that the selection is not exhaustive and depends on the policy environment. Key to read the table: positive ↑, negative ↓, uncertain →, (v) potential veto player.
Modified from Lah (2017a).

(avoid, shift, and improve) have synergies with other sustainable transport goals and societal actors. There are challenges with advancing collaborations between different societal actors like public authorities and business entrepreneurs. For example, in the case of initiatives for new business models or smart solutions, an uneven organisational capacity from local authorities limits the interaction with (disruptive) business models (e.g. Uber and Lyft).

Similarly, in interactions between private and public actors, there may be difficulties and privacy concerns in maintaining data openness. Therefore, absent sharing information makes difficult or impossible to run some new mobility interventions. This is the case, for example, when data are owned by one actor and are not being shared by another actor who may need these to implement a new mobility service. In the transport sector, there is a race to collect and mine the digital traces of data that people leave behind as they travel. This raises the problem that, for the companies pushing with new solutions, human mobility seems to have become a source of data provision to be mined and one that can be managed via algorithms. Public authorities are still in the position where there is a transport planning tradition that seeks to involve participants in the design of services that ultimately will impact the quality of life of the society at large. The raise of private companies harnessing data and suggesting new transport service applications (e.g. car sharing, Uber, MaaS, and bike sharing) is bringing new models and more opportunities for moving people and goods around. The task of evaluating and integrating the solutions is still in the hand of public authorities that have at best a short-handed technical capacity to assess and determine what can be the best solution in the face of multiple and competing goals to reach.

5 Conclusions

Nothing short than an unprecedented transformation of the systems supporting the movement of goods and people worldwide will deliver the necessary transport sector's contribution of GHG emissions reduction necessary to limit the increase in global average temperature to 1.5°C above preindustrial levels by 2050. Because transport is so central to all countries' economies and to achieve a fast and complete global decarbonisation, there are strong reasons to keep the management of transport energy demand at the centre of debate in research and policy action. This chapter highlights that a focus on energy use in transport attends only partially to the more comprehensive goals of sustainable transport concerning equity, accessibility, and contributions to a better quality of life. Given the pressing task of reducing emissions in transport, the focus of research and policy implementation has been on shifting fuels, avoiding travel, and improving vehicles, and without doubt, these are the three most direct forms of advancing transport decarbonisation. Still, the implementation of those related policies requires attentive focus on technologies and technical practices, and even with technological advances at hand, implementation has not proven easy. This is because the task requires a deeper understanding of the role of the human dimension to help managing transport energy demand. This chapter reviewed current interdisciplinary approaches spanning a variety of dimensions of transport sustainability. A collective response is necessary for a low-carbon transport transition to contribute to global goals. But a collective transition requires implementation processes that ensure that the proposed changes are acceptable and supported by local population. With the continuous growth of transport demand globally, the long-run feasibility of the low-carbon energy transition is at stake. The chapter highlights why there is a need to prioritise and support social innovation elements that can strengthen a collective transformation. The chapter also explored the most important current policy and technological alternatives of decarbonisation, emphasising how every alternative has issues that relate to human dimensions of sustainable mobility. The tension of providing low-carbon mobility services in cities and countries experiencing a fast-growing transport energy demand can be approached with consideration of cobenefits to assist policy implementation and help gain public acceptance of interventions. Cobenefits exist in areas such as health, reduced air pollution, noise and congestion, and increased quality of life in cities (McCollum et al., 2018). The question with alternative fuels and new vehicles is on affordability, availability, and the kind of instruments that facilitate scaling up the use of new technologies. The chapter highlights the need for addressing the human dimensions not only through policy planning but also through a combined role for intelligent models and organised civil society arrangements supporting forms of social innovation. The chapter reflects on the need for approaching transport policy planning and infrastructure appraisal that accounts for wider socio-economic cobenefits suggesting that taking an integrated approach to sustainability mobility is vital to cost-effectively reduce transport sector energy demand and GHG emissions (IPCC, 2014; Figueroa et al., 2014). The approach offered for minimising rebound effects is through the implementation of measures in package combinations that maximise synergies between solutions. Overall private motorised travel demand reduction and shifts to public modes would need to be substantially strong and accompanied by efficiency improvements within the vehicle fleet (Figueroa et al., 2014; Fulton et al., 2013). Reaching the full environmental potential of interventions makes indispensable a mind-set change, and this highlights the importance of advancing research on the human dimensions of the sustainable energy transition. The transition to sustainable energy for transport requires elements of policy integration and stakeholder cooperation that a pure reliance on data and information technology cannot by itself resolve. To achieve cooperative and integrative approaches, direct human understanding and interaction is needed to find synergies among the involved stakeholders. Aside from contributions to mitigate climate change, the transition to sustainable transport energy use has many positive impacts and entry points for engagement; therefore, civil society actors, businesses, and policy makers have a variety of dimensions where they can begin to intervene in this transition. Irrespective of who is proposing a solution, if it is new technologies from the market, a social innovation from organised civil societal actors, or a soft or hard policy intervention from government, actions and solutions will need to become acceptable, scaled up, and supported by people to be truly sustainable.

A rapid shift and build-up of the physical infrastructure to support low-carbon fuels and vehicles requires the allocation of societal resources. Concerted near-to-midterm actions, policy, and solution implementation by government, businesses, and civil society to succeed in addressing the major turn of individual and collective mobility behaviour into a low-carbon trajectory, hinges upon crafting an integrated approach to understanding the scope of the human dimensions for sustainable transport.

References

Ajanovic, A., Haas, R., Wirl, F., 2016. Reducing CO2 emissions of cars in the EU: analyzing the underlying mechanisms of standards, registration taxes and fuel taxes. Energy Efficiency 9 (4), 925–937. https://doi.org/10.1007/s12053-015-9397-4.

Ali, Q., Ahmad, N., Malik, A., Ali, G., et al., 2018. Issues, challenges, and research opportunities in intelligent transport system for security and privacy. Appl. Sci. 8 (10), 1964. https://doi.org/10.3390/app8101964.

Aljoufie, M., 2016. Exploring the determinants of public transport system planning in car-dependent cities. Procedia Soc. Behav. Sci. 216, 535–544. https://doi.org/10.1016/j.sbspro.2015.12.013.

Amini, M.H., Moghaddam, M.P., Karabasoglu, O., 2017. Simultaneous allocation of electric vehicles' parking lots and distributed renewable resources in smart power distribution networks. Sustain. Cities Soc. 28, 332–342. https://doi.org/10.1016/j.scs.2016.10.006.

Anable, J., 2005. 'Complacent Car Addicts' or 'Aspiring Environmentalists'? Identifying travel behaviour segments using attitude theory. Transp. Policy 12 (1), 65–78. https://doi.org/10.1016/j.tranpol.2004.11.004.

Anenberg, S.C., Schwartz, J., Shindell, D., Amann, M., et al., 2012. Global air quality and health co-benefits of mitigating near-term climate change through methane and black carbon emission controls. Environ. Health Perspect. 120 (6), 831–839. https://doi.org/10.1289/ehp.1104301.

Ang, B.W., 1990. Reducing traffic congestion and its impact on transport energy use in Singapore. Energy Policy 18 (9), 871–874. https://doi.org/10.1016/0301-4215(90)90067-E.

Azevedo, I.M.L., 2014. Consumer end-use energy efficiency and rebound effects. Annu. Rev. Environ. Resour. 39 (1), https://doi.org/10.1146/annurev-environ-021913-153558.

Bamberg, S., Fujii, S., Friman, M., Gärling, T., 2011. Behaviour theory and soft transport policy measures. Transp. Policy 18 (1), 228–235. https://doi.org/10.1016/j.tranpol.2010.08.006.

Banister, D., 2008. The sustainable mobility paradigm. Transp. Policy 15 (2), 73–80. https://doi.org/10.1016/j.tranpol.2007.10.005.

Banister, D., Anderton, K., Bonilla, D., Givoni, M., et al., 2011. Transportation and the environment. Annu. Rev. Environ. Resour. 36 (1), 247–270. https://doi.org/10.1146/annurev-environ-032310-112100.

Bardhi, F., Eckhardt, G.M., 2012. Access-based consumption: the case of car sharing: Table 1. J. Consum. Res. 39 (4), 881–898. https://doi.org/10.1086/666376.

Barth, M.J., Wu, G., Boriboonsomsin, K., 2015. Intelligent transportation systems and greenhouse gas reductions. Curr. Sustain. Renew. Energy Rep. 2 (3), 90–97. https://doi.org/10.1007/s40518-015-0032-y.

Bongardt, D., Scmid, D., Huizenga, C., Litman, T., 2011. Sustainable Transport Evaluation: Developing Practical Tools for Evaluation in the Context of the CSD Process.

Börjesson, M., Eliasson, J., Hugosson, M.B., Brundell-Freij, K., 2012. The Stockholm congestion charges—5 years on. Effects, acceptability and lessons learnt. Transp. Policy 20, 1–12. https://doi.org/10.1016/j.tranpol.2011.11.001.

Brownstone, D., Small, K.A., 2005. Valuing time and reliability: assessing the evidence from road pricing demonstrations. Transp. Res. A Policy Pract. 39 (4), 279–293. https://doi.org/10.1016/j.tra.2004.11.001.

Cajaiba-Santana, G., 2014. Social innovation: moving the field forward. A conceptual framework. Technol. Forecast. Soc. Chang. 82, 42–51. https://doi.org/10.1016/j.techfore.2013.05.008.

Caspersen, O.H., Konijnendijk, C.C., Olafsson, A.S., 2006. Green space planning and land use: an assessment of urban regional and green structure planning in Greater Copenhagen. Geogr. Tidsskr. 106 (2), 7–20.

Caulier-Grice, J., Davies, A., Patrick, R., Norman, W., 2012. Defining Social Innovation. Part One of Social Innovation Overview: A deliverable of the project: "The theoretical, empirical and policy foundations for building social innovation in Europe" (TEPSIE), European Commission – 7th Framework Programme, Brussels: European Commission, DG Research. Available at: https://youngfoundation.org/wp-content/uploads/2012/12/TEPSIE.D1.1.Report.DefiningSocialInnovation.Part-1-defining-social-innovation.pdf. (Accessed 18 July 2019).

Cervero, R., 2004. Transit-Oriented Development in the United States: Experiences, Challenges and Prospects.

Colville-Andersen, M., 2018. Copenhagenize: The Definitive Guide to Global Bicycle Urbanism. Island Press, Canada.

Creutzig, F., 2016. Evolving narratives of low-carbon futures in transportation. Transp. Rev. 36 (3), 341–360. https://doi.org/10.1080/01441647.2015.1079277.

Cuenot, F., Fulton, L., Staub, J., 2012. The prospect for modal shifts in passenger transport worldwide and impacts on energy use and CO2. Energy Policy 41, 98–106. https://doi.org/10.1016/j.enpol.2010.07.017.

D'Cruz, C., Sonia Fadrigo Cadornigara, S., Satterthwaite, D., International Institute for Environment and Development, 2014. Tools for Inclusive Cities: The Roles of Community-Based Engagement and Monitoring in Reducing Poverty. IIED, London.

Dahiya, B., 2012. Cities in Asia, 2012: demographics, economics, poverty, environment and governance. Cities 29, S44–S61. https://doi.org/10.1016/j.cities.2012.06.013.

Dahiya, B., 2016. ASEAN economic integration and sustainable urbanization. J. Urban Cult. Res. 13, https://doi.org/10.14456/jucr.2016.10. (Accessed 3 January 2019).

de Coninck, H., Revi, A., Babiker, P., Bertoldi, M., et al., 2018. Strengthening and implementing the global response. In: Global Warming of 1.5°C. Sustainable Development, and Efforts to Eradicate Poverty. Intergovernmental Panel on Climate Change, Cambridge University Press, Cambridge. An IPCC Special Report on the impacts of global warming of 1.5°C above pre-industrial levels and related global greenhouse gas emission pathways, in the context of strengthening the global response to the threat of climate change.

de Souza, A.M., Brennand, C.A., Yokoyama, R.S., Donato, E.A., et al., 2017. Traffic management systems: a classification, review, challenges, and future perspectives. Int. J. Distrib. Sens. Netw. 13 (4), https://doi.org/10.1177/1550147716683612. 1550147716683612.

den Boer, E., Van Essen, H., Brouwer, F., Pastori, E., et al., 2011. Potential of Modal Shift to Rail Transport. 119. Available at: http://www.cedelft.eu/publicatie/potential_of_modal_shift_to_rail_transport/1163?PHPSESSID=85969a496d79705462017a60f30353cc.

Dirgahayani, P., 2013. Environmental co-benefits of public transportation improvement initiative: the case of Trans-Jogja bus system in Yogyakarta, Indonesia. J. Clean. Prod. 58, 74–81. https://doi.org/10.1016/j.jclepro.2013.07.013.

EEA, 2011. Monitoring the CO_2 Emissions From New Passenger Cars in the EU: Summary of Data for 2010.

Estupinan, N., Gomez-Lobo, A., Munoz-Raskin, R., Serebrisky, T., 2007. Affordability and Subsidies in Public Urban Transport: What Do We Mean, What Can Be Done? SSRN eLibrary. Available at: http://papers.ssrn.com/sol3/papers.cfm?abstract_id=1073383. (Accessed 19 October 2011).

Ewing, R., Cervero, R., 2010. Travel and the built environment. J. Am. Plan. Assoc. 76 (3), 265–294. https://doi.org/10.1080/01944361003766766.

Fargione, J.E., Plevin, R.J., Hill, J.D., 2010. The ecological impact of biofuels. Annu. Rev. Ecol. Syst. 41 (1), 351–377. https://doi.org/10.1146/annurev-ecolsys-102209-144720.

Fellows, N.T., Pitfield, D.E., 2000. An economic and operational evaluation of urban car-sharing. Transp. Res. D Transp. Environ. 5 (1), 1–10. https://doi.org/10.1016/S1361-9209(99)00016-4.

Figueroa, M.J., Fulton, L., Tiwari, G., 2013. Avoiding, transforming, transitioning: pathways to sustainable low carbon passenger transport in developing countries. Curr. Opin. Environ. Sustain. 5 (2), 184–190. https://doi.org/10.1016/j.cosust.2013.02.006.

Figueroa, M.J., Nielsen, T.A.S., Siren, A., 2014. Comparing urban form correlations of the travel patterns of older and younger adults. Transp. Policy 35, 10–20. https://doi.org/10.1016/j.tranpol.2014.05.007.

Figueroa, M.J., Navrátil, J., Turrini, A., Krlev, G., 2019. Social innovation in environmental sustainability: promoting sharing public spaces for bicycle use. In: Anheier, H.K., Krlev, G., Mildenberger, G. (Eds.), Social Innovation: Comparative Perspectives. Routledge Studies in Social Enterprise & Social InnovationRoutledge, New York, NY, pp. 149–172.

Frantzeskaki, N., Dumitru, A., Wittmayer, J., Avelino, F., et al., 2018. To transform cities, support civil society. In: Elmqvist, T., Bai, X., Frantzeskaki, N., Griffith, C., et al. (Eds.), Urban Planet. first ed.. Cambridge University Press, pp. 281–302. https://doi.org/10.1017/9781316647554.016. (Accessed 18 June 2018).

Fridstrøm, L., Østli, V., 2017. The vehicle purchase tax as a climate policy instrument. Transp. Res. A Policy Pract. 96, 168–189. https://doi.org/10.1016/j.tra.2016.12.011.

Fu, M., Andrew Kelly, J., 2012. Carbon related taxation policies for road transport: efficacy of ownership and usage taxes, and the role of public transport and motorist cost perception on policy outcomes. Transp. Policy 22, 57–69. https://doi.org/10.1016/j.tranpol.2012.05.004.

Fulton, L., Lah, O., Cuenot, F., 2013. Transport pathways for light duty vehicles: towards a 2° scenario. Sustainability 5 (5), 1863–1874. https://doi.org/10.3390/su5051863.

Gillingham, K., Kotchen, M.J., Rapson, D.S., Wagner, G., 2013. Energy policy: the rebound effect is overplayed. Nature 493 (7433), 475–476. https://doi.org/10.1038/493475a.

Glotz-Richter, M., 2012. Car-Sharing—"Car-on-call" for reclaiming street space. Procedia Soc. Behav. Sci. 48, 1454–1463. https://doi.org/10.1016/j.sbspro.2012.06.1121.

Haase, D., 2013. Shrinking cities, biodiversity and ecosystem services. In: Elmqvist, T., Fragkias, M., Goodness, J., Güneralp, B., et al. (Eds.), Urbanization, Biodiversity and Ecosystem Services: Challenges and Opportunities: A Global Assessment. Springer Netherlands, Dordrecht, pp. 253–274. https://doi.org/10.1007/978-94-007-7088-1_12. (Accessed 3 January 2019).

Habermas, J., 1984. The Theory of Communicative Action: Reason and the Rationalization of Society. Beacon Press, Boston, MA.

Hawkins, T.R., Singh, B., Majeau-Bettez, G., Strømman, A.H., 2013. Comparative environmental life cycle assessment of conventional and electric vehicles: LCA of conventional and electric vehicles. J. Ind. Ecol. 17 (1), 53–64. https://doi.org/10.1111/j.1530-9290.2012.00532.x.

Hensher, D.A., 2008. Influence of vehicle occupancy on the valuation of car driver's travel time savings: identifying important behavioural segments. Transp. Res. A Policy Pract. 42 (1), 67–76.

Hensher, D.A., 2017. Future bus transport contracts under a mobility as a service (MaaS) regime in the digital age: are they likely to change? Transp. Res. A Policy Pract. 98, 86–96. https://doi.org/10.1016/j.tra.2017.02.006.

Hill, J., Polasky, S., Nelson, E., Tilman, D., et al., 2009. Climate change and health costs of air emissions from biofuels and gasoline. Proc. Natl. Acad. Sci. 106 (6), 2077–2082. https://doi.org/10.1073/pnas.0812835106.

Ho, C.Q., Hensher, D.A., Mulley, C., Wong, Y.Z., 2018. Potential uptake and willingness-to-pay for Mobility as a Service (MaaS): a stated choice study. Transp. Res. A Policy Pract. 117, 302–318. https://doi.org/10.1016/j.tra.2018.08.025.

Hüging, H., Glensor, K., Lah, O., 2014. Need for a holistic assessment of urban mobility measures—review of existing methods and design of a simplified approach. Transp. Res. Procedia 4, 3–13. https://doi.org/10.1016/j.trpro.2014.11.001. Sustainable Mobility in Metropolitan Regions. mobil.TUM 2014. International Scientific Conference on Mobility and Transport. Conference Proceedings.

Hymel, K.M., Small, K.A., Dender, K.V., 2010. Induced demand and rebound effects in road transport. Transp. Res. B Methodol. 44 (10), 1220–1241.

IEA, 2013. World Energy Outlook 2013. Organization for Economic Cooperation and Development, Paris.

IEA, 2018a. Global Electric Vehicle (EV) Outlook 2018 [Online]. Available at: https://www.iea.org/gevo2018/.

IEA (2018b) Global EV Outlook 2018 [Online]. Available at: www.iea.org.

IEA, 2018b. World Energy Outlook. OECD.

International Transport Forum, 2017. ITF Transport Outlook 2017. OECDhttps://doi.org/10.1787/9789282108000-en. (Accessed 7 January 2019).

IPCC, 2014. Climate change 2014—Mitigation of climate change, 5th assessment report. Intergovernmental Panel on Climate Change, Working Group III. Cambridge University Press. Available at http://www.ipcc-wg3.de/assessment-reports/fifth-assessment-report.

Johnstone, N., Karousakis, K., 1999. Economic incentives to reduce pollution from road transport: the case for vehicle characteristics taxes. Transp. Policy 6 (2), 99–108.

Kahn Ribeiro, S., Figueroa, M.J., 2012. Energy end-use: transportation. In: Global Energy Assessment—Toward a Sustainable Future. International Institute for Applied Systems Analysis and Cambridge University Press, Vienna/Cambridge/New York, NY, pp. 575–648.

Kamargianni, M., Matyas, M., 2017. The Business Ecosystem of Mobility-as-a-Service. Transportation Research Board, Washington, DC. Available at: http://discovery.ucl.ac.uk/id/eprint/10037890.

Koppenjan, J.F.M., Enserink, B., 2009. Public–private partnerships in urban infrastructures: reconciling private sector participation and sustainability. Public Adm. Rev. 69 (2), 284–296. https://doi.org/10.1111/j.1540-6210.2008.01974.x.

Lah, O., 2014. The barriers to vehicle fuel efficiency and policies to overcome them. Eur. Transp. Res. Rev.. Available at: http://link.springer.com/journal/12544.

Lah, O., 2015. The barriers to low-carbon land-transport and policies to overcome them. Eur. Transp. Res. Rev. 7 (1), 1–12.

Lah, O., 2017a. Factors of Change: the influence of policy environment factors on climate change mitigation strategies in the transport sector. Transp. Res. Procedia 25, 3495–3510. https://doi.org/10.1016/j.trpro.2017.05.265.

Lah, O., 2017b. Sustainable development synergies and their ability to create coalitions for low-carbon transport measures. Transp. Res. Procedia 25, 5088–5098. https://doi.org/10.1016/j.trpro.2017.05.495.

Litman, T., 2006. Parking Management: Strategies, Evaluation and Planning. Victoria Transport Policy Institute. http://www.vtpi.org.

Liu, Z., Li, C., Li, C., 2009. Traffic impact analysis of congestion charge in mega cities. J. Transp. Syst. Eng. Inf. Technol. 9 (6), 57–62. https://doi.org/10.1016/S1570-6672(08)60088-4.

Lockwood, M., 2015. Fossil fuel subsidy reform, rent management and political fragmentation in developing countries. New Polit. Econ. 20 (4), 475–494. https://doi.org/10.1080/13563467.2014.923826.

Loose, W., 2010. The State of European Car-Sharing. 129.

Lucas, K., Pangbourne, K., 2014. Assessing the equity of carbon mitigation policies for transport in Scotland. Case Stud. Transp. Policy 2 (2), 70–80. https://doi.org/10.1016/j.cstp.2014.05.003.

Maione, M., Fowler, D., Monks, P.S., Reis, S., et al., 2016. Air quality and climate change: designing new win-win policies for Europe. Environ. Sci. Pol. 6548–6557. https://doi.org/10.1016/j.envsci.2016.03.011.

Martin, E., Shaheen, S., 2011. The impact of carsharing on public transit and non-motorized travel: an exploration of North American carsharing survey data. Energies 4 (11), 2094–2114. https://doi.org/10.3390/en4112094.

McCollum, D.L., Echeverri, L.G., Busch, S., Pachauri, S., et al., 2018. Connecting the sustainable development goals by their energy inter-linkages. Environ. Res. Lett. 13 (3), 033006. https://doi.org/10.1088/1748-9326/aaafe3.

McLaren, D., Agyeman, J., 2017. Sharing Cities: A Case for Truly Smart and Sustainable Cities. MIT Press. Google-Books-ID: MBIxvgAACAAJ.

Moudon, A.V., Lee, C., 2003. Walking and bicycling: an evaluation of environmental audit instruments. Am. J. Health Promot. 18 (1), 21–37.

Moulaert, F., MacCallum, D., Mehmood, A., Hamdouch, A., 2013. The International Handbook on Social Innovation. Edward Elgar Publishing, UK. Available at: http://www.elgaronline.com.esc-web.lib.cbs.dk/view/9781849809986.xml. (Accessed 4 July 2014).

Mulley, C., Nelson, J.D., Wright, S., 2018. Community transport meets mobility as a service: on the road to a new a flexible future. Res. Transp. Econ. 69, 583–591. https://doi.org/10.1016/j.retrec.2018.02.004.

Newbery, D.M., 2005. Why tax energy? Towards a more rational policy. Energy J. 26 (3), 1–39.

Newman, P., Kenworthy, J., 1989. Gasoline consumption and cities: a comparison of U. S. cities with a global survey. J. Am. Plan. Assoc. 55 (1), 24–37.

Newman, P., Kenworthy, J., 2006. Urban design to reduce automobile dependence. Opolis 2 (1), 35–52.

Newman, P., Kenworthy, J., 2011. Evaluating the transport sector's contribution to greenhouse gas emissions and energy consumption. In: Technologies for Climate Change Mitigation— Transport Sector. UNEP Riso Center, pp. 7–23. Available at: http://www.uneprisoe.org/TNA-Guidebook-Series.

Nijland, H., van Meerkerk, J., 2017. Mobility and environmental impacts of car sharing in the Netherlands. Environ. Innov. Soc. Trans. 23, 84–91. https://doi.org/10.1016/j.eist.2017.02.001.

Noy, K., Givoni, M., 2018. Is 'smart mobility' sustainable? examining the views and beliefs of transport's technological entrepreneurs. Sustainability 10 (2), 422. https://doi.org/10.3390/su10020422.

Nykvist, B., Nilsson, M., 2015. Rapidly falling costs of battery packs for electric vehicles. Nat. Clim. Chang. 5 (4), 329–332. https://doi.org/10.1038/nclimate2564.

Pitsiava-Latinopoulou, M., Basbas, S., Papoutsis, K., Sdoukopoulos, E., 2012. Parking policies for supporting sustainable mobility. Procedia Soc. Behav. Sci. 48, 897–906. https://doi.org/10.1016/j.sbspro.2012.06.1067.

Prud'homme, R., Bocarejo, J.P., 2005. The London congestion charge: a tentative economic appraisal. Transp. Policy 12 (3), 279–287. https://doi.org/10.1016/j.tranpol.2005.03.001.

Pucher, J., Buehler, R., 2010. Walking and cycling for healthy cities. Built Environ. 36 (4), 391–414. https://doi.org/10.2148/benv.36.4.391.

Roy, J., Tschakert, P., et al., 2018. Sustainable development, poverty eradication and reducing inequalities. In: Global Warming of 1.5°C. Available at: https://www.ipcc.ch/sr15/chapter/chapter-5/. (Accessed 3 January 2019).

Ruzzenenti, F., Basosi, R., 2008. The rebound effect: an evolutionary perspective. Ecol. Econ. 67 (4), 526–537. https://doi.org/10.1016/j.ecolecon.2008.08.001.

Santos, G., 2008. London congestion charging. In: Brookings-Wharton Papers on Urban Affairs. pp. 177–234.

Santos, G., Behrendt, H., Teytelboym, A., 2010. Part II: Policy instruments for sustainable road transport. Res. Transp. Econ. 28 (1), 46–91. https://doi.org/10.1016/j.retrec.2010.03.002.

Schaeffer, R., et al., 2015. Who drives climate-relevant policies in Brazil. IDS Evidence Report 132 IDS, Brighton.

Schor, J., 1998. The overspent American: Upscaling, downshifting, and the new consumer, first ed. Basic Books, New York, NY.

Shaheen, S.A., Cohen, A.P., 2013. Carsharing and personal vehicle services: worldwide market developments and emerging trends. Int. J. Sustain. Transp. 7 (1), 5–34. https://doi.org/10.1080/15568318.2012.660103.

Shaheen, S., Cohen, A., 2018. Shared ride services in North America: definitions, impacts, and the future of pooling. Transp. Rev. 1–16. https://doi.org/10.1080/01441647.2018.1497728.

Short, J., Kopp, A., 2005. Transport infrastructure: investment and planning. Policy and research aspects. Transp. Policy 12 (4), 360–367. https://doi.org/10.1016/j.tranpol.2005.04.003.

Shoup, D.C., 2011. The High Cost of Free Parking. Planners Press, American Planning Association, Chicago.

Shove, E., Pantzar, M., Watson, M., 2012. The Dynamics of Social Practice: Everyday Life and How It Changes. Sage, London, NY. http://public.eblib.com/choice/publicfullrecord.aspx?p=880778_0. (Accessed July 24, 2019).

Sims, R., Schaeffer, R., Creutzig, F., Cruz-Núñez, X., et al., 2014. Transport. In: Edenhofer, O., Pichs-Madruga, R., Sokona, Y., Farahani, E., Kadner, S., Seyboth, K., … Minx, J.C. (Eds.), Climate Change 2014: Mitigation of Climate Change. Cambridge University Press, Cambridge, United Kingdom and New York, NY, pp. 599–670. Contribution of Working Group III to the Fifth Assessment Report of the Intergovernmental Panel on Climate Change.

Skovgaard, J., van Asselt, H., 2019. The politics of fossil fuel subsidies and their reform: implications for climate change mitigation. Wiley Interdiscip. Rev. Clim. Chang. 10 (4), e581. https://doi.org/10.1002/wcc.581.

SLoCaT, 2018. Transport and climate change global status report 2018. Available at: http://slocat.net/tcc-gsr.

Small, K.A., Verhoef, E.T., 2007. The Economics of Urban Transportation. Routledge, New York.

Smith, G., Sochor, J., Karlsson, I.C.M., 2018. Mobility as a Service: development scenarios and implications for public transport. Res. Transp. Econ. 69, 592–599. https://doi.org/10.1016/j.retrec.2018.04.001.

SOLUTIONS, 2016. Urban Mobility Solutions and Urban Electric Mobility Initiative. UEMI. Available at: http://www.uemi.net/. (Accessed 2 January 2019).

Sorrell, S., 2010. Energy, economic growth and environmental sustainability: five propositions. Sustainability 2 (6), 1784–1809.

Steg, L., 2005. Car use: lust and must. Instrumental, symbolic and affective motives for car use. Transp. Res. A Policy Pract. 39 (2–3), 147–162.

Steg, L., Gifford, R., 2005. Sustainable transportation and quality of life. J. Transp. Geogr. 13, 59–69.

Steg, L., Perlaviciute, G., van der Werff, E., 2015. Understanding the human dimensions of a sustainable energy transition. Front. Psychol. 6, 1–17. https://doi.org/10.3389/fpsyg.2015.00805.

Sterner, T., 2007. Fuel taxes: an important instrument for climate policy. Energy Policy 35 (6), 3194–3202. https://doi.org/10.1016/j.enpol.2006.10.025.

Tenenbaum, D.J., 2008. Food vs. fuel: diversion of crops could cause more hunger. Environ. Health Perspect. 116 (6), A254–A257.

Tirachini, A., Hensher, D.A., 2012. Multimodal transport pricing: first best, second best and extensions to non-motorized transport. Transp. Rev. 32 (2), 181–202. https://doi.org/10.1080/01441647.2011.635318.

Tiwari, G., 2016. Perspectives for Integrating Housing Location Considerations and Transport Planning as a Means to Face Social Exclusion in Indian Cities. https://doi.org/10.1787/4b3b9950-en. (Accessed 9 March 2019).

UN, 2014. World Urbanization Prospects. Population Division—United Nations. Available at: https://population.un.org/wup/. (Accessed 3 January 2019).

UN General Assembly, 2015. Transforming Our World: The 2030 Agenda for Sustainable Development. Available at: https://www.un.org/sustainabledevelopment/development-agenda/.

UN-Habitat, 2013. Planning and design for sustainable urban mobility—Global report on human settlements 2013. 348. Available at: http://www.unhabitat.org/content.asp?catid=555&typeid=19&cid=12336.

van der Have, R.P., Rubalcaba, L., 2016. Social innovation research: an emerging area of innovation studies? Res. Policy 45 (9), 1923–1935. https://doi.org/10.1016/j.respol.2016.06.010.

WHO, 2018. Global status report on road safety 2018. Available at http://apps.who.int/iris.

Wilson, C., 2018. Disruptive low-carbon innovations. Energy Res. Soc. Sci. 37, 216–223. https://doi.org/10.1016/j.erss.2017.10.053.

Woetzel, J., Remes, J., et al., 2018. Smart Cities: Digital Solutions for a More Livable Future. Available at: https://www.mckinsey.com/~/media/McKinsey/Industries/Capital%20Projects%20and%20Infrastructure/Our%20Insights/Smart%20cities%20Digital%20solutions%20for%20a%20more%20livable%20future/MGI-Smart-Cities-Full-Report.ashx.

Yang, S., Chen, B., Ulgiati, S., 2016. Co-benefits of CO2 and PM2.5 emission reduction. Energy Procedia 104, 92–97. https://doi.org/10.1016/j.egypro.2016.12.017.

Yap, M., Munizaga, M., 2018. Workshop 8 report: big data in the digital age and how it can benefit public transport users. Res. Transp. Econ. 69, 615–620. https://doi.org/10.1016/j.retrec.2018.08.008.

Zhang, F., Gallagher, K.S., 2016. Innovation and technology transfer through global value chains: evidence from China's PV industry. Energy Policy 94, 191–203. https://doi.org/10.1016/j.enpol.2016.04.014.

Further Reading

Anheier, H.K., Krlev, G., Mildenberger, G. (Eds.), 2019. Social Innovation: Comparative Perspectives. Routledge Studies in Social Enterprise & Social InnovationRoutledge, New York, NY.

Bongardt, D. (Ed.), 2013. Low-Carbon Land Transport: Policy Handbook. first ed.. Routledge, New York, NY.

Figueroa Meza, M.J., Lah, O., Fulton, L.M., McKinnon, A.C., et al., 2014. Energy for transport. Annu. Rev. Environ. Resour. 39 (1), 295–325.

The impact of the institutional context on the potential contribution of new business models to democratising the energy system

2.5

Ruth M. Mourik[a], Sylvia Breukers[a], Luc F.M. van Summeren[b],
Anna J. Wieczorek[b]

[a]*Duneworks, Eindhoven, The Netherlands*
[b]*School of Innovation Sciences, Eindhoven University of Technology, Eindhoven, The Netherlands*

1 Introduction: The incumbent energy system in transition

1.1 Technological and institutional lock-ins

'The energy transition' is becoming a mainstream term in energy policymaking, with multiple references to bottom-up and community energy projects in which households (consumers and prosumers) are assumed to take an active role. At the same time, this raises the question as to how and to what extent the organisation of the current energy system actually allows for such initiatives to become successful. How much room is there for community initiatives to collectively organise themselves to enable local generation, storage, and distribution of self-generated energy? Before explaining how we intend to discuss this question, we first provide a brief characterisation of the incumbent energy system.

The traditional incumbent energy supply system has for long been based on large-scale centralised energy generation by a limited number of (formerly state-owned) large companies, whereby energy was distributed to end-users via a centrally organised transmission and distribution grid, owned and managed via natural monopolies of transmission system operators (TSOs) and distribution system operators (DSOs). The sociotechnical nature of the energy system reflects a historical co-evolutionary path of physical objects, infrastructures, institutions, actors, networks, and social norms. By institutions, we mean the informal and formal rules that structure behaviour and organisation (Hall and Taylor, 1996).

Through a so-called path-dependent process, this particular co-evolutionary trail has become carved out deeper and deeper, whereby both physically and

Energy and Behaviour. https://doi.org/10.1016/B978-0-12-818567-4.00009-0

institutionally, our societies have become locked into a fossil fuel-based energy system (Unruh, 2000). These lock-ins, combined with the increasing economics of scale that the focus on large and centrally organised energy provision entailed, ensured the continuation of this particular developmental path even when it became clear that this system has negative societal impacts due to CO_2 emissions and environmental pollution and even when low-carbon alternatives were available (Geels, 2002; Verbong and Loorbach, 2012). Although change is occurring in our energy system, the current institutional settings of policies, regulation, codes, behavioural codes of conduct, and policy networks are still to a large extent predicated on this past. The physical energy infrastructures still reflect a centralised and hierarchically controlled organisation of energy supply (Smith, 2012). The consequences of this become visible when we examine how innovative initiatives that operate on a smaller scale and that focus on renewable energy generation, storage, and distribution try to find feasible organisational and business models to sustain a new more sustainable energy pathway.

So basically, despite this heavy path dependence, the large-scale, supply-oriented machinery is slowly adapting to changing realities. Recent decades have brought not only institutional changes such as the unfolding energy market liberalisation, deregulation, and privatisation processes (reflecting a dominant economic discourse of the 1980s and 1990s) but also an increase of European influence on both national energy and climate change policies. As such, concerns about environmental impacts and the climate change debate have started to influence changes in our energy system. In addition, three general trends play a particularly important role in further adjustments of the energy system. The first trend is the increasing electrification of both the energy and mobility systems, with more and more energy consuming functions being electrified, including vehicles. The second trend—related to the environmental awareness increase—is the rise of distributed generation from renewable sources, for example, photovoltaics (PV) and wind power, and in relation to this the emergence of a number of new actors. Furthermore, with the increase of distributed generation, new challenges have emerged which relate to the inability of the 'old' grid to deal with the intermittent nature of renewable energy production. This brings us to a third trend: developments in smart grid technologies supported by the rise of digital technologies and the strategic use of data, also accompanied by the emergence of new actors in the energy field (Bloomberg, 2017; IEA, 2017; Sharma and Mukherji, 2015; Statkraft, 2017). In smart grid developments, information and communications technology (ICT)-supported solutions are sought to enable more flexible supply and demand management options and to engage users–producers at all levels—energy intensive industries and households.

1.2 Community energy: Against all odds?

Community energy initiatives have been defined as initiatives that aim at contributing to a more sustainable energy system through community-based activities, for example, locally owned renewable energy generation, community-level energy saving initiatives, or any other kind of local collective energy-related initiatives. Generally, community energy initiatives face a range of challenges. Next to internal issues related to the need for further professionalisation, improvement of business cases, and advanced social organisation, they have to deal with the overwhelming market power and dominance of the established fossil fuel-based and renewable energy companies that reap the fruit of the economies of scale. Regardless of policy rhetoric in praise of bottom-up initiatives, in practice, these initiatives often face institutional and technological barriers that are the direct result of the lock-ins created in the path-dependent development of the 'old' energy system (Unruh, 2000). Obviously, possibilities for community energy initiative are shaped not only by the ideas, visions, skills, relations, and patience of their initiators but also as much by the physical (infrastructural), socio-political, cultural, and institutional characteristics of the context for which they are planned. In other words, the barriers in place distribute impacts such that they negatively influence the opportunities for community energy initiatives. This problem is investigated further in this chapter.

1.3 Conceptual approach, main questions

To clarify how distributive issues have occurred in a specific context and how the underlying structural (institutional, physical, and political) conditions have created this distribution, the environmental justice literature is helpful (Davoudi and Brooks, 2014; Schlosberg, 2004, 2013; Wolsink, 2013). The mutually reinforcing institutional and technological lock-ins in the still predominantly fossil fuel-based energy system raise the question what would be needed to get rid of obsolete institutions and/or to create new ones. It has been pointed out how institutional conditions are major determinants in the process of deployment of renewables (Breukers and Wolsink, 2007; Unruh, 2000; Wolsink, 2013). The challenge addressed is not only renewable energy sources deployment but also the role of local community members in generation, storage, and distribution of energy. This involves the use of innovative smart grid technologies to enable local optimisation of energy and financial flows. We investigate the ways in which institutional conditions—the informal and formal rules that structure behaviour—affect possibilities for such community-based initiatives. In addition, our inquiry also addresses the question which institutional conditions would be needed to enable a more inclusive energy transition—because the community energy initiatives reflect ambitions of community members to gain influence, control, and ownership over the energy system.

The explicit focus of this chapter is thus on the tension between a very distinct and new type of community energy initiatives, their business model, and the incumbent energy system. This new type of community energy is the delivery of electricity from community-based virtual power plants (cVPPs), a concept we discuss in more detail in Section 2.

1.4 Method

We build on insights from multiple research methods. First, a desk study including both academic and grey literature research was conducted to map the Irish and

Dutch institutional arrangements shaping community energy projects like cVPPs. Secondly, two Scopus-based literature reviews were conducted on 'virtual power plant' and 'community energy' to conceptualise cVPP. To answer the questions raised, we use empirical material from two cVPP initiatives in two different country contexts (in the villages of Loenen in The Netherlands and Templederry in Ireland). Between June 2018 and March 2019, two rounds of semistructured interviews were conducted with initiators of two cVPP projects that are being set in the Netherlands and Ireland. Interviews were recorded, transcribed, and coded using the software tool Nvivo. Finally, researchers participated in various face-to-face and virtual cVPP project meetings between December 2017 and December 2018. Data were gathered by making field notes and recordings of which relevant parts were transcribed and coded.

1.5 Reader's guide

First, in Section 2, we present what a cVPP is, we discuss the community logic which summarises the community values that may be crucial for those initiating community energy projects, and we present the Dutch and Irish cVPP empirical cases. Next, in Section 3, we elaborate on how we operationalise the business model concept using the canvas developed by Osterwalder and Pigneur (2013) as a heuristic. Using the overview/review of community values, we explore how energy project business models can be community centred.

As a next step, Section 4 takes an institutional lens to examine how business models for community energy projects such as cVPPs are influenced by the institutional context. Basically, we confront the models developed in Section 3 with the real-world institutional contexts. This is done in Section 5 by analysing how existing policy, regulation, and ways of doing influence the business models that cVPP projects develop, using the Dutch and Irish cVPP cases as illustrations. Finally, in Section 6, we discuss how the business models may harness potential for restructuring the energy system in the long run. And, based on this analysis, we discuss the implications in terms of justice and fairness using the environmental justice perspective to assess how the current setting works out for cVPP projects and how this arguably affects the democratic quality of the energy transition.

2 cVPPs, community energy projects and the community logic

2.1 Community virtual power plants

Of the many different types of community energy projects, we have taken two examples that entail a rather specific type: community virtual power plants (cVPP). A cVPP is a VPP adopted by a community. Van Summeren et al. (forthcoming) define a virtual power plant (VPP) as: 'A VPP is a software-based solution that aggregates distributed energy resources into one coordinated and controlled portfolio that operates as one single entity similar to a conventional power plant, which allows for performing roles in the electricity system related to managing and trading of electricity'. In contrast to a microgrid project that stands (almost) separate from the grid, a VPP makes use of the existing grid infrastructure (Asmus, 2010). By making distributed energy resources both visible and controllable to system operators and market players, a VPP enables their integration into the existing electricity grid and participation in the wholesale energy market. Existing VPPs and other smart grid applications have mainly been developed by large utilities or research organisations for the sake of providing balancing services for the grid (Verkade and Höffken, 2018). In these projects, there is little attention for active user participation in the design and operation of the VPPs, risking that they lose the users in the process (Verkade and Höffken, 2018). They 'involve' the prosumers by offering them remuneration, but neither individuals nor the communities that are part of the current VPPs have any say in how the benefits and risks are distributed, how the VPPs are organised and controlled, who in fact is involved, and who is left out (Van Summeren et al., forthcoming). Community-based VPP embodies a different, more distributed organisational logic.

2.2 Community energy projects and community logic

A community is often referred to as a place- or interest-based network of people that collectively engage in projects or initiatives (Klein and Coffey, 2016; Seyfang et al., 2013). A community energy project can include not only citizens but also organisations like municipalities and local SMEs. It is not only the involvement of a community that distinguishes cVPP from VPP but also the 'community logic' that inspires it (Van Summeren et al., forthcoming). This community logic implies that the community itself has a say in how the cVPP is developed and operated.

A literature review by Van Summeren et al. (forthcoming) on 'community energy' identified seven characteristics of community logic, for example, which characterise the way many community energy projects and initiatives operate:

1. Community energy projects are driven by a diversity of community needs and motivations, which often go beyond energy security and monetary gains and which may include strengthen the local economy, create local jobs, increase support for renewable energy, reduce carbon emissions, ensure appropriate siting and scales, become more energy independent, and improve social cohesion (Seyfang et al., 2013).
2. The community strives for a fair and transparent distribution of (collective) benefits, costs, and risks (Walker and Devine-Wright, 2008).
3. Community ownership of the assets, organisation, and/or the cVPP is important.
4. The organisation operating the cVPP has high degrees of community involvement in governance and decision-making (Hicks and Ison, 2018). For example, in the frequently used cooperative model, all members have equal voting rights regardless of the amount of shares they own (Šahović and Da

Silva, 2016). Cooperative members vote for long-term strategies and elect representatives that take place in the executive board, which is responsible for the daily operation of the cooperative (Šahović and Da Silva, 2016).

5. The community strives for local embeddedness and acceptance by actively engaging community members beyond the often small group of initiators (Hicks and Ison, 2018). Engagement may range from being informed to providing input or active participation in the development or operation of the initiative or project.

6. Community initiatives or projects tend to be open and inclusive towards the whole community, regardless of status and resources (Walsh, 2018). For example, the cooperative model is characterised by open membership, meaning that everyone from the community can become a member of an initiative or participate in a project (Šahović and Da Silva, 2016).

7. Communities aim for a scale of technology that reflects local energy demand and/or their needs and motivations. In contrast, many commercial project developers aim for profit maximisation by maximising energy generation (Hicks and Ison, 2018).

Adopting a VPP allows communities not only to generate energy and feed it back into the grid but also potentially to manage and distribute their self-generated energy within their community, trade energy on the energy market, and provide grid services to system operators. By enabling access to wholesale energy markets and/or the provision of grid services, a VPP potentially provides new business opportunities and revenue streams for communities, which would make them less dependent on feed-in tariffs and other funding schemes. Furthermore, VPP technology, in theory, allows communities to explore new ways of organising the generation and distribution of energy within their own community, more decentralised or distributed and more sustainable way, which better reflects community values. The earlier explanation of cVPP is based on what a cVPP ideally would look like, without considering institutional and physical barriers that exist in the incumbent energy supply system.

2.3 Introducing the two cases

2.3.1 cVPP in Loenen (Netherlands)

In 2014, the Dutch municipality of Apeldoorn invited inhabitants of Loenen, a village consisting of approximately 1300 households, and other villages, to participate in a contest to present an idea for a local energy project. The winning village would win 200.000 Euro to realise the project as part of a European project. Loenen won the contest and started the 'Loenen Energy Neutral' (LEN) foundation which, since 2014, governs a revolving fund. This fund initially started with activities supporting individual households to install energy saving measures (e.g. insulation) and renewable energy generation capacity in their homes, mainly solar photovoltaics. Between 2014 and 2018, LEN initiated 170 separate project activities aimed at energy saving and generation, investing approximately 1 million Euro.

The community aim is to continue activities to increase the amount of local renewable energy generation using the LEN revolving fund. In addition, the community has set up another foundation called 'Sustainable Projects Loenen' (DPL) and, at the time of writing, is founding an overarching energy cooperative that will govern LEN, DPL, and the cVPP. From here on, we refer to both LEN and DPL as cVPP Loenen. Next to the individual PV panels on the homes, cVPP Loenen aims at collective ownership of renewable energy generation on roofs of public buildings, industry, and SMEs. cVPP Loenen already has permission and agreement for a subsidy scheme for the development of a 900-kW solar photovoltaic farm on a roof of a local company, and they are planning to build two other solar farms of in total 1 MW in the near future. They are furthermore exploring the amount of controllable loads present in the community and industry and possibilities for installing energy storage systems like a power-to-gas storage system and a neighbourhood battery. In addition, they want to adopt a VPP to locally manage and potentially trade self-generated energy.

In the ideal future scenario, for example, neglecting institutional barriers, this cVPP would enable the community to store, manage, and distribute all self-generated renewable energy locally. The final cVPP design will be developed based on input gathered from the community during workshops and questionnaires regarding preferred objectives and activities, the availability of controllable distributed energy resources, and (future) institutional and physical barriers and opportunities.

Clearly, cVPP Loenen is still in the early stages of the process, and it is not at all clear what choices will be made exactly. Therefore, the community logic elements, as shown in Table 1, reflect the way these elements are currently dealt with, and this may change in the course of further developments.

2.3.2 cVPP in Ireland: Templederry wind farm and community power

Community owned energy generation is almost non-existent in Ireland. One of the few exceptions is Templederry Wind Farm. In 1999, a group of 28 inhabitants of the village of Templederry came together with the idea to set up a local energy project. After reviewing several options, they chose to go for wind energy. Templederry Wind Farm was developed with the aim to focus on values to the local community, improve the local economy, strengthen the local community, and take back power from big corporations and utilities. Each of the 28 inhabitants that invested and participated in setting up the wind farm owns one share and receives an equivalent of the revenues. Two additional shares are owned by the rest of the village, and the accompanying revenues go into a community fund, which is used to invest in the local community.

When the wind farm was realised, the community decided that they did want not only to generate energy locally but also to supply it to the community members. With the revenues, they set up their own 'Community Renewable Energy Supply' (CRES) company to indeed be able to supply the generated energy to their members. CRES became a licensed supply company, but for various reasons, further elaborated in Section 4, it was not possible for CRES to supply the energy generated by the wind farm to the community itself. However, being an energy supplier, CRES is now planning the development of collective generation projects together

Table 1 Community logic in Loenen cVPP (Van Summeren et al., forthcoming)

	Community logic in cVPP	Community logic in Loenen cVPP
1	Community needs and motivations driving the cVPP project	By focusing on a collective project, the initiative hopes to contribute to community building. cVPP Loenen focuses on values for the community such as lowering household energy bills; lowering carbon emissions by increasing the amount of renewable energy generation; improving integration of renewables into the grid, to allow for a larger amount of local renewable energy generation and increase control over the siting of renewable energy generation; and improving the local economy by, for example, hiring local companies to insulate homes and install solar photovoltaic panels
2	Distributional issues: Ideas about the distribution of benefits, costs, and risks within the community and between community members and others (and the decision-making about this distribution)	Households who own (shares) of solar panels receive financial revenues or face lower energy bills. It is not yet decided how revenues from collective solar farms and energy management will be divided between individual shareholders and the cooperative. This will be decided by the members of energy cooperative, who can vote for the preferred distributional mechanisms
3	Community ownership of assets and the organisation	The community adopted a cooperative ownership model in which community members collectively own the energy cooperative, solar farms, and the VPP. Domestic solar panels are owned by private house owners
4 5 6	Procedural issues: Organisation of the cVPP Engagement of community members and others in decision-making (and what this decision-making is about) Engagement with local noncommunity members	cVPP Loenen adopted a cooperative structure in which all community members have equal voting rights, allowing them to vote for changes in the bylaws and long-term strategies as well as on who represents them on a day-to-day basis in the board of directives In the early stages of designing the cVPP, all inhabitants of Loenen were invited to participate in workshops in which they could provide input regarding preferred activities and objectives that should guide the design of the cVPP
7	The way in which the scale and complexity of the technologies used matches with the needs, abilities, and ambitions of the local community members	The community would preferably generate and consume renewable energy as much as possible locally and sell potential surplus of energy for a good price. In theory, this could be done through a local energy market or by peer-to-peer energy trading. The underlying idea is that by allowing for local supply, the community would have an incentive to develop an energy generation mix that relates to (local) energy demand rather than aiming to maximise generation to maximise profits

with other Irish energy cooperatives and setting up a joint cVPP project—whereby CRES, which was renamed to 'Community Power', takes the role of a community-owned energy supply company on the national level. As such, this innovative initiative aims at enabling geographically dispersed Irish communities (e.g. at the Aran Islands, Tipperary, Limmerick, and Claremorris) to develop community-owned renewable energy, supply the energy to themselves through Community Power, and allow communities and citizens to receive revenues for their (surplus of) renewable energy generation.

The Community Power cVPP has not yet been materialised, and the search process to identify the best options is not yet finalised. However, some of the community logic elements have been strongly present from the start of the Templederry Wind Farm. Table 2 shows the community logic as it has become apparent at the time of writing and does not present a static picture.

Having introduced the illustrative Dutch and Irish cases and how a community logic is visible in both, we will now discuss how business models in theory could be matched to the community logic elements.

3 Business models for more decentralised community-based energy systems

3.1 The move to more user-centred business models in the energy sector

As discussed earlier, new sociotechnical configurations are emerging that aim for a more decentralised, distributed energy generation and consumption. These new configurations embody experiments with new types of business models as well. We follow the definition of a business model as the organisational logic of a firm (enterprise and cooperative) and its partners used to create and capture value, a mechanism to market a product or service, or, in the case of cVPPs, a new sustainable system of technologies and services (Boons and Lüdeke-Freund, 2013; Osterwalder and Pigneur, 2013; Zott et al., 2011). We use the business model canvas (Fig. 1) as developed by Osterwalder and Pigneur (2013) who state that essentially a business model consists of nine building blocks to ensure a good market delivery:

1. The involved partners and their role in the creation of the value proposition
2. The core activities required to realise the value proposition and how these related to each other
3. The resources necessary to secure the proposition
4. The value proposition: the offer provided and how it solves a problem, need or wish felt by the customer, the benefits that accrue to the user of the product or service
5. The quality of the relationship a business has with its customer
6. The channels used to interact with the customer
7. Which customer segments are targeted

Table 2 Community logic in Community Power (Van Summeren et al., forthcoming)

	Community logic in cVPP	Community logic in Irish case
1	Community needs and motivations driving the cVPP project	Before it became part of a cVPP project, the Templederry Wind Farm already addressed local community values and needs such as strengthening the local economy, create local employment, and take back control and ownership from large corporations and utilities. The communities involved in Community Power formulated five core values that drive the cVPP: local benefit, democracy and cooperation, clean energy, fair prices, and resilience
2	Distributional issues: Ideas about the distribution of benefits, costs, and risks within the community and between community members and others (and the decision-making about this distribution)	Templederry Wind Farm took risks by investing money in CRES, which will be compensated by other communities who will buy shares of Community Power Members of Community Power, who are also customers, will pay fair prices for energy that reflect the generation, overhead costs, and a share to be reinvested by Community Power. The latter will be reinvested in new generation projects, which should result in lower energy prices for all customers Community members who invested in generation project will receive a dividend that reflects their investments
3	Community ownership of assets and the organisation	Individual citizens and local energy communities involved will collectively own Community Power and energy generation technologies. In the future, the local communities might also own local energy generation while still able to sell energy through Community Power
4 and 5 and 6	Procedural issues: Organisation of the cVPP Engagement of community members and others in decision-making (and what this decision-making is about) Engagement with local noncommunity members	Community Power will act as a commercial actor. The executive board will consist of representatives of the different local communities involved in Community Power The exact ownership and governance structures are currently being developed in a cocreation process together with representatives of the different communities involved Each local community will be responsible for engaging with their own local community and for developing a local customer base for Community Power
7	The way in which the scale and complexity of the technologies used matches with the needs, abilities, and ambitions of the local community members	Templederry Wind Farm and Community Power originally aimed to match supply with the demand of the Templederry community. In contrast, Community Power aims, as a national energy supplier, to match its energy generation portfolio with the total energy demand of its customer base

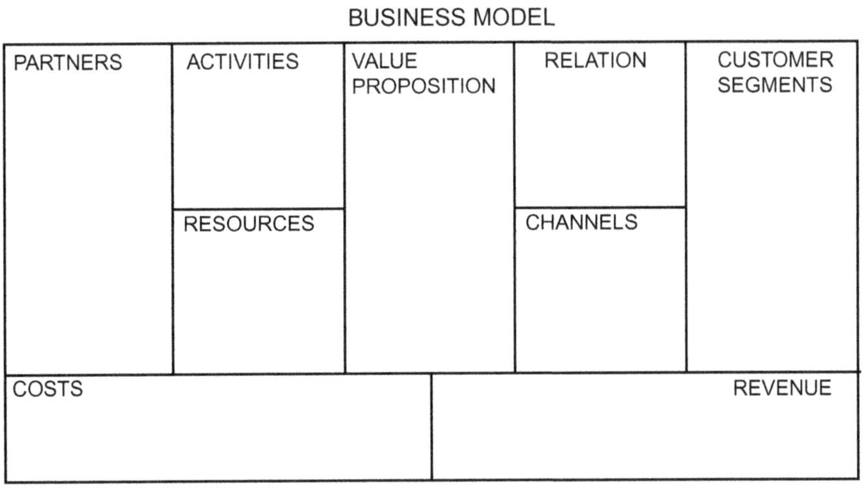

BUSINESS MODEL

PARTNERS	ACTIVITIES	VALUE PROPOSITION	RELATION	CUSTOMER SEGMENTS
	RESOURCES		CHANNELS	
COSTS				REVENUE

FIG. 1

The building blocks of the business model as designed by Osterwalder and Pigneur (2013).

8. Cost structure: all the costs and expenses incurred while operating the business model
9. Revenue streams: the answer to the question through what process money is made.

Recently, much research has been undertaken to understand the role of business models in reshaping our current energy system towards a more sustainable and democratic configuration that recognises that empowerment of users or customers should be part of the business model design. Tolkamp et al. (2018) claim that such a user-centred approach has the potential to be more effective in creating market uptake of new energy services, because the value being generated and the delivery mechanism for that value are better tailored to the actual problems, needs, and wishes of users. A shift to value creation for both users and suppliers and a clear focus on the user perspective and their unique reasons for buying energy services, for example, through the cVPPs, can be considered to be the next step in creating a larger market uptake for energy efficiency and renewable energy (Hienerth et al., 2011; Kindström et al., 2017; Nilsson et al., 2012; Vargo and Lusch, 2008).

Furthermore, sustainable business models are considered key in facilitating the shift of our current energy production and consumption system towards a more sustainable and/or democratic system because these business models explicitly aim at balancing multiple types of value being generated, that is, social, environmental, and economic, across the different stakeholders involved (Bocken et al., 2014; Boons and Lüdeke-Freund, 2013; Vezzoli et al., 2015). According to Hall and Roelich (2016), the traditional value proposition known to centralised energy systems and used in traditional business models is indeed changing into 'complex value' defined as 'the production of financial, developmental, social and environmental benefits which accrue to different parties, across multiple spaces and times, and through several systems'.

3.2 Community logic-informed business models

In the succeeding text, we discuss how the nine building blocks of the business model would look like when a community logic would inform the development of a business model for community energy.

3.2.1 Value proposition

Community values often go beyond the standard economic values. The community logic typically includes values in five categories: economic/financial, social, political, environmental, and technical/infrastructural, for example, not only at the economic value at individual household level with a lower energy bill but also at the economic value for the community or region in the form of rural regeneration, increased social capital, the development of community skills, creation of local employment, and any environmental value for both the community and society at large (Hicks and Ison, 2018; Seyfang et al., 2013). The community logic would entail that the business model is created by, for, and with the community members that initiate the cVPP and that the community has a say in its design, implementation, and operation. This implies that the value generated with the business model is driven by a diversity of needs and motivations of the community and that the collective benefits and the risks and costs are distributed accordingly. Values include not only democratisation of control, ownership, decision-making, and planning but also the creation of a system that provides a lower energy bill and tariff fairness.

3.2.2 Main activities

The setting up of a cVPP is not a value in itself but one of the activities that allow for the creation of a diversity of other community and societal values. Following the community logic, the main activities of the cVPP business model would be the community planning; implementation and operation of a cVPP, which includes local generation, supply, and, if necessary, storage of energy; and maintenance or outsourcing of maintenance as decided by the community, as well as the day-to-day administration and management of the cVPP. Engagement of the others not (yet) involved can take place in several ways, for example, by selling shares of the assets (Hoffman and High-Pippert, 2010; Seyfang et al., 2013).

To realise a good match between the local demand and local supply of energy, the activities around a cVPP would furthermore include the use of ICT: VPP technology and an energy management system to enable demand response. The scale and complexity of these technologies preferably would match with the local needs, ambitions, competencies, and capacities.

3.2.3 Resources

The main resources of this community logic-based cVPP business model would preferably consist of voluntary and paid contributions (based on various skills and expertise) by and of the community members acting as managers. Outsourcing would not be the first option for community led energy projects. Community owned land and property would be important resources as well.

3.2.4 Customer segment

A community logic-based business model aims to serve the community, comprising the households, enterprises, schools, municipal services, etc., but given the community centredness of the cVPP, these clients would most likely also be owners, shareholders, or operators of the cVPP, as well as partners. In addition, the business model would also consider selling the generated energy in case of surplus to an energy company that would thus become a customer but nothing more. The community customers would have full control and ownership, through different means, and the energy company customer would have not. On a more general level, the community logic-based cVPP business models would also consider society at large to be a customer, since the cVPP aims to contribute to an inclusive low-carbon energy transition.

3.2.5 Relations

Following the community logic, the relationship of the business developers and the customers would be face to face and personal, also because they would be one and the same to a certain degree, namely, the community members.

3.2.6 Channels

In terms of the channels used for interaction, these would be face-to-face meetings, for example, through the daily administration and management of the cVPP and through shareholder meetings.

3.2.7 Partners

The partners for this model following the community logic would first be the community members, both the directly involved and other residents, then the technology developers needed to develop the cVPP configuration, any supply chain partners needed and finally ethical financiers.

3.2.8 Costs

Costs incurred in the community logic-based type business model would include costs for a feasibility study and development of plans, costs for land use planning approval, capital/asset costs for construction and installation of the cVPP, costs of market participation, transaction costs, and costs of reinvesting in improving, replicating, or upscaling the cVPP.

3.2.9 Revenues

Revenues for this community logic-based type of business model would comprise a contribution from community members in return for shares, energy contracts with suppliers to whom the cVPP would be selling energy in surplus, subsidies received through financial support schemes (e.g. feed-in-tariff schemes or other long-term contracts between the government and renewable energy producers), and finally rent received for access to property where assets would be located.

The picture drawn in this chapter started from the community as a point of departure, not with the actual context that cVPP projects are confronted with in practice. In fact, initiators of cVPP projects struggle to find the 'right' sociotechnical configuration in terms of combinations of technologies and organisational structure including the business model that allow them to develop a cVPP project that is feasible in a challenging institutional and market setting, as the next chapter will show.

4 The shaping of cVPP business models by the institutional context

As mentioned before, the energy system in many countries, including the Netherlands and Ireland, is designed around a centralised market, where supply and demand are organised on a national and regional scale (Braunholtz-Speight et al., 2018; Ofgem, 2015). Institutional arrangements from the energy domain have worked to perpetuate the continuation of this system. By investigating energy-related policies and regulations (including tax incentives, rules, investment structures, regulations, and norms that structure the energy market developments), we try to assess the extent to which these dimensions of the institutional context provide room for alternative low-carbon organisational modes—that is, the business models for cVPP. We do not aim to be comprehensive, but focus on those institutional arrangements that proved to be the most important in shaping a viable business model.[a]

4.1 Local energy supply only is not allowed because of balancing and settlement codes

The first institutional context barrier is that cVPP projects are not allowed to supply energy only locally. In addition, the physically trade of energy between peers is not allowed either. Both in Ireland and the Netherlands, electricity producers are legally required to supply at least at regional or national levels. Because of this, all activity and most of the existing business models are centred on achieving and enlarging scale, by increasing unit sales volume and increasing throughput in the system (Hall and Roelich, 2016). Regulation effectively discourages smaller volumes ad scales of supply and trade.

4.2 Electricity balancing and settlement code

A second institutional context barrier is the electricity Balancing and Settlement Codes rule in place in many countries, including the Netherlands and Ireland (Commission Regulation (EU) 2017/2195). These codes require a balancing mechanism as a means for the transmission system operator to balance demand and supply

---

[a] For instance, the institutional conditions in the domain of spatial planning is not considered in this paper, but it will have significant impact on the eventual planning, decision-making, and siting of assets that are part of cVPP projects.

as real time (half hour) as possible, without using storage, thus reducing risks of blackouts and corrective actions. It does so by means of primary and secondary reserves that are controlled by the TSO (Ofgem, 2015). This means that energy communities that supply electricity need a large enough portfolio of energy to participate in this balancing. And having such a portfolio requires considerable capital.

4.3 Legal entities and local supply

To avoid individual financial liability, in case of going bankrupt or another financial crisis, cVPPs may need to become a legal entity. A third institutional barrier is the discrepancy caused by different institutional arrangements. If a cVPP project decides to become a legal entity, it may choose a form that will allow for the eligibility for certain financial benefits, for instance, in relation to tax exemptions. However, the requirement attached to these benefits is that the service provided by this cVPP is that the energy is distributed among its members only. However, this requirement goes against the rule that the cVPP needs to also provide energy nationally (Braunholtz-Speight et al., 2018). This issue plays out differently in different countries since the legal entities might have different rules. This needs to be further investigated.

4.4 Investment structures

A fourth institutional barrier has to do with investment structures. Since cVPP projects are new entrants, they have no credit history. As a consequence, their credit rating is low. The projects, however, do need investors to cover part of the high upfront capital investments in the assets (the RES generation installation) and other necessary technologies, as well as costs for planning, permission, customer acquisition, etc. Commercial investors also require energy initiatives to be investment ready, especially because they see this type of investment as risky due to the long payback time. This means that a process is to be undertaken to prepare the organisation to be able to successfully bid for repayable finance to fund the project, which is again very costly and difficult to acquire (Braunholtz-Speight et al., 2018). Large-scale initiatives also have to deal with many of these costs, but for cVPP projects, these have higher cost variability due to the lack of expertise and smaller-scale operation—so communities have fewer possibilities to spread their risk across multiple projects (Hall and Roelich, 2016; Braunholtz-Speight et al., 2018).To summarise, the (financial) complexity related to market rules is much harder for a cVPP project to comply with when compared with a large-scale energy sector company. The energy system is designed and organised such that the latter can participate using their in-house financial and legal expertise, spreading risks across projects, and being able to fund or find finance for upfront costs. For community projects, this is hardly possible, therewith undermining their possibilities for equal access to the energy market. In some cases, a public or charity/lottery grant funding or finance from incumbent energy sector actors can help out, but for smaller projects, this issue is a real barrier. Market participants—including cVPP projects—are furthermore required to have very high levels of liquidity to be able to trade outside of one's own portfolio to comply with the Balancing and Settlement Codes (European Commission, 2017). But cVPP projects usually do not have high liquidity levels once they are ready to supply energy, since they have a lot of investments to do upfront. In addition, their low rating means that they need even more collateral to be able to participate in energy trading.

4.5 Limited focus on community and system scale of implementation instead of single households and single technologies

A general characteristic of those supportive (financial) incentives that do target smaller-scale initiatives is that these tend to focus on single RES technologies rather than on sociotechnical configurations consisting of various technological components and that existing subsidies and incentive tend to target individual households rather than community owned or shared systems. The subsidy system for storage, for example, favours private household systems and not community owned or shared systems (Müller and Welpe, 2018).

4.6 Lack of institutional support through standardisation

Another institutional barrier is that no harmonisation of standards for system elements of (c)VPPs exists, because of the lack of maturity. This results in additional complexity that is posing challenges to cVPP initiatives [e.g. expertise and knowledge are required to understand all the industry codes in place with respect to the implementation of the technological system (Ofgem, 2015)].

4.7 Lack of institutional support due to new actors, new technologies and unclear direction

An additional issue adding to the complexity of becoming a player in the energy system is the current uncertainty because the energy system and market are in transition. This results with many new actors rising such as aggregators, new energy service companies, and new technologies such as blockchain, while at the same time, there are no clear directions, standards, or rules that give direction to these new innovations (Braunholtz-Speight et al., 2018; Buth et al., 2019).

To summarise, we see the continuation of path dependency and institutional voids.

Institutional arrangements clearly structure grid management and energy market developments in such a manner that they mainly perpetuate the distribution of opportunities such that large players with financial clout (these can be new or incumbent energy system actors) and large players with good access to expertise [often (linked to) incumbent energy system actors] have a more advantageous position compared with cVPP projects. In addition, policy support for smaller-scale energy initiatives is not providing the support needed by innovative projects like cVPPs. This, in combination with an absence of rules and standards, makes it very difficult for innovative cVPP projects that want to generate, store, distribute, and self-consume to plan and move forward. The next chapter will illustrate this using the examples of Loenen cVPP and Community Power cVPP.

5 The cVPP business models in the face of institutional conditions that are not conducive

In this section, we will now discuss the actual Dutch and Irish cVPP business models. We will discuss how these cVPP projects have had to adapt their business model to fit with the current institutional arrangements and how this has affected the community logic that these projects initially were based on.

5.1 Activities

As explained before, local generation and self-consumption as aimed for from a community logic perspective would mean peer-to-peer trade, which is as of yet not allowed in many European countries. cVPP Loenen has not decided what the course of action in reaction to this is going to be. When it decides to become a supplier itself, this would mean that its scale of operation has to become large (supralocal), limiting the possibilities for energy management to optimise local self-consumption. However, an option for cVPP Loenen could be to apply for the so-called experimental arrangement, which provides projects that meet certain requirements, temporary exemptions on the electricity bill, which means that, in these cases, it is allowed for energy cooperatives to manage their own production, distribution, and consumption of locally produced power 'as if they were behind a single meter connection' (De Graaf, 2017:25). However, there is no exemption from the tax regulation that would still apply to all the connections in such a project. Currently, the experimental arrangement is being revised, and new applications are not yet possible at the time of writing. A final remaining activity for cVPP Loenen consists of selling flexibility on the energy market to support the TSO in grid balancing and the suppliers in optimising their portfolio.

The initially planned activities of the Community Power cVPP aimed at matching local supply and demand, but just as in the Dutch situation, the Irish institutional context forced Templederry Wind Farm (and CRES) to shift to activities at a national scale and start delivering energy on a national scale through cooperation with a national electricity supplier. In addition, as a national energy supplier, Community Power had to implement IT systems and an administration structure that comply with the requirements of the regulator, thus moving further away from a situation where the technology matches the community scale. However, the initiators initiated new community-focused activities, using the wind farm revenues and their energy trading license as a vehicle to support other Irish energy cooperatives, including other cooperatives in the cVPP, thus becoming a communities' VPP instead of a single community VPP. In this way, the initiators aim to create a future in which they can enable other Irish communities to develop community-owned renewable energy, supply the energy to themselves through Community Power, and allow communities and citizens to receive revenues for their (surplus of) renewable energy generation. However, to be able to perform the activity of selling aggregated surplus energy back to the grid, a cVPP project would need to obtain a supply and distribution license. This license is costly and requires a certain aggregation ability of individual generation assets.

5.2 Value proposition

In practice, the value proposition of the Dutch and Irish cVPPs had to change towards providing greater value for the grid than anticipated, because the activities necessary to generate community values, such as keeping energy and economic flows locally, were not allowed. In terms of values, this means that the focus has to shift from local values and a community logic to delivering value to the electricity grid and the energy incumbents. This also means a focus on financial revenues rather than on nonmonetary community values.

The value proposition, therefore, would now have to include providing reliability, network services, grid investment deferral, primary and secondary reserve capacity, demand response flexibility, and/or balancing; reducing network congestion; improving load forecasting; and simplifying communication by having a single aggregator to communicate with (Burger and Luke, 2017; Ofgem, 2015).

The Irish cVPP also had to shift the business model focus from delivering local community values towards delivering values on a national scale to the national grid. However, the revolving fund they initiated to start developing other community cooperatives was a way of still focusing on the community values.

5.3 Resources

Local voluntary contributions and (time, relational, and intellectual) resources that are central in the community logic-based business model become less relevant when cVPP initiators are confronted with a situation in which the value proposition and activities have to change as set out above. Because of the increasing complexity, as mentioned earlier, other resources that often cannot be provided by local community members become more important, such as external funding, external expertise, and professional organisational skills.

Resources that cVPP Loenen and the Irish cVPP aimed to use consist of the revolving fund and investments of households who individually or collectively invest in renewable energy generation and/or energy conservation and of course the resource of a lot of voluntary time and intellect. In addition, funding from European projects is used to develop and implement the cVPP and provide room for experimentation. However, for both cVPPs, the necessity to grow beyond the community scale will also mean that an increasing number of resources are necessary, which go beyond the means available in the community and human resources would need to be outsourced, placing day-to-day operation outside of the community.

5.4 **Customer segment**

To participate in the market as energy traders, supplying their surplus energy, cVPPs would need to have a substantial size in terms of connected renewable sources, or work with a licensed supplier that thus would become a customer of the cVPP, receiving its generated energy and would need to have an aggregator as customer to sell their flexibility to. Most business models that currently include energy trading and provide flexibility are focused on large customers, not households, and can thus avoid this licensing partner construction (Burger and Luke, 2017, p. 12). In addition, because the value being delivered also shifts to include value to grid stakeholders such as DSOs and TSOs, these also become the customer segment targeted by the business model.

In the Dutch case, we indeed see a clear risk that the primary customer segment no longer is the (initial) local community (including also local industry and SMEs) itself but the grid stakeholders. Templederry Wind Farm also had to include a large national energy supplier as a customer, to sell their energy to. As such, the community is no longer the primary customer segment.

5.5 **Relation**

In a business model only informed by a community logic, the relationship between the cVPP and the community would revolve around involving the community in decision-making on renting or buying (access to) property, technology selection, and procurement (Müller and Welpe, 2018). However, the customer relation in both cVPP models is increasing in complexity because the customers include a wide variety of institutional stakeholders, from energy suppliers to whom surplus electricity is sold, to aggregators that buy the flexibility, to other grid stakeholders.

Apart from the risk of losing the connection with the community base, both the Irish and Dutch cVPPs risk a significant increase of the administrative burden to the extent that it may no longer be feasible for the community to handle the customer relationship itself. If this leads to a decision to outsource components of the cVPP, the business model is turned upside down: the community members become 'mere' shareholders and resources for aggregators and suppliers instead of the providers of the value proposition and owners of the ideas and assets. In line with that, they will lose the ability to influence and control how energy is generated, stored, distributed, and self-consumed within the community.

5.6 **Channels**

When cVPP projects change their proposition to be able to arrive at feasible business model, the customer interactions and relations change as well, as explained earlier. There is consequently the risk that the direct, personal, face-to-face channels of interaction within the community become much less important as the cVPP initiators need to turn towards a variety of other stakeholders that they depend on for realising the cVPP initiative. This implies that the Dutch and Irish channels run the risk of becoming much more professional and traditional, resembling the current mass-market type of interactions between energy suppliers and their customers. The community-focused type of channels would fade into the background.

5.7 Partners

Following the community logic, the cVPP would be owned and operated by, for, and with the community members, making them the main partner. However, the current institutional setting drives cVPP initiators towards partnering with incumbent, licensed, and large energy companies to make use of their license, under their flag—creating a dependent relationship with these partners, which makes the ownership and participation of the community more challenging. Such partnering also is the only way to avoid the costly and organisational complex consequences of having to comply with the balancing mechanism regulation, which requires either a very large portfolio in terms of the amount of electricity available to feed into the grid at any given time or a very large sum of capital to be able to buy that amount of electricity quickly. Because most cVPP projects lack the expertise and financial clout, they need an intermediary for this, whether it is an aggregator or a large incumbent supplier (Ofgem, 2015). In addition, a cVPP may also need to outsource most of their management to an aggregator that centralises all data, loads, and activities (Wainstein et al., 2017). Partnerships with incumbent partners and other intermediary actors can take different forms, but in any case, it will be an asymmetrical relationship that risks undermining the role (influence and autonomy) of the community—contrary to the main values of the energy community logic.

Initially, cVPP Loenen partnered with local actors like volunteers, the village counsel that could act as a supervisory board, local industry that provides space on their roofs for community-owned solar panels, and local SMEs taking care of insulating houses and installing solar panels. However, to stay up and running, cVPP Loenen now has to partner with the DSO rather than collaborating mainly with local community actors. This DSO supports the integration of renewable energy sources and possesses information regarding the status of the grid.

In an attempt to maintain its community character, Community Power is transforming from a small-scale energy supplier into a nationwide energy supplier co-owned by multiple energy cooperatives and communities, including Templederry Wind Farm. These other energy communities not only are groups of customers but also are seen as important partners in developing community-owned energy generation. Another important partner is Friends of the Earth (https://www.foe.ie/), who, together with the involved communities, is lobbying for regulatory change that makes the Irish context more suitable for community-owned renewable energy generation and trading.

5.8 Costs

The costs accompanying the changes to the business model, as discussed earlier, are very different from the community logic type of costs, which would be manageable for the community to bear.

For cVPP Community Power, to cover the costs for meeting the requirements involved in becoming a national supplier, it needs to either grow its customer base significantly or find investors. For cVPP Loenen, this will also be the case if it decides to become a national level supplier or otherwise participate in the electricity market.

5.9 Revenue

All of the changes to the business model also result in a different type of revenue. When the community value that was central in the community logic-based models was replaced by value for the grid stakeholders, that value disappeared from the business model only to be financed, if sufficient funds remained, through the revenue (Burger and Luke, 2017; Kerr et al., 2017; Müller and Welpe, 2018). However, the question is if revenue is enough to indeed create the community value that Loenen cVPP and Community Power cVPP aimed for. Other values could include community building, lowering household energy bills, increasing control over the siting of renewable energy generation, and improving the local economy by, for example, hiring local companies to insulate homes and install solar photovoltaic panels. A community logic-based type of revenue aimed at serving the community is not yet achievable in practice, unless community members accept really long payback times (de De Graaf, 2017). However, evidence so far suggests that the business model for a cVPP that has adapted to the institutional realities is unlikely to generate sufficient surplus revenue to pay for the desired community values beyond some direct economic value to individual participants (Braunholtz-Speight et al., 2018; Hall and Roelich, 2016). This is indeed what the Community Power cVPP fears. The fact that cVPP initiatives need to achieve financial viability to survive means that, in the current institutional setting, they will opt first and foremost for a VPP that aims for financial optimisation (rather than an optimum based on CO_2 reduction or maximising self-consumption).

6 Discussion

In this chapter, we focused explicitly on the tension between initiatives, their business model, and the incumbent energy system, taking a very distinct and new type of community energy project as example: community-based virtual power plants. We first set out the community logic based on a literature review, to be able to assess how a business model based on such a logic ideally would look like. The cases of the Dutch Loenen cVPP and the Irish Community Power cVPP project allowed us to explore how the institutional conditions shaped the Dutch and Irish cVPP business models.

Both sociotechnical and environmental justice literatures point out how important institutional conditions are in shaping opportunities for (social) innovation. For the deployment of distributed and renewable energy resources, in the past decades, major impediments did not result from a lack of ideas, enthusiasm, or available technologies, but rather from the incumbent institutions that favour the fossil fuel-based energy resource deployment, distribution, and use. The institutional analysis in Section 4 revealed path dependency and institutional voids. Institutional arrangements that structure grid management and energy market developments work to reproduce unequal opportunities in such a manner that large players with financial clout and with good access to expertise are privileged over cVPP projects. In addition, the voids consist of a lack of policy support for more complex social and technological configurations such as cVPPs. This, in combination with an absence of rules and standards for these new configurations, creates enormous challenges for innovative cVPP projects that want to generate, store, distribute, and self-consume.

The Irish and Dutch cVPP examples demonstrate that the institutional context in practice requires that cVPPs partly abandon their community logic-based business model. The community logic of partnering with the community, that is, the community being involved, engaged, and included as key element of the model, is problematic. The design, ownership, and operation of the cVPP is not only going to be decided by the community, with community benefits, values being centre stage, at present also the incumbent system is at least as influential in directing the configuration of the cVPP. The community logic of a technology use being matched to the local demands and needs is replaced by a configuration so complex, technically, organisationally, and financially, that partnering with the incumbent system stakeholders is the only option. Activities will not follow the community logic of being driven by community needs and motivations, but become focused on the national scale delivery of electricity and delivering values for the grid stakeholders next or even first, instead of for the community. Costs are very high, while revenues are low. In general, the necessary business model demonstrates a trend towards more scale and more complexity, both technologically and organisationally, on the administrative level, etc.

What we can conclude is that, although cVPPs are only just emerging, the tensions between their initial community logic-based intentions and the institutional realities are such that their business models, scale of operation, and community values are not tenable if they want to survive in the current institutional context. In other words, an economically viable roll-out of community logic-informed type of business models is not possible in the current situation.

However, things are changing. At the time of writing of this chapter, the clean energy for all Europeans package—Electricity Directive (European Commission, 2019)—was published, and this directive states that: 'Citizens energy communities should not face regulatory restrictions if they apply existing or future ICT technologies to share electricity from generation assets within the community between its members or shareholders based on market principles, for example by offsetting the energy component of members using the generation available within the community, even over the public network, provided that both metering points belong to the community'. This directive also states that communities should have access to the electricity markets: 'Member States shall ensure that citizens energy communities can access all electricity markets either directly or through aggregation in a non-discriminatory manner'. However, how this would be designed and how this will influence the cVPP business models is yet to be seen.

6.1 Suggestions for further research and practice

More and more communities are setting up cooperatives to create a local market for renewable energy, with the accompanying business models. These business models face great problems with the institutional and regulatory national systems, problems similar to those described for the cVPPs in Ireland and The Netherlands. On the one hand, communities developing cVPPs, a new and thus challenging sociotechnical configuration, need to be aware of the prevailing energy systems. As such, they can aim to learn about this interplay to successfully develop strategies to implement sustainable business models (Ceschin, 2013; Geels, 2002; Huijben et al., 2016; Smith and Raven, 2012).

However, the business models now being developed are part of an energy system under change and may have the potential to contributing to a more inclusive energy transition. These communities are developing business models that to some extent fit the current system, but some key elements making their cVPP community based are under great danger of being obliterated in light of the need to comply with context rules and regulations. Some exemptions are in place to help initiative experiment with models that would not fit the existing regulatory, institutional, legal system such as sandboxes in the United Kingdom or the experimentation arrangement in the Netherlands. These sandboxes or experimentation arrangements allow business developers to test their product or service in a controlled environment, with a temporarily customised regulatory environment so that the business developers can learn about its viability in a relatively protected environment (De Graaf, 2017). However, for initiatives that want to enter the market and be commercially viable, this experimentation is no option since no structural changes are made to the context itself. Burger and Luke (2017, p. 3) even state that: 'the regulatory and policy environment is a larger driver of business model structure than technological differences or other factors'.

Thus, if the aim is to change our current energy system to allow for more bottom-up community energy, more distributed and decentralised generation, and a more democratic energy system, what needs to be considered is how current energy market infrastructures, regulation, and support mechanisms in different countries influence the newly emerging sociotechnical configurations that are composed of social, technical, physical, and organisational dimensions. In addition, the question of what the consequences for the future energy system are if this context remains needs to be addressed. Cross-cultural learning and experimentation such as those that are taking place in the Netherlands and Ireland are of main importance for business model development aimed at mass uptake of new energy services (McGrath, 2010; Mullins and Komisar, 2009). Another important step would be to further investigate how to innovate the regulatory system to allow for better regional development of renewable energy initiatives aimed at local generation and supply, driven by community logic. This type of community energy projects are potential sociotechnical configurations for a more inclusive and democratic development, and this implies that their development and decisions made need to be evaluated in light of their impact on the distribution of control and power over the energy system (Burke and Stephens, 2017).

Another future research avenue concerns the role of intermediaries that can take over the role now often played by incumbent energy supply companies, which are not necessarily contributing to the ideals of a more inclusive or democratic energy system. Seyfang et al. (2014) argue that the question is indeed not if intermediaries are needed, they already exist and are needed because they can support the much-needed learning at the level of community energy. The question to be researched is what these intermediaries can achieve and how they can contribute to the ideals of energy democracy. That raises the question what the ideal type intermediary for cVPP would then look like. Would the ideal type intermediary have to be able to combine and contextualise multiple technologies into an innovative integrated system, in line with regulation and ensuring financial viability? Potentially, this intermediary should also have competencies in terms of technology, system integration, institutional and techno-financial requirements, translating between needs and motivations from community and the energy system, and mobilisation skills. Or would these intermediaries in particular have a role to play in identifying the many tensions between new business models such as those for cVPPs and the existing context and use this to initiate a dialogue about the existing path dependencies and lock-ins? Efforts at dialogue, discussion, and action are needed in any case, focusing on the question as how to create a more conducive environment for the type of business models and sociotechnical configurations such as cVPPs that have the potential to change our current energy system into a more inclusive and democratic energy system.

Acknowledgements

This text was written as part of the IEA DSM TCP Task 25 project 'Energy Service Supporting Business Models and Context' and the Interreg project cVPP 'Community-based Virtual Power Plant: a novel model of radical decarbonisation based on empowerment of low-carbon community driven energy initiatives'. We would also like to thank the editors for their review of earlier versions of the chapter.

References

Asmus, P., 2010. Microgrids, virtual power plants and our distributed energy future. Electr. J. 23 (10), 72–82.

Bloomberg, 2017. Market for Digitalization in Energy Sector to Grow to $64BN by 2025 [Online]. Available at https://about.bnef.com/blog/market-digitalization-energy-sector-grow-64bn-2025/. Accessed 23 April 2018.

Bocken, N.M.P., Short, S.W., Rana, P., Evans, S., 2014. A literature and practice review to develop sustainable business model archetypes. J. Clean. Prod. 65, 42–56.

Boons, F., Lüdeke-Freund, F., 2013. Business models for sustainable innovation: state-of-the-art and steps towards a research agenda. J. Clean. Prod. 45, 9–19.

Braunholtz-Speight, T., Mander, S., Hannon, M., Hardy, J., McLachlan, C., Manderson, E., Harmina, M., 2018. The evolution of Community Energy in the UK. UKERC, London.

Breukers, S., Wolsink, M., 2007. Wind power implementation in changing institutional landscapes: an international comparison. Energy Policy 35 (5), 2737–2750.

Burger, S.P., Luke, M., 2017. Business models for distributed energy resources: a review and empirical analysis. Energy Policy 109, 230–248.

Burke, M.J., Stephens, J.C., 2017. Energy democracy: goals and policy instruments for sociotechnical transitions. Energy Res. Soc. Sci. 33, 35–48.

Buth, M.C., Wieczorek, A.J., Verbong, G.P.J., 2019. The promise of peer-to-peer trading? The potential impact of blockchain on the actor configuration in the Dutch electricity system.. Energy Res. Social Sci. 53, 194–205.

Ceschin, F., 2013. Critical factors for implementing and diffusing sustainable product-Service systems: insights from innovation studies and companies' experiences. J. Clean. Prod. 45, 74–88.

Davoudi, S., Brooks, E., 2014. When does unequal become unfair? Judging claims of environmental injustice. Environ. Plann. 46 (11), 2686–2702.

De Graaf, F. (2017) New Strategies For Smart Integrated Decentralised Energy Systems [Online]. Available at https://www.metabolic.nl/wp-content/uploads/2018/09/SIDE-Systems-Report.pdf (Accessed: 2 January 2019).

European Commission, 2017. (EU) 2017/2195 of 23 November 2017. Establishing a guideline on electricity balancing. Off. J. Eur. Union.

European Commission, 2019. Analysis of the final compromise text with a view to agreement (11 January 2019), 15150/1/16 REV 1. In: Proposal for a DIRECTIVE OF THE EUROPEAN PARLIAMENT AND OF THE COUNCIL on common rules for the internal market in electricity (recast)—2016/0380(COD), Brussels, 2019.

Geels, F.W., 2002. Technological transitions as evolutionary reconfiguration processes: a multi-level perspective and a case-study. Res. Policy 31 (8–9), 1257–1274.

Hall, S., Roelich, K., 2016. Business model innovation in electricity supply market: the role of complex value in the United Kingdom. Energy Policy 92, 286–298.

Hall, P.A., Taylor, R.C.R., 1996. Political science and the three new institutionalisms. Polit. Stud. 44 (5), 936–957.

Hicks, J., Ison, N., 2018. An exploration of the boundaries of "community" in community renewable energy projects: Navigating between motivations and context. Energy Policy 113, 523–534.

Hienerth, C., Keinz, P., Lettl, C., 2011. Exploring the nature and implementation of user-centred business models. Long Range Plann. 44, 344–374.

Hoffman, S.M., High-Pippert, A., 2010. From private lives to collective action: recruitment and participation incentives for a community energy program. Energy Policy 38 (12), 7567–7574.

Huijben, J.C.C.M., Verbong, G.P.J., Podoynitsyna, K.S., 2016. Mainstreaming solar: stretching the regulatory regime through business model innovation. Environ. Innov. Soc. Trans. 20, 1–15.

IEA (2017) Digitalization & Energy [Online]. Available at http://www.iea.org/publications/freepublications/publication/DigitalizationandEnergy3.pdf (Accessed 10 September 2018).

Kerr, S., Johnson, K., Weir, S., 2017. Understanding community benefit payments from renewable energy development. Energy Policy 105, 202–211.

Kindström, D., Ottosson, M., Thollander, P., 2017. Driving forces for and barriers to providing energy services—a study of local and regional energy companies in Sweden. Energ. Effic. 10 (1), 21–39.

Klein, S.J.W., Coffey, S., 2016. Building a sustainable energy future, one community at a time. Renew. Sustain. Energy Rev. 60, 867–880.

McGrath, R.G., 2010. Business models: a discovery driven approach. Long Range Plann. 43 (2–3), 247–261.

Müller, S.C., Welpe, I.M., 2018. Sharing electricity storage at the community level: an empirical analysis of potential business models and barriers. Energy Policy 118, 492–503.

Mullins, J., Komisar, R., 2009. Getting to Plan B: Breaking Through to a Better Business Model, first ed. Harvard Business Press, Boston, US.

Nilsson, H., Bångens, L., Govén, B., Andersson, B., 2012. We are lost if we don't develop new business-models. In: ECEEE 2012 Summer Study—Energy Efficiency in Industry, pp. 363–368.

Ofgem (2015) Non-traditional business models: supporting transformative change in the energy market [Online]. Available at: https://www.ofgem.gov.uk/ofgem-publications/93586/non-traditionalbusinessmodelsdiscussionpaperpdf (Accessed 21 September 2018).

Osterwalder, A., Pigneur, Y., 2013. Business Model Generation: A Handbook for Visionaries. Game Changers, and Challengers. Wiley.

Šahović, N., Da Silva, P.P., 2016. Community Renewable Energy—Research Perspectives. Energy Procedia 106, 46–58.

Schlosberg, D., 2004. Reconceiving environmental justice: global movements and political theories. Environ. Polit. 13, 517–540.

Schlosberg, D., 2013. Theorizing environmental justice: the expanding sphere of the discourse. Environ. Polit. 22, 37–55.

Seyfang, G., Hielscher, S., Hargreaves, T., Martiskainen, M., Smith, A., 2014. A grassroots sustainable energy niche? Reflections on community energy in the UK. Environ. Innov. Soc. Trans. 13, 21–44.

Seyfang, G., Park, J.J., Smith, A., 2013. A thousand flowers blooming? An examination of community energy in the UK. Energy Policy 61, 977–989.

Sharma, R.R., Mukherji, S., 2015. Organizational transformation for sustainable development: a case study. In: Management of Permanent Change. Springer Fachmedien Wiesbaden, Wiesbaden, pp. 195–216.

Smith, A., 2012. Civil society in sustainable energy transitions. In: Verbong, G., Loorbach, D. (Eds.), Governing the Energy Transition: Reality, Illusion or Necessity? Routledge Studies in Sustainability Transitions. Routledge Taylor & Francis Group, New York, pp. 180–202.

Smith, A., Raven, R., 2012. What is protective space? Reconsidering niches in transitions to sustainability. Res. Policy 41 (6), 1025–1036.

Statkraft (2017) Digitalization in the Energy Industry—Statkraft Summer Project Blog [Online]. Available at: https://statkraftsummerproject.wordpress.com/2017/08/07/digitalization-in-the-energy-industry/ (Accessed 23 April 2018)

Tolkamp, J., Huijben, J.C.C.M., Mourik, R.M., Verbong, G.P.J., Bouwknegt, R., 2018. User-centred sustainable business model design: The case of energy efficiency services in the Netherlands. J. Clean. Prod. 182, 755–764.

Unruh, G.C., 2000. Understanding carbon lock-in. Energy Policy 28 (12), 817–830.

Van Summeren, L.F.M., Wieczorek, A.J., Bombaerts, G.J.T., Verbong, G.P.J., Community energy meets smart grids: reviewing goals, ownership, and roles in community-based Virtual Power Plants in Belgium, Ireland, and the Netherlands. Energy Res. Soc. Sci. (forthcoming).

Vargo, S.L., Lusch, R.F., 2008. Service-dominant logic: continuing the evolution'. J. Acad. Market. Sci. 36, 1–10.

Verbong, G.P.J., Loorbach, D. (Eds.), 2012. Governing the energy transition: reality, illusion or necessity? Routledge Studies in Sustainability Transitions. Routledge Taylor & Francis Group, New York.

Verkade, N., Höffken, J., 2018. The design and development of domestic smart grid interventions: Insights from the Netherlands. J. Clean. Prod. 202, 799–805.

Vezzoli, C., Ceschin, F., Diehl, J.C., Kohtala, C., 2015. New design challenges to widely implement "Sustainable Product-Service Systems". J. Clean. Prod. 97, 1–12.

Wainstein, M.E., Dargaville, R., Bumpus, A., 2017. Social virtual energy networks: Exploring innovative business models of prosumer aggregation with virtual power plants. In: 2017 IEEE Power & Energy Society Innovative Smart Grid Technologies Conference (ISGT). pp. 1–5.

Walker, G., Devine-Wright, P., 2008. Community renewable energy: What should it mean? Energy Policy 36 (2), 497–500.

Walsh, B., 2018. Community: A powerful label? Connecting wind energy to rural Ireland. Comm. Develop. J. 53 (2), 228–245.

Wolsink, M., 2013. Fair distribution of power-generating capacity: justice, microgrids and utilizing the common pool of renewable energy'. In: Bickerstaff, K., Walker, G., Bulkeley, H. (Eds.), Energy Justice in a Changing Climate: Social Equity and Low Carbon Energy. Zed Books, London, pp. 116–138.

Zott, C., Amit, R., Massa, L., 2011. The business model: recent developments and future research. J. Manage. 37 (4), 1019–1042.

Further reading

Energy4All (n.d.) ENERGY4ALL Green Co-operative Energy [Online]. Available at https://energy4all.co.uk/ (Accessed: 2 January 2019).

Energy, human activity, and knowledge: Addressing smart city challenges

2.6

Sarah J. Darby

Environmental Change Institute, University of Oxford, Oxford, United Kingdom

1 Introduction: City evolution and the arrival of smart technology

The first cities were established over 6000 years ago, but it is only quite recently that there have been mass movements worldwide from the countryside into cities. More than half the world's people now live in cities, and the United Nations estimate that around two-thirds of us will be city dwellers by 2050.[a] The future of cities therefore matters to everyone, including all the people and organisations who produce food, water, fuel, and power for city dwellers. When a 'smart city' appears, what does that mean for life in the city and beyond?

This chapter begins with a short reflection on what cities are and how they evolve, defining some smart city concepts and discussing what they may imply in terms of 'energy and behaviour'. It analyses how humans and technologies interact in smart systems, showing the importance of different actors and types of communication, then outlines some challenges we might expect if such concepts are put into practice. Outcomes will depend on *how* we deploy the new combinations of technology, people, and skills and on what we want to achieve in social and environmental terms. Realistic expectations of smart cities will prepare us best for using smart technology effectively.

Each city is unique, a complex organism with its own geography, ecology, people, buildings, public and private spaces, technologies, knowledge and expertise, governance, industries, and services. A city is a place where we find people carrying out activities that need specialisation and training: somewhere that is 'well organised socially with a very developed culture and way of life' and with 'a relatively high level of cultural and technological development'.[b] It is a site or theatre where people

---

[a] https://population.un.org/wup/.
[b] These definitions come from the Oxford Advanced Learners' English Dictionary and the Merriam-Webster Dictionary, respectively.

Energy and Behaviour. https://doi.org/10.1016/B978-0-12-818567-4.00010-7

'perform' city life each day. As with theatre or live concert performances, cities are never quite the same from day to day, even though there may be a lot of predictable routine. The performance of city life each day reflects how people will be doing some different activities from those they did yesterday, or similar activities at different times or speeds, or with different people. When we add accidents and emergencies to the list of possible changes, along with road and rail maintenance, public events, and seasonal shifts in activity, the complexity of life becomes awe-inspiring, even in a stable, mature city. Shaping and upholding this complexity are a crucial part of city governance and energy systems that play a central role here.

Electrification has been a factor in the growth of cities, adding to their complexity by making new activities possible in industry, commerce, public services, and everyday life. An electrified city is also more complex in terms of timing than its pre-electric version: one where work and leisure can continue day and night without the risks, smells, and inconvenience of candles, kerosene, or gas lights. Photographs of the world at night, taken from space, show vividly where people are lighting up their environment, and these bursts of light also show where we are likely to find the most intense and diverse economic activity.

The arrival of information and communications technology (ICT) is already adding to the complexity of life—more data, more sophisticated services, more rapid interactions, and transactions. It also lends itself to new ways of thinking about 'smart' city design and activity, some utopian, some the opposite. Before going any further, it is probably time to stop and think about definitions. What is smart? And what might it mean to be a smart city?

1.1 The meaning and purpose of 'smart' devices and systems

The first uses of 'smart' to mean 'quick, active, clever' date from the 14th century; however, it was not until the 1970s, with the dawn of artificial intelligence, that 'smart' was applied to inanimate things 'behaving as though guided by (human) intelligence'[c]. A current definition states, more explicitly, that a smart device has been 'programmed so as to be capable of independent action'[d]. In other words, a smart device or system is activated by human intelligence but then becomes capable of acting without human intervention, through machine learning. This second definition is probably the most useful when we think about smart cities. It tells us that humans plan them and design the algorithms that control activities in the city, so the designers' aims, biases, knowledge, and risk perceptions are built into those algorithms. It also tells us that smart systems will at times act independently of direct human control.

Designers of smart systems have traditionally emphasised technical and economic possibilities, with the prospect of reduced human effort and increased convenience. There is a huge body of research literature and marketing material based

---

[c] Online Etymological Dictionary, 2019.
[d] The Oxford Dictionaries, 2019.

on 'prospective' studies of the potential outcomes from smart systems—far greater than the literature that evaluates actual performance in real life. Gram-Hanssen and Darby (2018) discuss this mismatch between desk studies and empirical evaluation where smart homes are concerned, and there is a similar mismatch in the literature on smart cities.

The literature on smart technology in everyday life mostly assumes such technology will meet people's individual needs as well as contributing to system management goals. These assumptions, though, raise questions about what people need, how they have met their needs in the past, and how they might respond to living with devices that are remotely monitored, accessed, and controlled. It can be difficult to persuade people that they want or need remote sensing and control for actions that they have already carried out themselves, without difficulty, for many years (Poulsen et al., 2002; Darby, 2018). A growing body of research finds that smart technologies in the home can have unexpected and sometimes negative outcomes. Users have to make an effort to understand and use them in the way the designers intended and are not always willing or able to do this (Hargreaves et al., 2017; Darby et al., 2018); they can alter the balance of power and control in the household (Wilson et al., 2015), and they may lead to increased consumption (Nyborg and Røpke, 2013).

Where cities are concerned, similar questions arise about needs and about how people may adapt to smart technology or resist it. While some smart functions may be relatively simple, for example, making it easier to find out when the next bus or tram is available, there will also be interactions between different infrastructures and services to take into account, for example, water, wastewater and energy, and transport and police support for traffic management and passenger safety. If smart technologies in a home can lead to unexpected outcomes and unwelcome outcomes, it is reasonable to 'expect the unexpected' and to assess risks, when ICT becomes part of something as complex as a city with thousands or millions of people and all their diverse activities, priorities, and resources.

1.2 Smart technology in the city

We have no single agreed definition of a smart city, although the term has been in use since the 1990s. Ramaprasad et al. (2017) offer a brief history of the smart city concept and analyse how it has evolved. Identifying over 25,000 potential components of a smart city from their literature review, they produce a complex definition that incorporates structures, functions, focus, semiotics (data, information, and knowledge), stakeholders, and outcomes. This is not very useful for non-specialists, perhaps, but it does reflect the complex nature of city life and the huge range of possibilities for cities of the future.

Two broad types of definition have emerged: one concerned mostly with infrastructure control and management and the other with social and (sometimes) sustainability outcomes (Ahvenniemi et al., 2017). The more 'humanist' interpretations of smart cities stress the importance of social learning, education, and social capital. But both agree that ICT connectivity is what makes a city smart, giving it a new nervous system that is continually sensing, processing, and transmitting information to districts, neighbourhoods, individuals, and devices.

Another approach is to ask, what does a smart city *do*? Is it something new, or is it simply a city with added smartness, a city that somehow performs better than other cities, or better than it used to do? Is smart city governance a matter of developing 'new forms of human collaboration through the use of ICTs to obtain better outcomes and more open governance processes' (Meijer and Rodríguez Bolivar, 2016, p. 392) or, more neutrally, something that involves the use of digital technology to *inform* city managers, *control* city processes and services, and *promote* innovation? (e.g. Liu and Peng, 2014). Based on these and similar commentaries, I suggest the following definition, one that assumes nothing about the quality of outcomes that

> *A smart city is one where ICT is applied extensively in living and working environments, and in city management and government.*

That is, 'smartening' means the incorporation of ICT into city processes and functions, no more and no less. We should not therefore assume that smarter cities are going to be more sustainable than less-smart cities. In frameworks to assess and evaluate smart cities, the focus is usually on social and economic indicators. If sustainability is included, it tends to be low down the list of desirable outcomes (Ahvenniemi et al., 2017). So, we can only safely assume that the process of introducing ICT will bring about different types of connectedness, control, and planning that will lead to new ways of organising, investing, operating, and being accountable. A smart city is therefore more than a city where people just use ICT for their own purposes, however, intensively; it relies on ICT to function as a complex system.

A smart city does not have to be a new city—just one where ICT shifts activity to new levels in a controlled and coordinated way and where citizens with internet-connected devices can connect with city infrastructure. For example, public transport vehicles can be monitored continually so that users can easily find out when to expect the next bus, tram, or train; sanitation workers can locate blocked wastewater pipes more easily; electric vehicle charging can be managed in ways that do not threaten local network stability; and engineers can locate leaks in low-pressure water distribution systems. In a 'smartened' city, the ICT nervous system will typically cover a wide range of functions and services: a review of 70 smart city initiatives identified six domains and 28 subdomains, from renewable energy through to cultural heritage management (Neirotti et al., 2014). Barcelona offers an example of a well-established and prosperous city that decided to adopt smart city technologies and services relatively early and sent through a process of adapting human activities and responsibilities; Bakici et al. (2014) outline how existing physical, knowledge, and administrative structures have been adapted, and 'intermediary organisations' have been recruited to carry out a top-down programme of smartening. What we mostly see in smart city planning is an extension or development of our existing cities, so they incorporate more ICT-enabled control and communication.

At the other end of the scale, though, stand 'new build' smart cities such as Masdar in Abu Dhabi and the hugely ambitious Dholera in India. Planned from afar, these experiments are typically presented to the world in utopian language, with their proponents showing great faith in the ability of financiers, technology developers, planners, and the construction industry to produce viable, liveable, and sustainable settlements, even in inhospitable and sparsely populated parts of the world. (See Cugurullo (2016), for an analysis of how easily such ambitious experimental city planning can fail.) It is reasonable to ask whether a smart city 'solution' has been proposed because it meets an actual need or because the developer of a particular technology or service is determined to sell more of it and is trying to persuade potential buyers through 'corporate storytelling' (Söderström et al., 2014).

An electricity infrastructure, entwined with a communication infrastructure, is central to all smart city concepts, whether the cities are old or new. ICT depends on reliable electricity supply, and our energy systems increasingly depend on reliable ICT: we cannot talk for long about smart cities without referring to energy-related challenges. So, it is time to think about what we know about energy and society. The next section of the chapter looks at how energy systems are collaborations between people and things that provide services for specific activities in particular places. It notes how they require different types of knowledge and skill, the centrality of communications, the potential for unexpected outcomes, and the need for resilience.

2 Energy and society: What have we learned?

It is useful to review some of the knowledge about energy systems we have gained through carrying out research into their social dimensions, to understand better the energy-related challenges posed by smart cities. The material in this section draws upon research relating to everyday activities, decisions, policy, and governance. Each subsection looks at a feature of energy systems.

2.1 Energy systems are social

Energy demand is highly social. We share access to electricity and gas grids, heat networks, street lighting, and public transport infrastructure. We use energy for work, travel, and recreation; more and more, we use it to communicate with each other and with appliances, devices, and systems. Yet this social dimension is often overlooked: even energy efficiency researchers can think of people in a very limited way, as individual (usually domestic) consumers (Moezzi and Janda, 2014). A more rounded view of how we live as social beings gives us more insight into the nature of energy demand and into how decisions relating to energy use are made. It also opens up new possibilities for change such as collaborative learning, promoting climate-appropriate architecture rather than 'international' building styles, and modelling buildings with reference to how they perform when people are using them, not just as collections of physical material (ibid.).

Energy supply, too, is social. Even if power and heat come from large, seemingly impersonal corporate suppliers, these contain people who are constantly making strategic and operational decisions and carrying out activities that include designing, installing and maintaining physical infrastructure, generating power, buying and storing fuel, marketing, metre reading, billing, accounting, trading, and lobbying. If energy is supplied on a smaller, more local scale—for example, municipal district heating or a cooperatively owned wind farm—the social aspect may be more obvious.

Adding a new piece of energy infrastructure—a tramway, electricity substation, smart meters, new buildings (because buildings, too, are infrastructures through which energy flows)—will involve numerous actors to plan, specify, install, advise, and maintain. Developing a workable business model means bringing together an effective combination of people, things, and processes. And the rules that shape energy supply and use, such as building standards, electricity tariffs, and speed limits, are made and applied through social processes.

It is for reasons such as these that we can only describe and analyse energy systems comprehensively by understanding them in social, organisational, and physical terms. When looking ahead to low-carbon systems, the same considerations apply (Eyre et al., 2018).

2.2 **Energy issues are local and global**

Patterns of energy demand and supply differ around the world, between and within countries: geography matters. Most of the research literature on both demand and supply comes from the wealthier countries of the world, mostly members of the Organisation for Economic Cooperation and Development (OECD). Yet most of the world's people live elsewhere, and conditions can vary a lot between regions and local areas, even within OECD countries. So, it is important, when thinking about city resources and energy demands, to consider where the city is located. This will influence factors such as the mix of fuels for generation and heating, which will depend on availability, infrastructure, and local rules such as air pollution standards, and whether electricity supply is reliable and constant, reliable but patchy, or erratic.

Within cities, too, geography matters. Some cities are tightly zoned, with distinct industrial, commercial, and residential districts, while others have more mixed-use neighbourhoods. A densely populated city is likely to require less energy for transport than a sprawling one. Some areas will be wealthier than others, with different types of building and different patterns of mobility. Useful examples of how local factors shape energy demand can be seen in a detailed analysis of residential electricity and gas use in districts of California (Lutzenhiser and Bender, 2008), a socio-technical analysis of household electricity use in Lahore (Khalid and Sunikka-Blank, 2017), and an account of energy transition in a district of Scotland that used to rely on mining and heavy industry (Darby, 2017).

Three trends in electricity systems make 'place' even more significant than it has been in the past. One is the move from centralised systems to highly distributed generation. The second is the arrival of substantial new demand such as that from heat pumps, electric vehicles, air conditioners, and IT servers (with or without the network operator being aware of this). The third is the growing need for regulatory and commercial arrangements to make smart grids viable, at international, national, and local levels (Darby et al., 2013). The value of smart technology in a city will depend on how well it fits with the other elements of the city, far more than on the market readiness of individual technologies.

An example from a recent smart technology project may help to illustrate this last point. The aim was to demonstrate the demand response value of smart thermal storage in several European locations, by aggregating the heating demand and introducing 24-h direct load control, subject to customer preferences (Darby et al., 2018). Customers who participated in the trial lived in homes with a range of thermal characteristics; they also showed a range of temperature and timing preferences. Some had no difficulty programming their heating with the new digital controls, while others struggled and needed help. It became clear that different households would have different potential for demand response, due to these differences, and that customer support and advice could improve that potential. The study also showed how physical and regulatory environments shaped the possibilities for demand response. Climate and housing quality both influenced the overall demand for heating, while European, national, and local rules for housing, heating systems, metering, and tariffs set boundaries for what aggregators, retailers, and network operators could do to establish more flexible demand. In all, the 3-year trial showed how effective demand response could be achieved after painstaking work to bring together a collaboration between appliances, aggregator 'backend' platforms, electricity suppliers, customers, and various 'middle actors' (Parag and Janda, 2014).

2.3 **Energy provides services so that people can act**

Energy is commonly defined as the ability to do work. This is not the whole story, but it is good enough for everyday use and worth remembering when thinking about energy system design. Energy provides services that allow us to work and to make changes in our environment, services such as rapid transit, cleanliness, entertainment, warmth or 'coolth', and lighting. Electricity and fuel use will vary depending on whether it is needed for heating and ventilating a library or concert hall, food processing, office work, hairdressing, steelmaking, or domestic activities. All have their own rhythms and environmental impacts, and, as electricity supply comes to rely more on variable generation from wind and sun, flexible demand becomes more desirable, and the rhythms of different services become more significant (Torriti et al., 2015).

Daily routines in the home are influenced by shared social norms (e.g. about how warm a home or workplace should be, how often clothes need to be washed and how they are dried, and how much ICT equipment is purchased and left switched on). However, these conventions are interpreted in many ways, which helps to explain why similar numbers of people in identical buildings can vary so much in the amounts of electricity and fuel they use—for example, by a factor of ~3 for heating and ~5 for electrical appliances (Gram-Hanssen, 2013)—and why the timing of electricity use can also vary according to what energy services we expect, how we achieve them, and which technologies we use (McKenna et al., 2018).

Cities are not only places where people use energy services for their own purposes. They also provide public services that may not be available in smaller communities. So, a smart city is one where we would expect public services to be carefully integrated and publicised with the help of ICT, to maximise citizens' access.

2.3.1 Energy poverty is about poor access to energy services

Poverty in general is a state of deprivation, and energy poverty is an inability to access the energy services needed to live in a society (Bouzarovski and Petrova, 2015). Fuel poverty, a related concept that is usually applied to cool climates, has three elements: the unit cost of fuel or electricity, household income, and the energy efficiency of the home (Boardman, 2009). If a home is energy efficient enough, the household can be 'fuel poverty proof' because it will have affordable warmth and other energy services even if the unit cost of fuel rises and the household income is low.

The crucial word here is 'services'. People do not want kilowatt-hours of electricity or cubic metres of gas, as such: they want the services that these can provide. Also, energy services do not always imply the use of electricity or fuel. For example, a passive house in a cool temperate climate can supply all heating needs, by containing heat from solar radiation and occupants' bodies and some waste heat from their appliances; a protected outdoor space for drying clothes may take away the need for a laundry dryer; a local health centre, pharmacy, and library will reduce people's need to travel long distances to access health and information services. If there is over-reliance on energy service provision through technology alone, the first to suffer may be citizens who are isolated and vulnerable because of age, illness, or disability. The terrible history of the 1995 Chicago heatwave, during which over 700 citizens died (many at home, alone), shows the central significance of human connections and care in making services available (Klinenberg, 2002). Some authors point to the potential use of data from smart meters and sensors to diagnose fuel or energy poverty (Gouveia et al., 2018), and such data could have shown where the most overheated apartments lay, but diagnosis is no substitute for cure.

The main point here is that smart technologies alone are unable to solve poverty-related problems that stem from some combination of poor housing, inefficient appliances, low incomes, or high unit energy prices. Indeed, they could make things worse if they lead to services that are hard to access for people without broadband or digital literacy.

2.4 Energy systems call for knowledge and skills

Energy supply and use have always called for some knowledge and skill. At a very basic level, it has been useful to know what types of wood burn best and how to use fire efficiently for cooking or smelting metal. With industrialisation, the need for more specialist knowledge has grown, and it continues to grow as we move from demand-led to supply-led electricity systems, where demand must match available supply from distributed and renewable sources. We are also seeing a shift in the financing of electricity, from systems where 'end-users' pay a supplier for units of energy to those where they become generators themselves and may even trade their own generation, storage, and demand to supply ancillary services to balance demand and supply in real time.

Such changes add to system complexity. Smart energy systems are also more technical than traditional methods of accessing energy services, and they can require highly specialised, knowledge and the ability to install and operate advanced technology. (For more on the technical and human implications of this shift towards microgeneration, new forms of demand and 'active demand', see Sauter and Watson, 2007; Caird and Roy, 2010; Morstyn et al., 2018.)

It is still quite unusual for householders and small businesses to have the knowledge and ability to adopt and adapt smart energy technology themselves (e.g. Mennicken and Huang, 2012; Woo and Lim, 2015). Most continue to rely on trained engineers to install and maintain solar photovoltaic panels (PV), batteries, and devices such as vehicle chargers and building management systems or home hubs. In addition to basic qualifications, these engineers are very likely to need in-service training so that they can keep up with new developments. Not just engineers but the growth of conferences, training events, and webinars on all aspects of smart energy shows how energy professionals around the world are hungry for knowledge and understanding.

There is also a less recognised need for widespread, general, 'ordinary' knowledge and skill in the provision of energy services. There are consumer issues: what products and services are available, at what cost, and how reliable are they? If something goes wrong, who is responsible for compensating a customer? There are also operational issues: how do the new products and services work and what are they compatible with? How are they maintained or upgraded? What sort of controls are available? Without this sort of knowledge, neither energy users nor systems can realise potential benefits from smart energy. For example, it was only possible to gain effective demand response during a trial of smart residential thermal storage once the customers understood the basic proposition, how to operate their controls, and why it was important not to switch their heaters off at the walls (to maintain connectivity). The suppliers, network operators, and demand aggregators also needed to understand the customers' situation, and the 'middle actors'—project coordinators, landlords, call centre advisory staff—became valuable teachers, enabling the whole project to achieve its aims (Darby et al., 2018).

2.5 Effective energy systems need good communications

The parts that make up a system have to be connected and able to communicate; otherwise, there is no system—just a collection of elements. Energy analysts express this in different ways, according to their priorities. For example, engineering diagrams typically show the physical elements of an energy system; operational diagrams show how one process is related to another; policy diagrams illustrate the links between organisations.

All these system representations can be useful, but they only give a partial picture of system structure and communications. They often underestimate the significance of people as actors in systems, with their individual and collective activities and decisions. Crucially, they often miss out how people, through their knowledge, ability to act, and *connections* with others, become agents of change. As systems grow more complex, this agency becomes even more significant. For example, Janda and Parag (2013) show how 'middle actors' or 'middle agents' shape the practices through which energy services are delivered and used, contributing to professional knowledge and the ability to communicate it to others; Chapter 5 in this book also explores the role of middle actors in building refurbishment.

The need for major, rapid change in energy systems, to integrate supply with storage and demand in real time, increases the need for effective communication. There is a lot to learn, at all levels, and there is probably most to learn where the end-users ('bottom actors') and middle actors are concerned. People-focussed research can show, for example, the conditions in which customers may be willing to allow external control of some of their appliances by a network operator or demand aggregator (Fell et al., 2015), or how energy planners think about energy users and their ability to contribute to smart energy designs (Skølsvold and Lindqvist, 2015), or what actually happens when people adopt smart technologies or energy services with new forms of communication and control, and how this differs from what the designers imagined would happen (e.g. Nyborg and Røpke, 2013; Hargreaves et al., 2017).

We can summarise how communication in energy systems may take place under three headings: connectivity, control, and care (Darby et al., 2018). *Connectivity* is the communication between technical devices, something that requires high levels of skill to achieve and maintain, especially when software upgrades take place or components have to be replaced. *Control* is the interaction between people and technologies or the programming of automated control/design of algorithms by people. Control implies responsibility for outcomes: even if artificial intelligence is involved, people are ultimately liable for the consequences of using it. *Care* is the person-to-person communication that explains, advises, discusses, and generally contributes to understanding how systems work and how to operate and develop them.

The argument here is that all three types of communication are needed for effective energy systems. Without connectivity, the technologies fail to interact properly. Without user-friendly controls, the human–technology interface fails. And without effective person-to-person communications, it is hard if not impossible to prepare effectively for change in sociotechnical systems by involving people in system specification (what is it for? who will it serve and how?), transferring knowledge of how new technology works, and troubleshooting when things go wrong.

2.6 Energy systems are not always predictable

This heading will probably not surprise readers, but serves as a reminder of risk (quantifiable) and uncertainty (unquantifiable) in complex sociotechnical systems. For example, we know well by now that new or refurbished buildings will not 'perform' as predicted, once real people are living and working in them

(Bordass et al., 2001; Majcen et al., 2013).We know that wind turbine or hydropower outputs will fluctuate according to weather conditions and that from time to time there will be shocks such as natural disasters, market malfunctions, damage from conflicts and ICT failures, or security breaches. When these fluctuations and shocks occur, systems have to adapt or fail.

In practice, this means that people have to adapt. Where fluctuations in supply or demand are relatively predictable, they can adapt through demand response in which people alter their service expectations, activities, and/or technologies (McKenna et al., 2018). This may be quite easily managed through some voluntary load switching and automated control, or it may be painful, as when utilities are forced to 'load shed' and institute rolling blackouts to provide services for at least part of the time (Lodhi and Malik, 2013). But a major, unexpected shock to an energy system tests a society's resilience in an extreme way. The more people have depended on that system, the greater the disruption.

There are some helpful recent analyses of how people have reacted to system shocks. These include the account by Murakoshi et al. (2013) of how Japanese citizens reduced their electricity demand following the earthquake, tsunami, and flooding of the Fukushima nuclear power plants; the account by Leighty and Meier (2011) of how the people of Juneau, Alaska, reduced electricity demand immediately and in the longer term, following an avalanche that cut the transmission cables; and a detailed econometric analysis of public and policy responses to the Californian electricity supply crisis in 2000-01, showing short- and long-term demand reductions of around 13% and 4%, respectively (Reiss and White, 2008). All these studies show how electricity customers responded to the crises not simply as rational economic actors but as members of sociotechnical networks who had some concern for the public good: as citizens who, because of the sudden shortage of supply, became aware of being part of a shared system with limits. They made behavioural changes to reduce their demand although there was, at the time, no economic incentive to do so, and many also invested in more efficient equipment. Reiss and White (2008) comment that 'although economists tend to be dismissive of public appeals that run contrary to private incentives…one cannot dismiss [the] striking reduction in average household electricity consumption… The empirical facts here indicate that consumers do respond to voluntary appeals to modify their consumption behaviour, provided (i) the costs of a collective-action failure are tangible (here, involuntary blackouts for some) and (ii) the public is well aware of it' (p. 658).

Such findings are significant for the future design and governance of grids and networks. They highlight the need to think in terms of social and individual potential, but, at an even more fundamental level, they point to unpredictability as a fact of life, even in modern industrialised societies.

3 Addressing energy challenges in smart(er) cities

The previous section set out some characteristics of energy systems, with an emphasis on their social and technical nature: energy services in cities cannot be conceptualised and planned simply in terms of supplying physical infrastructures and kilowatt-hours. We can expect smart city energy challenges to be social, political, educational, and organisational as well as technical. For example, they may well involve adapting institutions and governance to carry out new functions, such as training people to work with the new technologies, altering market rules, and setting standards for data privacy, security, and consumer rights (Darby et al., 2013).

The need for new expertise in smart cities is recognised by many planners, who see the smart city concept as a way of attracting highly qualified and enterprising people to live and work in the city. But this will only be part of the story. The following section attempts to build on the previous one by looking at some of the more challenging aspects of energy services and city life, as ICT is absorbed into both.

3.1 Local issues

ICT has connected technologies and people to an extent that was almost unbelievable only 50 years ago, and electricity is available—at least some of the time—to most city dwellers around the world. It can seem as though a single vast 'global village' is not far ahead in which everything and everyone is interconnected seamlessly through an Internet of Everything. But the image of a single connected community is somewhat misleading. In reality, there are striking differences between cities and between the neighbourhoods, groups, and individuals within each city. New technologies will produce locally specific outcomes, because they will interact with local infrastructure and devices, and they will be installed and operated by people with diverse knowledge, ability, resources, and goals. So, a network of things and people, however large, can only be understood properly by studying how it works locally.

Energy system studies help us understand smart city challenges better, as they require multi- or interdisciplinary approaches to describe and analyse characteristics of demand and supply in a given place, and diversity between and within places. Such studies can show how effective smart city service provision will require careful preparation and testing in situ. For example, the Bristol Smart Energy City Collaboration, launched in April 2015, aimed to have 'the capabilities and capacity to make the most of new opportunities to capture, analyse and act on smart energy data for the benefit of people and businesses across the city' (BSECC, accessed 2019). Collaborators came from the public, private, and voluntary sectors: this was not a corporate, business-driven venture. They first identified opportunities to develop smart system activities and benefits throughout the city, paying attention to specific issues such as fuel poverty, economic prosperity, and the integration of locally generated electricity. The collective experience of testing these opportunities led to a conclusion that 'the ability of actors within the city to realise (a range of benefits) will be in part dependent on them…taking initiative…and in part on how market access rules, settlement systems, charging methodologies and licensing regimes create space and offer due reward for such initiative-taking' (CSE, 2017). That is, local actors with local knowledge were seen as essential but in need of support from central government and market regulators.

This locally based approach to energy, action, and knowledge in the city offers a model of how to test smart city concepts, taking seriously the aim of improving services and the quality of life in the city. By starting from local concerns, involving a range of stakeholders and taking time to gather evidence carefully, it arguably offers a much better prospect of valuable outcomes than an approach that is technology led. It is through local initiatives such as this that a useful body of knowledge is being built up that can act as a guide to what is possible in different circumstances.

3.2 Service provision, equity, and energy poverty

Some smart-enabled services—for example, distribution grid, traffic, or wastewater management—can work for everyone in a city, although only a few people may actually understand and manage the technologies. Others, such as smart electric vehicle charging, or smart storage, will only work for citizens who can afford the 'right' technologies and know how to use them. That is, they will need finance plus practical skills and social networks through which they can access services and learn how to benefit from them. Some citizens will be in a better position to benefit from smart services simply because of their greater access to technology and greater knowledge of how to use it (Lytras and Visvizi, 2018).

Fuel poverty and energy poverty are both concepts used to indicate an inability to afford energy services, and both describe a situation that can be complex in which the people most in need of services may be least able to access them, not simply because they lack money or live in poor housing but because they are in poor health or socially isolated (Baker et al., 2018).

Making a city smart, if by that we simply mean introducing ICT to many services, is therefore not likely to make energy poverty go away. Indeed, there is a risk that it could increase inequality if there is a serious divide between citizens who find it easy to understand and participate in new systems and those who do not. In an extreme scenario, the most marginalised people in a city could disappear from view and from care, if city administrators were to rely entirely on smart sensing and algorithms for information and decision-making. The complex nature of energy poverty strongly indicates that human observation and interaction are also needed (ibid.), along with a willingness to adapt systems to particular situations.

An approach based on local knowledge and stakeholder involvement can deal explicitly with the political processes of deciding priorities and allocating resources. It can be a sound basis for deciding, first of all, what functions a smart system is needed for, before working out which resources are needed to carry out those functions. It can also help avoid situations in which 'solutions' are decided before problems have been properly defined. Once problems have been agreed and defined by relevant stakeholders, decision-makers can move on to the question of whether smart or non-smart measures, or some combination, will be the best way of improving city life and avoiding environmental damage. Where energy is concerned, priorities for the poorest citizens are likely to be good-quality housing (to reduce the need for heating/cooling); gainful employment; and affordable access to workplaces, health services, shops, parks, and other amenities. Smart energy services do not insulate walls or roofs or provide public facilities. As a recent analysis of fuel and travel poverty in the United Kingdom points out, there is a strong need for a joined-up approach to energy affordability (Matteoti et al., 2017), and this applies also to the adoption of smart energy initiatives in cities.

3.3 Knowledge and skills for smart energy

Smart grid developments call for a shift in thinking at many levels and by many people, from regulators and network operators to customers. They also call for new knowledge and skills and bring new actors into energy systems. Professional actors include software engineers to develop algorithms for network management and market operations; demand aggregators to make a business out of aggregating and selling demand response services; and lawyers to deal with new types of contract between market actors, including contracts to provide ancillary system services and developments in consumer protection law. In addition, there are engineers who design, install, and maintain technologies and utility call centre staff and energy advisers who offer support to customers.

Smart energy systems therefore involve training and learning. At a minimum, the people involved in design and development will need to learn the necessary skills, while those who are directly involved in operating devices and systems will have to know how they work. There is also the need to exchange knowledge and practical experience between specialists and nonspecialists, so that members of the general public have a better chance of understanding new technologies and professionals can develop their understanding of how technologies are used in different contexts. For example, all smart meter installers in Great Britain have been required to take part in communication skills training so that they can explain to customers the purpose of the new metre and how to use the in-home display and can offer basic energy advice (Ofgem, 2013). They are also able to pass customer feedback to their employers, to improve the understanding of how a technology operates in real-life conditions (Darby and Liddell, 2015). This combination of smart meter plus in-home display plus trained installer is unusual (probably unique), in smart meter rollouts, in realising benefits from transferring technology together with some knowledge of how to use it effectively. While the British smart meter rollout has been plagued with technical difficulties, levels of customer satisfaction have remained high (see Darby et al., 2015, and customer surveys carried out for Smart Energy GB).

Successful introduction of smart energy in cities therefore relies on human and machine intelligence. Technologies can easily fail to live up to 'technical potential', partly due to design that doesn't match the ways in which people actually use devices or controls—the 'user logic'—and partly due to the lack of know-how among installers and users (Caird and Roy, 2010; Stevenson and Rijal, 2010). But when users *are* involved in design, experimentation, adoption, and use, the outcomes can be more productive (e.g. Ornetzeder and Rohracher, 2006; Mennicken and Huang, 2012). The need for specialist technical knowledge for smart energy is well recognised, as can be seen in the level of funding for research and development in this area; the need for practical and communication skills is often not sufficiently acknowledged and provided for.

3.4 **The challenge of adequate information and communications**

ICT has always required an electricity supply, and modern energy systems (electricity, gas, and heat) rely increasingly on effective ICT. People therefore increasingly rely for their energy services on two interlocked systems, each of which needs the other, whereas in the past, they were able to access some of these services, at least (lighting, heating, and ventilation), without the help of either electricity or ICT. This issue is explored further in the succeeding text, in the section on robustness and resilience. Here, the question is how to make sure a smart system is well enough informed and connected.

As noted earlier, there are three vital types of communication in sociotechnical systems, binding the human with the non-human: *connectivity* between devices and systems; human *control* of devices and systems (control), which requires effective interfaces; and *care,* constructive communication between people.

The first two of these, connectivity and control, are now branches of engineering and design, with a large body of research literature dealing with technical and conceptual issues. Most of it lies beyond the scope of this chapter, but it is worth saying a little about feedback on energy use to customers, via metering systems. There is a sizeable body of research into different modes of feedback (e.g. Karlin et al., 2014) and their effectiveness, showing a range of outcomes in terms of customer responses and impact on energy use. This variety of outcomes reflects the number of variables in play, such as demographics, locations, activities and usage patterns, feedback types, and customer–utility relationships. Broadly, findings support the case for

- better feedback from a mix of sources, for example, informative bills/printed reports based on measured (not estimated) consumption plus in-home display (IHD) plus alerts if consumption levels are exceeding normal limits plus comparisons with customers in similar situations (Ehrhardt-Martinez et al., 2010; VaasaETT, 2011),
- developing feedback in the context of social learning in which people do not simply receive information but apply it to develop, together, their understanding of energy and to improve their living conditions (Glad, 2012).

Feedback is valued mainly because, if carried out well, it raises awareness and helps promote action to reduce demand, because it can help build knowledge and trust, and because it can boost the effectiveness of demand response (e.g. VaasaETT, 2012; Faruqui and Sergici, 2013).

As cities incorporate more smart technology, person-to-person communications are still likely to be essential to maintain services, for at least three reasons. The first is advisory—assisting people as they adopt and adapt to new technologies. The second reason is to provide reality checks, where professionals and city managers can learn from the experience of technology users in real-life conditions. And the third is the double-loop process of learning from project and programme evaluations. As the climate crisis intensifies, it becomes even more important to avoid making the too-familiar mistakes from overspecified systems, poorly designed buildings, and inadequate market arrangements or regulatory processes. This risk can be addressed systematically by policymakers and practitioners, given the will, resources, and time to evaluate and learn (Stern et al., 2016; Bordass and Leaman, 2005).

3.4.1 Security, privacy, trust and accountability

Smart systems are data driven, and data sharing has become an extremely sensitive issue, along with security concerns. ICT opens up massive opportunities for connectedness—and for malicious hacking. Every Internet-connected device makes an electricity network a little more vulnerable to cyberattack than it already is. This is not the place to go into data security in any detail, only to point out that smart systems are always likely to be limited by security considerations in at least two ways: the technical or operational risks attached to huge numbers of interconnected devices and the willingness or unwillingness of citizens to tolerate high levels of data gathering about their activities.

As Smale et al. (2019) point out, the ICT component of smart grids 'shrinks the distance' between households, workplaces, and the grids that supply them. In principle, we might expect this new closeness to increase trust. But the ICT is also the vehicle for data harvesting, and a utility or city that is constantly collecting data about citizens' whereabouts and activities is likely to face challenges in gaining and maintaining trust. Their residents can justifiably ask what data gatherers need their data for, who they share it with, where they store it and who decides what is deleted, and when. A city that relies heavily on 'big data' may be unacceptably intrusive into the lives of its citizens (Kitchin, 2014). Big data has to be at the service of city life, not a beast that must be continually fed through intrusive monitoring. Automated data gathering, in which data are generated routinely through a device or system, is a particular cause for concern (ibid.); it is one thing to offer personal data in return for a service but quite another to have it taken from you without your knowledge. We can expect levels of trust to be influenced by factors such as rules for privacy and data sharing and the quality and reliability of services (Fell et al., 2015).

Trust can be an issue when customers are needed to cooperate in more active demand. For example, electricity users are unlikely to alter the timing of their washing-machine use or vehicle charging, or to hand over control of their water heater to someone else, unless they believe there is a valid reason for doing so. They may also object to sharing data about their appliance usage. Safeguards against 'oversharing' of data can be built into information systems: for example, high-resolution data from smart meters can remain within the household or business as a default arrangement, only going to the utility with customers' explicit consent (McKenna et al., 2012). But this sort of arrangement does have to be designed into a demand response programme, after careful preparation. A review of nine major smart-enabled demand response and efficiency programmes carried out by VaasaETT (2012) shows how developing good customer relations over time can improve effectiveness.

In a smart system, it needs to be clear who is responsible for service delivery and for commissioning and designing algorithms. As machine learning becomes more commonplace, accountability for decisions made by machines will increasingly become a live issue for people seeking compensation for disrupted services and for their lawyers.

3.5 Robustness and resilience: Responding to shocks and fluctuations

Section 2.6 pointed out how energy systems are not always predictable and how a reliable electricity supply can be wiped out in seconds as a result of an accident or natural disaster. To appreciate the impact of a long and unplanned electricity blackout on city life, including communications, it is worth reading Kemp's (2017) forensic analysis of how citizens reacted to the flooding of a substation in North West England, during a severe winter storm. This shows in detail how disruptive this event was and describes the network operator's mighty effort to restore services to a population of ~100,000 within a few days. The analysis also tells how people experienced the blackout differently depending on where they lived; the quality of their housing; whether they had 'back-up' resources such as fireplaces, candles, and camping stoves; and whether they were able to work without an electricity supply. It shows how the wired telephone system (powered from batteries) mostly continued to function while the mobile phone systems (powered from the local supply) failed and how businesses, schools, the university and railway station faced major difficulties as cashless payments, swipe cards, messaging, and safety controls ceased to work. The author raises important questions about how a 'smart' society should prepare to cope with shocks like these. As smart devices and processes become more commonplace, people come to rely on them more heavily, and system failure can be extremely serious. Increasing the number of connected devices can add to the risk of system failure through hacking or malfunction.

One of the crucial smart city challenges is therefore to make sure that energy systems are robust enough and well-enough understood to cope with shocks and fluctuations. Introducing ICT to city life brings new risks and uncertainties, including connectivity failures, overconcentration of control in a few experts, and less ability to carry out everyday activities without uninterrupted power and data streaming. More generally, there is a risk that attention and resources may shift towards new smart developments and away from persistent problems caused by inadequate buildings and infrastructures. The claims made for smart cities are typically focused on technology and positive, especially those coming from the ICT industry. For example, the head of strategy and marketing in the software business unit of Ericsson is quoted as claiming that a smart city is one that 'can utilise Internet of Things (IoT) sensors, actuators, and technology to connect components across…every layer of a city, from the air to the street to underground. It's when you can derive data from everything that is connected and utilise it to improve the lives of citizens and improve communication between citizens and the government that a city becomes a smart city'.[e] Yet, at the time of writing this chapter, Ericsson and O_2 had to issue a joint apology to millions of customers for a shutdown to their data services, attributed to an expired certificate in the software versions installed with these customers; O_2 were estimating the cost of such a shutdown as ~€110m[f]. Some customers, unable to access the

---

[e] https://www.techrepublic.com/article/smart-cities-the-smart-persons-guide/
[f] https://www.bbc.co.uk/news/business-46499366

Internet on their phones for a while, may only suffer minor inconvenience when something like this happens. But some will rely on 24-h data streaming to run their businesses or to manage illnesses such as diabetes. Are they entitled to compensation and, if so, to how much? More importantly, are there ways of working around such shutdowns: having a plan B or, in the language of adaptation, having resilient systems with several pathways to the same goal?

These uncertainties and risks matter greatly when the major ICT companies are striving to become leaders in a huge emerging smart market. Compromises around open operating standards seem a long way off when there is so much to be gained commercially by locking customers into a single firm's products and services. As argued by Viitanen and Kingston (2014), highly technical 'intelligent systems' are likely to leave cities less resilient than before, relying heavily on expert knowledge and equipment that can only be supplied by commercial partners.

In summary, a viable smart city has to be able to handle risk and uncertainty: it has to have people and other resources to mobilise 'plan B' when one or more systems or processes fail. A focus on functionality seems like the best place to start. As an example, the Future Systems Power Architecture project in the United Kingdom has defined 35 functions that a future system will need to perform. These range from 'black start' capability—to get a system up and running from a single generator after a shutdown, without relying on an external network—to metering and settlement that allows for flexible tariffs (FPSA, 2017). Such an approach can develop resilience and reliability, especially if it is coupled with knowledge sharing, so that no system relies heavily on the expertise of a small group of people.

4 Summary: Meeting smart city energy challenges, energetically

This chapter has argued that to achieve good environmental, energy, and social outcomes, we need to take a critical view of smart city proposals, especially those that are based on technological 'solutions' to generic problems. The concept of a smart city can be utopian, promising fully integrated control over city life along with high-quality services for citizens at low environmental cost. But such visions typically rely on unrealistic assumptions about the nature of energy systems and social systems and how benefits will be distributed. We have no evidence that such a city can work reliably for the good of all citizens and plenty of evidence that smart technologies fail at times. The size and scale of corporate interests in pushing smart city technologies, if nothing else, should make us wary that they may be products in search of a market, rather than practical ways of addressing social need and ecological crisis.

Transitioning to low-carbon energy is not simply a matter of changing fuel sources: it requires people not only to alter the way they live from year to year but also to change how they invest in buildings, vehicles, and appliances and how they plan and regulate land use, taxation, markets, and infrastructures. Fortunately, energy studies can already provide plenty of knowledge relevant to the future of our cities—especially the more interdisciplinary studies that take in social, ecological, and political aspects of energy. They show how energy systems, like cities, are physical *and* ecological, social, *and* technical. It is not only individual technologies or individual behaviours and choices that make a system function but also interactions between people and technology, people and people, and people and their environment.

The challenge for city dwellers, in a growing climate crisis and an energy transition, is to make cities liveable in the long run: for ecosystems, people, technology, and all. So far, the reality of new build smart cities has fallen a long way short of the utopian promises, and it would be surprising if it were otherwise. Most cities are not new build, though, and here, the challenge is one of integrating ICT selectively, in ways that incorporate local characteristics, actors, knowledge, and skills to provide the services that citizens need.

Medium-scale trials around the world are now showing practical ways of testing smart technologies in near-real-life interactions with people, built environment, and market conditions. Through such trials, it becomes possible to identify the materials, actors, connections, risks, and uncertainties that matter most, so that smart city proposals can be assessed under headings such as

- services needed;
- how technology will be used to meet them;
- citizens' access to services: what technologies and know-how are required?
- types of expertise needed for reliable services, including person-to-person, situation-specific advice;
- power and raw material requirements, with carbon and other environmental impacts;
- people needed to introduce and operate new technologies, to negotiate their use in different situations, and to troubleshoot when they fail;
- types of control needed and how that control is distributed among citizens, organisations, and government;
- governance and accountability;
- long-term reliability, for example, in relation to connectivity, human–technology interfaces, and human oversight;
- types of data transmitted and safeguards against misuse;
- resilience: when power, hardware, or software fail, how can a city keep functioning?
- adaptability: how much scope is there to learn and change, avoiding harmful lock-ins to particular devices, technologies, or processes?

Experience to date shows how hard it is to predict accurately what a smart city will achieve in terms of energy performance or citizen welfare. Cities rely heavily on services such as waste disposal, sewage treatment, emergency services, and chilled food storage, and disrupting any of these can have severe impacts on well-being. If the running of these services is taken away from human agents and invested in machines, specialised, expensive knowledge will be needed to repair the damage when a system fails; therefore, there are strong arguments for making sure that many people are trained to understand the intricacies of city life and to improvise and adapt when necessary to maintain services.

Ultimately, the value of 'smart' in a smart city will depend on local conditions, civic values, and the distribution of useful knowledge between people and technologies. This poses some exciting challenges to researchers. I would argue that we need more researchers who are prepared to move beyond their specialism and think broadly, treating environmental, technical, and social issues with equal seriousness. We need to study the connections between social expectations, activities and technologies, and the unforeseen effects that will emerge. We need to be critical of 'drawing-board' claims made for smart cities and put more effort into evaluating how smart technologies work out in real-life conditions. Perhaps most importantly, we need to keep asking basic questions when considering smart city initiatives, such as: what is the problem that this is intended to solve? How will it solve the problem? What sort of actors and skills are needed to make this work? Who will gain and lose, and what will they gain and lose? Where will the power and data be flowing? Who faces the most and least risk and uncertainty? And who is accountable for the outcomes?

Acknowledgement

The author gratefully acknowledges funding from the UK Energy Research Centre; editorial guidance from the guest editors; and many conversations with colleagues over the years, especially Ellen Christensen, Kirsten Gram-Hanssen, Susse Georg, Katy Janda, Marina Topouzi and Sarah Higginson.

References

Ahvenniemi, H., Huovila, A., Pinto-Seppä, I., Airaksinen, M., 2017. What are the differences between sustainable and smart cities? Cities 60, 234–245.

Baker, K.J., Mould, R., Restrick, S., 2018. Rethink fuel poverty as a complex problem. Nature Energy 3, 610–612.

Bakici, T., Almirall, E., Wareham, J., 2014. A Smart City initiative: the case of Barcelona. J Knowl Econ 4, 135–148.

Boardman, B., 2009. Fixing Fuel Poverty: Challenges and Solutions. Routledge, London.

Bordass, B., Leaman, A., 2005. Making feedback and post-occupancy evaluation routine 1: a portfolio of feedback techniques. Building Research and Information 33 (4), 347–352.

Bordass, B., Leaman, A., Ruyssevelt, P., 2001. Assessing building performance in use 5: conclusions and implications. Building Research and Information 29 (2), 144–157.

Bouzarovski, S., Petrova, S., 2015. A global perspective on domestic energy deprivation: Overcoming the energy poverty–fuel poverty binary. Energy Research & Social Science 10, 31–40.

BSECC, 2019. Bristol Smart Energy City Collaboration. Taking action to realise the benefits of smart energy data. accessed March 2019.

Caird, S., Roy, R., 2010. Adoption and use of household microgeneration heat technologies. Low Carbon Economy 1 (2), 61–70.

CSE, 2017. A Smart, Flexible Energy System: Call for Evidence. CSE's Response to BEIS and Ofgem. Centre for Sustainable Energy, Bristol. https://www.cse.org.uk/downloads/file/CSE-response-smart-flexible-energy-system-consultation-jan-2017.pdf. (Accessed March 2019).

Cugurullo, F., 2016. Urban eco-modernism and the policy context of new eco-city projects: where Masdar City fails and why. Urban Studies 53 (11), 2417–2433.

Darby, S.J., 2017. Coal fires, steel houses and the man in the moon: local experiences of energy transition. Energy Research & Social Science 31, 120–127.

Darby, S.J., 2018. Smart technology in the home: time for more clarity. Building Research and Information 46 (1), 140–147.

Darby, S.J., Liddell, C., 2015. Communicating 'smartness': smart meter installers in UK homes. In: Proceedings, ECEEE Summer Study, pp. 1991–2001.

Darby, S.J., Strömbäck, J., Wilks, M., 2013. Potential carbon impacts of smart grid development in six European countries. Energy Efficiency 6, 725–739. https://doi.org/10.1007/s12053-013-9208-8.

Darby, S.J., Liddell, C., Hills, D., Drabble, D., 2015. Smart Metering Early Learning Project: Synthesis report. For the Department of Energy and Climate Change, London. https://www.gov.uk/government/uploads/system/uploads/attachment_data/file/407568/8_Synthesis_FINAL_25feb15.pdf.

Darby, S.J., Higginson, S., Topouzi, M., Goodhew, J., Reiss, S., 2018. Getting the Balance Right: Can Smart Thermal Storage Work for Both Customers and Grids?. Consumer Impact study, RealValue project Horizon 2020, Environmental Change Institute, University of Oxford. http://www.realvalueproject.com/images/uploads/documents/RealValue_Consumer_Impact_Report_-_FINAL_%28Compressed_spread%29.pdf.

Ehrhardt-Martinez, K., Donnelly, K.A., Laitner, J.A., 2010. Advanced Metering Initiatives and Residential Feedback Programmes: A Meta-Review for Household Electricity-Saving Opportunities. American Council for an Energy Efficient Economy, Washington DC.

Eyre, N., Darby, S.J., Grünewald, P., McKenna, E., Ford, R., 2018. Reaching a 1.5C target: Socio-technical challenges for a rapid transition to low carbon electricity systems. Philosophical Transactions of the Royal Society A: Mathematical, Physical and Engineering Sciences 376, 20160462.

Faruqui, A., Sergici, S., 2013. *Arcturus:* international evidence on dynamic pricing. The Electricity Journal 26 (7), 55–65.

Fell, M., Shipworth, D., Huebner, G.M., Elwell, C.A., 2015. Knowing me, knowing you: the role of trust, locus of control and privacy concern in acceptance of domestic electricity demand-side response. In: Proceedings, ECEEE Summer Study, Hyères, France, June 2015, pp. 2153–2163.

FPSA, 2017. Future Power System Architecture Project, Phase 2 Synthesis Report. Institution of Engineering and Technology with the Energy Systems Catapult, London, UK. https://www.theiet.org/sectors/energy/resources/fpsa/fpsa-future-system-challenges.cfm.

Glad, W., 2012. Housing renovation and energy systems: the need for social learning. Building Research & Information 40 (3), 274–289.

Gouveia, J.P., Seixa, J., Long, G., 2018. Mining households' energy data to disclose fuel poverty: Lessons for Southern Europe. Journal of Cleaner Production 178, 534–550.

Gram-Hanssen, K., 2013. Efficient technologies or user behaviour, which is the more important when reducing households' energy consumption? Energy Efficiency 6, 447–457.

Gram-Hanssen, K., Darby, S.J., 2018. 'Home is where the smart is'? Evaluating smart home research and approaches against the concept of home. Energy Research and Social Science 37, 94–101.

Hargreaves, T., Wilson, C., Hauxwell-Baldwin, R., 2017. Learning to live in a smart home. Building Research and Information 46 (1), 127–139.

Janda, K.B., Parag, Y., 2013. A middle-out approach for improving energy performance in buildings. Building Research and Information 41 (1), 39–50.

Karlin, B., Ford, R., Squiers, C., 2014. Energy feedback technology: a review and taxonomy of products and platforms. Energy Efficiency 7 (3), 377–399.

Kemp, R., 2017. Electrical system resilience: a forensic analysis of the blackout in Lancaster, UK. In: Proceedings of the Institution of Civil Engineers—Forensic Engineering. https://doi.org/10.1680/jfoen.16.00030.

Khalid, R., Sunikka-Blank, M., 2017. Homely social practices, uncanny electricity demands: Class, culture and material dynamics in Pakistan. Energy Research & Social Science 34, 122–131.

Kitchin, R., 2014. The real-time city? Big data and smart urbanism. GeoJournal 79, 1–14.

Klinenberg, E., 2002. Heatwave: A social Autopsy of Disaster in Chicago. University of Chicago Press.

Leighty, W., Meier, A., 2011. Accelerated electricity conservation in Juneau, Alaska: a study of household activities that reduced demand 25%. Energy Policy 39 (5), 2299–2309.

Liu, P., Peng, Z., 2014. China's smart city pilots: a progress report. Computer 41 (10), 72–81.

Lodhi, R.N., Malik, R.K., 2013. Impact of electricity shortage on daily routines: a case study of Pakistan. Energy & Environment 24 (5), 701–709.

Lutzenhiser, L., Bender, S., 2008. The "average American" unmasked: social structure and differences in household energy use and carbon emissions. In: Proceedings, ACEEE Summer Study. pp. 7-191 to 7-20.

Lytras, M.D., Visvizi, A., 2018. Who uses smart city services and what to make of it: toward interdisciplinary smart cities research. Sustainability 10 (6). https://doi.org/10.3390/su10061998.

Majcen, D., Itard, L., Visscher, H., 2013. Actual and theoretical gas consumption in Dutch dwellings: What causes the differences? Energy Policy 61, 460–471.

Matteoti, G., Lucas, K., Marsden, G., 2017. Transport poverty and fuel poverty in the UK: from analogy to comparison. Transport Policy 59, 93–105.

McKenna, E., Richardson, I., Thomson, M., 2012. Smart meter data: balancing consumer privacy concerns with legitimate applications. Energy Policy 41, 807–814.

McKenna, E., Higginson, S., Grunewald, P., Darby, S.J., 2018. Simulating residential demand response: Improving socio-technical assumptions in activity-based models of energy demand. Energy Efficiency 11, 1583–1597.

Meijer, A., Rodríguez Bolivar, M.P., 2016. Governing the smart city: a review of the literature on smart urban governance. International Review of Administrative Sciences 82 (2), 392–408.

Mennicken, S., Huang, E.M., 2012. Hacking the natural habitat: an in-the-wild study of smart homes, their development, and the people who live in them. In: Kay, J., Lukowicz, P., Tokuda, H., Olivier, P., Krüger, A. (Eds.), Pervasive Computing. 7319. Springer, Berlin and Heidelberg, pp. 143–160.

Moezzi, M., Janda, K., 2014. From "if only" to "social potential" in schemes to reduce building energy use. Energy Research & Social Science 1, 30–40.

Morstyn, T., Farrell, N., Darby, S.J., McCulloch, M., 2018. Using peer-to-peer energy trading platforms to incentivise prosumers to form federated power plants. Nature Energy 3, 94–101.

Murakoshi, C., Hirayama, S., Nakagami, H., 2013. Analysis of behaviour change due to electricity crisis: Japanese household electricity consumer behaviour since the earthquake. In: Proceedings, ECEEE Summer Study, pp. 29–39.

Neirotti, P., de Marco, A., Cagliano, A.C., Mangano, G., Scorrano, F., 2014. Current trends in Smart City Initiatives: some stylised facts. Cities 38, 25–36.

Nyborg, S., Røpke, I., 2013. Constructing users in the smart grid—insights from the Danish eFlex project. Energy Efficiency 6, 655–670.

Ofgem, 2013. SMICoP: Smart Metering Installation Code of Practice. Office of Gas and Electricity Markets, London.

Online Etymological Dictionary, 2019. https://www.etymonline.com/word/smart. referred to March 2019.

Ornetzeder, M., Rohracher, H., 2006. User-led innovations and participation processes: lessons from sustainable energy technologies. Energy Policy 34, 138–150.

Oxford Dictionaries, 2019. https://en.oxforddictionaries.com/definition/smart. referred to March 2019.

Parag, Y., Janda, K., 2014. More than filler: middle actors and socio-technical change in the energy system from the "middle-out". Energy Research & Social Science 3, 102–112.

Poulsen, D., Nicolle, C.A., Galley, M., 2002. Review of the current status of research on smart homes and other domestic assistive technologies in support of the TAHI trials. Prepared for the Department of Trade and Industry, Loughborough University Institutional Repository. https://dspace.lboro.ac.uk/2134/1030.

Ramaprasad, A., Sánchez-Ortiz, A., Syn, T., 2017. A unified definition of a smart city. Chapter 2, In: Janssen, M., Axelsson, K., Glassey, O., Klievink, B., Krimmer, R., Lindgren, I., Trutnev, D. (Eds.). Electronic Government, Proceedings, 16th IFIP WG 8.5, International Conference, EGOV 2017, St Petersburg, Russia, September 4–7, 2017.

Reiss, P.C., White, M.W., 2008. What changes energy consumption? Prices and public pressures. RAND Journal of Economics 39 (3), 636–663.

Sauter, R., Watson, J., 2007. Strategies for the deployment of micro-generation: Implications for social acceptance. Energy Policy 35 (5), 2770–2779.

Skølsvold, T.M., Lindqvist, C., 2015. Ambivalence, designing users and user imaginaries in the European smart grid: insights from an interdisciplinary demonstration project. Energy Research and Social Science 9, 43–50.

Smale, R., Spaargaren, G., van Vliet, B., 2019. Householders comanaging energy systems: space for collaboration? Building Research & Information 47 (5), 585–597.

Söderström, O., Paasche, T., Lauser, F., 2014. Smart cities as corporate storytelling. City 18 (3), 307–320.

Stern, P.C., Janda, K.B., Brown, M.A., Steg, L., Vine, E.L., Lutzenhiser, L., 2016. Opportunities and insights for reducing fossil fuel consumption by households and organisations. Nature Energy 1. https://doi.org/10.1038/NENERGY.2016.43.

Stevenson, F., Rijal, H.B., 2010. Developing occupancy feedback from a prototype to improve housing production. Building Research and Information 36, 549–563.

Torriti, J., Hanna, R., Anderson, B., Yeboah, G., Druckman, A., 2015. Peak residential electricity demand and social practices: deriving flexibility and greenhouse gas intensities from time use and locational data. Indoor and Built Environment 24 (7), 891–912.

VaasaETT (2011) Empower Demand. The potential of smart meter enabled programmes to increase energy and systems efficiency: a mass pilot comparison. (Lead authors Strömbäck, K, Dromacque, C, and Yassin, MH). Report for the European Smart Metering Industry Group. Vaasa Energy Think Tank, Helsinki.

VaasaETT, 2012. Empower Demand 2: Energy Efficiency through Information and Communication Technology—Best Practice Examples and Guidance. VaasaETT Global Energy Think Tank, Finland. https://esmig.eu/resource/empower-demand-report-phase-ii.

Viitanen, J., Kingston, R., 2014. Smart cities and green growth: outsourcing democratic and environmental resilience to the global technology sector. Environment and Planning A 46, 803–819.

Wilson, C., Hargreaves, T., Hauxwell-Baldwin, R., 2015. Smart homes and their users: a systematic analysis and key challenges. Personal and Ubiquitous Computing 19 (20), 463–476.

Woo, J., Lim, Y., 2015. User experience in do-it-yourself-style smart homes. ACM Press, pp. 779–790. https://doi.org/10.1145/2750858.2806063.

Modelling energy behaviour

3

Energy and enjoyment: The value of household electricity consumption

3.1

Phil Grunewald, Marina Diakonova

Environmental Change Institute, University of Oxford, Oxford, United Kingdom

1 Introduction

For the needs of engineers and economists, household energy consumption is relatively easy to measure and quantify. Electricity is measured in kilowatt-hour and valued in euros or dollars. These are established units in engineering and economic studies and have become the dominant units of energy policy narratives globally. In broad terms, the policy aims in many countries with liberalised energy markets are therefore a tension between two aims: (1) driving GDP ($) growth through economic activity using more electricity (kWh) and (2) reducing expenditure ($) on energy with more efficient generation and use of kilowatt-hours.

More recently, the timing of electricity use has attracted attention. In low-carbon systems, the availability and cost of electricity generation could vary significantly over time. Depending on the availability of renewable resources, such as wind and solar, prices can rise or fall to new extremes. The natural policy response is to change the timing of use with time-varying prices, especially penalising peak-time consumption or enticing increase (or better shifts) of demand to times with lower prices using time-of-use tariffs (Darby and McKenna, 2012).

At this point—where price seeks to reshape the patterns of daily life—a good understanding of the dynamics that shape these patterns and their potential flexibility becomes important. While some time-of-use tariffs have proved effective in shaping and reducing peaks (Schofield et al., 2014; CER, 2011; Bulkeley et al., 2014), the full potential and inhibitors to residential flexibility are still poorly understood.

What purely price-based policies fail to recognise is that for end-users, both money and kilowatt-hour are abstract concepts at the time of using electricity. When performing energy-intensive activities, the very fact that energy is involved is well shielded from the user. Appliances are specifically and carefully designed to make the performance of their tasks appear effortless. The electricity used by these appliances is neither visible nor easy to quantify, making the implicit cost of the performed task an even more remote concept. Fig. 1 seeks to illustrate the conceptual distance between what individuals do (their activities) and some of the implications.

Energy and Behaviour. https://doi.org/10.1016/B978-0-12-818567-4.00011-9

FIG. 1

Degrees of abstraction between individuals and policy instruments.

Often, the most remote concepts, such as cost, CO_2 emissions, and climate change, are used as motivators intended to change individual's activities.

Exposing end-users to time-varying prices directly may therefore not be the most effective means to achieve the ultimate aim of demand reshaping and reduction. And it may not be necessary, either. The commercial incentive to change demand patterns could stay with suppliers and other service providers, who would benefit from having to supply less energy at times when they are exposed to higher wholesale market prices. Instead of passing the price through directly, they may engage in more subtle incentive schemes, grounded in behavioural economics, marketing, or measures that help their customers in practical ways to be more flexible. For this long anticipated 'energy service' model to be successful, a good understanding of the dynamics of demand could make an important contribution. The units of energy and money alone do not suffice in this effort.

This chapter deals with quantities that are more difficult to quantify yet much closer to the everyday experience of people: activities and their enjoyment. In Section 2, we critically review tools and methods to attempt to capture these quantities and relate them to electricity use. Section 3 presents new data on activities, enjoyment, and electricity use and discusses methods and limitations to their analysis. Results on relationships between these data are shown in Section 4. The final section discusses how a perspective that is more closely aligned with the everyday experience of energy users may inform policies in more constructive ways than dollars and kilowatts alone.

2 Quantifying life

2.1 Measuring electricity consumption

New and sophisticated means to measure electricity use have become available in recent years. For the time being, however, it is worth noting how crude electricity measurements are for most households. Electricity readings rely on utility employees visiting their customers in person. In countries with internally fitted metres, they must gain access to the property and read values from an often poorly accessible display. The process is inconvenient and costly enough not to be performed frequently (less than twice per year) nor regularly (different months each year). The result is that these readings are very difficult to compare and interpret. Seasonal variations in demand mean that one would struggle to know whether 10 kWh per day from July to October is better or worse than 12 kWh the next year between August and December. What part is seasonal, which is fluctuation, and which is an increase? Add to this is that most bills—the main means by which households receive any feedback about their energy consumption—often add another layer of obscurity. Readings are often mere estimates. If these estimates are too high in one year, they will be reduced the next to compensate. Thus, the billed cost is an even less reliable indicator of consumption.

Smart meters could improve this state of affairs considerably (Darby et al., 2015). They enable reliable feedback on electricity use and cost via in-home displays and other platforms of analysis. The higher temporal resolution (half hourly or better) can begin to reveal causes of high electricity use (DECC, 2012). For instance, a simple night-time baseload reading can give a helpful indication of the extent of permanent and stand-by appliances in use. This is not possible from random six readings.

Moving from 1–2 readings per year to 17,530 half hourly readings per household holds significant potential for a better understanding of the underlying demand patterns (Jin et al., 2017), enough, indeed, for privacy concerns about such data to arise (Véliz and Grunewald, 2018; Murrill et al., 2012).

The default assumption in the design of smart meters has long been that they transmit all readings to somewhere outside the home, where the data are stored and processed. This approach is consistent with the spirit of 'surveillance capitalism' where organisations assume ownership of private data (Zuboff, 2015). However, in strictly technical terms, there is no need for these data to leave the property at all. Even a time-varying tariff could theoretically be transmitted to the smart meter where in each settlement period the consumption (kilowatt-hour) is multiplied by the rate ($/kWh) and aggregated over weeks or even months. After this time, a single billing amount ($) is transferred from the user to the supplier for invoicing. No use data need to be transmitted in this model. If organisations wish to have these (valuable and insightful) data, they would need to pay users for it explicitly (which could take the shape of a lower-cost electricity supply).

However, for a specific understanding of what is behind this demand, even smart meter data may not be sufficient. The resolution of half an hour blurs many uses of electricity into an indistinguishable average. Boiling a kettle 10 times in half an hour looks the same as someone using an eclectic loan mower.

Two solutions have been explored in energy research: (1) higher temporal resolution and (2) appliance-specific measurements. The power of readings with high temporal resolution (>50 Hz) to disaggregate individual appliances in use is still emerging. Machine learning and artificial intelligence (AI) solutions may have a lot to contribute in this space. More and more data can train such tools to detect appliance patterns. Activity sequences can be introduced to these approaches to illuminate behavioural patterns underlying their use (De Lauretis et al., 2017; Armel et al., 2012; Verv, 2018).

The alternative approach of measuring individual appliances directly can be used by households themselves using smart plugs. This approach was deployed intensively by the Household Electricity Survey (HES, 2013), where appliances, including ovens and light circuits, were measured directly in 250 homes. Valuable insights into the relative share of consumption between appliance types and intraday patterns can be gained from these data (Terry et al., 2014; Zimmermann et al., 2012).

What neither of these approaches is able to establish, however, is the social function these appliances perform (see Shove et al., 2012; Walker, 2014). We argue that the timing of appliance use is not always (in fact, rarely) the choice of the appliance itself, but rather governed by the patterns dictated by life—the fixtures in our daily routines and responses to spontaneous choices and pressures. In the following section, we therefore present data that were collected alongside electricity recordings measured using a current clamp attached to the household's mains electricity metre. This electricity recorder collects power readings with 1-second resolution. We will explore how the observed load patterns can be understood in the context of additional activity data.

2.2 Quantifying everyday life

The most appropriate method to understand household activities would be through direct observation (see Higginson et al., 2013). However, such processes are intrusive, labour intensive, and therefore difficult to scale up. Scale matters in this context, because activities are highly diverse and small samples are difficult to scale up to representative levels.

A scalable approach to collecting activity information has been developed by the time-use research community (Gershuny and Sullivan, 2017; Gershuny et al., 2017; Eurostat, 2014). Paper diaries are sent to several thousand households, to be completed for 1–2 days. Activities are manually coded using the Harmonised European time-use codes, which allow for comparison of data across countries.

Here, we present data that were collected in a modified form of this process. Two significant changes to the standard time-use survey apply (more detail on the methodological differences can be found in Grünewald (2017)). Firstly, data are collected using an interactive app, instead of paper diaries. Secondly, activities are reported at the time of the participant's choice, rather than for 144 prescribed 10-minute intervals for a given day.

The latter constitutes a material difference in the meaning of the data and its possible uses. The data presented here are not a time budget, meaning it does not account for how long certain activities were performed for. Each reported activity has a specific time at which it was reported to have taken place. The next reported activity may or may not overlap in time. This distinction is important for the interpretation of data. People can (and do) report several things in close proximity in time (busy or active periods) at certain times of day, while at others, very little are reported. Reporting frequency and types of activity reported do themselves gain greater significance, given that they were the reporting choice of the participant, rather than the result of a prescribed 10-minute reporting grid.

The app, activity lists, and selection structure are open sourced (Grünewald, 2017).

Several biases apply to the data. Firstly, the study has been recruited on an opt-in basis. In addition to the aforementioned affluence, participants tend to be more energy literate and engaged than the general population (Grunewald and Diakonova, 2018). Furthermore, during observational studies, the participant's behaviour may be modified. Virtuous activities are overreported (and overperformed), while some stigmatised or undesirable activities are underrepresented. These effects have been observed in time-use research and are discussed by Gershuny and Sullivan (2017) and Gershuny et al. (2017).

The entry sequence options are illustrated as shown in Fig. 2. Each of these steps offers 6 options, which differentiate over 140 activities with often fewer than 3 selections. This final screen always asks about the enjoyment of this activity (Fig. 3).

2.3 Quantifying enjoyment

The most recent UK time-use survey (2015) included a column about enjoyment for each reported time period. This approach has been adopted in the app with the question at the end of each activity entry 'How much did you enjoy this?'. If the meaning of this question is not clear to participants, the additional explanation under 'help' is 'Think how this made you feel'.

The responses are coded numerically as 1–5, where 1 means 'Not at all' and 5 is 'Very much'. If the question was not answered or skipped, the value is 0 and will be excluded from further analysis here.

Enjoyment is a subjective measure. Some participants have a higher default choice than others or use wider ranges, which may not necessarily translate into experiencing greater or more volatile enjoyment, but is merely a different interpretation of the question.

A unique advantage of the app-based approach over conventional paper diaries is that the time of reporting can be recorded alongside the time at which the activity is said to have taken place. In this way, it is possible to distinguish between the experienced and the remembered enjoyment (Kahneman and Tversky, 2003; Kahneman and Riis, 2005).

It has not been possible yet to validate specific biases with enjoyment data collection. Some potential biases can be envisaged, specifically when participants complete their entry while in the company of others. Being seen to report 'not enjoying' the company of others may be socially problematic.

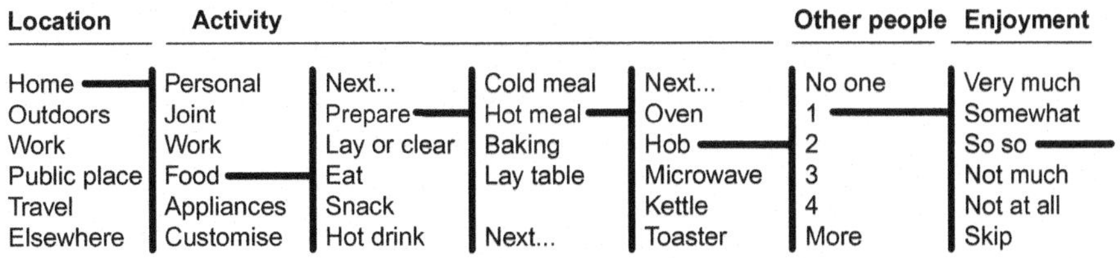

Location	Activity					Other people	Enjoyment
Home	Personal	Next...	Cold meal	Next...		No one	Very much
Outdoors	Joint	Prepare	Hot meal	Oven		1	Somewhat
Work	Work	Lay or clear	Baking	Hob		2	So so
Public place	Food	Eat	Lay table	Microwave		3	Not much
Travel	Appliances	Snack		Kettle		4	Not at all
Elsewhere	Customise	Hot drink	Next...	Toaster		More	Skip

FIG. 2

Guided six choice reporting sequence for hot meal preparation. Every activity entry ends with questions about other people and enjoyment.

FIG. 3

Screen to select the level of enjoyment.

3 Activity, enjoyment, and electricity use data

The data collection is ongoing. The results presented here were collected between 2016 and 2018. Summary statistics are shown in Table 1. Some of the sociodemographic biases of this approach are discussed in Grunewald and Diakonova (2018) and include a propensity towards affluent participants. Each participant provides 28-hour worth of data. The 28-hour period spans from 5 p.m. on the first day to 9 p.m. on the second day. The additional 4 hours capture a second peak demand period (typically between 5 p.m. and 7 p.m. in the United Kingdom) and allow for comparisons between otherwise identical times of day. A longer period of observation has been found to lead to fatigue effects and a reduction in activity reporting.

Table 2 gives an illustrative example of an activity record. Rows in Table 3 combine activity codes with the electricity reading for the time period when any member of a household reported this activity. For this analysis, the unit of enquiry is the household, since electricity readings are collected at household level

Table 1 Sample sizes of records

Data	Records
Activity records	15,939
Electricity readings (h)	10,020
Individuals	576
Households	319

Table 2 Illustrative extract of activity data

idDevice	dt_activity	tuc	Activity	Location	Enjoyment	People
3799	2018-11-18 19:11:50	214	Hot drink	1	5	1
3799	2018-11-19 13:49:17	9110	Travel (work)	21	4	0
3799	2018-11-19 12:44:47	3912	Being with others	4	5	5
3799	2018-11-15 20:15:19	8215	Watching TV	1	5	1
3799	2018-11-18 21:43:19	111	In bed	1	5	1

Table 3 Illustrative extract of activity and electricity data (tuc, time-use code; location, 1 = at home)

idHousehold	tuc	Location	Enjoyment	Date/time	Watt
7929	3910	1	4	2016-04-27 17:00	1581
7929	8551	1	3	2016-04-27 19:00	720
7929	3810	1	4	2016-04-27 20:00	611
7929	8551	1	3	2016-04-27 20:00	611
7929	8215	1	5	2016-04-27 22:00	703
7929	111	1	3	2016-04-27 23:00	92

and cannot easily be attributed to individuals, as discussed by Grunewald and Diakonova (2018). Activities are attributed to the electricity use for the hour surrounding the reported activity time. Some activities show more pronounced dependencies if a smaller window is selected (such as 10 minutes for kettles), but 1 hour is sufficiently long to allow comparison of most activities, without excessive averaging effects.

Any activities with a location other than 'home' are excluded from the electricity analysis, since they are unlikely to directly affect load. For greater explanatory power of the electricity signatures, the baseload consumption for each household is subtracted from the data. Baseload is defined here as the minimum demand during the

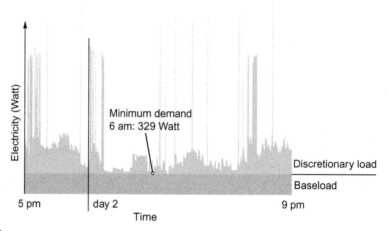

FIG. 4

Discretionary load is defined as load above the minimum value observed during the 28-hour recording for each household.

observation period. It is assumed to be largely activity independent, and the household load without baseload is therefore a better representation of discretionary, that is, activity-dependent, electricity use. The division of load into baseload and discretionary load is illustrated in Fig. 4.

4 Relationships between electricity consumption, activities, and enjoyment

4.1 Activity patterns

Figs. 5 and 6 give examples of the distribution of activities throughout the day. Washing machines are reported to be used throughout the day but with a slight preference for the morning. Hot meals are considerably more common in the evening. This raises the question of whether hot meals might be a significant contributor to peak demand. In the absence of parallel collection of electricity and activity data, as performed here, such a link had to be assumed. With these new data, the relationship can be tested.

4.2 Household electricity consumption

Electricity use is highly variable in numerous ways. It varies by household and by time. The sample of 319 households has diverse average consumption, shown as a power histogram in Fig. 7. The histogram in Fig. 8 shows the distribution limited to

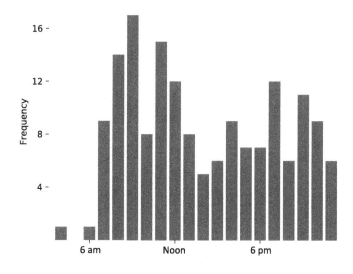

FIG. 5

Histogram of reported washing machine use. $N = 196$.

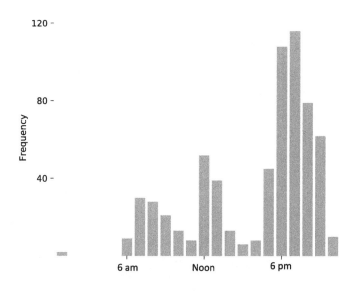

FIG. 6

Histogram of hot meal reporting. $N = 999$.

the period 6 p.m. to 7 p.m. Average load is 30% higher, but even during system peak demand, over half (53%) of households still are below the all-day average (>614 W). The other 47% do use 3.5 times as much electricity, and understanding the contributing factors to their high use could help in achieving considerable system savings (Ofgem, 2015).

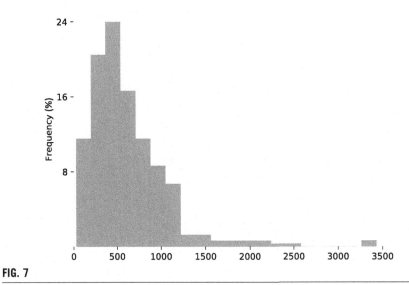

FIG. 7

Histogram of household mean power over 28 hours. $N=319$, mean $=614$ W, StdDev $=461$.

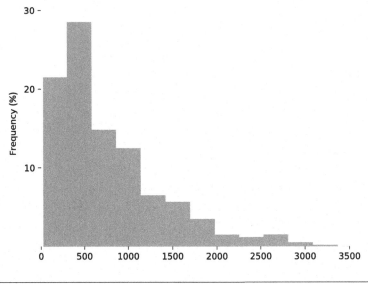

FIG. 8

Histogram of household mean power between 6 p.m. and 7 p.m. $N=568$, mean $=800$ W, StdDev $=699$.

4.3 Activity patterns

Fig. 9 shows the most commonly reported activities and their associated electricity consumption. Only activities that make up at least 1% of total activities reported are included. This depiction of demand is fundamentally different from conventional appliance-based data. Even activities labelled by appliance (such as 'TV') are not the

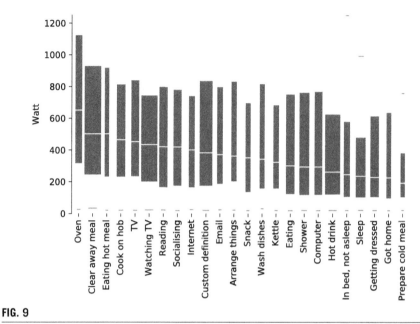

FIG. 9

Activities ranked by the median electricity use at the time of performance. Bars span the 50th percentiles. Width corresponds to the frequency of reporting.

consumption of the TV per se. It is the household consumption at the time when at least one household member reported that this appliance was in use. Many other uses of electricity are likely to happen at the same time, and it would not be meaningful to attribute them directly to activities. Lighting and heating are examples of demand that are nonrival in this sense.

This association is therefore best thought of as temporal, rather than causal. If, for instance, 'reading' was routinely done while the washing machine is running, 'reading' would appear to have a somewhat higher electricity consumption than what one would intuitively attribute to the activity itself in the most narrow sense (a reading light and an e-reader charger at most). Suspending the activity of reading is unlikely to reduce demand by much, but the timing of reading alongside other activities can help us understand the nature of household electricity use better. It is therefore helpful to think of these activities as parts of household life patterns, which can overlap, coincide, and even correlate.

Of particular interest is the discrimination of system critical periods, such as peak demand. In the United Kingdom, this tends to fall between 5 p.m. and 7 p.m. Fig. 10 presents the subset of data for this time window.

Compared with the all-day data in Fig. 9, this graph is even more dominated by activities related to hot meals. Cold meals, which are more common at earlier times of day, no longer feature as prominently as contributors to peak demand. The activity 'eating a hot meal', unlike the appliance 'oven', is itself not electricity intensive, but it constitutes an important marker in the rhythm of daily life that can help to explain when and why household electricity use peaks.

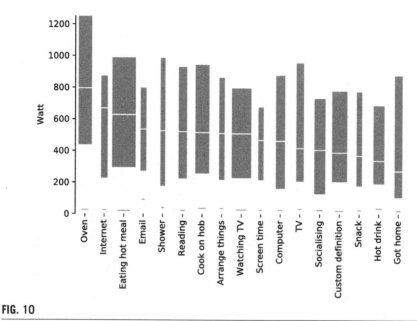

FIG. 10

Activities during peak demand (5–7 p.m.) ranked by their electricity intensity. Bars span the 50th percentiles. Width corresponds to the frequency of reporting.

4.4 Enjoyment

Alongside each reported activity, participants rate their enjoyment on a scale from 1 (not at all) to 5 (very much). These numerical values allow some comparisons and trends to be deduced from the data. Which activities score most highly and lowly on this metric? We exclude activities that have been reported less than 100 times, to avoid uncommon and extreme activities to distort the analysis. (All four reported instances of 'Swimming in river' scored 5.)

After we established in the previous section that hot meals play a central role in evening peak-time electricity use, Table 4 further confirms that this activity is among the most enjoyable. Many of the most enjoyable activities are consistent with conventional expectations. It is perhaps surprising that a supposedly unconscious activity like 'sleeping' could score so highly on enjoyment. Obviously, the activity was not reported while sleeping, such that the score is more of a reflection of how sleep (and any other activity) is perceived relative to other activities.

The distribution of enjoyment scores is skewed towards the more enjoyable end. Scores 1 and 2 are only assigned in extreme cases, such that the averages at the lower end are in fact closer to mid-range (Table 5). On this end of the enjoyment spectrum, we find many household chores, such as clearing up and laundry-related activities. In contrast to 'hot meals', it is worth noting that cold meal preparation itself scores lowly.

Table 4 Most enjoyable activities (by decreasing enjoyment)

Activity	Enjoyment	Instances reported
Socialising	4.64	548
Reading books	4.63	386
Food eat hot	4.60	902
Care rest	4.56	109
Exercise	4.52	245
Play with child	4.51	120
Food eat snack	4.48	256
Sleep	4.47	476
Food drink hot	4.43	832
Screen time	4.40	115
Appliances TV	4.37	337
In bed not asleep	4.36	289
Leisure TV	4.34	755
Being with others	4.34	183
Food eat cold	4.33	453
Got home	4.33	177
Me time	4.30	171
Gardening	4.21	151

Table 5 Least enjoyable activities (by increasing enjoyment)

Activity	Enjoyment	Instances reported
Disposal of waste	2.97	158
Wash dishes	3.19	191
Washing machine	3.19	136
Clear away meal	3.26	175
Hygiene/beauty	3.42	165
Work house clear up	3.48	215
Wash	3.50	138
Brush teeth	3.51	154
Travel purpose work	3.55	655
Care wash dress	3.59	403
Work	3.60	247
Work (computer)	3.75	741
Lay table	3.76	114
Travel purpose friend	3.77	135
Travel purpose household	3.80	205
Email	3.81	223
Shopping	3.90	151
Food preparation cold	3.91	169

4.5 **The consumption–enjoyment frontier**

Having explored electricity consumption and enjoyment during activities independently, the linked nature of the data allows to combine all three. Can we identify activities that are both high in enjoyment and low in electricity demand? Conversely, are there activities with high-energy footprint and little enjoyability? Such clusters may be candidates for activities that want to be encouraged or discouraged.

Fig. 11 provides a map of activities by electricity use and enjoyment. The top left is high consumption with low enjoyment; the bottom right constitutes enjoyable activities with low electricity use. The space is divided along the diagonal. Many of the least enjoyable household chores from Table 5 are located in the top left quadrant. Some of the most enjoyable activities, on the other hand, are associated with relatively low consumption. Among them are reading, sleeping, and socialising, which

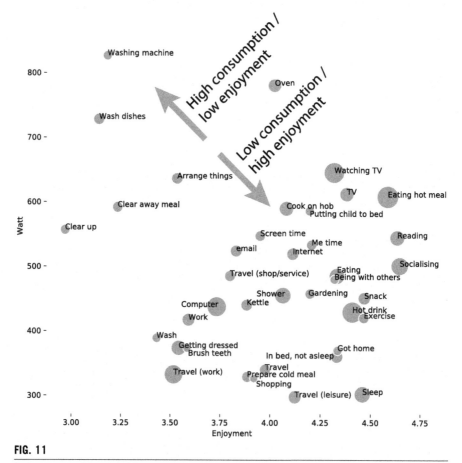

FIG. 11

Enjoyment and household electricity consumption in the hour of performing a given activity. Size represents frequency of an activity (minimum 100 reported instances).

form something of a low-energy/high-enjoyment frontier. In the following section, we will discuss how such findings may be used to shape demand and policy.

5 Discussion: An activity perspective of demand

5.1 Attribution of load and activities

The data presented here are the first of their kind. While many attempts have been made to infer electricity use from time-use data to the author's knowledge, this is the first study collecting both empirically at the same time, such that the data can be linked.

The nature of the data is materially different from a technical measurement of appliances. Applying a power metre to a kettle makes a directly attributable reading. It can be safely assumed that when the kettle is on, it will consume a certain amount of power. When it is off, it won't.

For activities, this attribution is different. Activities can be performed in parallel and may overlap in time with the electricity use associated with them to a greater or lesser extent. Some activities are preceded by load (going to bed); others are near instantaneous (hot drink) or lead to an increase afterwards (coming home). Some electricity footprints can be the result of activity sequences (reading after having started the dishwasher).

However, with sufficiently large datasets, it is possible to identify trends and differences between activities that are statistically significant. It may further be possible to use machine learning techniques to distil activity-specific load signatures.

An additional help for such attribution can come from interventions, which can explore how the suppression of certain activities affects the load profile for better or worse and which activity (sequences) are more or less flexible in response to interventions. The ongoing data collection now includes such interventions, and future analysis will focus on more direct attributions.

5.2 Looking at demand differently

Electricity use has many negative connotations: waste, emissions, and cost. However, this should not come as a surprise; electricity underpins many highly enjoyable activities and produces great value for people. Even household chores would arguably be less enjoyable without the help of electricity. Yet, the fact that much of the highest electricity uses are reportedly so much less enjoyable could open up new opportunities for energy solutions.

As many social practice theorists argue, energy use is shaped by social norms, skills, and the material world that surrounds us (Shove, 2010; Walker, 2014). The private ownership of washing machines is the results of such processes with many decades of technology evolution. These have led to practices that are neither energy nor time efficient. Millions of washing machines need to be individually loaded and unloaded. They feed on cold water, which has to be heated electrically. One way to alleviate this low-enjoyment activity could be a laundry service that collects laundry and processes it centrally using waste heat from other processes. Such more centralised services would further be able to optimise the timing of operation better, for instance, by providing an overnight service. While the upsides are clearly apparent, this approach is poorly aligned with current social practice, and many deeply held cultural norms around laundry. With both enjoyment and energy at stake, it may be worth challenging some of these conventions.

The second practice we have explored in this chapter is hot meal preparation. The data show a clear trend for hot meal preparation to coincide with UK peak demand and for it to be a potentially significant contributor. As with laundry, hot meals are deeply embedded in cultural norms and meaning, and unlike laundry, hot meals happen to be enjoyable. Thus, they are less likely to be shifted or suppressed without resistance. In this context, it may be informative to compare data with other cultures where different practices have established themselves. In Germany, for instance, hot lunches are more common, and evening meals are more often cold. This pattern is supported by a strong provision of hot lunches at work places, including subsidised canteens. Co-benefits could include better workplace cohesion, productivity improvements, and other health benefits. These could be worth exploring as part of an energy policy that extends well beyond energy alone.

Lastly, a group of pleasurable activities with very low consumption include reading, sleeping, and socialising. It can be argued that modern Western society is suffering from a lack of all three. Any policy or other measure to promote more sleep or time to read and socialise would not only benefit society; based on our data, it could also lead to a reduction in electricity consumption.

To deliver the benefits of flexibility for energy systems and wider society, three steps may lead to effective solutions:

1. Identify activity patterns that dominate periods of high consumption.
2. Test these patterns for flexibility.
3. Decide on the appropriate measures to engage this flexibility.

These steps are equally valid for policy and business model development. The second step should address questions, such as the following:

- What measures lead to the rearrangement of patterns in time?
- How sustainable are these responses?
- What obstacles inhibit their flexibility?

Where price signals prove effective policies could encourage markets to deliver on these. Some of the obstacles to flexibility under (2) may, however, require wider regulation and policy support, such as product standards and even work-time standards and measures to address potential equality issues arising from the inability of poorer and vulnerable households to participate.

6 Conclusions

We have shown with a new class of data that activity patterns can hold important information to explain and potentially reshape electricity load profiles. While cost-reflective price signals are a vital enabler for markets to have the right incentives for system cost minimisation, these may not necessarily be the most effective way to engage end-users. Storage and some fully automated loads may be capable to respond to fast-changing real-time prices, and some users will be able to adapt use patterns in response to price changes. For more fundamental changes to the patterns that shape electricity use, we conclude that an understanding of the activities may hold the key to more effective and meaningfully targeted policies.

The relationships between reported activities and electricity use have shown which activities coincide with high and low electricity consumption throughout the day and during peak demand in particular. Hot meals have been identified as one of the key activities mapping onto high consumption during this period.

Enjoyment data have added an extra dimension of interest to these considerations. It is now possible to differentiate activities not only by high and low electricity consumption but also by the perceived enjoyment of people performing them. Our data show some clear distinction between a group of high-energy low-enjoyment activities relating to household chores and other, more enjoyable activities. Of particular note are reading, sleeping, and socialising, which are among the most enjoyable activities, while having particularly low electricity consumption footprints. Future energy policy may wish to consider some of the opportunities for societal well-being where this aligns well with other energy policy goals. In some cold-temperate climates, this could take the form of nonenergy energy policies that focus on better provision of hot meals in schools and work places, thereby reducing the contribution of hot evening meal preparation towards peak demand. It may also be worth considering whether some of the least enjoyable chores, such as laundry, could be avoided with more efficient and time-saving laundry service offerings. Since much resistance to such services is deeply culturally rooted, such a campaign may take time and careful nudges by policy makers.

References

Armel, K.C., Gupta, A., Shrimali, G., Albert, A., 2012. Is disaggregation the holy grail of energy efficiency? The case of electricity. Energy Policy. https://doi.org/10.1016/j.enpol.2012.08.062.

Bulkeley, H., Bell, S., Lyon, S., Powells, G., Judson, E., Lynch, D., 2014. Customer-Led Network Revolution. In: Social Science Research report CLNR-L052 CLNR-L052. Durham University.

CER, 2011. Electricity Smart Metering Customer Behaviour Trials Findings Report. Information Paper CER11080a. The Commission for Energy Regulation.

Darby, S., McKenna, E., 2012. Social implications of residential demand response in cool temperate climates. Energy Policy 49, 759–769. Special Section: Fuel Poverty Comes of Age: Commemorating 21 Years of Research and Policy.

Darby, S., Liddell, C., Hills, D., Drabble, D., 2015. Smart Metering Early Learning Project, Synthesis report Department of Energy and Climate Change.

De Lauretis, S., Ghersi, F., Cayla, J.-M., 2017. Energy consumption and activity patterns: an analysis extended to total time and energy use for French households. Appl. Energy 206, 634–648. https://doi.org/10.1016/j.apenergy.2017.08.180.

DECC, 2012. Smart Meter Roll-Out for the Domestic Sector (Gb). Impact Assessment, Department of Energy and Climate Change.

eurostat, 2014. Harmonised European Time Use Surveys. 2008 guidelines, Office for Official Publications of the European Communities.

Gershuny, J.I., Sullivan, O., 2017. United Kingdom Time Use Survey, 2014-2015. [data collection]. SN: 8128 , UK Data Service. https://doi.org/10.5255/UKDA-SN-8128-1.

Gershuny, J., Harms, T., Doherty, A., Thomas, E., Milton, K., Kelly, P., Foster, C., 2017. CAPTURE24: Testing Self-Report Time-Use Diaries Against Objective Instruments in Real Time. Working paper Centre for Time Use Research, University of Oxford.

Grünewald, P., 2017. Meter App. Source code repository, GitHub. https://github.com/PhilGrunewald/MeterApp.

Grunewald, P., Diakonova, M., 2018. The Electricity Footprint of Household Activities—Implications for Demand Models. Energy and Buildings 174, 635–641. https://doi.org/10.1016/j.enbuild.2018.06.034.

HES, 2013. Light Foot Energy Services. Household Energy Services.

Higginson, S., Thomson, M., Bhamra, T., 2013. For the Times They Are a-Changin': The Impact of Shifting Energy-Use Practices in Time and Space. Int. J. Justice Sustain. https://doi.org/10.1080/13549839.2013.802459. https://doi.org/10.1080/13549839.2013.802459.

Jin, L., Lee, D., Sim, A., Borgeson, S., Wu, K., Spurlock, C.A., Todd, A., 2017. In: Comparison of Clustering Techniques for Residential Energy Behavior Using Smart Meter Data. AAAI Workshops. https://aaai.org/ocs/index.php/WS/AAAIW17/paper/view/15166.

Kahneman, D., Riis, J., 2005. Living, and thinking about it: two perspectives on life. In: The Science of Well-Being, pp. 285–304.

Kahneman, D., Tversky, A., 2003. Experienced utility and objective happiness: a moment-based approach. In: The Psychology of Economic Decisions. vol. 1, pp. 187–208.

Murrill, B.J., Liu, E.C., Thompson II, R.M., 2012. Smart Meter Data: Privacy and Cybersecurity. . CRS Report for Congress 7-5700. Congressional Research Service.

Ofgem, 2015. Making the Electricity System More Flexible and Delivering the Benefits for Consumers. Position paper, Office of Gas; Electricity Markets.

Schofield, J., Carmichael, R., Tindemans, S., Woolf, M., Bilton, M., Strbac, G., 2014. Residential Consumer Responsiveness to Time-Varying Pricing. In: Report A3 for the "Low Carbon London" LCNF project. Imperial College, London.

Shove, E., 2010. Beyond the ABC: climate change policy and theories of social change. Environ. Plan. A 42 (6), 1273.

Shove, E., Pantzar, M., Watson, M., 2012. The Dynamics of Social Practice. SAGE Publications.

Terry, N., Palmer, J., Godoy, D., Firth, S., Kane, T., Tillson, A., 2014. Further Analysis of the Household Electricity Survey: Lighting Study. Final Report Ref No. 475/09/2012, Cambridge Architechural Research Limited; Loughborough University.

Véliz, C., Grunewald, P., 2018. Protecting Data Privacy Is Key to a Smart Energy Future. Nature Energy 1. https://doi.org/10.1038/s41560-018-0203-3. Nature Publishing Group.

Verv, 2018. Unlocking Unique Energy Insights with AI. https://verv.energy/. https://verv.energy/.

Walker, G., 2014. The dynamics of energy demand: change, rhythm and synchronicity. Energy Res. Soc. Sci. 1, 49–55. Elsevier.

Zimmermann, J.-P., Evans, M., Griggs, J., King, N., Harding, L., Roberts, P., Evans, C., 2012. Household Electricity Survey a Study of Domestic Electrical Product Usage. Report R66141, Intertek.

Zuboff, S., 2015. Big other: surveillance capitalism and the prospects of an information civilization. J. Inform. Technol. 30 (1), 75–89. Springer.

Further reading

Grünewald, P., Diakonova, M., Zilli, D., Bernard, J., Matousek, A., 2017. In: What We Do Matters—a Time-Use App to Capture Energy Relevant Activities. *Eceee 2017 Summer Study Proceedings,* pp. 2085–2093.

Developing quantitative insights on building occupant behaviour: Supporting modelling tools and datasets

3.2

Tianzhen Hong, Jared Langevin, Na Luo, Kaiyu Sun

Building Technology and Urban Systems Division, Lawrence Berkeley National Laboratory, Berkeley, CA, United States

1 Introduction

The energy-related behaviours of building occupants constitute a key factor influencing building performance; accordingly, realistic representation of occupant behaviour in building performance simulations is essential to ensuring that such simulations yield accurate guidance for building design and operation decisions. For the purpose of this chapter, building occupant behaviour (OB) refers to (1) occupant presence in spaces and movement between spaces, (2) occupant interactions with building systems, and (3) occupant adaptations (e.g. changing clothing and having hot/cold drinks).

Occupant behaviours such as adjusting a thermostat for comfort, switching lights on/off, using appliances, opening/closing windows, pulling window blinds up/down, and moving between spaces can have a significant impact on energy use and occupant comfort in buildings. Depending on the building type, climate, and degree of automation in operation and controls, existing studies have found that occupant behaviour may increase or decrease energy use by a factor of up to three for residential buildings (Andersen, 2012) and increase energy use by up to 80% or reduce energy use by up to 50% for single-occupancy offices (Hong and Lin, 2013); another study (Sun and Hong, 2017a) estimates that occupant behaviour measures have a 41% energy savings potential for office buildings. Developing a deeper understanding of occupant behaviour and further quantifying its impact on the use of building technologies, occupant satisfaction, and building performance using simulation tools is crucial to the design and operation of low-energy buildings with high indoor environmental quality (IEQ).

Energy and Behaviour. https://doi.org/10.1016/B978-0-12-818567-4.00012-0

Nevertheless, in most building design, construction, operation, and retrofit practices, the influence of occupant behaviour remains underrecognised and oversimplified. In the most widely used building performance simulation programs, for example, the representation of occupant behaviour is limited to predefined static schedules or fixed settings and rules (Cowie et al., 2017), leading to deterministic and homogeneous simulation results that fail to capture the stochasticity, dynamics, and diversity of occupants' energy behaviour in buildings. Meanwhile, available models of occupant behaviour have been developed across different researchers and have showed inconsistencies, precluding arrival at a consensus within the research community on how to approach experimental design and modelling methodologies. Given the issues earlier, a strong need has emerged in recent years for researchers to work together on devising a consistent research framework for occupant behaviour definition and simulation.

Whole-building performance simulation (BPS) programs such as EnergyPlus (BTO, 2017), ESP-r (Hand, 2015), IDA-ICE (Equa, 2017), DeST (Yan et al., 2008), and TRNSYS (2017) have recently been applied to quantitatively evaluate the effects of occupant behaviours on the performance of building technologies and energy systems, with the aim of reducing energy use in buildings and associated greenhouse gas emissions. Half of current BPS programs include built-in stochastic occupant behaviour modelling capabilities; however, this functionality is far from consistent across different BPS tools and generally lacks flexibility for user customisation (Cowie et al., 2017). In these programs, prescribed schedules and rule-based control are frequently used to represent building occupants and their energy-related behaviours. Overall, the stochastic representation of occupants within BPS programs is much less ubiquitous than deterministic modelling capabilities (Cowie et al., 2017).

A recent study (Hong et al., 2017) provides a thorough overview of OB implementation approaches in the current BPS tools, which are as follows: (1) *Direct input or control*—it refers to the case when occupant-related inputs are defined using the semantics of BPS programs, just as other model inputs are defined (building geometry, construction, internal heat gains, and HVAC systems); (2) *Built-in OB models*—users can choose deterministic or stochastic OB models already implemented in the BPS program, which are originally data-driven and use functions and models such as linear or logit regression functions. These models typically include occupant movement models, window operation models, and lights switching on/off models; (3) *User function or custom code*—users can write functions or custom code to implement new or overwrite existing or default building operation and supervisory controls; and (4) *Cosimulation approach*—it allows simulations to be carried out in an integrated manner, running modules developed by different programming languages or in different physical computers. For a building energy modeller, the choice of which implementation approach to select is a difficult one. All of these approaches have their advantages and disadvantages, such as precision, calculation time, and input model development time.

There is a strong need to homogenise and stimulate wider uptake of stochastic occupant modelling capabilities in BPS programs. The development of a BPS program-independent cosimulation platform could address the gaps by centralising functionality, allowing models to be implemented within the platform and then applied in a consistent way amongst different BPS tools. There is also a significant need for developing a suite of new occupant behaviour modelling tools to improve the building performance simulation by (1) providing a standard representation of occupant behaviour models, enabling the exchange and use of occupant behaviour models between BPS programs, applications, and users to improve the consistency and comparability of simulation results, and (2) generating realistic occupancy schedules. These tools capture the diversity, stochastics, and complexity of occupant behaviour in buildings to improve the simulation and evaluation of behavioural measures, as well as of the impact of occupant behaviour on technology performance and energy use in buildings.

In this chapter, state-of-the-art methods and tools that enable more sophisticated occupant behaviour simulation in BPS programs are reviewed, along with the most prominent datasets available to support quantitative behaviour model development. A particular focus is placed on the OB tools yielded by the recently concluded International Energy Agency (IEA) Annex 66: Definition and Simulation of Occupant Behaviour in Buildings. Four advanced occupant behaviour modelling tools that allow for a rapid and widespread integration of OB models in various BPS programs are introduced: (1) obXML, an XML schema representing OB models using the drivers-needs-actions-systems (DNAS) ontology; (2) obFMU, an OB model solver using the functional mockup unit; (3) Occupancy Simulator, an agent-based Markov chain model of occupant presence and movement in buildings; and (4) Buildings.Occupants, an open-source package of occupant behaviour models implemented in Modelica, an equation-based, object-oriented language. Next, an overview of application areas for OB modelling tools and datasets across the building life cycle is presented, and example applications are demonstrated through three case studies. The chapter concludes by identifying emerging opportunities and challenges surrounding the use of occupant behaviour simulation to support the design and operation of low-energy buildings that foster greater occupant satisfaction.

2 Occupant behaviour modelling methods, datasets, and simulation tools

This section reviews the state-of-the-art methods and tools that enable more sophisticated occupant behaviour simulation, along with the data collection approaches and the most prominent datasets available to support quantitative behaviour model development.

2.1 State-of-the-art occupant behaviour modelling approaches

Various mathematical methodologies have been used in occupant behaviour modelling. Classical statistical models such as general and generalised linear models have been applied extensively, while for time-dependent data, Markov and Hidden Markov chains (Dong and Lam, 2014; Liisberg et al., 2016; Andersen et al., 2014; Richardson et al., 2008) have proved to be useful tools. Mixed-effects and agent-based models have been applied to capture the diversity amongst occupants (Haldi, 2013; Langevin et al., 2015b), and machine learning and data mining techniques such as clustering (Pan et al., 2017; Ren et al., 2015) and decision trees have followed from the improved availability of large occupant behaviour datasets (Hong et al., 2015b). In this section, the use of different modelling approaches in the literature is described, with the models organised by the different behaviour types for which they were developed. As aforementioned, the occupant behaviours referred to in this chapter are (1) occupant presence/absence in spaces and movement between spaces, (2) occupant interactions with building systems (e.g. opening the windows and operating the HVAC system), and (3) occupant personal adaptations (e.g. changing clothing and having hot/cold drinks). This section mainly focuses on the first two occupant behaviour types, as few existing modelling studies cover occupants' personal adaptations.

2.1.1 Representative modelling approaches

Markov chains assume that future system states (e.g. of occupancy or of a building control) are dependent only on the current system state together with the probabilities of the state changing. A Markov chain consists of a set of transition probability matrices that describe the transition between states in each time step. The matrix entries can be estimated from the source data using maximum likelihood estimation. A hidden Markov model (HMM) consists of a Markov chain whose states are not directly observed, and information is derived about the unobserved entity from a series of related observations. For a detailed description of Markov chains, refer to Zucchini et al. (2016). Time series in which quantities take a finite number of states can be modelled using Markov chains. In practice, Markov chains are employed to model (1) occupancy (presence, absence, and people count), (2) window states over time (open or closed), (3) blind usage (open, closed, and fraction of opening), and (4) activity level (working, sleeping, resting, laundry, cooking, or absent).

The *general linear model* (classical GLM) is a classical statistical model that assumes normally distributed response variables and a linear relationship between the explanatory variables and the response variable. For example, ordinary linear regression and the analysis of variance (ANOVA) and mixtures thereof are classical examples of GLM. Let $Y=(Y_1,\ldots,Y_n)$ be a vector of n observations of a response variable. We assume that Y follows a multivariate normal distribution $N(\mu,\Sigma)$. In the classical GLM, it is assumed that the vector of mean values $\mu=(\mu_1,\ldots,\mu_n)$ can be expressed as a linear combination of some explanatory variables expressed by column vectors $X_1, \ldots, X_k$.

Generalised linear models (GLM) relax the assumption of normally distributed errors, relating a linear predictor $X\beta$ to the expected response $E(Y)$ via a link function g where $g(E(Y))=X\beta$. In a binary logistic regression modelling a dichotomous response variable, for example, the link is defined as $\ln\left(\frac{\pi}{1-\pi}\right)=X\beta$, where π is the expected probability of a response $Y=1$ and model errors are assumed to follow a logistic distribution. A further generalisation of the linear model adds random effects U to the fixed effects of predictor variables X, or in the case of the binary logistic regression, $\ln\left(\frac{\pi}{1-\pi}\right)=X\beta+ZU$; random effects account for unobserved heterogeneity in the data. This class of approaches is termed generalised linear mixed models (GLMM).

Bayesian network models (BNs) are directed acyclic graphs or belief networks that are used to represent the relationships amongst a predefined group of discrete and continuous variables (Xi). BNs consist of a graphical model and an underlying conditional probability distribution. The nodes of the graph represent the variables, and the dependencies between variables are depicted as directional links corresponding to conditional probabilities. Hence, the construction of a BN consists of determining the structure and the probability distribution associated with these relations. The relationships between nodes can be explained by employing a family metaphor: a node is a parent of a child if there is an arc from the former to the latter. For instance, if there is an arc from X_1 to X_3, then node X_1 is a parent of node X_3. The Markov property of the BNs implies that all probabilistic dependencies are identified via arcs and that child nodes only depend on the parent nodes.

Agent-based models (ABMs) represent systems from the bottom-up, simulating individual actors or 'agents' with personal attributes and behavioural possibilities, as well as rules for interacting with other agents and their surrounding environment; macrolevel or group-level behaviours emerge from the microlevel behaviours of individual agents. For more guidance on the agent-based modelling approach, refer to (Macal and North, 2010) and the chapter in this book titled *Agent-Based Modelling of the Social Dynamics of Energy End-Use* by Chappin et al.

2.1.2 Modelling studies of building occupancy

Occupancy is defined in existing studies as either the presence or absence of an occupant or the occupant count (the number of occupants) in a given space. One of the most typical occupancy modelling approaches is Markov chains. The occupancy models of Richardson et al. (2008) and Page et al. (2008) are the earliest published examples of first-order Markov chains being used to generate stochastic synthetic occupancy patterns. This first-order Markov chain technique has since been widely adopted in the development of occupancy models in office buildings (Wang et al., 2011; Liao et al., 2012; Andersen et al., 2014). In certain studies, presence/absence at the space level is modelled alongside the number of occupants—for example, in (Hong et al., 2013), which uses the occupancy models to determine the lighting and heating requirements of a building. More recently, Wilke et al. (2013) used first- and higher-order homogeneous Markov processes to represent building occupancy, where a higher-order Markov process extends the first-order Markov case by including multiple past values of occupancy state. This approach is coupled with a survival analysis method in which a Weibull distribution is used to estimate occupant presence durations at greater time lags before the present simulation step. Hence, information about the next time step is based not only on current occupancy state but also on past occupancy values through the survival function that also captures the durations of occupant presence and absence coherently.

2.1.3 Modelling studies of occupant interactions with building systems

In naturally ventilated buildings, window opening and closing behaviour is an important control mechanism used by building occupants to regulate the indoor thermal environment and air quality. It is crucial to have window operation models that create realistic patterns for use in building performance simulations. Accordingly, models of window use are particularly prevalent within the existing behaviour literature.

The most common modelling approach for window operations is logistic regression as a special case of GLMs. In some cases, interaction terms between several predictors are considered. Time dependencies are modelled by Markov chains (Fabi et al., 2014; Calì et al., 2016), and survival analysis has been applied to model opening durations (Haldi and Robinson, 2009).

More recently, GLMMs have been used to model the diversity in window opening behaviour across occupants (Schweiker et al., 2012); this application of GLMMs has also been suggested by Haldi (2013). Here, the inclusion of random effects in the GLMM approaches allows interindividual variability to be described—that is, the diversity in behaviour amongst different occupants, where fixed effect models only capture an average occupant's behavioural tendencies. Hence, the GLMMs separate the variability in the data corresponding to occupants' diversity from other sources of uncertainty. These kinds of models are especially useful for Monte-Carlo simulations, because an occupant is randomly drawn from a population in every simulation run, resulting in a spread of behaviour that reflects reality.

In another recent study, Barthelmes et al. (2017) used a Bayesian network (BN) to model window control behaviour in the residential sector. Their study addressed five key research questions related to modelling window control behaviour: (1) variable selection for identifying the key drivers of window control behaviour, (2) correlations between key variables for structuring a statistical model, (3) target definition for finding the most suitable target variable (window control actions rather than window states), (4) development of a BN model with the ability to treat mixed data, and (5) validation and demonstration of the high predictive power of stochastic BN models.

In addition to the window opening and closing behaviour, light switching behaviour is also considered as a major factor influencing the electricity use in domestic homes and office buildings. Studies on the modelling of lighting switch behaviour have mostly focused on small office and residential buildings, with the research findings greatly dependent on building layout and daylight control systems. The first report of a stochastic approach to manual lighting control was by Newsham et al. (1995), who developed a regression model called Lightswitch that simulated user activities of turning lights on/off in the workplace based on measured field data from an office building in Ottawa, Canada. The probability of turning on lights has also been modelled as dependent on natural/daylight level and occupant movement (Widén et al., 2009).

A small number of studies attempt to integrate prediction of multiple types of human–building interactions in a single modelling package. For example, the agent-based modelling approach reported by Langevin et al. (2015a, b) predicts the probability of several interdependent behaviours, including window opening and closing, adjustment of thermostats, use of personal heating and cooling devices, and the adjustment of personal clothing levels. In this framework, each agent represents an individual office occupant that acts adaptively based on the simulated distance between current thermal sensation and a thermal acceptability range, where both are modelled probabilistically based on occupant comfort field data from the ASHRAE RP-884 database (https://sydney.edu.au/architecture/staff/homepage/richard_de_dear/ashrae_rp-884.shtml).

2.2 Occupant behaviour datasets

2.2.1 Data collection approaches

The collection of datasets for developing and validating behaviour models is an essential driver of improved understanding and representation of behaviour in BPS. To capture occupants' energy-related behaviours in buildings, researchers may collect two types of information: (1) reported information using surveys and/or (2) monitored information from sensing and data acquisition technologies.

Surveys are a cost-effective means of achieving a large sample size and can measure phenomena that would be difficult or impossible to measure with sensors (e.g. thermal comfort sensation and clothing level, social interactions, and attitudes). Several recent studies (Becerik-Gerber et al., 2011; Konis, 2013; Haldi and Robinson, 2008) have relied on custom technological survey solutions for polling occupants more frequently than a telephone, paper, or online survey would allow. Surveys have also been used to develop models (e.g. Haldi and Robinson, 2009; Langevin et al., 2015b). Despite the aforementioned benefits to using surveys in occupant research, a number of established psychological biases, including the Hawthorne effect and social desirability bias, suggest that self-reported behaviour may not always match observed behaviour (McCambridge et al., 2014). In addition, a lack of understanding of different building services systems or the misinterpretation of questions may cause occupants to unknowingly report certain variable states incorrectly. Relative to other in situ and laboratory monitoring approaches, surveys typically do not lend themselves to frequent sampling because they rely on occupants' active input; therefore, their use in longitudinal studies may be limited to targeted periods of study, with the goal of limiting occupant fatigue (Langevin et al., 2015a).

Outside of surveying techniques for behaviour data collection, previous studies have used sensing and data acquisition technologies to yield more granular occupant information (including both occupancy and activities). Sensor data collection may be conducted either in situ (e.g. in a field setting) or in a laboratory condition. Data are typically acquired passively through sensors that feed into the building automation system (BAS). Such sensors measure variables that include occupants' presence; adaptive actions (e.g. changing window or door state and turning on/off personal heating and cooling devices); energy use (through submetering); and environmental variables such as temperature, humidity, air velocity, lumen level, and CO_2 concentration (Haldi and Robinson, 2010; Duarte et al., 2013). For in situ studies, the measurement sample size may be constrained by the small number of willing participants in a given building, though frequent measurement intervals (1–5minutes) may still yield large amounts of data for a small occupant sample. Additionally, the lack of flexibility in sensor placement to avoid interfering with occupants' activities or to prevent the measurements being disturbed by the occupants can reduce the accuracy of measurements (Reinhart and Voss, 2003; Andersen et al., 2013).

On the other hand, laboratory facilities for occupant research are typically costly to build and operate, and experiments are often significantly more expensive than in situ studies because of the human resources required. Another downside to laboratory studies is that the short-term and potentially unnatural characteristics of laboratory environments may yield occupant response data that are unrepresentative of a field setting; new experimental techniques propose using augmented reality technologies to allow laboratory settings for occupant behaviour research to more closely mimic occupants' experiences in real buildings over long time periods (Saeidi et al., 2017).

To summarise, occupant behaviour data collected through survey instruments may reveal important insights on the rationales and motivations for behaviour and cover variables that would be difficult to measure using sensor and data acquisition instruments; yet, survey techniques can prompt occupant response fatigue, and they often rely on occupant recall of behaviour, which may be inaccurate. Meanwhile, data collected by sensing instruments may yield richer and more granular insights on variables such as occupant presence, certain actions, and environmental conditions across long time periods; yet, measurement sample size may be constrained by the number of occupants who are willing to be monitored and suboptimal sensor placement in the field can compromise the accuracy of collected data.

Ultimately, the most favourable behaviour data approaches are likely to involve some combination of both survey and sensor measurements, enabling one data source to be cross-referenced with the other and supporting the compilation of a comprehensive set of information on the physiological, psychological, and social aspects of occupant behaviour.

2.2.2 Existing dataset resources

An increasing number of datasets on building occupant behaviour are being developed and shared through a growing variety of channels.[a] Here, we list a few of the most prominent existing datasets concerning occupant behaviour.

- *ASHRAE Global Thermal Comfort Database I and II*
  The ASHRAE Global Thermal Comfort Database project (Földváry Ličina et al., 2018) was launched in 2014 under the leadership of the University of California at Berkeley's Center for the Built Environment and The University of Sydney's Indoor Environmental Quality (IEQ) Laboratory. The exercise began with a systematic collection and harmonisation of raw data from the last two decades of thermal comfort field studies around the world. The final database is composed of field studies conducted between 1995 and 2015 from around

---

[a] Google recently developed a database search tool (https://toolbox.google.com/datasetsearch), which includes occupant behaviour–related datasets.

the world, with contributors releasing their raw data to the project for wider dissemination to the thermal comfort research community. After the quality-assurance process, there was a total of 81,846 rows of data of paired subjective comfort votes and objective instrumental measurements of thermal comfort parameters. An additional 25,617 rows of data from the original ASHRAE RP-884 database are included, bringing the total number of entries to 107,463. The database is intended to support diverse inquiries about thermal comfort in field settings. To achieve this goal, two web-based tools were developed to accompany the database:

a. Interactive visualisation tool (https://cbe-berkeley.shinyapps.io/comfortdatabase/): provides a user-friendly interface for researchers and practitioners to explore and navigate their way around the large volume of data in ASHRAE Global Thermal Comfort Database II.

b. Query builder tool (http://www.comfortdatabase.com/): allows users to filter the database according to a set of selection criteria and then download the results of that query in a generic comma-separated values (.csv) file.

- *Library of occupant behaviour models*
Within the effort of Annex 66, energy-related OB literatures have been reviewed to identify and compile a list of 127 commonly-used OB models in the field that cover the following categories: (1) *Behaviour types*—occupant movement and different types of occupant interactions with windows, doors, shading, blinds, lighting systems, thermostats, fans, HVAC systems, plug loads; making hot/cold beverages; and adjusting clothing levels. (2) *Building types*—office, residential, and school buildings. In this list, those models with clear documentation were considered for library inclusion and were processed and implemented using the drivers-needs-actions-systems (DNAS) framework, presented in a standardised way called obXML (occupant behaviour eXtensible Markup Language, see later). In addition to the obXML library (Belafi et al., 2016), a library of occupant behaviour models in another language, Modelica was also recently developed (Wang et al., 2018). This Modelica package of occupant behaviour models could be more conveniently integrated into other Modelica-based building system models.

- *Surveys on building simulation practices and human–building interactions*
As aforementioned, surveys are a cost-effective means of achieving a large sample size and can measure phenomena that would be difficult or impossible to measure with sensors. Two surveys were conducted under Annex 66 (https://annex66.org/).

The first survey seeks to understand the current practices and attitudes of current building simulation users towards occupant modelling (O'Brien et al., 2017b). In total, 274 valid responses were collected from BPS users (practitioners, educators, and researchers) from 37 countries. The results of this 36-question international survey indicate that occupant assumptions made by simulation users vary widely and are considerably simpler than what has been observed in reality. Most participants cited the lack of time or understanding as their primary reason for not delving deeply into occupant modelling but responded that they are receptive to further training.

The second survey is a cross-country questionnaire based on theories and insights from building physics and social psychology. This survey aims to investigate the building–user interaction in the workspace and the degree to which this interaction impacts comfort provision, energy use, and operating costs in diverse office settings and cultural contexts worldwide (D'Oca et al., 2015). A total of 37 questions were devised by an interdisciplinary team having architecture, engineering, and social science backgrounds, and responses were collected from administrative staff and faculties at 14 universities and research centers across six countries spanning the United States, Europe, China, and Australia. The outcomes highlight the correlation between perceived behavioural control and perceived comfort, satisfaction, and productivity in office spaces.

- *OpenEI.org*
OpenEI provides a free platform for sharing datasets specifically in the area of renewable energy and energy efficiency. Currently, there are several datasets available on this platform that relate to occupant behaviour. For instance, Langevin et al. (2015a, b) published a 1-year longitudinal dataset (15-minute interval) on local thermal conditions, related occupant behaviours, and comfort of 24 occupants of a medium-sized office building between July 2012 and August 2013 in Philadelphia, PA. The long-term data were collected via online daily surveys and data logger measurements of the local thermal environment and behaviour.

- *Zenodo*
Zenodo (Zenodo.org) is a web platform that promotes open data for open science. Datasets, software, and other materials can be deposited and shared through the Zenodo platform, which includes datasets related to occupant behaviour. For instance, a recently published dataset on Zenodo contains movement behaviour (head, eye, and torso) and electroencephalogram (EEG) signals (a recording of the electrical activity of the brain from the scalp) of 21 young normal-hearing (11 males and 11 females, mean age 25 +/− 3.6 years) and 19 elderly normal-hearing subjects (9 males and 12 females, mean age 69 +/− 5.4 years) measured in virtual audiovisual environments in the laboratory. The virtual audiovisual environments that were used are a living room, a lecture hall, a cafeteria, a street, and a train station. The video and audio material for the environments is also available on the website (Hendrikse et al., 2018).

- *Nature Scientific Data*
Nature Scientific Data is a peer-reviewed, open-access journal for descriptions of scientifically valuable datasets and research that advances the sharing and reuse of scientific data. The journal was launched by Nature Research to enable the discoverability, reproducibility, and reuse of valuable data. Scientific Data primarily publishes Data Descriptors, a new type of publication that combines the narrative content characteristic of traditional journal articles with structured, curated metadata that outline experimental workflows and point to publicly archived data records. Currently, there are around 20 datasets in this resource

that relate to occupant behaviour or energy consumption measurements. For instance, Makonin et al. (2016) collected long-term measurements of electric and water consumption, energy use behaviour, and HVAC operational parameters for a residential house in Canada between 2012 and 2014.

2.2.3 Advanced occupant behaviour simulation tools
A suite of computational tools has been developed to standardise the representation of OB models and enable their use via cosimulation with BPS programs.

2.2.3.1 obXML: An occupant behaviour XML schema
obXML (Hong et al., 2015a, b) is an XML schema that standardises the representation and exchange of occupant behaviour models for building performance simulation. obXML builds upon the drivers-needs-actions systems (DNAS) ontology to represent energy-related occupant behaviour in buildings. Drivers represent the environmental and other context factors that stimulate occupants to fulfil a physical, physiological, or psychological need. Needs represent the physical and nonphysical requirements of occupants that must be met to ensure satisfaction with their environment. Actions are the interactions with systems or activities that occupants can perform to achieve environmental comfort. Systems refer to the equipment or mechanisms within the building that occupants may interact with to restore or maintain environmental comfort. A library of obXML files, representing typical occupant behaviour in buildings, was developed from the literature (Belafi et al., 2016). These obXML files can be exchanged between different BPS programs, different applications, and different users. Fig. 1 shows the four key elements of the obXML schema and their subelements.

2.2.3.2 obFMU: An occupant behaviour functional mockup unit
obFMU (Hong et al., 2016) is a modular software component represented in the form of functional mockup units (FMUs), enabling its application via cosimulation with BPS programs using the standard functional mockup interface (FMI). FMU is a file (with extension .fmu) that contains a simulation model that adheres to the FMI standard. obFMU reads the occupant behaviour models represented in obXML and functions as a solver. A variety of occupant behaviour models are supported by obFMU, including (1) lighting control based on occupants' visual comfort needs and availability of daylight, (2) comfort temperature set points, (3) HVAC system control based on occupants' thermal comfort needs, (4) plug load control based on occupancy, and (5) window opening and closing based on indoor and outdoor environmental parameters. obFMU has been used with EnergyPlus and ESP-r via cosimulation to improve the modelling of occupant behaviour. Fig. 2 shows the workflow of cosimulation using obFMU and EnergyPlus.

2.2.3.3 Occupancy simulator: A web-based occupancy app
Occupancy Simulator (Chen et al., 2018; Luo et al., 2017) is a web-based application running on multiple platforms to simulate occupant presence and movement in buildings. The application can also generate subhourly occupant schedules for each

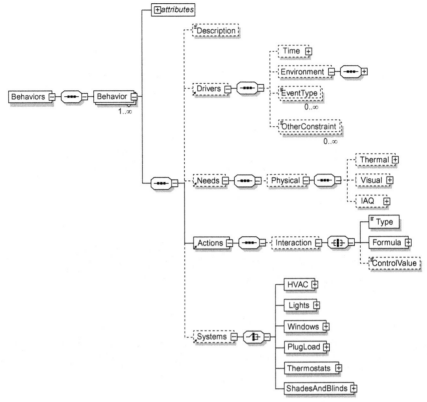

FIG. 1

Overview of the obXML schema showing the DNAS ontology.

Credit: Hong, T., et al., 2016. An occupant behavior modeling tool for co-simulation. Energy Build., 117, 272–281.

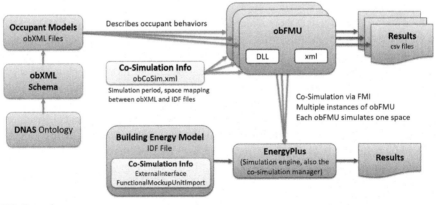

FIG. 2

Cosimulation workflow of obFMU with EnergyPlus.

Credit: Yan, D., Hong, T., 2018. Fina Report of IEA EBC Annex 66: Definition and simulation of occupant behavior in buildings, May 2018.

space and individual occupants in the form of CSV files and EnergyPlus IDF files for building performance simulations. Occupancy Simulator uses a homogeneous Markov chain model (Wang et al., 2011; Feng et al., 2015) and performs agent-based simulations for each occupant. A hierarchical input structure is adopted, building upon the input blocks of building type, space type, and occupant type to simplify the input process while allowing flexibility for detailed information capturing the diversity of space use and individual occupant behaviour. Users can choose an individual space or the whole building to see the simulated occupancy results.

2.2.3.4 Buildings.Occupants: An occupant behaviour model package in Modelica

To simulate the continuous and dynamic interaction between occupants and building systems, Buildings.Occupants (Wang et al., 2018) can be used. The Buildings. Occupants package, as part of the Modelica Buildings Library (Wetter et al., 2014), supports fast prototyping by seamlessly integrating occupant behaviour models with Modelica models from existing libraries for building dynamics. Additionally, the structure of the package has been designed to allow for the flexible implementation of user-defined models by tuning the parameters and calling functions defined in the BaseClasses package. The Buildings.Occupants package includes reported occupant behaviour models in the literature that are more commonly used and well documented in terms of the data source, mathematical equation, independent variables, parameter values, etc. The models are categorised into subpackages based on the building types and systems. There are 34 occupant behaviour models for office and residential buildings that are included in the first release of the Buildings.Occupants package. The office building models include eight models on windows operation, six models on window blind operation, four models on lighting operation, and one model on occupancy.

3 Application of occupant behaviour models across the building life cycle

This section brings an overview of application areas for occupant behaviour modelling tools and datasets across the building life cycle. It summarises Annex 66's 32 case studies of building occupant behaviour modelling applications from around the world and then introduces three example applications through case studies focusing on the building design-stage, operation, and control-stage, as well as the retrofit-stage.

3.1 Fit-for-purpose occupant behaviour modelling in the building life cycle

As suggested earlier in this chapter, occupant behaviour is an important source of uncertainty when dealing with BPS (Clevenger and Haymaker, 2006; Hoes et al.,

2009), and an increasing number of models have accordingly attempted to represent occupant behaviour in a more realistic manner within BPS. Such models can be classified according to their complexity—here defined as the amount of detail in a model, which in turn results from its size and resolution (Zeigler and Oren, 1979).

At the lowest spectrum of complexity are the diversity factors, or schedules: hourly fractions from 0 to 1, which are multiplied for a maximum amount of, for example, heat gains due to people, equipment, lighting, etc. Schedules are commonly employed to represent occupant presence and occupant behaviour in current BPS tools, due to their ease of use and to the incentives from the building code (Yan et al., 2015). However, it is argued that simple schedules are not representative of actual occupant behaviour, which is typically stochastic and influenced by a high number of variables. Moreover, schedules neglect occupants' diversity (O'Brien et al., 2017a).

For this reason, researchers have developed nonprobabilistic, probabilistic, and agent-based models, which give a more accurate representation of people's behaviour (Gaetani et al., 2016; Gunay et al., 2013). Here, it is important to note that the required confidence in the building performance prediction depends on the purpose of the simulation. For example, Gaetani et al. (2016) argue that a more complex behaviour model is needed when energy usage for a single building is assessed (design/retrofit), but such complexity may not be necessary or feasible when aggregating predictions across the scale of a district with a collection of buildings. Furthermore, different buildings and performance metrics may be affected in a diverse way by the various aspects of occupant behaviour: some cases are extremely sensitive to the way a particular aspect is modelled, while others may be little affected.

In practice, BPS users may not understand the details of available behaviour models and may not use them as intended. Above all else, the modeller must justify the chosen behaviour modelling approach on a case-by-case basis to ensure that it is fit-for-purpose. Fig. 3 illustrated the fit-for-purpose framework for occupant behaviour model selection and application (Gaetani et al., 2016). First, there is a need to select an appropriate tool for the given system design complexity. Then, information on the design parameters should be commensurate with the level of detail of the model. The characteristics of BPS tools that incorporate occupant behaviour should therefore vary according to application context. Highly complex behaviour modelling software may not be of much use when simple energy use estimations are required. In contrast, for a building design phase that calls for detailed modelling, additional behaviour modelling sophistication may be warranted.

3.2 Summary of 32 case studies for occupant behaviour models and data in Annex 66

IEA EBC Annex 66: Definition and simulation of occupant behaviour in buildings (Yan et al., 2017), an international collaborative project involving more than 120 researchers from 20 countries working together from November 2013 to May 2018, collected a set of 32 case studies (Clinton et al., 2017) of modelling occupant behaviour in buildings using various computational decision support tools.

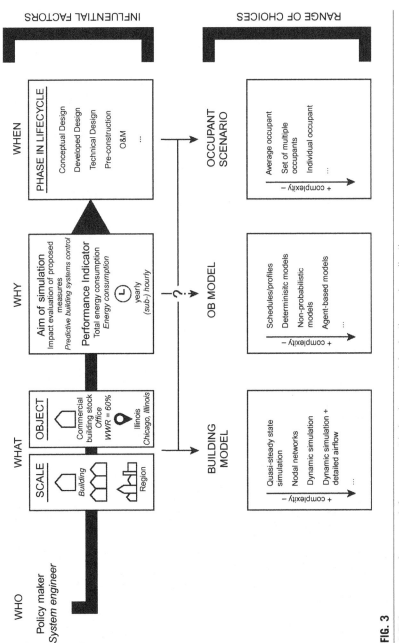

FIG. 3

A fit-for-purpose framework for occupant behaviour model selection and application.

Credit: Gaetani, I., Hoes, P.-J., Hensen, J.L.M., 2016. Occupant behavior in building energy simulation: towards a fit-for-purpose modeling strategy. Energy Build., 121. 188–204.

Motivation to accurately represent occupant behaviour in these case studies comes from BPS practitioner beliefs that occupant behaviour is a major source of discrepancy between simulated and measured building energy performance and that current modelling practice with regard to occupant behaviour is overly simplistic (O'Brien and Gunay, 2016). Indeed, the previously cited review of nine current BPS programs by Cowie et al. (2017) identified "a widening gap between knowledge and implementation in the field of occupant behaviour modelling."

Accordingly, the purpose of the case study review was to illustrate the range and types of occupant behaviour modelling applications, to contribute to a framework for classifying these applications, and to explore which behaviour modelling approaches are most appropriate for which contexts. Essential elements of this framework answer the canonical journalistic questions about any story: who, what, why, when, and where. To determine which model is most fit for which context, three dimensions emerge as being particularly important: the stakeholder and their problem (who and why); the building type, services, and provisions (what); and the process stage and relevant tools (when).

The most innovative cases of occupant behaviour modelling provide a "demand-pull" view, as seen by the users of such tools, to counterbalance the "supply-push" perspective that many who create such models bring to the subject (Godin, 2017). The case studies collectively provide a framework for thinking about (1) when occupant behaviour becomes important for making decisions about buildings, (2) which tools are most appropriate for specific applications, and (3) what insights emerge from practical experience with these tools.

3.3 Three representative case studies: Design-stage application, operation and control-stage application, and retrofit-stage application

3.3.1 Case study 1—the impact of occupant behaviour modelling assumptions on energy-efficiency measure performance

To improve energy efficiency—during new building design and building retrofit—evaluating the energy saving potentials of energy conservation measures (ECMs) is critical. ECMs refer to building technologies (e.g. LED lights), control strategies (e.g. daylighting and dimming control of lights), and behaviour changes (e.g. occupants turning off lights when leaving an office) that improve upon the per-unit energy use of comparable incumbent or "business-as-usual" technologies or approaches. Occupant behaviours significantly impact building energy use and raise uncertainty when estimating the effectiveness of ECMs. This case study presents a simulation framework of quantifying the impact of occupant behaviour on ECM savings.

3.3.1.1 Methodology

The ECM savings are influenced by many factors such as the building type, weather data, building operation, and occupant behaviours. The estimated ECM savings would vary with different model input assumptions. Traditional ECM evaluation methods adopt deterministic inputs and generate a static single result of energy savings, which neglects the uncertainty of the ECM savings. However, estimating the uncertainties of the ECM savings is critical, especially during risk analysis and decision making for ECM investment (Heo et al., 2011). Decision makers should be aware of the potential risks of implementing ECMs before selecting the most appropriate ECMs for a specific building.

In this case study, a simulation framework was proposed to evaluate ECM savings considering the variations of occupant-related inputs and their influence on the ECM energy savings (Fig. 4). This proposed framework includes the following steps: (1) defining the three occupant behaviour styles representing people with different levels of energy consciousness (austerity, normal, and wasteful), using quantitative occupant behaviour models; (2) developing three baseline models using each of the predefined three occupant behaviour styles and other same model inputs such as weather data, internal heat gains, and energy system efficiencies; (3) calculating the energy uses of the three baseline models; (4) applying the ECMs to each baseline model to create the alternate models for each ECM; and (5) simulating the ECM energy models to calculate their energy use.

The simulated ECM saving results using the proposed framework are a range of values instead of a single fixed value, which reflects the possible variations of the ECM savings due to different occupant behaviours in the building. Therefore, the framework can be adopted to evaluate ECM savings in a more comprehensive and robust way, giving decision makers the information they need to recognise and assess the potential risks of investing in ECMs in buildings with different occupant behaviours. ECMs with consistent large energy savings can be prioritised for investment compared with those ECMs with savings that are sensitive to occupant behaviour style.

This framework was demonstrated in a real office building to quantify the influence of occupant behaviours on ECM savings. Fig. 5 shows the overall workflow of

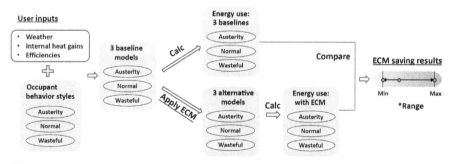

FIG. 4

A framework to quantify the impact of occupant behaviour on performance of ECMs.

Credit: Sun, K., Hong, T., 2017a. A framework for quantifying the impact of occupant behavior on energy savings of energy conservation measures. Energy Build., 146, 383–396. Available at: http://dx.doi.org/10.1016/j.enbuild.2017.04.065..

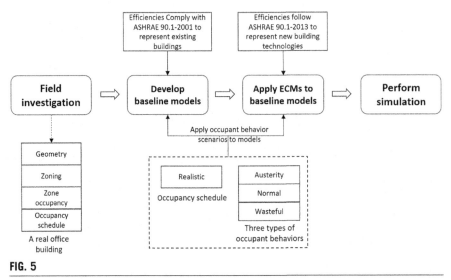

FIG. 5

The workflow of the pilot study.

Credit: Sun, K., Hong, T., 2017a. A framework for quantifying the impact of occupant behavior on energy savings of energy conservation measures. Energy Build., 146, 383–396. Available at: http://dx.doi.org/10.1016/j.enbuild.2017.04.065.

the pilot study. Field investigation was conducted in the building to gather information for creating the baseline energy model, including the geometry, zoning, number of occupants in each zone, and occupant schedules. Three occupant behaviour styles, representing the proactive energy savers, average (norm) occupants, and the energy spenders, respectively, were adopted to represent different levels of energy consciousness and the boundaries of either extreme (as in energy savers and spenders). Occupant schedules, generated by the Occupancy Simulator with inputs from the site survey of the case building, were used in the energy models.

Seven individual ECMs and one packaged ECM were evaluated in this study, including reducing lighting power density (LPD), reducing plug-in electric equipment power density (EPD), improving envelope performance, improving HVAC system efficiency, daylighting control, variable refrigerant flow system, and natural ventilation coupled with the variable refrigerant flow (VRF) system. These ECMs were chosen considering their application to a 15-year-old building designed to comply with ASHRAE Standard 90.1-2001 standards, which were adopted in the baseline models to represent existing buildings. The efficiencies of the ECMs in this study were obtained from the more recent ASHRAE 90.1-2013 standards. The impact of occupant behaviour on ECM energy savings was evaluated in four different climates—Chicago, Fairbanks, Miami, and San Francisco—so that the potential sensitivity to climate could be studied as another dimension. The selected cities represent the four typical climate types in the United States: humid continental, subarctic, tropical (subtropical), and Mediterranean, respectively.

Whole-building simulation using EnergyPlus was used to evaluate the impact of occupant behaviours on the ECM savings. Baseline models were developed in EnergyPlus to represent the investigated office building. EnergyPlus is an open-source program that models heating, ventilation, cooling, lighting, water use, renewable energy generation, and other building energy flows (Crawley et al., 2001) and is the flagship building simulation engine supported by the Department of Energy. It includes innovative simulation capabilities (e.g. subhourly time steps, natural ventilation, thermal comfort, cosimulation with external interfaces, renewable energy systems, and user customisable energy management systems). Some of the innovative capabilities, such as natural ventilation, daylighting, external schedules, and energy management systems, were used in this pilot study.

3.3.1.2 Results

Fig. 6 shows an example of the pilot study results, which illustrates the ECM energy saving percentages compared with the baseline models under the three behaviour styles in Chicago. The simulation results indicate that the saving percentages of LPD, EPD, envelope, system efficiency, and daylighting control are minimally affected by occupant behaviour styles. This is because they are all purely technology-driven ECMs, which don't rely on the interactions with the occupants to save energy. On the other hand, the saving percentages of the VRF system, natural ventilation, and integrated ECM are significantly affected by occupant

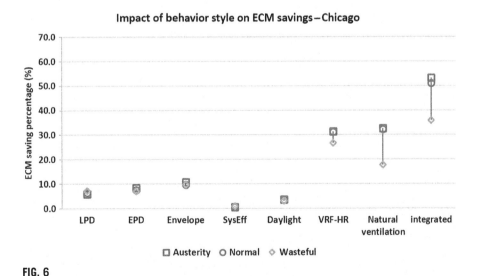

FIG. 6

ECM saving percentages compared with the baseline models with different behaviour styles in Chicago.

Credit: Sun, K., Hong, T., 2017a. A framework for quantifying the impact of occupant behavior on energy savings of energy conservation measures. Energy Build., 146, 383–396. Available at: http://dx.doi.org/10.1016/j.enbuild.2017.04.065.

behaviour styles, because the energy performance in these ECMs is closely related to how the occupants interact with the ECM. For example, once the VRF system is installed, which allows zonal control, the occupants have decisions to make on how to control their indoor units: the austerity occupants only turn on the indoor units when they feel hot, normal occupants turn on the indoor units as long as they are in the room, but the wasteful occupants keep the indoor units on during the entire working hours. Also, cooling and heating set points are different amongst the behaviour styles. Therefore, the energy performance of such ECMs heavily depends on how the occupants behave.

3.3.1.3 Conclusion
The main findings from this study are as follows:

(1) The occupant behaviour style has significant influence on building energy use. Buildings occupied by energy spenders could consume more than twice the energy of the energy savers.
(2) For occupant-independent ECMs, which are purely technology-driven and have little interaction with the occupants, such as reducing LPD, reducing EPD, improving envelope properties, and improving HVAC system efficiency and daylighting control, energy saving percentages are minimally (less than 2%) affected by occupant behaviour styles. For occupant-dependent ECMs, which have strong interaction with the occupants, such as the VRF system and natural ventilation, energy saving percentages are significantly (up to 20%) affected by occupant behaviour styles.
(3) The wasteful behaviour style generally achieves the greatest absolute energy savings, while its saving percentages are close to or even lower than those of the austerity and normal behaviour. This is important information for decision makers in retrofit planning.
(4) The occupant schedule has certain impacts on the simulated results of ECM savings, especially for the occupant-dependent ECMs coupled with the austerity behaviour style. Adopting realistic occupant schedules rather than normalised ones would help improve the accuracy of ECM saving evaluation.

The zero-net energy (ZNE) technologies are successful and growing today as energy performance requirements are becoming more and more stringent. ZNE technologies, such as natural ventilation, HVAC control, and demand response, tend to need more interaction with occupants. They are more sensitive to occupant behaviours and reactions to stimulations, which makes occupant behaviour a significant uncertainty factor for the technology's performance. In other words, occupant behaviour may significantly change the way technologies are designed and expected to perform. The proposed framework provides a novel simulation approach enabling energy modellers to calculate the ECM savings as a range rather than a single fixed value considering the variations of occupant behaviours in buildings, which provides a critical input to the risk analysis of ECM investments.

3.3.2 Case study 2—Simulating the dynamic feedbacks between individual-level behavioural adaptations and building operations

Real office building occupants interact with and adapt to their surrounding environments in deliberate and meaningful ways that affect both energy consumption and indoor environmental quality. As suggested throughout this chapter, numerous studies have estimated the magnitude of these effects, establishing the high degree of influence that occupant behaviour exerts on building energy use and thermal comfort relative to other potentially significant factors (Hong and Lin, 2013; Haldi and Robinson, 2011; Bourgeois et al., 2006).

Given the importance of occupants' environmental adaptations to building energy and comfort outcomes, this case study presents a Human and Building Interaction Toolkit (HABIT) that cosimulates building energy and office occupant behaviour. The toolkit uses a field-validated, agent-based model that estimates both individual and group-level comfort/behaviour outcomes; accommodates whole building-level analyses; and yields comprehensive outputs on energy, behaviour, and indoor environmental quality (IEQ) that can guide the design and operation of low-energy, high-quality office building environments.

The case study begins by describing the HABIT cosimulation exchange and its underlying agent-based model of thermally adaptive behaviours. The usefulness of the toolkit is then demonstrated through a series of case study simulations that explore a range of occupant behaviour scenarios, including multiple cases where wider thermostat set point ranges are paired with the provision of efficient local heating and cooling options for occupants. The relative merits of each scenario are assessed by comparing resulting energy use intensities alongside occupant thermal unacceptability and productivity outcomes.

3.3.2.1 Methodology

Cosimulation overview. HABIT pairs a previously published ABM of office occupants' thermal comfort and adaptive behaviours (Langevin et al., 2016) in MATLAB with whole-building energy simulations of office buildings in EnergyPlus (BTO, 2017) using the Building Controls Virtual Test Bed (BCVTB) cosimulation program (Wetter, 2011).

The EnergyPlus/MATLAB information exchange runs as follows: EnergyPlus simulates zone-level thermal conditions and passes these as inputs to the MATLAB comfort/behaviour model; the MATLAB model predicts thermal comfort and related behaviour outcomes for each occupant (i.e. fan on, window open, etc.) and aggregates these outcomes across all agents in the zone; the aggregated behaviour outcomes are passed back to EnergyPlus and used to adjust appropriate zone schedules (i.e. heater/fan equipment gains and thermostat set points) for the next time step; the process repeats until a simulation end time is reached.

The BCVTB negotiates single runs of the above MATLAB/EnergyPlus exchange. However, the MATLAB comfort/behaviour model contains probabilistic elements. Thus, the exchange must be rerun multiple times to assess a range of possible outcomes.

Agent-based behaviour model overview. In the default HABIT setup, each office occupant is represented in the MATLAB comfort/behaviour model as a simulated agent that acts adaptively based on the scheme described in Langevin et al. (2015a, b). Under this scheme, behaviour is considered to be the by-product of a negative feedback loop in which an agent acts to bring its current thermal perception into line with a reference range of seasonally acceptable ASHRAE thermal sensations, despite environmental disturbances.

An agent's current thermal sensation and seasonally acceptable thermal sensation range are both modelled probabilistically using the distributions developed in Langevin et al. (2013); daily occupant arrival/lunch/departure times are also sampled from a normal distribution around user-defined means. Agent behaviour choices may be constrained by the building management or by other agents in the space that share a given control. The reader is referred to Langevin et al. (2015a, b) for full details on the HABIT ABM and its validation.

Table 1 presents the full set of behavioural adaptations simulated by the HABIT ABM and shows how their feedback on the thermal environment/comfort is represented on both the MATLAB and EnergyPlus sides of the BCVTB cosimulation.

Table 1 Behaviours simulated in the HABIT framework and their default local/zone-level feedbacks in BCVTB cosimulation

Behaviour	Impact	Feedback increments[a]	
		MATLAB ABM (agent-level)	EnergyPlus (zone level)
Clothing adjustment (minor)	Adds/subtracts clothing insulation from morning level	−0.08 Clo[b]	
Clothing adjustment (major)		+/− 0.30 Clo	
Fan on	Increases local air velocity; adds zone-level electric equipment gain	+0.75m/s	+15 W equipment load
Heater on	Increases local and zone level ambient/radiant temperature; adds zone-level electric equipment gain	+2°C	+1200W equipment load
Thermostat up/down	Increases/decreases zone level ambient temperature		+/−1 set point adjustment
Window open	Increases local air velocity; zone-level ambient temperature moves towards outdoor temperature	+0.25m/s	25X design infiltration rate

[a] Refer to Langevin et al. (2016) for details on sourcing of these assumptions.
[b] The clo unit expresses the thermal insulation provided by garments and clothing ensembles; 1 clo=0.155m²°C/W (0.88ft²h°F/Btu).
Credit: Langevin, J., Wen, J., Gurian, P.L., 2016. Quantifying the human-building interaction: considering the active, adaptive occupant in building performance simulation. Energy Build.

Case study simulations. To demonstrate the usefulness of the simulation framework earlier, case study simulations were performed for a medium-sized office building. The simulated building has three stories with 5600m^2 of total floor space and one core/four perimeter thermal zones per floor; masonry construction with 20% glazing; variable-air-volume air-handling units with hot water reheat; an occupant density of 0.05person/m^2; and a baseline occupied infiltration rate of 2.4CFM/m^2 applied to perimeter zones (see Hendricken et al. (2012) for more details).

Case study simulations are performed on all zones of the case study building for the months of January and July in the Philadelphia climate.[b] These simulations each cover 15 zones (3 core; 12 perimeter) and 297 occupants. These case study simulations are ultimately intended to yield a complete picture of the link between behaviour, energy, and IEQ across multiple thermal zones with different orientations and locations within the general building geometry.

In each of the simulations, multiple behaviour scenarios are run to test the influence of behaviour modelling assumptions on energy and IEQ outcomes. As shown in Table 2, behaviour scenarios range from a "Baseline" case (B) where no thermally adaptive actions are possible, to a "Fully Unrestricted" (UR) case where a full range of actions are possible, finally to a series of "Wider Set Points" (WSP) cases where zone thermostat set point ranges are progressively widened and occupants are provided with more efficient local heating/cooling options at their desks. The "Wider Set Points" scenarios are of particular interest in testing the degree to which localised heating/cooling devices can save energy while maintaining or improving occupant thermal comfort levels (as proposed, e.g. in Hoyt et al. (2014)).

Simulation outcomes are evaluated from both the energy and IEQ perspectives. For energy, an energy use intensity is calculated; for IEQ, the percentage of occupants that the behavioural model indicates are outside their acceptable thermal sensation range without any behavioural remedy is recorded at each time step and averaged across the entire simulation period, yielding an overall percentage thermally unacceptable outcome (a thermal comfort indicator). Relative work underperformance percentage (a productivity indicator) is also evaluated using the polynomial relationship presented in Jensen et al. (2009), which describes relative performance in terms of thermal sensation:

$$\text{Relative work performance} = -0.0069tsv^2 - 0.0123tsv + 0.9945 \qquad (1)$$

where *tsv* is an occupant's thermal sensation vote on the ASHRAE sensation scale. Relative underperformance is then simply 1—relative work performance.

3.3.2.2 Results
Key results for the case study simulations are summarised through Fig. 7A–C. Fig. 7A shows that HVAC energy use intensity increases slightly in both January and July from the "Baseline" behaviour scenario through the "Fully Unrestricted

---

[b] Refer to Langevin et al. (2016) for additional case study simulations that explore the sensitivity of outcomes to climate and location.

Table 2 Behaviour scenarios for case study simulations. Note: cell values indicate EnergyPlus settings associated with each behaviour

Name	Clothing[a]	Fans	Heaters	Thermostat[b]	Window[c]
Baseline (B)					
Fully restricted (R)	–	+15W	+1200W	21–24°C +/−1°C	+25X infiltration
Fully unrestricted (UR)	–	+15W	+1200W	21–24°C +/−1°C	+25X infiltration
Wider set points—base (WSP)[d]	–	+15W	+800W	20–27°C +/−1°C	+25X infiltration
Wider set points—base+educate (WSPe)[e]	–	+15W	+600W	20–27°C +/−1°C	+25X infiltration
Wider set points—moderate (WSP2)	–	+15W	+800W	19–28°C +/−1°C	+25X infiltration
Wider set points—aggressive (WSP3)	–	+15W	+800W	17–30°C +/−1°C	+25X infiltration

■ Unavailable/not possible.

Restricted by management; 50% of occupants assumed to adhere to management restrictions.

Restricted by management and others in the space; 30% of occupants assumed not to take shared action if it is predicted to make >50% of other occupants in the space uncomfortable.
[a] Clothing level has no direct impact on simulated energy use in EnergyPlus. Simulated clothing level must remain between 0.3 and 1.3.
[b] Shown are the heating set point in January and cooling set point in July ±1°C occupant adjustment capability.
[c] Windows remain closed if outdoor running mean temperature is greater than indoor operative temperature.
[d] In all WSP scenarios, it is assumed that simulated occupants are given more efficient halogen heaters (max power use=+800W).
[e] In the WSPe scenario, half of the occupants are assumed to use halogen heaters on a low setting (+400W) due to energy education efforts.
Credit: Langevin, J., Wen, J., Gurian, P.L., 2016. Quantifying the human-building interaction: considering the active, adaptive occupant in building performance simulation. Energy Build.

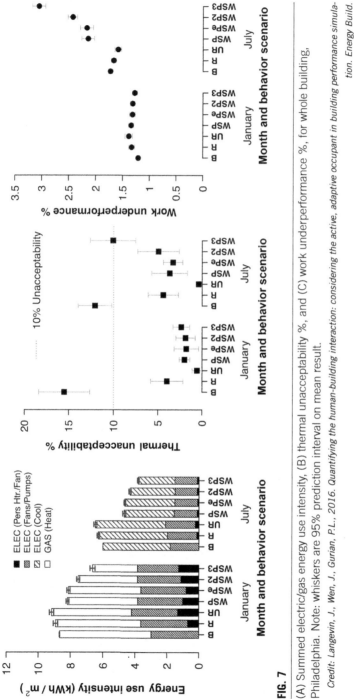

FIG. 7

(A) Summed electric/gas energy use intensity, (B) thermal unacceptability %, and (C) work underperformance %, for whole building, Philadelphia. Note: whiskers are 95% prediction interval on mean result.

Credit: Langevin, J., Wen, J., Gurian, P.L., 2016. Quantifying the human-building interaction: considering the active, adaptive occupant in building performance simulation. Energy Build.

Behaviour" scenario; the energy use intensity then moves back down across the "Wider Set Points" scenarios, which range from an initial $\pm 1°$ C widening of the set point range ("Wider Set Points") to a $\pm 4°$C widening of this range ("Wider Set Points (Aggressive)").

Energy end use breakdowns in Fig. 7A allow a more specific examination of these trends. In January, for example, significant reductions in HVAC energy use for the "Wider Set Points (Aggressive)" scenario ($\sim$24%) result from a decrease in gas space heating consumption that more than offsets an associated increase in electrical equipment energy from more frequent use of personal heaters. Similarly, in July, significant reductions in HVAC energy use by the "Wider Set Points (Aggressive)" scenario ($\sim$37%) result from a decrease in electric space cooling consumption that more than offsets a small associated increase in electrical equipment energy from more frequent use of personal fans.

Regarding IEQ, Fig. 7B shows a significant decrease in percentage of time thermally unacceptable relative to the baseline through the "Wider Set Points (Aggressive)" scenario in January and through the "Wider Set Points (Moderate)" scenario in July, with the lowest thermally unacceptable percentage occurring as in the zone-level simulations for the "Fully Unrestricted Behaviour" scenario in both months. In July, however, the "Wider Set Points (Aggressive)" scenario yields a thermally unacceptable percentage for July that is close to that of the baseline, with the prediction intervals for the two overlapping. This result also violates the 10% thermally unacceptable threshold used in thermal comfort standards, suggesting that a "Wider Set Points (Aggressive)" strategy for a Philadelphia office in the summer will yield substantial warm discomfort amongst occupants. Work underperformance results for this scenario in Fig. 7C reinforce this conclusion, moving up to 3% underperformance as a result of the high cooling set point. Ultimately, a "Wider Set Points (Moderate)" strategy is the better option for achieving significant HVAC energy savings ($\sim$28%) while maintaining good IEQ in the Philadelphia summer months.

3.3.2.3 Conclusion

Taken together, the results from this case study suggest that building managers can pair the use of more efficient local heating and cooling options with strategic thermostat set point adjustments as a simple way of saving substantial amounts of energy (up to 24%/28% in heating/cooling months, respectively) while also improving occupant thermal acceptability. Care must be taken to consider additional outcome metrics, however: managers who value occupant productivity above all else, for example, may view the small productivity decrements predicted from raising cooling set points as unacceptable. Moreover, in the heating cases, the potential disadvantages of trading natural gas for electric heating fuel in terms of energy costs and greenhouse gas emissions must also be taken into consideration.[c]

---

[c] These cost tradeoffs are further quantified in Langevin et al. (2016).

3.3.3 Case study 3—Modelling and evaluating the energy savings potential of behaviour-focused retrofit measures

Occupant behaviour in buildings is a leading factor influencing building energy use. Low-cost behavioural solutions have demonstrated significant potential energy savings. Estimating the behavioural savings potential is important for a more effective design of behaviour change interventions, which in turn will support more effective energy-efficiency policies. This case study introduces a simulation approach to quantify the energy saving potentials of occupant behaviour measures.

3.3.3.1 Methodology

This case study investigated the energy saving potentials of occupant behaviour measures by (1) conducting field investigation on a real office building (including the geometry, zoning, occupancy schedule, lighting schedule, and plug load power density and schedule); (2) developing the baseline models based on the information earlier; (3) defining five occupant behaviour measures, including lighting, plug load, thermal comfort criteria, HVAC control, and window control; and (4) running simulation to calculate the energy saving potentials of the occupant behaviour measures across four typical US climates (Chicago, Fairbanks, Miami, and San Francisco) and two vintages (1989 and 2010). Overall methodology is illustrated in Fig. 8.

The case building has two above-ground stories with a total conditioned floor area of 1723 m^2. Main room functions include office, conference room, classroom, and lounge (corridor). Smaller corridors are merged into office zones for simplification.

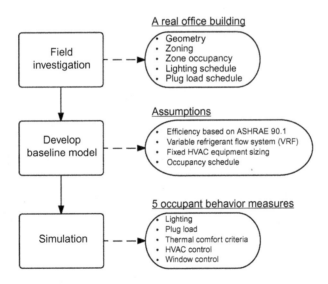

FIG. 8

Overall methodology.

Credit: Sun, K., Hong, T., 2017b. A simulation approach to estimate energy savings potential of occupant behavior measures. Energy Build., 136, 43–62.

The perimeter zones have operable windows, which allow the occupants to open windows for cooling or ventilation. The total number of occupants in the case building is 63. Figs. 9 and 10 show the floor plan of the first and second floors, indicating the room functions. Baseline models representing the case building were developed in EnergyPlus.

Five occupant behaviour measures implemented in this case study are as follows: (1) Lighting control—lights are only on if a space is occupied and occupants feel too dark. The conditional probability of turning on/off the lights follows a

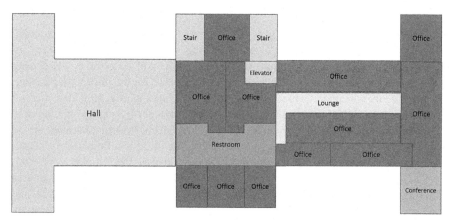

FIG. 9

The first floor plan.

Credit: Sun, K., Hong, T., 2017b. A simulation approach to estimate energy savings potential of occupant behavior measures. Energy Build., 136, 43–62.

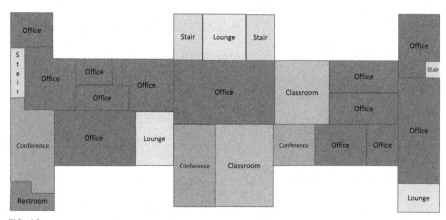

FIG. 10

The second floor plan.

Credit: Sun, K., Hong, T., 2017b. A simulation approach to estimate energy savings potential of occupant behavior measures. Energy Build., 136, 43–62.

three-parameter Weibull distribution, defined in Wang's paper (Wang et al., 2015). In this study, we referred to the parameters' values in Wang's paper; (2) Plug load control—30% of power is turned off if unoccupied; (3) Thermal comfort criteria—two thermal comfort criteria were considered: one is the ASHRAE standard 55 comfort zone limits (ANSI/ASHRAE, 2013), where the upper temperature limit of the ASHRAE 55 comfort zone was taken as the cooling set point in the simulation, while the lower limit was taken as the heating set point. The other one is the adaptive comfort model (Brager and De Dear, 2001) with 80% acceptability limits to calculate a dynamic comfort range based on ambient temperature, which was then used as dynamic cooling/heating set points in simulation; (4) HVAC control—HVAC is turned on if a space is occupied and occupants feel hot (in cooling mode) or cold (in heating mode). Ren's model (Ren et al., 2014) was adopted to estimate the time-step HVAC control status in our study, which used a three-parameter Weibull distribution function to describe different air conditioning usage patterns; and (5) Window control—the control logic is illustrated in Fig. 11. The Weibull distribution functions describing the conditional probability of turning on/off the lights and HVAC, which were defined by Wang (Wang et al., 2015) and Ren (Ren et al., 2014), were adopted in this study. Other than the individual measures, all the five measures were integrated as well and their integral energy savings were simulated.

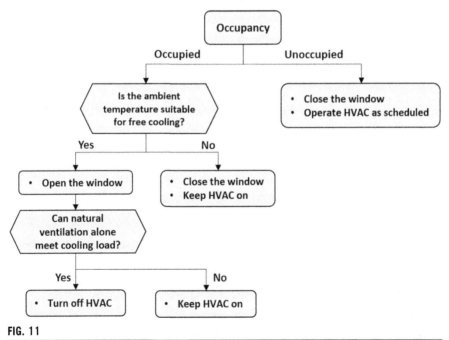

FIG. 11

Control logic of window Measure.

Credit: Sun, K., Hong, T., 2017b. A simulation approach to estimate energy savings potential of occupant behavior measures. Energy Build., 136, 43–62.

The stochastic occupant schedules were generated by the Occupancy Simulator. Compared with the normalised identical occupant schedule in all spaces, the generated schedules can reflect the variation, diversity, and stochastic characteristic of the realistic occupant movements. To make it consistent for all the studied measures, the same set of generated schedules is applied to both the baseline model and the five occupant measures.

3.3.3.2 Results

Fig. 12 shows an example of the breakdown end uses of the baseline model, the five individual measures, and the integrated measure. Each measure has its different impact on energy consumption: (1) the lighting measure and the plug load measure reduce the internal heat gains, which cut the cooling load but raise the heating load; (2) the comfort criteria measure reduces the heating/cooling load by enlarging the comfort boundary; (3) the HVAC measure and the window measure reduce the energy consumption by decreasing the HVAC operation time. When they are integrated, the effect of (3) is relatively weakened due to a lower cooling load level resulting from (1) and (2) and due to the higher heating load resulting from (1).

Based on the simulation results, the occupant behaviour measures can achieve considerable energy savings as high as 22.9% for individual measures and up to 41.0% for the integrated measures. The main energy savings captured by the occupant behaviour measures come from the avoidance of energy waste in unoccupied rooms especially for their lighting, plug load, and HVAC systems.

It should be noted that if the static occupant schedules in ASHRAE standard 90.1 were used, the behavioural measures savings will be significantly reduced by up to 50%. The occupant schedule makes a significant difference on the energy savings of occupant-based measures. Therefore, when estimating the potential energy savings of occupant-related measures, it is crucial to apply realistic occupant schedules that reflect occupancy variations in each room.

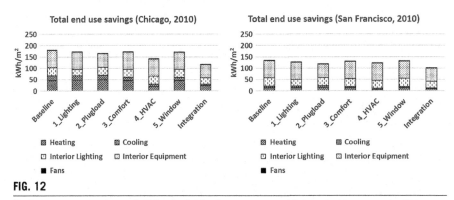

FIG. 12

End-use energy savings of all five individual measures and integrated measure in Chicago and San Francisco for the baseline model of vintage 2010.

Credit: Sun, K., Hong, T., 2017b. A simulation approach to estimate energy savings potential of occupant behavior measures. Energy Build., 136, 43–62.

Although energy savings of behaviour measures would vary depending upon many factors, the presented simulation approach in this case study is robust and can be adopted for other studies aiming to quantify occupant behaviour impact on building performance.

4 Future perspectives on data and computational tools for occupant behaviour modelling

Although significant progress on occupant behaviour modelling and simulation has been made through international collaboration under IEA EBC Annex 53 (Yoshino et al., 2017) and Annex 66, as well as the associated development of an occupant behaviour research community, future work will be challenged to leverage the interdisciplinary nature of occupant behaviour research, communicate its contribution to a building industry that is moving towards zero-net energy or zero-net emissions, and embrace the adoption of supporting technologies for behaviour research like Internet of Things (IoT), big data, machine learning, and exascale computing. A few of the most pressing needs for future occupant behaviour research are enumerated later.

Data: (1) develop low-cost and reliable methods and tools to collect large-scale, high-quality occupant data covering all behaviour types (i.e. presence and movement, adaptive behaviour, and comfort preference), occupant types, building types (commercial and residential), cultures, and climates, through in situ smart sensing and online surveys and/or lab settings that can better mimic field settings through innovative use of virtual reality technologies; (2) employ machine learning algorithms and stochastic modelling techniques to extract knowledge and establish mathematical models of occupant behaviour from the collected data; and (3) develop or adopt data sharing protocols that standardise variable types and response formats and substantively address behavioural data privacy and security concerns.

Modelling and simulation: (1) develop representation of complex occupant effects and interactions at various scales (e.g. group behaviour, social dynamics, interoccupant behaviour diversity, multiple behaviour choice hierarchies, and aggregation to the whole-building or grid-level); (2) establish a global open-source repository of occupant behaviour models using standardised representation schema to enable interoperability between tools, users, and applications; (3) implement a rigorous and transparent process of model creation, evaluation, and verification to ensure model validity and applicability; (4) integrate occupant modelling with building information modelling in the building design and operation workflow; and (5) develop a synthetic population of building occupants with representative behaviour types to support agent-based modelling and simulation.

Application: (1) develop guidelines and tools for fit-for-purpose application of an occupant behaviour model suite in commonly used building energy simulation tools; (2) improve occupant-related assumptions and increase the use of occupant behaviour models in critical decision points of the building life cycle including load calculation, evaluation of energy conservation measures, selection of equipment and system types, code compliance, performance rating, and occupant-responsive model predictive control; (3) quantify key building performance uncertainties related to occupant behaviour and choose design strategies that are robust to these uncertainties in fostering energy-positive human–building interactions; and (4) establish clear communications with customers or stakeholders regarding the influence that the diversity and stochasticity of occupant behaviour has in determining a range of possible outcomes for key building performance metrics.

Interdisciplinary collaboration: occupant behaviour is diverse and stochastic, requiring an interdisciplinary approach to gain a deeper understanding that spans building science, environmental engineering, social and behavioural science, data science, and computer science. Through several IEA EBC annexes (i.e. Annex 53, 66 and 79), social scientists and psychologists contributed to the design of surveys that explore the link between social and contextual factors, such as culture, gender, age, and habits, and occupants' energy use behaviours. This contribution is important and complementary to the engineering approach of using measured variables, such as indoor and outdoor environmental parameters, to formulate mathematical algorithms that yield accurate probabilities of occupants' behavioural actions. Fostering an interdisciplinary approach that engages building designers and operators is also crucial to the practical integration of occupant behaviour insights (e.g. occupant needs and human–building interactions) across the building life cycle (design, construction, operation, and retrofit), supporting the achievement of building performance goals through leveraging both the technological and human dimensions (D'Oca et al., 2018).

To address the aforementioned research needs, it is crucial to foster and sustain an occupant behaviour research community through continuous international collaboration such as the on-going effort of IEA EBC Annex 79 (http://annex79.iea-ebc.org/), IEA Task 24 Phase II: Behaviour Change in DSM (http://www.ieadsm.org/task/task-24-phase-2/), and professional organisations such as ASHRAE Multidisciplinary Task Group on occupant behaviour in buildings (MTG.OBB). Dedicated conferences such as Behaviour, Energy, and Climate Change (BECC; beccconference.org) and BEHAVE European Conference on Behaviour and Energy Efficiency (information on the 2018's edition is available at https://www.zhaw.ch/en/about-us/news/events/behave/) also provide excellent venues for researchers and practitioners to exchange and share knowledge, experience, success stories, and important lessons learned about the growing field of occupant behaviour research.

References

Andersen, R., 2012. The influence of occupants' behaviour on energy consumption investigated in 290 identical dwellings and in 35 apartments. In: Proceedings of Healthy Buildings. Brisbane, Australia.

Andersen, R., et al., 2013. Window opening behaviour modelled from measurements in Danish dwellings. Build. Environ. 69, 101–113.

Andersen, P.D., et al., 2014. Dynamic modeling of presence of occupants using inhomogeneous Markov chains. Energy Build. 69, 213–223.

ANSI/ASHRAE, 2013. Thermal environmental conditions for human occupancy. Standard 55-2013 2013.

Barthelmes, V.M., et al., 2017. Exploration of the Bayesian Network framework for modelling window control behaviour. Build. Environ. 126, 318–330.

Becerik-Gerber, B., et al., 2011. Application areas and data requirements for BIM-enabled facilities management. J. Constr. Eng. Manag. 138 (3), 431–442.

Belafi, Z.D., Hong, T., Reith, A., 2016. A library of building occupant behaviour models represented in a standardised schema. Energy Effic. 1–15.

Bourgeois, D., Reinhart, C., Macdonal, I., 2006. Adding advanced behavioural models in whole building energy simulation: a study on the total energy impact of manual and automated lighting control. Energy Build. 814–823.

Brager, G.S., De Dear, R., 2001. Climate, comfort, & natural ventilation: a new adaptive comfort standard for ASHRAE standard 55.

BTO, U., 2017. EnergyPlus.

Calì, D., et al., 2016. Analysis of occupants' behavior related to the use of windows in German households. Build. Environ. 103, 54–69.

Chen, Y., Hong, T., Luo, X., 2018. An agent-based stochastic occupancy simulator. In: Building Simulation. Springer, pp. 37–49.

Clevenger, C.M., Haymaker, J., 2006. The impact of the building occupant on energy modeling simulations. In: *Joint International Conference on Computing and Decision Making in Civil and Building Engineering, Montreal, Canada*. pp. 1–10. Citeseer.

Clinton, A., et al., 2017. Annex 66: Definition and Simulation of Occupant behavior i buildings. In: Technical Report: Occupant Behavior Case Study Sourcebook.

Cowie, A., et al., 2017. Usefulness of the obFMU module examined through a review of occupant modelling functionality in building performance simulation programs. In: *IBPSA Building Simulation Conference*.

Crawley, D.B., et al., 2001. EnergyPlus: creating a new-generation building energy simulation program. Energy Buildings 33 (4), 319–331.

D'Oca, S., et al., 2015. Annex 66, Definition and simulation of occupant behavior in buildings. In: Technical Report: An International Survey of Occupant Behavior in Workspaces. (November). Available at: http://www.annex66.org/.

D'Oca, S., Hong, T., Langevin, J., 2018. The human dimensions of energy use in buildings: a review. Renew. Sust. Energ. Rev. 81, 731–742.

Dong, B., Lam, K.P., 2014. A real-time model predictive control for building heating and cooling systems based on the occupancy behavior pattern detection and local weather forecasting. In: Building Simulation. Springer, pp. 89–106.

Duarte, C., Van Den Wymelenberg, K., Rieger, C., 2013. Revealing occupancy patterns in an office building through the use of occupancy sensor data. Energy Build. 67, 587–595.

Equa, A.B., 2017. IDA Indoor Climate and Energy.

Fabi, V., et al., 2014. Occupants' behaviour in office building: stochastic models for window opening. In: *8th windsor conference*. Network for Comfort and Energy Use in Buildings.

Feng, X., Yan, D., Hong, T., 2015. Simulation of occupancy in buildings. Energy Build. 87, 348–359.

Földváry Ličina, V., et al., 2018. Development of the ASHRAE Global Thermal Comfort Database II. Build. Environ..

Gaetani, I., Hoes, P.-J., Hensen, J.L.M., 2016. Occupant behavior in building energy simulation: towards a fit-for-purpose modeling strategy. Energy Build. 121, 188–204.

Godin, B., 2017. Models of Innovation: The History of An Idea. MIT Press.

Gunay, H.B., O'Brien, W., Beausoleil-Morrison, I., 2013. A critical review of observation studies, modeling, and simulation of adaptive occupant behaviors in offices. Build. Environ. 70, 31–47.

Haldi, F., 2013. In: A probabilistic model to predict building occupants' diversity towards their interactions with the building envelope. *Proceedings of the international IBPSA conference*, Chambery, France. pp. 147.

Haldi, F., Robinson, D., 2008. On the behaviour and adaptation of office occupants. Build. Environ. 43 (12), 2163–2177.

Haldi, F., Robinson, D., 2009. A comprehensive stochastic model of blind usage: theory and validation. In: *Proceedings of the Eleventh International IBPSA Conference*. pp. 529–536.

Haldi, F., Robinson, D., 2010. Adaptive actions on shading devices in response to local visual stimuli. J. Build. Perform. Simul. 3 (2), 135–153.

Haldi, F., Robinson, D., 2011. The impact of occupants' behaviour on building energy demand. J. Build. Perform. Simul..

Hand, J.W., 2015. Strategies for Deploying Virtual Representations of the Built Environment (aka The ESP-r Cookbook). Energy Systems Research Unit, Department of Mechanical and Aerospace Engineering. University of Strathclyde, Glasgow, UK.

Hendricken, L., et al., 2012. Capital Costs and Energy Savings Achieved by Energy Conservation Measures for Office Buildings in the Greater Philadelphia Region. In: SimBuild 2012. International Building Performance Simulation Association, Madison, WI. Available at: https://www.researchgate.net/publication/266553922.

Hendrikse, M.M.E., et al., 2018. Database of movement behavior and EEG in virtual audiovisual everyday-life environments for hearing aid research. Available at: https://doi.org/10.5281/zenodo.1470805#.W_3MOboHRG0.mendeley. (Accessed November 27, 2018).

Heo, Y., Augenbroe, G., Choudhary, R., 2011. Risk analysis of energy-efficiency projects based on Bayesian calibration of building energy models. Build. Simul. 2579–2586.

Hoes, P., et al., 2009. User behavior in whole building simulation. Energy Build. 41 (3), 295–302.

Hong, T., Lin, H.-W., 2013. Occupant Behavior: Impact on Energy Use of Private Offices. Berkeley, CA. Available at: https://www.researchgate.net/publication/258246691.

Hong, T., Chang, W.-K., Lin, H.-W., 2013. A fresh look at weather impact on peak electricity demand and energy use of buildings using 30-year actual weather data. Appl. Energy 111, 333–350.

Hong, T., D'Oca, S., Taylor-Lange, S.C., et al., 2015a. An ontology to represent energy-related occupant behavior in buildings. Part II: Implementation of the DNAS framework using an XML schema. Build. Environ. 94, 196–205.

Hong, T., D'Oca, S., Turner, W.J.N., et al., 2015b. An ontology to represent energy-related occupant behavior in buildings. Part I: Introduction to the DNAs framework. Build. Environ. 92, 764–777.

Hong, T., et al., 2016. An occupant behavior modeling tool for co-simulation. Energy Build. 117, 272–281.

Hong, T., et al., 2017. Occupant behavior models: a critical review of implementation and representation approaches in building performance simulation programs. In: Building Simulation. Springer, pp. 1–14.

Hoyt, T., Arens, E., Zhang, H., 2014. Extending air temperature setpoints: Simulated energy savings and design considerations for new and retrofit buildings. Build. Environ. 88, 89–96.

Jensen, K.L., Toftum, J., Friis-Hansen, P., 2009. A Bayesian Network approach to the evaluation of building design and its consequences for employee performance and operational costs. Build. Environ. 44 (3), 456–462.

Konis, K., 2013. Evaluating daylighting effectiveness and occupant visual comfort in a side-lit open-plan office building in San Francisco, California. Build. Environ. 59, 662–677.

Langevin, J., Wen, J., Gurian, P.L., 2013. Modeling thermal comfort holistically: Bayesian estimation of thermal sensation, acceptability, and preference distributions for office building occupants. Build. Environ. 69, 206–226.

Langevin, J., Gurian, P.L., Wen, J., 2015a. Tracking the human-building interaction: a longitudinal field study of occupant behavior in air-conditioned offices. J. Environ. Psychol. 42, 94–115.

Langevin, J., Wen, J., Gurian, P.L., 2015b. Simulating the human-building interaction: development and validation of an agent-based model of office occupant behaviors. Build. Environ. 88, 27–45.

Langevin, J., Wen, J., Gurian, P.L., 2016. Quantifying the human-building interaction: considering the active, adaptive occupant in building performance simulation. Energy Build. 117, 372–386.

Liao, C., Lin, Y., Barooah, P., 2012. Agent-based and graphical modelling of building occupancy. J. Build. Perform. Simul. 5 (1), 5–25.

Liisberg, J., et al., 2016. Hidden Markov Models for indirect classification of occupant behaviour. Sustain. Cities Soc. 27, 83–98.

Luo, X., et al., 2017. Performance evaluation of an agent-based occupancy simulation model. Build. Environ. 115, 42–53.

Macal, C.M., North, M.J., 2010. Tutorial on agent-based modelling and simulation. J. Simul. 4, 151–162.

Makonin, S., et al., 2016. Electricity, water, and natural gas consumption of a residential house in Canada from 2012 to 2014. Sci. data 3, 160037.

McCambridge, J., Kypri, K., Elbourne, D., 2014. Research participation effects: a skeleton in the methodological cupboard. J. Clin. Epidemiol. 67 (8), 845–849.

Newsham, G.R., Mahdavi, A., Beausoleil-Morrison, I., 1995. Lightswitch: a stochastic model for predicting office lighting energy consumption. In: *Proceedings of right light three, the third european conference on energy efficient lighting, Newcastle-upon-Tyne.* pp. 60–66.

O'Brien, W., Gunay, H.B., 2016. Implementation of the occupant behaviour and presence models in OpenStudio. . Final Report Submitted to Natural Resources Canada. Carleton University. Google Scholar.

O'Brien, W., et al., 2017a. A preliminary study of representing the inter-occupant diversity in occupant modelling. J. Build. Perform. Simul. 10 (5–6), 509–526.

O'brien, W., Gaetani, I., Gilani, S., Carlucci, S., Hoes, P.-J., Hensen, J., 2017b. International survey on current occupant modelling approaches in building performance simulation. J. Build. Perform. Simul. 10, 653–671.

Page, J., et al., 2008. A generalised stochastic model for the simulation of occupant presence. Energy Build. 40 (2), 83–98.

Pan, S., et al., 2017. Cluster analysis for occupant-behavior based electricity load patterns in buildings: a case study in Shanghai residences. In: Building Simulation. Springer, pp. 889–898.

Reinhart, C.F., Voss, K., 2003. Monitoring manual control of electric lighting and blinds. Light. Res. Technol. 35 (3), 243–258.

Ren, X., Yan, D., Wang, C., 2014. Air-conditioning usage conditional probability model for residential buildings. Build. Environ. 81, 172–182. Available at: https://doi.org/10.1016/j.buildenv.2014.06.022.

Ren, X., Yan, D., Hong, T., 2015. Data mining of space heating system performance in affordable housing. Build. Environ. 89, 1–13.

Richardson, I., Thomson, M., Infield, D., 2008. A high-resolution domestic building occupancy model for energy demand simulations. Energy Build. 40 (8), 1560–1566.

Saeidi, S., et al., 2017. Application of Immersive Virtual Environment (IVE) in Occupant Energy-Use Behavior Studies Using Physiological Responses. In: Computing in Civil Engineering 2017.

Schweiker, M., et al., 2012. Verification of stochastic models of window opening behaviour for residential buildings. J. Build. Perform. Simul. 5 (1), 55–74.

Sun, K., Hong, T., 2017a. A framework for quantifying the impact of occupant behavior on energy savings of energy conservation measures. Energy Build. 146, 383–396. Available at: https://doi.org/10.1016/j.enbuild.2017.04.065.

TRNSYS, 2017. TRNSYS Transient Sytem Simulation Tool. Available at: http://www.trnsys.com. (Accessed August 5, 2017).

Wang, C., Yan, D., Jiang, Y., 2011. A novel approach for building occupancy simulation. In: Building Simulation. Springer, pp. 149–167.

Wang, C., Yan, D., Ren, X., 2015. Modeling individual's light switching behavior to understand lighting energy use of office building. In: *CUE National Conference*, pp. 1–14.

Wang, Z., Hong, T., Jia, R., 2018. Buildings. Occupants: a Modelica package for modelling occupant behaviour in buildings. J. Build. Perform. Simul. 1–12.

Wetter, M., 2011. Co-simulation of building energy and control systems with the Building Controls Virtual Test Bed. J. Buildi. Perform. Simul. 4 (3), 185–203.

Wetter, M., Zuo, W., Nouidui, T.S., Pang, X.P., 2014. Modelica buildings library. J. Build. Perform. Simul. 7 (4), 253–270.

Widén, J., Nilsson, A.M., Wäckelgård, E., 2009. A combined Markov-chain and bottom-up approach to modelling domestic lighting demand. Energy Build. 41 (10), 1001–1012.

Wilke, U., et al., 2013. A bottom-up stochastic model to predict building occupants' time-dependent activities. Build. Environ. 60, 254–264.

Yan, D., et al., 2008. DeST—An integrated building simulation toolkit Part I: Fundamentals. In: Building Simulation. Springer, pp. 95–110.

Yan, D., et al., 2015. Occupant behavior modeling for building performance simulation: Current state and future challenges. Energy Build. 107, 264–278.

Yan, D., et al., 2017. IEA EBC Annex 66: Definition and simulation of occupant behavior in buildings. Energy Build. 156 (April 2018), 258–270.

Yoshino, H., Hong, T., Nord, N., 2017. IEA EBC annex 53: Total energy use in buildings—Analysis and evaluation methods. Energy Build. 152 (March 2013), 124–136. Available at: https://doi.org/10.1016/j.enbuild.2017.07.038.

Zeigler, B.P., Oren, T.I., 1979. Theory of modelling and simulation. IEEE Trans. Syst. Man Cybernet. 9 (1), 69.

Zucchini, W., MacDonald, I.L., Langrock, R., 2016. Hidden Markov Models for Time Series: An Introduction Using R. Chapman and Hall/CRC.

Further reading

Behavior, O., 2017. Technical Report: Surveys to understand current needs, practice and capabilities of occupant modeling in building simulation.

Sun, K., Hong, T., 2017b. A simulation approach to estimate energy savings potential of occupant behavior measures. Energy Build. 136, 43–62.

Yan, D., Hong, T., 2018. Fina Report of IEA EBC Annex 66: Definition and simulation of occupant behavior in buildings. May 2018.

Agent-based modelling of the social dynamics of energy end use

3.3

Émile J.L. Chappin[a], Igor Nikolic[a], Neil Yorke-Smith[b,c]

[a]Faculty of TPM, Delft University of Technology, Delft, The Netherlands
[b]Faculty of EEMCS, Delft University of Technology, Delft, The Netherlands
[c]Olayan School of Business, American University of Beirut, Beirut, Lebanon

1 Introduction

Socio-technical complexity is conspicuous in the domain of energy. From day-ahead pricing markets to government policy to end-user behaviour, the energy domain is a manifest of intertwined technical challenges and social factors. This chapter considers agent-based modelling and simulation as a tool for understanding behaviours and for advising policy in the energy domain.

As recognised in many chapters of this book, the 21st century sees energy transition as the pressing driver and consequence of energy technology, policy, and systems in much of the world. The grand challenge around today's energy infrastructures is to reach the energy transition in a timely fashion, while meeting other requirements such as security of supply and affordability (Armaroli and Balzani, 2007).

Energy transition is an example of a multistakeholder *wicked problem* (Rittel and Webber, 1974), characterised by the involvement of a variety of stakeholders and decision makers with conflicting values and interests and diverging ideas for solutions. Even the problem formulation itself is contested and, since cause and effect are intertwined with each other, a clear analytical separation between the problem formulation and the solution is not possible.

The consequences of any decision on wicked problems can be profound, difficult if not impossible to reverse, and may result in lock-ins for future decision making. Consider, for example, the decision to construct or decommission a nuclear power plant. Kwakkel et al. (2016) argue that planning and decision making in wicked problem situations should, therefore, be understood as an argumentative process, where three technical constructors—the problem formulation, a shared understanding of system functioning and how this gives rise to the problem, and the set of promising solutions—emerge gradually through debate among the involved decision makers and stakeholders.

Such a socio-technical systems (STSs) perspective offers a conceptual basis for dealing with wicked problems, grounded in systems science, engineering, social sciences, and complex adaptive systems thinking. Adopting an STS perspective implies that we understand and model parts of the man-made world as systems composed of two deeply interconnected subsystems: a social network of actors and a physical network of technical artefacts (Van Dam et al., 2013).

Together, these intertwined systems form a complex adaptive system: a multiactor network determines the development, operation, and management of the technical network, which in turn affects the behaviour of the actors buy setting physical boundaries and shaping the dynamics of change. The interactions within and between technical systems are defined by causal relationships which are governed by laws of nature, while the actors in the social system develop intentional relationships, based on habits, norms, formal and informal cultural arrangements, and individual psychological mechanisms and biases, to accomplish their individual goals. At multiple hierarchical levels, the technical network is shaped by the social network and vice versa, with feedback loops running across multiple levels and time scales. These elements and interactions form a self-organising, hierarchical, open system with a multiactor, multilevel, and multiobjective character (Kay, 2002). Fig. 1 presents this in a schematic representation (Chappin and van der Lei, 2014).

Granted this perspective, agent-based modelling (ABM) is a paradigm for simulating the actions and interactions of autonomous heterogeneous agents, which do not need to be perfectly rational or perfectly informed, in order to study the emergent system-level effects of collective agents' behaviour within a certain environment, overtime. As we will discuss in the following section, the agents can represent actors

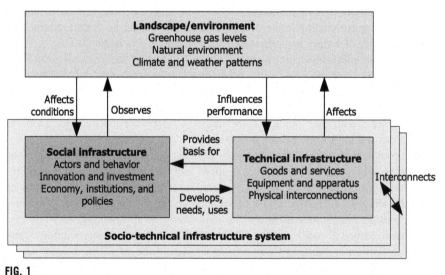

FIG. 1

Socio-technical systems, embedded in a broader ecological system.

From Chappin, E.J.L., van der Lei, T., 2014. Adaptation of interconnected infrastructures to climate change:
a socio-technical systems perspective. Util. Policy 31, 10–17.

at different levels of abstraction and hierarchy: for example, individual end users or collective entities such as households or businesses.

The remainder of the chapter is structured as follows. Section 2 introduces more fully the methodology of ABM, placing its methodological approach in the larger STS context. We summarise the steps at a high level in building an ABM, and the challenges and drawbacks of the methodology. Section 3 focuses on the energy domain, and in particular end-user energy use. Section 4 makes the discussion specific by examining three case studies. As a prelude, we point out important considerations when reading a case study of ABM. Section 5 summarises the chapter in the context of the book and points our directions for future research and application of ABM in the energy domain.

2 **Agent-based models**

Socio-technical systems are defined by simultaneously possessing interrelated social and technical aspects. These interrelated aspects manifest the interaction between human behaviours and relationships, and societal infrastructure and organisational processes. STSs are thus characterised by a set of factors that readily lead to complexity:

- many social and technical components (Hughes, 1987);
- parallel, distributed self-organisation with reflective downward causation (Holland, 1996, 1997; Kauffman, 1993; Kroes, 2009);
- evolution overtime (Dennet, 1996);
- requiring multiple formalisms to understand fully the system (Mikulecky, 2001); and
- loaded with values, emotions, and norms (Funtowicz and Ravetz, 1993; Roeser, 2012; Van den Hoven et al., 2012).

Given the complex nature of STSs, if we are to understand—and further, manage or influence—an STS, then we must be aware of Ashby's Law of Requisite Variety (Ashby, 1991): "A model system or controller can only model or control something to the extent that it has sufficient internal variety to represent it" (Heylighen and Joslyn, 2001). A key term in Ashby's Law is 'sufficient': the expressiveness needed of the model. Note that this needed sufficiency can be defined by the modeller, that is, we can choose a modelling paradigm and design the model with sufficient expressiveness for that which we are modelling.

Nevertheless, 'good' (and 'bad') models come in all shapes and sizes, and should be well positioned in what they do and do not cover. We return to this topic of model quality in Section 4.

When modelling STSs, one can loosely separate the modelling approaches and associated tools into top-down and bottom-up approaches, and across the dimensions of design, analysis, and exploration, as visualised in Fig. 2.

Sitting at one corner point of the figure, agent-based modelling finds itself as a bottom-up approach typically used for analysis and exploration, which, as we shall

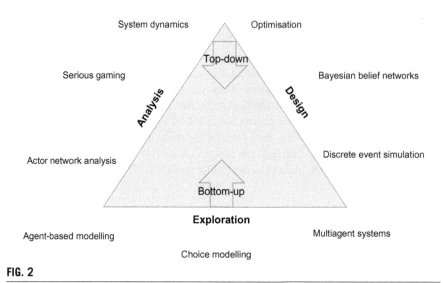

FIG. 2

Dimensions and approaches to modelling STSs.

see, fits its generative nature (Epstein, 1999). At other corners of the figure are multiagent systems (MAS), which we contrast below with ABM in Section 2.1.1; and traditional top-down approaches such as system dynamics and mathematical optimisation. MAS is, like ABM, a bottom-up approach emphasising exploration, but more strongly towards design rather than analysis. Top-down approaches can be used for both analysis and design but are not orientated for exploration, as we discuss in Section 2.1.2.

Anticipating this chapter's discussion of ABM for the energy domain, Table 1 summarises existing approaches for energy scenarios and system models, in studies of energy policy. There is a focus on equilibrium models, in various shapes and forms, which study energy policy questions from the perspective of how they influence the lowest cost/equilibrium solution, either static or under lurching dynamics. The other strand of popular models is system models that study pathways or end states, mostly to find those paths or end states that are assumed to have the lowest system costs. In all these types of models, many of the uncertainties that affect real-world behaviour are ignored or underrepresented. Here, ABM may be complementary, as uncertainties are explicitly studied. Further, elements that do not fit within the economic framing of the modelling are often hard to represent and are typically not in the focus of attention in the modelling practices that have emerged (e.g., see the remainder of the chapter).

2.1 Definition, philosophy, and purposes of agent-based modelling

The modern origins of ABM arise from the work of Conway (1970), Schelling (1971), Axelrod and Hamilton (1981), and Reynolds (1987). Modelling tools became widely available in the 1990s, as the object-oriented paradigm in software development

Table 1 Existing approaches to study energy (and climate) policy.

Element	Computational general equilibrium	Partial equilibrium	Socio-technical energy system models
Time	Equilibrium (static, recursive-dynamic, or dynamic)		Pathway or end state (of least cost)
Scope	Economy as aggregate whole; details for particular sectors	Individual sector in detail	Individual sector with specific technology details
Uncertainty	Some included; in general ignore uncertainties caused by system out of equilibrium, and dynamics from policy uncertainty		

Adapted from Chappin, E.J.L., de Vries, L.J., Richstein, J., Bhaghwat, P., Iychettira, K., Khan, S., 2017a. Simulating climate and energy policy with agent-based modelling: the energy modelling laboratory (EMLab). Environ. Model. Software 96, 421–431.

became more mainstream. Since then, ABM grew as a methodological approach for the social sciences, and evolved into fields such as computational social science and social simulation. Text-books on ABM were written by 2000, and the field had journals and academic societies. For a systematic review and overview, see Bonabeau (2002), Edmonds and Meyer (2013), and Wilensky and Rand (2015).

2.1.1 Different schools of agent-based thinking

Before proceeding further, it is worth recognising that in the broad literature dealing with agent-based perspectives, we can differentiate at least three related, yet distinct schools (besides agent-based methodologies in specific application domains). These schools are:

- AA: Autonomous agents;
- MAS: Multiagent systems; and
- ABM(S): Agent-based modelling (and simulation).

We could understand the AA perspective as asking the *What is?* questions using agents. AA is a part of field of artificial intelligence (AI) field, which itself is a part of computer science. Here, agents are understood as autonomous reasoning and problem-solving entities, and research focuses on representing general reasoning processes in artificial entities. Shoham (1993) provides a background on the AI perspective on agents, and Lee et al. (2014) provide an example of a typical application of agents in AI.

MAS could be understood as asking the *How to?* question. It is related to AA but is closer to engineering in its focus and tools. The goal of MAS is to design distributed control and management systems, that interact, often in real time, with real-world complex systems, such as the electricity grids, traffic control, etc. Čaušević et al. (2017) present a typical example of an MAS applied to managing a decentralised energy production and consumption system.

Finally, ABM(S) could be understood as asking the *What if?* question. This approach is mainly used in engineering and social science fields. Here, agents are understood as relevant and descriptive representations of real-world entities, whose interactions are studied under a range of different conditions to answer questions about consequences of interventions. In this chapter, we focus on this perspective.

2.1.2 Philosophical orientation of ABMs

The mantra of ABM is well expressed by Epstein (1999): "If you did not grow it, you did not explain it!" This expression aligns with questions of interest in social science, where explanation is as important as the 'answer' (Wallach, 2018). Computationally, agent-based models (ABMs) are pattern generators based on state machines, computational devices that systematically simulate interactions of entities overtime, in some environment. The interactions that lead to a new state of an agent and the system are based on the previous agent and system states. Time progresses in discrete steps, and everything within a time step is assumed to happen at once, in parallel.

Unlike traditional modelling techniques which focus on describing the essence of the system, entity, or pattern being modelled, ABMs focus on the essence of the *process* that gives rise to that system, entity, or pattern. Hence, ABMs 'build understanding from the bottom up' we could say.

Western scientific tradition is overwhelmingly based on Aristotle's so-called substance philosophy. Agent-based modelling belong to what can be called *generative science* (Epstein, 1999), which is based on Heraclitus of Ephesus (~560 B.C.E.) and his famous quote 'Panta Rei'—everything flows, the only constant is change. In contrast to substance philosophy, process philosophy thus:

> *...has full systematic scope: its concern is with the dynamic sense of being as becoming or occurrence, the conditions of spatio-temporal existence, the kinds of dynamic entities, the relationship between mind and world, and the realization of values in action.*

> **(Seibt, 2017)**

Epstein (1999) goes on to define the central principle of generative science as "phenomena can be described in terms of interconnected networks of (relatively) simple units. Deterministic and finite rules and parameters of natural phenomena interact with each other to generate complex behaviour."

Will Wright, the creator of Sim City (Wright, 1996), which was arguably the first large-scale ABM that entered popular culture, says the following about generative science (O'Reilly, 2006):

> *Science is all about compressing reality to minimal rule sets, but generative creation goes the opposite direction. You look for a combination of the fewest rules that can generate a whole complex world which will always surprise you, yet within a framework that stays recognizable. It's not engineering and design, so much as it is gardening. You plant seeds.*

2.1.3 ABM as a modelling approach

With the earlier understanding of the generative, process-focused philosophy behind ABM, Shalizi (2006) defines an agent and agent-based models as:

> *An agent is a persistent thing which has some state we find worth representing, and which interacts with other agents, mutually modifying each others states. The components of an agent-based model are a collection of agents and their states, the rules governing the interactions of the agents and the environment within which they live.*

Fig. 3 depicts the agents interacting within the environment as an agent-based model.

ABM thus simulates system behaviour as the emergent result of the (inter)actions of situated heterogeneous individuals overtime. This simulated system behaviour makes the ABM methodology very suitable for the analysis of complex adaptive systems such as we find in the energy domain. An agent, its states, its actions, and the environment it 'lives' in are the formalisation of the modeller's simplified, formalised, and (software) codified representation of how an individual perceives the world, thinks, works, and interacts with his or her environment (Fig. 3). It is, therefore, often said that *The agent is the theory*.[a] If the agent is a basis of the generative

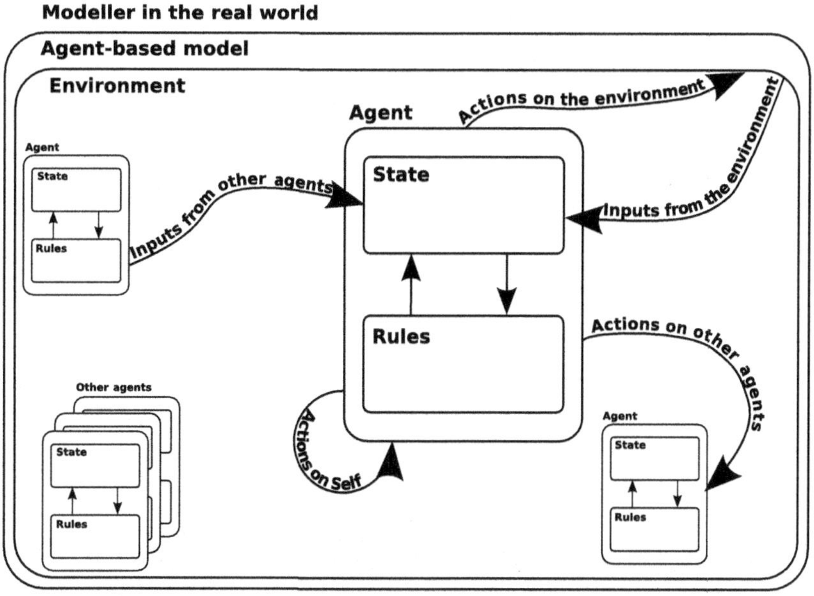

FIG. 3

Agents interact in the environment, in an agent-based model, designed by a modeller.

From Van Dam, K.H., Nikolic, I., Lukszo, Z. (Eds.), 2013. Agent-Based Modelling of Socio-Technical Systems. Springer, Dordrecht.

---

[a] Saying often attributed to Prof. Nigel Gilbert.

theory, simulated overtime to explore a system's emergent properties, what then are the specific kinds of questions, or purposes, that this tool can be used for? We address purposes of ABM next.

2.1.4 Purposes of ABM

Based on the work of Edmonds (2017) and our experience in the energy domain, we can identify at least six purposes relevant to this chapter.

Theory identification 'Agent is the theory' can more specifically be defined as 'Agent is the hypothesis'. By formalising a particular hypothesis about what an agent is, and how she reasons we can test our generative hypothesis by growing emergent patterns and comparing them with observed reality. A typical example of this type of use is presented by Schelling (1971).

Theoretical exposition ABMs can be thought of as 'a theory with a run button'. If we have a specific theory of the individual in mind, ABMs are excellent tools to present that theory and explore its consequences. As agency is the primary concepts behind ABM, 'things that want to do things to other things', theory exposition done through ABMs can provide intuitive and visual understanding. A typical example is presented by Janssen and Jager (2001).

Explanation When explanation is the main purpose of an ABM, the goal of modelling is to provide a consistent, realistic description of what is going on in a particular situation, regardless whether a description is based on the theoretical concepts, observed heuristics, or empirical data. We are looking for a set of realistic, plausible mechanisms that provide a generative account of the phenomenon. A typical example is presented by Knoeri et al. (2014).

Prediction In situations where the agents' behavioural rules, interactions, and environment are precisely known, ABMs can be used to predict future system states. The most common examples of this type of application are in traffic modelling, pedestrian flow simulations, and evacuation behaviour. A typical example is presented by Hedo and Martínez-Val (2010).

Exploration and discovery As depicted in Fig. 2, ABMs are ultimately suitable for explorative purposes. Asking the 'what if' question amounts to running a model repeatedly under (a large number of) varying initial conditions, parametrisations and external scenarios, and mapping out the space of outcomes. We observe changes in outcomes across the parameter space, and from these changes infer insights about the effects of interventions upon it. A typical example is presented by Richstein et al. (2015).

2.2 Methodology and caveats of agent-based modelling

In this section, we outline a suggested methodology of how to build an agent-based model, and we point out caveats that are important when designing ABMs for domains such as energy and climate.

2.2.1 ABM development process

In the following section, we present an extended model development process, based on the method presented by Van Dam et al. (2013), consisting of 11 steps. The method follows a usual software modelling cycle, adapted to the needs of ABM and the participatory stakeholder involvement. The steps are presented as higher-level questions that need to be addressed during each step, rather than a procedural description of these steps. Note the nuance in vocabulary: actors are the real-world entities involved with the system we are studying, and agents are model representations of actors.

Step 1: Problem formulation and actor identification

The main question here is: *What is the problem?* It leads to clarifying questions, such as *What is the exact lack of insight that we are addressing?*, *What is the observed emergent pattern of interest to us?*, *Is there a desired emergent pattern, and if so, how is it different from the observed emergent pattern?*, and *What is the initial hypothesis on how the emergent patterns emerge, or, why do the observed and desired emergent patterns differ?* Once the problem to be addressed is clarified, the social dimension of the problem field needs to be identified. Questions we, as modellers, can ask are: *Whose problem are we addressing?*, *Which other actors are involved in this problem?*, and finally, *What is our role?*

Step 2: System identification and decomposition

Once the lack of insight has been formulated, we can start to systematically define the system boundaries and the relevant level of abstraction and detail within those boundaries. The questions relevant here are: *Who is there, that affects the emergent pattern?* This questions identifies the relevant agents, based on relevant real-world actors. *What do they know or have?* identifies the relevant states and properties of the agents. *What is here, and is being acted upon by the relevant actors in the system?* This identifies the model objects, the technical components of the system. *What do these actors and objects do?* identifies the actions that agents take upon themselves, others, environment and the objects. The question of *How are these actions performed?* will be asked in step 4. The question *What is the nature of these actions?* identifies the types of flows and exchanges between agents, being money, information, influence, etc. Finally, the two question defining the agents' environment need to be asked. *What are the relevant things/properties of the world that all actors know about or have access to, but does not belong to any of them?* defines the environment. *What are all the relevant things that affect the actors and objects, but are considered too large and too far away to be affected by them, but do affect them?* identifies the system boundaries and determines the exogenous model variables and dynamics.

Step 3: Concept formalisation

In this step, the qualitatively defined agents, objects, their states and properties, actions, the environment, and the external world are formalised in a computational form. The are two main question to be asked here. First, *Is the formalisation commensurable to the involved stakeholders?* If the stakeholder cannot recognise or relate to the formalised concepts, model loses its authoritativeness from the get go. The second question is *Will this formalisation be practical for the kinds of actions and behaviours that we need to describe?* As ABMs are at one level 'just algorithms', choices for data structures have a large impact on the algorithm (and thus the model narrative) used.

Step 4: Model formalisation

At this step, the question asked is, if we will: *The agent wakes up in the morning, has a cup of coffee and does...?* This step explicitly defines the detailed narrative of each agents 'theory', thus answering the question *Who does what, with whom, when, and how?* All agents, objects, their states, environmental components, and the external world must be included in this narrative. This step is the final model design step that can be communicated in practice with stakeholders, before the modeller descends into software development details. It is therefore essential that this narrative is fully understood, internalised and accepted by the stakeholder, because all interpretation of model outcomes and the use of the model will be based on their mental model of this narrative.

Step 5: Software implementation

In this step, no specific question is asked, rather, the narrative is converted into a software artefact. As example of this process for an ABM in energy domain is Eppstein et al. (2011). There are a number of popular software tools used to develop agents-based models, such as NetLogo (Wilensky, 1999), RePast (North et al., 2013), GAMA (Taillandier et al., 2010), MASON (Luke et al., 2004), and SWARM (Terna et al., 1998). The tools differ in their topic focus, intended user group, programming language, user licence, etc. Elaborating on these differences falls outside the scope of this chapter. An up-to-date overview can be found at the Wikipedia page: https://en.wikipedia.org/wiki/Comparison_of_agent-based_modeling_software.

Step 6: Model verification

The main guiding question at this step is *Have we built the thing right?* Like software development meant for products, where the specific implementation details are not important to the end user (as long as software performs the required functions), we do specifically care about how the agent performs its decision-making processes and actions/interactions. In this step, we are building an evidence file on the correctness of the artefact's construction, so that we can have confidence in its outputs when we start using it.

Step 7: Model parametrisation

Now that we have confidence on the suitability of the model artefact for its purpose of generating emergent patterns from theories of individuals and their actions, the question that we face is *What are the right inputs?* There are two questions hidden behind the notion of inputs. First, *What is the state of the world, agents, the properties, and objects that is relevant to the problem?* Based on and limited by the formalisation choices, we need to translate real-world properties into parameter values for the model. The second question is *What else could the world be, and what would we like to do to it?* This identifies the parameter space within which alternative but relevant situations can be represented, and the scenario space, a set of possible states and dynamics of exogenous parameters, under which we wish to explore the model behaviour.

Step 8: Experimentation

Guiding question for the experimentation is *How do I run this thing?* How do we translate the lack of insight defined in step 1 into computational experiments that can generate emergent patterns that can provide that insight. A wealth of detailed choices have to be made, from the how to sample and explore the parameter and scenario space, how to get significant outcomes, given the inherent stochasticity of ABMs, how long to run the simulations in order to get the emergent pattern of interest and do we have the computational resources to perform these simulations. If not, how can we reduce the number of experiments while maintaining their usefulness.

Step 9: Data analysis

What happened? is the main question driving the analysis of the model outputs. It consists of two subquestions. First is *What is interesting?* and the second is *How to show it?*, and both are surprisingly nontrivial. What is a relevant emergent pattern arising from the generative process of agent interactions? Is the absence of a pattern that is interesting? When does it appear/disappear? Which pars of the parameter and scenario spaces lead to these patterns? Is the actual patter/the absence of pattern interesting, or is the difference between two parts of the parameter/scenario space interesting? The second question on visualisation is about identifying the most suitable way to represent an often highly dimensional datasets to a stakeholder in a way that it usefully relates to the lack of insight and their understanding of the model narrative. Data visualisation is a specialised field, and discussing the various options, considerations, and tools falls outside the scope of this work.

Step 10: Model validation

The guiding question behind validation is *Did we build the right thing?* Model validation is a process that is very well understood in the natural science. A model, based on the representation of laws of nature is constructed, it provides a prediction, which is then checked by empirical observation. In the case of socio-technical models, and especially in participatory setting where questions are asked about possible futures states, classical validation is not possible. This is exacerbated by the fact that, in cases where the ABM contains a description of the stakeholder, any model output may change the stakeholder and this renders the model invalid. Several alternative ways of validation are available and will be discussed in Section 2.2.3.

Step 11: Model use

The driving question for the final step is *What can I do with this?* and *How useful is it?* The model outcomes may corroborate or conflict the stakeholders intuitive understanding of the problem at hand. By comparing model outcomes with their own mental models and reasoning about the reality based on these differences, insight about the problem is gained. Whether this is useful, is completely observer dependent and subjective to the stakeholder. It is often reported by the stakeholders that the modelling process was highly insightful, sometimes more that the model outcomes themselves.[b] The model can in that case be considered to be useful, even when incomplete or coarse, as it has provided insights that would not be possible otherwise. It is often said that the usefulness of the model is measured by the speed with it is replaced: a model that is quickly abandoned and replaced with a new one is useful as it has provided so much insight and understanding that it makes its own limitations obvious, and drives the desire for a more complete insight.

2.2.2 Problem-driven strategy to agent-based models

A key strategy for developing an ABM—or any other model for that matter—is to put primary focus into formulating the problem that is addressed, and for that problem, argue how modelling can actually contribute insights towards a solution for that problem. This strategy is particularly relevant for the energy domain. Focusing on the problem and insights from its modelling is particularly relevant for ABM, because of the large number of possibilities in model purposes, in conceptualisation, in implementation, and in experimentation. Complementary, being explicit about the modelling purpose helps to identify the way in which modelling helps in solving the problem. For example, does the model help clarify the arguments, perform a complicated calculation, find problematic conditions, etc. Altogether, the modelling process may be as relevant as the model outcome. In the end, the reasoning of findings that stem from a modelling study should be robust enough to stand alone, irrespective of the model they lead to.

It is also important to acknowledge that good modelling does not often come from a linear approach: following one step at a time. Indeed modelling can help in the creative process by exploring possible ideas, explicating an intuition, as well as in the justification and underpinning of a particular reasoning. The only risk is mixing up these perspectives.

2.2.3 Challenges to agent-based models

As any other modelling approach, ABMs have a number of challenges associate with their creation and use. We will not discuss generic challenges faced by all simulation models, but will solely focus on those specific to agent-based modelling. These challenges stem from:

- the types of data required;
- the algorithmic nature; and
- the types of use

that are typical for ABMs in the energy domain.

Types of data required

For creating ABMs, two key types of inputs are needed. The first is theories or heuristics describing agents, and their behaviours; the second is facts parameterising those agents their behaviours and the environment.

---

[b] This is the personal experience of the authors when working on ABMs for various topics in the energy domain.

ABMs of STSs obviously must contain knowledge and facts from social sciences. A given in social science is the 'incoherency problem': the lack of a single fundamental understanding of what motivates human beings to cooperate and what their capacities are (Watts, 2017, p. 1). Natural sciences have a basic unifying ontology, the laws of nature, that are universally applicable and transferable across all natural science domains. Therefore, there can never be a universally accepted and objective choice for which theory of the individual must be used. Do we approach it from the sociological or psychological perspective? Or maybe an economic base for agent description is more suitable? What would the outcomes have been if we had chosen a completely different one over other? This leads to explicit structural uncertainty in ABMs, which may be difficult to systematically explore, both due to effort required to conceptualise multiple agents, thus creating effectively different models, and can be very computationally expensive. There is no closed-form solution here other than being very explicit in the choices made and transparent and systematic in describing the reasons behind those choices.

What about the facts describing the agents? As data are needed at a very disaggregate individual level, it is often either not available, or sufficiently sensitive that it may not be used due to individual privacy or industrial trade secrets or intellectual property issues. Some ways to deal with this is to sample from stochastic distributions describing the individuals data, in case where a large population is involved. If a single, or very few individuals are modelled, where it is clear who that party is, even when anonymised, once can resort to asking the stakeholder to provide data which have been modified with unknown random noise. For example, this situation occurs when modelling production facilities, such as oil refinery, where the specific operational parameters and product composition are highly sensitive, but the general information is not, as all refineries are very similar to each other.

Finally, we can create fact free, but knowledge-rich models. Here, we either randomly assign values to the parameters, or use a best guess method for describing 'reality'. This shifts the burden from data gathering to large uncertainty exploration. In some use cases, this is sufficient for the purpose, such as when we are doing theory identification.

Algorithmic nature

ABMs attempt to describe the parallel, instantaneous actions, and agents' interaction in some environment. We are, therefore, simulating parallel action in a serial processing computer. Even when we have access to heavily multiprocessing machines, there are limits to how much parallelism can be put into the model, even when we ignore the very significant effort in creating scalable parallel processing algorithms. As we simulate time steps, which all agents are assumed to act in parallel, we constantly have to randomise the iteration order of our simulated agents, in order to avoid first mover artefacts. For example, if there is a scarce supply of a good that all agents want to have, if at each time step that same agent gets to buy first, we would severely distort the model outcomes. As agents actions are iterated every step, and their states and actions almost always depend on the previous states and actions, we necessarily have a chaotic system. Chaotic systems are highly sensitive to initial conditions, and a random selection of the iteration order at the first step may be sufficient to produce unstable final outcomes.

This means that model analysis always has to be performed on an ensemble of model runs at identical parameter settings, and outcomes must include a statistical analysis of that ensemble. Luckily, chaotic systems have robust attractors (regardless whether they are stable, unstable, or metastable) and many techniques from dynamical systems mathematics exist to identify and analyse those attractors. This of course involves more effort than a linear model would require and makes model outcomes more difficult to communicate.

Types of use
ABMs, certainly in the energy domain, are usually used for 'what if' scenario exploration types of questions. As is often observed, 'Prediction is difficult, especially if it's about the future'. This truism is especially the case when exploring potential futures of complex, adaptive STSs, whose evolutionary paths are intractable. Further, as already mentioned, STSs only have limited physical science components for which predictable laws of nature hold true. To make matters worse, an agent-based model often contains behavioural models of the involved stakeholder. As the stakeholder uses the model to understand possible futures, her behaviours will change due to this knowledge, however, slightly. In a chaotic, evolving reality this loop—strictly speaking—necessarily invalidates the model, since it changes the agents' state and behaviour.

Further, since many model users are still unfamiliar with notions of emergence, self-organisation, chaos, and notions of generative science, and therefore, tend to expect 'single truths' as model outcomes (Hiteva et al., 2018), ABMs may face issues with model authoritativeness and acceptance of its outcomes.

3 Applying ABM to the energy domain
This section brings together agent-based modelling and the problems in the energy domain. We first identify distinguishing aspects of the domain, discuss, energy-related behaviour, the uptake of energy technology, perceptions, and expectations, interconnectedness, and then consider data, and finally purpose and validation in connection with policy.

3.1 Aspects of the energy domain
In Section 1, we saw the grand challenge around today's energy infrastructures is to reach the energy transition in a timely fashion, while meeting other requirements such as security of supply and affordability (Armaroli and Balzani, 2007). Meeting this challenge requires producing energy carriers from renewable resources, reducing the carbon emissions, and lowering energy demand.

Energy infrastructures provide the service of connecting energy suppliers and users, with the purpose of providing access to affordable, clean energy, with a secure supply. The various energy carriers that each require their own infrastructure; indeed, energy infrastructure is the basis for many of the functions of today's society: health, economy, transport, information technology, etc. Because of its capital-intensive nature, energy infrastructures emerged and evolved over decades and a broad set of actors with different roles, rules—but also markets and policies—has emerged. Any technology, device, fuel, or system is regulated in one way or another, for example, through taxes, tax reliefs, subsidies, concession restrictions, voluntary sector agreements, information instruments, bans, and many of these policies interact across timescales and borders. Altogether we see a complex multi-stakeholder system.

The standard framework for users in liberalised energy markets is a freedom to choose suppliers and a basic right to affordable energy. Nevertheless, climate goals have also translated into policy targets on the user side, in terms of energy efficiency (lowering energy consumption for the same functionality), lowering energy demand, shifting energy demand in order match supply (i.e., demand response), and increasing local energy production [e.g., solar photovoltaic (PV)]. The question is what policies, institutions, markets, and market conditions we need in order to meet the policy objectives. ABM fits well with studying all these user-related processes, since agent-based models can explicitly study under what conditions we may reach our policy targets.

3.2 Studying energy-related behaviour

Many of the relevant aspects of study in the energy domain translate into examining behaviour: simulating end users and their decisions. ABM can explicitly study the consequences of behavioural barriers and policies that may help to overcome them (Hesselink and Chappin, 2019): structural barriers (e.g., split incentives), economic barriers (e.g., lack of capital), behavioural barriers (e.g., bounded rationality), and social behavioural barriers (e.g., lack of trust). Related to but distinct from the study of end-user behaviours, Chapter 2.3 addresses barriers to energy transition in industry.

The barriers identified earlier have been shown to have large effects on why particular energy efficiency objectives were not met; indeed, the barriers are often the reason why particular policies are developed (Hesselink and Chappin, 2019). ABM can, in theory, explore all of these barriers for the energy domain. However, it is often the case that the barriers themselves are not explicitly studied with the models used to develop the policies, whether ABM or other types of models.

Further, the fact that there are so many different barriers, but also end-user groups, appliance types, etc. has led to a large variety of policies, often very detailed. Think for instance of the detailed discussions on energy labels and how they are revised. Because policies interact with each other, much needs to be considered in order to develop specific, coherent, and feasible policy advice.

In the context of liberalised markets, policies are often developed to move end users away from habit-driven behaviour, and towards developing more reasoned actions. A key example is in-home displays and smart meters, which, if supported by policies, inform people on their energy demand with the aim to induce more energy-efficient practices. Other decisions are characterised by the frequency in which they are considered. For instance, heating systems are capital intensive and only replaced infrequently. In particular, many noneconomic factors play a role in those decisions, for instance who's advice is followed and what infrastructure is available there.

Agent's behaviour, formalised as decision rules, may incorporate many of the factors involved with empirical data, in the form of utility functions. Although these utility functions may be heterogeneous and may be placed in a dynamic context of a simulation of interacting agents, the well-known limitations of utility functions still apply (Aleskerov and Monjardet, 2002).

3.3 Exploring the uptake of new energy technology

A purpose for which ABM is well suited is the study of the uptake of new technology, that is, the development of new types of demand and the adoption of new technologies for existing demand (Chappin et al., 2017b). What, for example, may we expect from demand-side management? In order for security of electricity supply, will end users accept giving up some of their autonomy? How can they be persuaded to participate in such a scheme? What infrastructure is needed? These are questions that can be explicitly explored with ABM. It requires an explicit understanding of people behaviour in their homes and the way they decide on their energy contracts, including the terms for financial renumeration (Vasiljevska et al., 2017).

A highly relevant example is how we change the (residential) heating infrastructure in the coming decades, so that sustainable energy sources are used for heating. In the Netherlands, for example, the decision has been made to replace natural gas. But there are many alternatives, and it is a wicked multiactor design problem to explore what technology would work where (heat pumps, district heating, biogas, full electric, etc.), with many fundamental uncertainties, and distributed decision making.

3.4 Understanding perceptions and expectations

Related to behaviour of end users adopting energy technologies, is often the perceptions and expectations of people have about those technologies, which is a key source for policies. The perception that resulted from the performance of energy-saving bulbs from the 1980s and 1990s affected the adoption since and led to a ban on incandescent and halogen bulbs in the European Union (see Chapter 4.3 for discussion). The undeniable factor that perceptions played here has been shown very explicitly with ABM (Chappin and Afman, 2013).

Similarly, expectations play a huge role in decisions in which significant uncertainties may be involved. A prominent example is how uncertainties in future CO_2

prices have been ignored in most modelling studies (Chappin et al., 2017a), where the expectations have affected the value of waiting (delaying the decisions to invest significantly beyond modelled expectations). These processes are explicitly studied with ABM and lead to unique insight in policy side effects and robustness (Chappin et al., 2017a).

3.5 Accounting for interconnectedness

The energy transition drives new interconnections between what were, in the past, separated services. For example, electric mobility interconnects energy, transport, and IT. If electric vehicles (EVs) penetrate the market, they create significant demand for electricity, which, depending on how it will be organised in contracts or regulated, may alter significantly the demand profile for households. Electric mobility is, therefore, a topic that is much broader than the adoption of EVs alone: it also involves design of new roles, new infrastructure, and question about what EV owners will find acceptable. ABM can study hypothetical arrangements of all these interconnections. As a second example, the so-called 'smart' infrastructures accompanied by 'smart' contracts are a means to balance energy demand and supply. These require interconnections with information infrastructures. Again, ABM can study the interconnections.

A second form of interconnectedness is, as we saw earlier, that energy is instrumental for many of today's societal functions. The consequence is that many energy-related decisions are habit driven or are dominated by other aspects than energy itself. This is vital for understanding energy-related behaviour and also for simulating it. It also provides key levers: how can we get end users out of their habit-driven behaviour and into a more reasoned type of action? This suggests the usefulness of in-home displays, smart meters and also of home automation, which essentially bypasses the habit-driven decisions by households. Questions to all these aspects can be simulated and explored with ABM, and the performance policies stimulating particular behaviour may be tested.

A third form of interconnectedness is that households are known to observe and influence each other in their decisions—and this can affect energy behaviour significantly. This influence differs for various appliances and functions: think of the visibility of cars in streets versus light bulbs in homes. For ABM, it is typical to use a constructed social network between agents and agents taking into account properties, opinions of others in their network in their own decisions.

3.6 Inputs (aka 'Data')

The data that goes into ABMs are a combination of data on rules, algorithms, and/or heuristics on how people make the particular decisions which are being modelled, as well as the parameters or properties that go into such decisions.

In order to structure the model of the agents' reasoning and decision-making algorithms, theory may be used. A review on energy efficiency models showed the popularity of the theory of planned behaviour, and of utility theory (Hesselink and Chappin, 2019). The theory employed gives a frame for what kind of data needs to be gathered. The agent-based modelling paradigm enables enormous flexibility in terms of theory and how particular theory can be operationalised. This flexibility has consequences for validation, but also for the data that needs to be gathered. For many elements of modelling the decision making of end users, there is no data and no accepted way of gathering it. For instance, how to get a representative sample of perceptions on brands, technologies, etc.

The data needed for ABM of energy end users include household preferences (which may well be rather specific to location, and particular decisions). In addition, some form of infrastructure data may be needed to be able to describe neighbourhoods and interactions adequately. There are many challenging aspects in collecting the appropriate data: if we think about agent interactions, for example, how do we find out how many people in the network to ask/observe? To what extent does it differ between households?

It is known that small differences may have large effects in complex systems; hence, errors in data could have large consequences. It is a challenge, but also a necessity, to systematically evaluate the sensitivity of a simulation model to possible errors (Van Dam et al., 2013).

Lastly, depending on the modelling scope, national statistics (e.g., census data) may be useful. A key challenge may be to disaggregate data into properties of individual agents. Therefore, such data may be more useful for validation tests than as inputs.

3.7 Purpose and validation

Validation of agent-based models is a particularly challenging task for ABMs of energy behaviour. This is not only because many aspects of 'the system' are relevant, but also because of the forward-looking nature of the questions we have. In the context of energy behaviour, the hypothesis is that if we anticipate the effects of policy decisions with proper use of agent-based modelling, we may be able to improve those policy decisions and influence the efficiency and effectiveness of the energy transition. This implies the need to impact the 'problem owner'.

Considering the role of the problem owner in relation to the purpose of the modelling study, agent-based models have the potential to be developed in participation with, or in close contact with, stakeholders. If energy policy makers, for instance, are involved in the model development process, they may be drawn into the 'way of thinking' with ABM (Section 2.1): how do individuals make their energy-related decisions and how may policy makers influence them efficiently and effectively? This process is less strong if the policy maker is less involved, or not involved at all.

More generally, how insights from models, such as ABM, can aid policy debates around wicked problems is the subject of the literature on the science-policy interface. The importance and challenge of this interface is summarised, for instance, by Watson (2005). Assessing the interface in practice, Brugnach et al. (2007) conclude that "the integration of information derived from models into policy is far away from being trivial or the norm. Part of the difficulties of this integration is rooted in the lack of confidence policy makers have on the incorporation of modelling information into policy formulation." These authors go on to pinpoint as critical factors how uncertainty is modelled and how the model is communicated to the public.

Going further, and writing a decade later, Strachan et al. (2016) find a similar disconnect between models, model outcomes, policy makers, and policy. The authors identify what they call 'short comings' of current practice in modelling in the energy domain: "Policy makers continue to struggle to assess insights from competing models that give alternative findings, or respond when different commentators interpret results to support their arguments."

This relates to the issue of model validity. When validation is defined as fit for purpose, it becomes unfeasible to consider traditional criterion of validity of predictive capability: predictions can never be verified. But in particular, if prediction is not the purpose in the first place, this particular test does not make sense. This important point relates strongly to the difficulties of gathering the right kinds of data, as we discussed earlier.

Although some other modelling approaches (besides ABM) are intended for prediction (recall Fig. 2), the question is whether they succeed at doing so (Chappin, 2018). Energy scenario studies are often supported by models that calculate lowest-cost pathways (Chappin et al., 2017a), which are 'validated' by default because the lowest-cost solution is guaranteed in the modelling method. In addition, models of physical energy systems may be validated in the traditional sense, as they are rooted in laws of physics and there are many known validation tests from electrical engineering. However, the challenge for the energy domain is not only to find lowest-cost solutions (the what?), but what policies may be needed to get there (the how?). In other words, when the scope increases to include social aspects, behaviour, policy, and—in particular with longer time scales—transition, no real validation data are available. Paradoxically, it is that in this wider context, prediction is perceived to be quite important (Chappin, 2018).

Designers of ABMs need to consider the purpose of the model carefully (Section 2.1.4), and 'users' of the modelling work (e.g., modeller, reviewer, policy maker, and colleague) need to be confident that the model fits to this purpose, in terms of (1) what is in the model and what is not for the purpose defined, (2) whether the way in which components are modelled make sense given that purpose, and (3) whether the outcomes can be understood, given those assumptions. If these three points are met and the purpose is useful for the energy domain, then the modelling results can be useful in the discussion around energy policy, understanding energy behaviour, and hopefully, accelerating the energy transition.

4 Case studies

This section looks in detail at three case studies of ABM in the energy domain: energy efficiency in domestic heating; EV adoption; and energy management in smart grids.

First, picking up the discussion with which we concluded the last section, we take the opportunity to reflect on *how* to approach a case study of ABM and the questions to ask about it. We divide these questions into two categories: the model and its use.

Regarding the agent-based model itself: Does the model have

- a clear modelling problem statement?
- a clear definitions of agents, their states, interactions, and the environment?
- a logical and relevant narrative?
- a clear choice for 'KISS' or 'KIDS' methodologies and approach to data (Edmonds and Moss, 2004)?
- sufficient heterogeneity and adequate freedom for agents to act?
- only stochastic decision making by agents?
- a clearly stated list of assumptions?
- any hard-coded choices for parameters only due to expediency?
- behavioural elements whose only justification is that they are easy to model?
- an explicit verification presented (Van Dam et al., 2013, Chapter 3)?

Regarding the use of the model: Does the model have

- a clear rationale for decision on parametrisation?
- a clear rationale why computational experiments are carried out, and under which conditions?
- multiple simulation runs performed?
- descriptive statistics of the outcomes?
- outcomes that are used to reason about reality?
- outcomes that are unwarrantedly used as prediction?
- an explicit statements about limitations?
- an explicit guide to interpretation of the results?

4.1 Energy efficiency

Our first case study reviews 23 agent-based models of energy efficiency (Hesselink and Chappin, 2019). We discuss this paper in this section.

One of the formulated energy goals, for instance in Europe, is to improve energy efficiency. A document from the European Union defines as follows: "Technically, 'energy efficiency' means using less energy inputs while maintaining an equivalent level of economic activity or service; 'energy saving' is a broader concept that also includes consumption reduction through behaviour change or decreased economic activity" (European Commission, 2011). This differs from other key policy objectives (e.g., how the energy input is supplied, i.e., with renewables). It has been shown that energy efficiency targets set at national and international levels are difficult to achieve, despite the fact that there are many energy efficiency policies in place. The potential added value of ABM here is to explicitly simulate the effects of policies through decisions of individuals. In this context, many of the aspects of the energy domain we noted in Section 3 are relevant and studied explicitly.

This case study describes the ongoing efforts in studying energy efficiency with ABM. Although energy efficiency applies both to small and large end users—households and small and large enterprises—Hesselink and Chappin (2019) focus on households.

The energy behaviour dimensions studied include the response of agents to policies, specific to particular types of devices, given various behavioural barriers, as explained in detail here.

4.1.1 Appliances

The following types of appliances have been studied with ABM in relation to energy efficiency: lighting technologies [incandescent, light emitting diode (LED), compact fluorescent lamp (CFL), halogen], residential heating systems [direct electric heating, wood pellet heating stoves, heat pump, aircon, micro-combined heat and power (CHP)], wall insulation, solar PV systems, EVs, CO_2 emission measurement devices, and behaviour changing feedback devices.

4.1.2 Barriers and policies

In the literature, many barriers for energy efficiency have been identified in a variety of studies and many energy efficiency policies have been proposed. Only a rather small set of the barriers have, to date, been explicitly modelled with ABM, and here is a huge opportunity because, in principle, ABM could simulate all of the barriers identified. The same holds for the possible energy efficiency policies. Fig. 4 illustrates which barriers have been studied.

4.1.3 Theory and data

The ABMs surveyed have different roots for their decision logic: most prominent is the theory of planned behaviour, which prescribes that the behaviour of an individual is determined by the intention and perceived behavioural control of this individual. The intention in turn is influenced by the attitude and subjective norm of the individual. Some models are inspired by network or social network theory and the diffusion of innovations.

For energy efficiency, a variety of studies use utility functions to structure the decision-making logic of agents. The decision of households to adopt a technology is modelled by the agent calculating the decision with the maximum expected utility. This approach provides a systematic and potentially valid means of collecting data and allows multiple aspects of the adoption decision to be expressed as utilities. However, it is also limited by what data one can actually obtain, and by other general limitations such as the choice experiments that are the basis for the utility functions used (Hodgson, 2012).

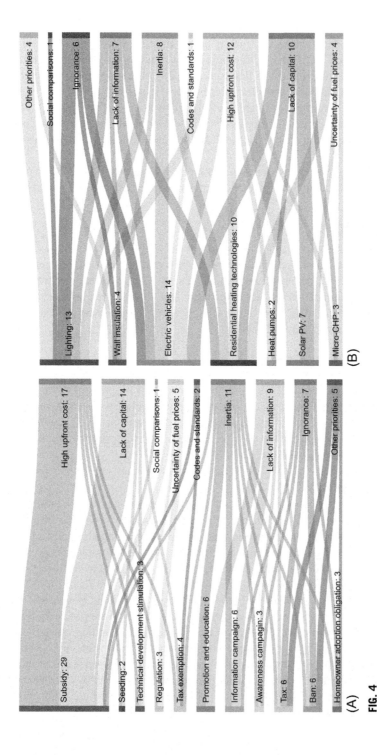

FIG. 4

Connections between policies, technologies, and barriers visualised in Sankey diagrams. (A) Links between policies and technologies. (B) Links between technologies and barriers.

From Hesselink, L., Chappin, E.J.L., 2019. Adoption of energy efficient technologies by households—barriers, policies and agent-based modelling studies. Renew. Sustain. Energy Rev. 99, 29–41.

Empirical data in agent-based models are normally used to provide for specifics on properties of barriers, technologies, households, and policies. It can also be used to structure the decision-making logic of agents: models that do so appear highly specific to a technology, barriers, and a population of households and lead to explicit policy-relevant findings.

4.1.4 Findings

Despite the fact that many types of policies are possible, there still is a focus on economic and financial instruments, such as subsidies or tax exemptions. These focus on barriers that relate to financing: high upfront costs and a lack of capital. There is not a consistent conclusion across the studies surveyed, mostly because policy conclusions differ for appliance types.

Key findings are that subsidies help to stimulate the adoption of EVs and alternative heating technologies, that banning incandescent lamps is the most effective policy to increase the adoption of efficient lighting, that an obligation for new homeowners to insulate their houses effectively helps to increase the adoption of wall insulation. Further, Hesselink and Chappin (2019) find that informational instruments are not as effective as other policies to stimulate EV adoption unless combined with a subsidy scheme.

The existing ABMs are specific to one or a few of the appliance types. This makes sense as the behaviour that end users have in relation to appliances may be very different, in terms of how they are purchased and how they are used. This implies that, for now, the field studies energy efficiency on the basis of cases, one by one, for a set of appliances, a set of barriers, and a set of policies.

ABM could also simulate other actors more explicitly, for example, the role of shops and other intermediary parties in the interconnected STS and a more elaborate institutional setting. This enables one to study more complicated policies, such as tradable 'white certificates' and energy efficiency tenders.

4.2 Consumer adoption of hybrid EVs

Our second case study simulates the market penetration of plug-in hybrid EVs (Eppstein et al., 2011). Eppstein et al.'s (2011) work was one of the first uses of agent-based modelling for studying in detail end-user adoption of such EVs, which combine a petrol engine with an electric engine. The petrol engine is refuelled at traditional petrol stations, while the electric engine is recharged at EV charging stations.

The context of Eppstein et al.'s (2011) study is the US end-user market, which, at the time of the study, showed low awareness of hybrid EVs, and which for many end users required relatively long-distance travel, beyond the range of pure EV technology at the time. Barriers hindering the uptake of EVs include reluctance to adopt unfamiliar technologies, uncertainty about battery life, recharging time and replacement costs, and difficulty in predicting cost of ownership (especially bearing in mind uncertainty about petrol prices). Studies of the US end users indicated that nonfinancial reasons were dominated in the decision to purchase a hybrid EV: for example, concern about CO_2 emissions. Altogether these comprise the energy behaviour dimensions studied by the paper.

The aims of Eppstein et al. (2011) are three:

> *present a framework for a novel agent-based vehicle consumer choice model, illustrate how such a model could be used by policy-makers and vehicle manufacturers to help prioritize investments influencing (hybrid EV) adoption, and identify additional empirical evidence that will be necessary to improve the predictive power of such a model.*

Thus, the case study explicitly identifies the purpose and limitations of the model, and treats the issues of data and policy implications. The model is simplified "due in large part to low model sensitivity to specific details or a lack of empirical data that could justify a more complex model."

Agents model end users, with attributes such as demographics (age, income, residential location), propensity to rationality, openness to EV adoption, years of car ownership, annual distance travelled by car, and attributes of their current car including traditional, hybrid, or pure EV. Each agent has a spatial neighbourhood which defines their geographical locality, and a social neighbourhood which corresponds to their nongeographical locality.

The decision-making procedure for end-user agents is a simple rule-based flow-chart. Each year, the agent decides whether or not to buy a (new) car. If the agent decides to buy a car, it weighs up the relative costs and benefits of the available vehicles, and buys the most 'desirable' vehicle within its range of financial affordability.

The paper pays a lot of attention to calibrating the model based on the US data. Various sources are used. Attention is given to a heuristic parameter for each agent, which models her attitude to the desirability of reducing petrol consumption. This parameter is influenced by media coverage (of EVs) and the agent's susceptibility to media influence, and by decisions of agents in her spatial and social networks regarding EVs and the agent's susceptibility to social influence. The latter networks thus express geographic and socioeconomic homophily as depicted in Fig. 5.

Based on the simulation results, the paper identifies six possible levers relevant for policy around petrol and electrical vehicles, noting the interplay of policies at local, regional/state, and national levels:

- purchase price of EV;
- battery range of EV;
- petrol price relative to electricity price;
- ability of end users to accurately assess fuel costs for vehicle types;
- comfort level of vehicle end users in adopting new (EV) technology; and
- relative weight that end users place on rational-financial versus other reasons to save petrol.

Concluding, Eppstein et al. (2011) suggest that data are lacking on the proportion of end users that are comfortable enough with EV technology to be willing to

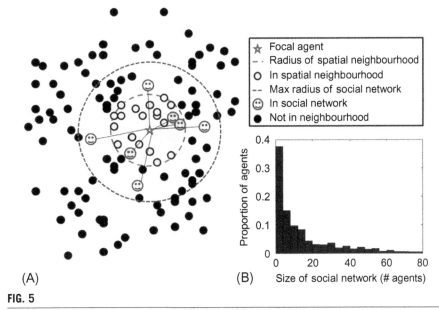

FIG. 5

(A) Agents' spatial and social neighbourhoods. (B) Long-tailed distribution of agents' social network sizes.

From Eppstein, M.J., Grover, D.A., Marshall, J.S., Rizzo, D.M., 2011. An agent-based model to study market penetration of plug-in hybrid electric vehicles. Energy Policy 39, 3789–3802.

consider adopting early (Leigh and Yorke-Smith, 2011). Follow-on ABM and non-ABM studies reflect upon the model and results of the paper.

4.3 Energy management in smart grids

Our third case study explores scenarios around more end users becoming local producers of energy, such as through roof-top PV. Klaimi et al. (2017) study the integration of such *prosumers* into the electricity grid. Their interest is to maximise savings for end users, while retaining power supply balance, and allowing energy storage devices. Further, the authors aim to accommodate end users' preferences. We discuss the paper as a case study of a typical agent-based approach to electricity pricing and coordination in a smart grid, with focus on the end user.

Klaimi et al. (2017) consider four classes of agents, power generation companies, storage devices, prosumers, and traditional end users, and for each present an algorithm for the agent's behaviour. The agents negotiate in a cooperative, distributed manner in order to agree local electricity prices and distribution.

Klaimi et al. (2017) do not distinguish between the traditional power generation company and the distributor of electricity, often called the distribution system operator (DSO). In deregulated jurisdictions, such as most of Europe, these entities are separated. The focus of the paper is rather on agent-based modelling including energy storage devices. These devices can be used in nodes of the smart grid to store energy when supply exceeds demand and dispatches it when needed. Such a device can be a battery or even an EV. Optimal use of energy storage devices improves energy efficiency, acts as an arbitrage instrument, and maintains the storage device life.

The proposed approach is named ANEMAS (Agent-based eNErgy MAnagement in Smart grids). The aim is "resolve the generation intermittency [sic] problem and to optimize in real time the end-user bill by integrating a storage system and using multiagent algorithms (including negotiation)."

In addition to the distributed agents, the approach assumes a centralised system operator (implicitly, a DSO). The roles of the four agent classes are:

- Grid agent: satisfies the energy lack (by producing nonrenewable energy, or purchasing renewable or stored energy) and buys the excess of energy produced.
- Storage agent: controls the energy storage ('batteries').
- Production agent (including prosumer): controls the distribution of the energy which he produces.
- Consumption agent: negotiates the energy purchase with other end users and prosumers.

These decisions of the respective agents—energy purchase, storage, and distribution decisions—constitute the energy behaviour dimensions studied by the paper.

The approach follows a two-step protocol. In the first step, the 'proactive layer', the system operator predicts energy consumption and production for the next 24-hour period. In the second step, the 'reactive layer', the storage, production, and consumption agents plan and negotiate consumption for the next 1-hour period. Agents communicate by sending messages.

Consumption agents model the consumption of a household in a smart homes. Consumption is prioritised: the highest-priority demand in each 1-hour period must be met; the medium-priority demand can be delayed for a small time; the lowest-priority demand can be delayed for a longer time.

Production agents model sources of renewable energy production. Production agents may also be end users (i.e., prosumers); if so, the agent firsts aims to satisfy its own demand, and sells any excess energy. If a pure producer, the agent sells all its production: either directly to end users or to a storage agent.

Various synthetic scenarios are considered, although with predicted values of consumption and production taken from the French electricity network. The largest scenario has 30 production agents, 30 consumption agents, 1 storage agent, and 1 grid agent. Simulation results indicate that ANEMAS increases the use of renewable energy, reduces grid use beyond the local smart grid node, and reduces prices for end users.

Open questions about the paper are the origin and rationale for the algorithms proposed for each class of agent; the source and calibration of the utility functions of the consumption and production agents, and the extent to which they model end-user preferences; the negotiation protocols; multiple storage agents; and competition as well as cooperation. The last point leads to the topic of mechanism design in smart grids (Espana et al., 2018).

5 Directions

This chapter surveyed the opportunities and challenges in agent-based models for the social dynamics of energy end use. A pivotal question in the energy domain is what policies, institutions, markets, and market conditions we need in order to meet policy objectives in the context of the energy transition. ABMs fit well with studying all these user-related processes, as they can explicitly be studied under what conditions we may reach our policy targets. Thus, ABMs are found effective in informing policy making and in identifying possible energy transition pathways by exploring what-if scenarios. We considered in detail three case studies, which address domestic heating, EV adoption, and energy management in smart grids.

ABMs relate to many other chapters of this book. Chapter 4.1 looks at context of energy behaviour change while Chapter 1.1 discusses the psychological factors around energy conservation. Regarding building occupancy models, Chapter 3.2 notes the representation of complex effects and interactions at various scales—which suggests ABMs. Indeed, cross-fertilisation of ABM with the social sciences can enrich how energy behaviours are modelled.

The flexibility and utility of agent-based modelling leads to several directions for future research. First, multimodelling (Camus et al., 2015). Strachan et al. (2016) argue for dialogue between multiple models specifically to narrow the science-policy interface gap in the energy domain.

The second direction is serious games of energy use. These kinds of games expose stakeholders to the roles and positions of other stakeholders in the STS. Example of successful agent-based computer-supported serious games includes Bourazeri and Pitt (2014) and Kurahashi and Jager (2017).

Third, model reuse. It is still the norm in agent-based modelling that each problem studied is model afresh from the ground up; there is limited reuse of previous model parts. The literature analyses a number of barriers to model reuse (Rouchier et al., 2008).

The fourth direction is using large volumes of data and machine learning techniques. We have seen the use of data in agent-based modelling for the energy domain (Section 3.6). Advances in data availability and data science tools generate new opportunities and, perhaps, necessity, in data-driven ABMs. The literature reflects a debate on this point (Cao et al., 2015; Nye, 2013).

Fifth, ABM for urban planning. We noted in this chapter the interconnectedness of modern society and the centrality of energy. If 'energy is everything', then the utility of ABM for exploring energy use and behaviour suggests utility for urban planning (Perez et al., 2016), and successful examples already exist (Ghavami et al., 2016).

References

Aleskerov, F., Monjardet, B., 2002. Utility Maximization, Choice and Preference. Springer Verlag, Heidelberg, Germany.

Armaroli, N., Balzani, V., 2007. The future of energy supply: challenges and opportunities. Angew. Chem. Int. Ed. 46 (1–2), 52–66.

Ashby, W.R., 1991. Requisite variety and its implications for the control of complex systems. Facets of Systems Science, Springer, Boston, MA, pp. 405–417.

Axelrod, R., Hamilton, W.D., 1981. The evolution of cooperation. Science 211 (4489), 1390–1396.

Bonabeau, E., 2002. Agent-based modeling: methods and techniques for simulating human systems. Proc. Natl. Acad. Sci. USA 99 (suppl 3), 7280–7287.

Bourazeri, A., Pitt, J.V., 2014. An agent-based serious game for decentralised community energy systems. In: Proceedings of 17th International Conference on the Principles and Practice of Multi-Agent Systems (PRIMA'14), pp. 246–253.

Brugnach, M., Tagg, A., Keil, F., de Lange, W.J., 2007. Uncertainty matters: computer models at the science-policy interface. Water Resour. Manag. 21 (7), 1075–1090.

Camus, B., Bourjot, C., Chevrier, V., 2015. Considering a multi-level model as a society of interacting models: application to a collective motion example. J. Artif. Soc. Soc. Simul. 18 (3), 7.

Cao, L., Zeng, Y., An, B., Symeonidis, A.L., Gorodetsky, V., Coenen, F., Yu, P.S. (Eds.), 2015. Proceedings of 10th International Workshop on Agents and Data Mining Interaction (ADMI'14). In: vol. 9145. Springer, New York, NY.

Čaušević, S., Warnier, M., Brazier, F.M.T., 2017. Dynamic, self-organized clusters as a means to supply and demand matching in large-scale energy systems. In: Proceedings of 14th International Conference on Networking, Sensing and Control (ICNSC'17), IEEE, pp. 568–573.

Chappin, E.J.L., 2018. Escaping the modelling crisis. Rev. Artif. Soc. Soc. Simul. https://roasss.wordpress.com/2018/10/12/ec/. Accessed 11 December 2018.

Chappin, E.J.L., Afman, M.R., 2013. An agent-based model of transitions in consumer lighting: policy impacts from the E.U. phase-out of incandescents. Environ. Innov. Soc. Trans. 7, 16–36.

Chappin, E.J.L., van der Lei, T., 2014. Adaptation of interconnected infrastructures to climate change: a socio-technical systems perspective. Util. Policy 31, 10–17.

Chappin, E.J.L., de Vries, L.J., Richstein, J., Bhaghwat, P., Iychettira, K., Khan, S., 2017. Simulating climate and energy policy with agent-based modelling: the energy modelling laboratory (EMLab). Environ. Model. Software 96, 421–431.

Chappin, E.J.L., Hesselink, L.X.W., Blok, K., Mueller, A., Forthuber, S., Braungardt, S., Fries, B., 2017. Changing energy efficiency technology adoption in households (CHEETAH)—working paper on modelling and survey [d3.2]. Technical Report. Fraunhofer ISI.

Conway, J., 1970. The game of life. Sci. Am. 223 (4), 4.

Dennet, D.C., 1996. Darwin's Dangerous Idea: Evolution and the Meanings of Life. Simon & Schuster, New York, NY.

Edmonds, B., 2017. Different modelling purposes. In: Edmonds, B., Meyer, R. (Eds.), Simulating Social Complexity: A Handbook. Springer, Cham, Switzerland, pp. 39–58.

Edmonds, B., Meyer, R., 2013. Simulating Social Complexity. Springer, Berlin, Germany.

Edmonds, B., Moss, S., 2004. From KISS to KIDS—an 'anti-simplistic' modelling approach. In: Lecture Notes in Computer Science. Proceedings of the Joint Workshop on Multi-Agent and Multi-Agent-Based Simulation, vol. 3415. Springer, New York, NY, pp. 130–144.

Eppstein, M.J., Grover, D.A., Marshall, J.S., Rizzo, D.M., 2011. An agent-based model to study market penetration of plug-in hybrid electric vehicles. Energy Policy 39, 3789–3802.

Epstein, J.M., 1999. Agent-based computational models and generative social science. Complexity 4 (5), 41–60.

Espana, G.M., Lorca, A., de Weerdt, M., 2018. Robust unit commitment with dispatchable wind power. Electr. Pow. Syst. Res. 155, 58–66.

European Commission, 2011. Energy efficiency plan 2011. European Commission. https://eur-lex.europa.eu/LexUriServ/LexUriServ.do?uri=COM:2011:0109:FIN:EN:PDF. Accessed 15 November 2018.

Funtowicz, S.O., Ravetz, J.R., 1993. Science for the post-normal age. Futures 25 (7), 739–755.

Ghavami, S.M., Taleai, M., Arentze, T.A., 2016. Socially rational agents in spatial land use planning: a heuristic proposal based negotiation mechanism. Comput. Environ. Urban. Syst. 60, 67–78.

Hedo, J.M., Martínez-Val, R., 2010. Computer model for numerical simulation of emergency evacuation of transport aeroplanes. Aeronaut. J. (1968) 114 (1162), 737–746.

Hesselink, L., Chappin, E.J.L., 2019. Adoption of energy efficient technologies by households—barriers, policies and agent-based modelling studies. Renew. Sustain. Energy Rev. 99, 29–41.

Heylighen, F., Joslyn, C., 2001. The law of requisite variety. http://pespmcl.vub.ac.be/REQVAR.html. Accessed 14 November 2018.

Hiteva, R., Ives, M., Weijnen, M., Nikolic, I., 2018. A complementary understanding of residential energy demand, consumption and services. In: Foulds, C., Robison, R. (Eds.), Advancing Energy Policy. Springer International Publishing, Cham, pp. 111–127.

Hodgson, G.M., 2012. On the limits of rational choice theory. Econ. Thought 1, 94–108.

Holland, J.H., 1996. Hidden Order: How Adaptation Builds Complexity. Addison-Wesley, Reading.

Holland, J.H., 1997. Emergence: From Chaos to Order. Addison-Wesley, Reading.

Hughes, T.P., 1987. The evolution of large technological systems. In: Bijker, W.E., Hughes, T.P., Pinch, T. (Eds.), The Social Construction of Technological Systems: New Directions in the Sociology and History of Technology. MIT Press, Cambridge, MA, pp. 45–76.

Janssen, M.A., Jager, W., 2001. Fashions, habits and changing preferences: simulation of psychological factors affecting market dynamics. J. Econ. Psychol. 22 (6), 745–772.

Kauffman, S.A., 1993. The Origins of Order: Self-Organization and Selection in Evolution. Oxford University Press, New York, NY.

Kay, J.J., 2002. On complexity theory, exergy and industrial ecology: some implications for construction ecology. In: Kibert, C., Sendzimir, J., Guy, B. (Eds.), Construction Ecology: Nature as the Basis for Green Buildings. Spon Press, London, UK, pp. 72–107.

Klaimi, J., Rahim-Amoud, R., Merghem-Boulahia, L., et al., 2017. Energy management in the smart grids via intelligent storage systems. In: Alonso-Betanzos, A., Sánchez-Maroño, N., Fontenla-Romero, O., Polhill, J.G., Craig, T., Bajo, J. (Eds.), Agent-Based Modeling of Sustainable Behaviors. Springer, Cham, Switzerland, pp. 207–225.

Knoeri, C., Nikolic, I., Althaus, H.-J., Binder, C.R., 2014. Enhancing recycling of construction materials: an agent based model with empirically based decision parameters. J. Artif. Soc. Soc. Simul. 17 (3), 10.

Kroes, P., 2009. Technical artifacts engineering practice and emergence. In: Functions in Biological and Artificial Worlds: Comparative Philosophical Perspectives, MIT Press, Cambridge, MA, pp. 277–292.

Kurahashi, S., Jager, W., 2017. An electricity market game using agent-based gaming technique for understanding energy transition. In: Proceedings of 9th International Conference on Agents and Artificial Intelligence (ICAART'17), pp. 314–321.

Kwakkel, J.H., Walker, W.E., Haasnoot, M., 2016. Coping with the wickedness of public policy problems: approaches for decision making under deep uncertainty. J. Water Resour. Plan. Manag. 142 (3), 01816001.

Lee, J., Padget, J., Logan, B., Dybalova, D., Alechina, N., 2014. Run-time norm compliance in BDI agents. In: Proceedings of 13th International Conference on Autonomous Agents & Multi-Agent Systems (AAMAS'14), IFAAMAS, pp. 1581–1582.

Leigh, L., Yorke-Smith, N., 2011. An adaptation of the Bass new product diffusion model for multiple purchases of capital items. Proceedings of 10th International Marketing Trends Conference. http://archives.marketing-trends-congress.com/2011/index.htm. Paris, France.

Luke, S., Cioffi-Revilla, C., Panait, L., Sullivan, K., 2004. Mason: a new multi-agent simulation toolkit. Proceedings of the 2004 SwarmFest Workshop. pp. 316–327.

Mikulecky, D.C., 2001. The emergence of complexity: science coming of age or science growing old? Comput. Chem. 25 (4), 341–348.

North, M.J., Collier, N.T., Ozik, J., Tatara, E.R., Macal, C.M., Bragen, M., Sydelko, P., 2013. Complex adaptive systems modeling with Repast Simphony. Complex Adapt. Syst. Model. 1 (1), 3.

Nye, B.D., 2013. What's between KISS and KIDS: a keep it knowledgeable (KIKS) principle for cognitive agent design. Proceedings of the 2013 AAAI Fall Symposium on Integrated Cognition, pp. 63–70. Arlington, VA.

O'Reilly, T., 2006. Generative creation. https://web.archive.org/web/20120119094929/http://radar.oreilly.com/2006/06/generative-creation.html. Accessed 5 December 2018.

Perez, P., Banos, A., Pettit, C., 2016. Agent-based modelling for urban planning current limitations and future trends. Agent Based Modelling of Urban Systems—First International Workshop, ABMUS 2016, Held in Conjunction with AAMAS, Singapore, Singapore, May 10, 2016, Revised, Selected, and Invited Papers, pp. 60–69.

Reynolds, C.W., 1987. Flocks, herds and schools: a distributed behavioral model. In: Proceedings of 14th Annual Conference on Computer Graphics and Interactive Techniques (SIGGRAPH'87), ACM, pp. 25–34.

Richstein, J.C., Chappin, É.J., de Vries, L.J., 2015. The market (in-)stability reserve for EU carbon emission trading: why it might fail and how to improve it. Util. Policy 35, 1–18.

Rittel, H.W.J., Webber, M.M., 1974. Wicked problems. Man-Made Futures 26 (1), 272–280.

Roeser, S., 2012. Risk communication, moral emotions and climate change. Risk Anal. 32, 1033–1040.

Rouchier, J., Cioffi-Revilla, C., Polhill, J.G., Takadama, K., 2008. Progress in model-to-model analysis. J. Artif. Soc. Soc. Simul. 11 (2), 8.

Schelling, T.C., 1971. Dynamic models of segregation. J. Math. Sociol. 1 (2), 143–186.

Seibt, J., 2017. Process philosophy. In: Zalta, E.N. (Ed.), The Stanford Encyclopedia of Philosophy. https://plato.stanford.edu/archives/spr2018/entries/process-philosophy/.

Shalizi, C.R., 2006. Methods and techniques of complex systems science: an overview. Complex Systems Science in Biomedicine, Springer, New York, NY, pp. 33–114.

Shoham, Y., 1993. Agent-oriented programming. Artif. Intell. 60, 51–92.

Strachan, N., Fais, B., Daly, H., 2016. Reinventing the energy modelling-policy interface. Nat. Energy 1, 16012.

Taillandier, P., Vo, D.-A., Amouroux, E., Drogoul, A., 2010. GAMA: a simulation platform that integrates geographical information data, agent-based modeling and multi-scale control. International Conference on Principles and Practice of Multi-Agent Systems, Springer, New York, NY, pp. 242–258.

Terna, P., et al., 1998. Simulation tools for social scientists: building agent based models with swarm. J. Artif. Soc. Soc. Simul. 1 (2), 1–12.

Van Dam, K.H., Nikolic, I., Lukszo, Z., 2013. Agent-Based Modelling of Socio-Technical Systems. Springer, Dordrecht.

Van den Hoven, J., Lokhorst, G.-J., Van de Poel, I., 2012. Engineering and the problem of moral overload. Sci. Eng. Ethics 18 (1), 143–155.

Vasiljevska, J., Douw, J., Mengolini, A., Nikolic, I., 2017. An agent-based model of electricity consumer: smart metering policy implications in Europe. J. Artif. Soc. Soc. Simul. 20 (1). Article No. 12.

Wallach, H.M., 2018. Computational social science ≠ computer science + social data. Commun. ACM 61 (3), 42–44.

Watson, R.T., 2005. Turning science into policy: challenges and experiences from the science-policy interface. Philos. Trans. R. Soc. Lond. B Biol. Sci. 360 (1454), 471–477.

Watts, D.J., 2017. Should social science be more solution-oriented? Nat. Hum. Behav. 1 (1), 0015.

Wilensky, U., 1999. NetLogo. Center for Connected Learning and Computer-Based Modeling, Northwestern University, Evanston, IL. http://ccl.northwestern.edu/netlogo/.

Wilensky, U., Rand, W., 2015. Introduction to Agent-Based Modeling: Modeling Natural, Social and Engineered Complex Systems With NetLogo. MIT Press, Cambridge, MA.

Wright, W., 1996. SimCity. Maxis Software, Inc., Emeryville, CA.

Preference elicitation approaches for energy decisions

3.4

Gabriela Oliveira[a,c], Luís C. Dias[a,c], Luís Neves[b,c]

[a]CeBER, Faculty of Economics, University of Coimbra, Coimbra, Portugal
[b]School of Technology and Management, Polytechnic Institute of Leiria, Leiria, Portugal
[c]INESC Coimbra, DEEC, University of Coimbra, Coimbra, Portugal

1 Introduction

Decision-making concerning energy encompasses a broad spectrum of problems, from simple ones such as selecting an electricity contract for our home to highly complex decisions involving conflicting stakeholders, with multiple, competing objectives, such as siting a new power plant or developing a national energy plan. When the consequences at stake are deemed important by the decision-maker (DM), the use of formal decision-aiding processes contributes to better understand the issues, to organise information, to generate and analyse alternatives, and to decide in a sound and auditable way. In a group decision-making context, the formal models of decision aiding also contribute to focus the discussion and help the group members to coconstruct their decisions.

Decision aiding entails elicitation, which may be defined as the facilitation of the quantitative expression of subjective judgement, whether about matters of fact or matters of value (Dias et al., 2018b). Elicitation is needed to build a model of the preferences of the DMs (and possibly other stakeholders) in a way that allows them to think clearly about these preferences.

Energy-related problems usually involve economic, social, environmental, and technical aspects. Thus, the models and methods of multicriteria decision analysis or multicriteria decision aiding (MCDA) (Belton and Stewart, 2002; Ishizaka and Nemery, 2013; Roy, 1996) are particularly adequate and are widely applied (Antunes and Henriques, 2016). MCDA encompasses many models and methods to appraise policies, projects, or any other type of alternatives in a decision process. Its most distinctive feature is that it explicitly uses multiple evaluation dimensions (the criteria or objectives), recognising that decisions involve compromises between objectives in conflict (e.g. costs vs. negative impacts vs. benefits). This allows an analysis of decision problems that is more transparent than putting a price on all kinds of social, technical, and environmental benefits or negative impacts to reduce them to monetary units. MCDA assesses the performance of each alternative on each criterion separately and then aggregates these assessments to derive a recommendation.

MCDA comprises four steps, among which Steps 1–3 require elicitation of preferences:

1. Problem structuring
2. Single-criteria assessment
3. Aggregation
4. Recommendation

The first stage (Step 1) is structuring the problem, which is the basis for all analyses that ensue. This includes defining what the problem is and identifying who the DM and the other relevant actors are. In MCDA, it also encompasses defining the evaluation criteria set, as it is well known that DMs find it difficult to articulate their objectives (Bond et al., 2008). This step can be supported by the use of problem structuring methods (PSM), discussed and illustrated in Section 2. PSMs are a set of methods or methodologies developed independently since the 1960s; aimed at giving structure to ill-defined or unstructured problems; and characterised by the existence of multiple actors, multiple perspectives, incommensurable and conflicting interests, important intangibles, and key uncertainties (Mingers and Rosenhead, 2004; Rosenhead and Mingers, 2008).

The second stage (Step 2) is to assess the performance of each alternative on each evaluation criterion. Criteria such as costs, emissions, populations, or areas are usually measured on quantitative scales, whereas criteria such as degree of opposition of the population or aesthetic impact on the landscape will usually be assessed using qualitative levels (e.g. negligible, weak, and strong). Complex decisions (e.g. locating a dam) usually involve experts in different domains (biology, engineering, economics, etc.) who evaluate the alternatives on the criteria pertaining to their expertise. The viewpoints of the affected populations and other stakeholders can also be elicited at this stage.

The third stage (Step 3) is to aggregate the evaluations obtained in Step 2 to derive a recommendation. There are essentially four strategies to aggregate multiple criteria assessments. The most common strategy is to synthetise these assessments into a single figure for each alternative, denoting its overall worth (value, utility, or priority). In some methods, for example, multiattribute value or utility methods (Keeney and Raiffa, 1993), the overall value of an alternative aggregates its value on the different criteria, independently of other alternatives. In other methods, for example, the analytic hierarchy process (Saaty, 2008), the overall value of an alternative aggregates how it compares relatively to the other alternatives being evaluated. A second strategy is to measure how near an alternative is to some reference. For instance, the TOPSIS method (Yoon and Hwang, 1995) defines an ideal reference and an antiideal reference, and then, it obtains scores that indicate how near each alternative is to the ideal and how far it is from the antiideal. A third strategy is to define outranking relations among the alternatives. In a typical outranking method such as the ELECTRE method (see Roy, 1991, 1996), an alternative a outranks another alternative b if the majority of the criteria agrees that a is at least as good as b (taking into account the criteria weights), and there is no criterion in which a is so much worse than b that it would 'veto' the outranking. Finally, a fourth strategy is to derive rules learnt from examples of preferences of the DM, as rough sets approaches do (Greco et al., 2001). Step 3 requires eliciting preferences concerning the importance of the criteria (and possibly other parameters), which is addressed in Section 3.

The final stage (Step 4) consists in deriving a recommendation based on the outputs of the previous stages. Recommendations are typically from one of the following types, depending on the 'problématique' (Roy, 1996):

- Selection (or choice) consists in identifying the best alternative (or the best k alternatives), for example, developing an energy plan for a country, selecting the best photovoltaic technology, or choosing the location for a wind farm.
- Ranking (or prioritisation) consists in ordering the alternatives from best to worst, for example, prioritising the order by which several projects will be implemented or ranking the environmental performance of energy producers.
- Sorting (or classification) consists in assigning alternatives to categories, which are typically defined a priori and ordered. Examples include product labelling and the applications presented in Section 3.

Besides the aforementioned situations in which one seeks to elicit the preferences of a DM, there are situations in which one wishes to elicit the preferences of a population. In the energy field, this typically concerns knowing more about how consumers behave. Conjoint analysis (CA) has become the most frequently used method to assess consumer preferences among the preference elicitation methods (Green et al., 2001; Sattler and Hensel-Borner, 2007; Scholz et al., 2010), and it has been considered a major contribution to understand the purchase behaviour of consumers (Netzer et al., 2008). CA was developed within the conjoint measurement area, in mathematical psychology, by Luce and Tukey (1964) and was later extended to consumer behaviour and marketing research (Green et al., 2001; Kuhfeld, 2010). The term 'conjoint' reflects the need to assess each product in its integrity, by taking into account the product's attributes jointly (Eggers and Sattler, 2011). Through the analysis of the trade-offs between attributes, CA exploits the consumers' decision process by defining which are the most determinant attributes and the most preferred combinations of attribute levels (Green et al., 2001; Kuhfeld, 2010).

CA is able to analyse why consumers choose one product, brand, or service over another one (Green et al., 2001), offering a good approximation of a purchase process in which consumers face a range of products, which they have to screen and select (Orme, 2009a). The introduction of efficient and user-friendly software has simplified not only the usually complex survey design but also the estimation procedures at an individual level (Halme and Kallio, 2011). The third topic covered in this chapter is then the elicitation of a population's preferences with CA based on a stated preferences survey (Section 4), with an application to study preferences concerning electric vehicles.

This chapter is thus structured as follows. Section 2 discusses PSMs and their use to support MCDA. It presents a case study regarding the promotion of energy efficiency, providing also general guidelines to use a specific PSM, soft systems methodology, to build a MCDA model. Section 3 addresses the elicitation of preferences in the scope of MCDA applied to energy decisions, demonstrated with two examples: the assessment of energy efficiency initiatives and the evaluation of policies to foster the development of smart grids. Section 4 addresses how to assess the preferences of a population using CA. It presents a case of eliciting preferences concerning electric vehicles, based on a stated preferences survey. The final section provides concluding comments.

2   Using problem structuring methods to elicit preferences

2.1   Brief review of problem structuring methods

Problem structuring methods are qualitative approaches for making progress with ill-structured problems (Smith and Shaw, 2019). PSMs provide different ways to describe problems, using a model to clarify stakeholders' visions of the problem, to converge on a potentially actionable mutual problem or issue within it (Mingers and Rosenhead, 2004), and to agree on commitments that will at least partially resolve it, which implies the following:

- Enabling several alternative perspectives to be brought into conjunction with each other
- Being cognitively accessible to actors with a range of backgrounds and without previous training, to allow a participative process of problem structuring
- Operating iteratively, so that the problem representation adjusts to reflect the state and stage of discussion among the actors and vice versa
- Permitting partial or local improvements to be identified and committed to, rather than requiring a global solution.
  The main PSMs include the following:
- Strategic options development and analysis (SODA) (Ackermann and Eden, 2010; Eden, 1989): a general problem identification method that uses cognitive mapping as a modelling device for eliciting and recording individuals' views of a problem situation. The merged individual cognitive maps (or a joint map developed within a workshop session) provide the framework for group discussions, and a facilitator guides participants towards commitment to a portfolio of actions.
- Soft systems methodology (SSM) (Checkland, 1989, 1999; Friend, 1989): a general method for system redesign. Participants idealise conceptual models, one for each relevant world view. They compare them with perceptions of the existing system to generate debate about what changes are culturally feasible and systemically desirable.
- Strategic choice approach (SCA) (Friend, 1989, 2011; Friend and Hickling, 2012; Hickling, 1989): a planning approach centred on managing uncertainty in strategic situations. Facilitators assist participants to model the

interconnectedness of decision areas. Interactive comparison of alternative decision schemes helps them to bring key uncertainties to the surface. On this basis, the group identifies priority areas for partial commitment and design exploration and contingency plans.

- Drama theory (Bennett et al., 1989): an interactive method of analysing cooperation and conflict among multiple actors. A model is built from perceptions of the options available to the various actors, and how they are rated. Drama theory looks for the 'dilemmas' presented to the actors within this model of the situation.

PSMs involve identifying issues; questions of concern; interesting observations; and motivations to frame a problem, usually involving interaction with individuals to record their opinions and interpretations regarding the situation. Following this phase, problems are described regarding what is to be used as inputs, desired outputs, decisions to make, and relationships between all these variables. Problem structuring usually ends with a model definition, which can include a mathematical model (Keisler, 2012).

2.2 Application of problem structuring methods to define MCDA models

The need for a structuring phase as a first step to develop a decision support model has been recognised by different authors (Belton et al., 1997; Bana e Costa et al., 1999; Keeney, 1992; Rodrigues et al., 2017), as real-world problems are frequently poorly defined, and it is uncommon to find starting points as clear as the ones from academic examples.

Effective problem structuring is critically important for MCDA as the subsequent phases of analysis are strongly influenced by the structuring process. Historically, this started to be addressed in the late 1990s with an increased focus on effective problem structuring for MCDA reflected in the publication of *Value-Focused Thinking* (Keeney, 1992) and applications that sought to integrate PSMs with MCDA (Belton et al., 1997; Bana e Costa et al., 1999; Ensslin et al., 2000; Montibeller et al., 2008).

Marttunen et al. (2017) reviewed the use of structuring problems for MCDA in practice, reporting that MCDA is increasingly seen as a powerful approach to support collaborative processes and that stakeholder involvement in MCDA is now common, particularly in environmental decision-making. As examples of coupling PSMs with MCDA, the following were identified:

- Cognitive maps and group maps have been coupled with the following MCDA methods: multiattribute value theory [e.g. MACBETH, SMART; (Bana e Costa et al., 2006; Myllyviita et al., 2014), multicriteria portfolio analysis (Franco and Lord, 2011), and the analytic network process (Wolfslehner and Vacik, 2011)];
- Combined uses of SSM and MCDA have enabled the analysts/facilitators to handle complex decision problems characterised by many stakeholders, variables, and a high level of uncertainty and to develop dynamic evaluation processes with the aim of identifying joint gains and compromises (Bana e Costa et al., 2014; Cerreta et al., 2012; Coelho et al., 2010; Neves et al., 2009).

Other cases that can also be considered PSM included SWOT analysis, scenario planning, and stakeholder analysis (Marttunen et al., 2017).

Combining PSMs and MCDA produces a richer view of the decision situation and provides a methodology that can better handle the various phases of decision-making.

2.3 Case study—Appraisal of initiatives to promote energy efficiency

The following example is based on works published by Neves et al. (2004, 2008, 2009), regarding the development of an MCDA model aimed at appraising actions or programs to promote energy efficiency, here named energy efficiency initiatives. These authors used SSM to help defining the decision problem context and the main actors involved and to unveil the relevant objectives for each stakeholder. Keeney's (1992) value-focused thinking approach was also used to refine and structure the list of objectives according to the perspective of the main evaluators identified. In addition to the description of this particular study, this section also provides some general guidelines on how SSM may facilitate the emergence of objectives for MCDA models.

2.3.1 Background

Energy efficiency (EE) has been promoted for years in different countries. Right after the 1970s' oil crisis, utilities implemented large-scale demand side management (DSM) initiatives, that is, initiatives aimed at shaping demand according to the interests of electric utilities, then still vertically integrated. Strategic conservation and load management were part of the strategies included in a portfolio usually not only financed by the ability to recover costs through general rate increases under a regulated monopoly framework but also motivated by high marginal costs of generation and capacity and regulatory pressure. However, with the liberalisation of the electricity markets, utilities changed the way the investment in DSM initiatives was made. In part, business-oriented DSM was replaced by legislative initiatives implemented by governments, aimed at transforming the markets, such as the imposition of standards and mandatory labelling of EE. But DSM is still seen as an important asset, namely, for the transmission and distribution businesses, which in most cases are still regulated businesses and subject to specific stimuli to maintain these practices.

The need for financing DSM/EE initiatives with general rate increases or taxes led to the formulation of rules to qualify those initiatives. A well-known methodology, which became a general reference, defined cost–benefit tests for different perspectives of analysis, concerning different stakeholders, namely, the participants, the utility, the ratepayers, and the society, although not stating how these perspectives should be used to qualify initiatives. Other approaches, more recent, included the consideration of impacts usually not quantifiable or impossible to measure in monetary units, therefore unsuitable for a cost–benefit analysis, and that should be listed and used only to help the analysis of the test results on a qualitative basis, implicitly suggesting a shift from cost–benefit analysis to MCDA.

A preliminary reflection about the deficiencies of the existing EE evaluation methodologies and the identification of relevant stakeholders suggested the choice of Checkland's SSM (Checkland, 1989; Checkland and Scholes, 1990) to tackle the ill-structured nature of the problem and to help structuring, or at least unveiling, a hierarchy of fundamental evaluation objectives for MCDA. SSM was used to generate a 'cloud of objectives' for MCDA, which was then structured using value-focused thinking. This process involved a group of experts, which included representatives from an electrical transmission company, an energy services company, an association of consumers, the Portuguese electrical sector regulator, and the Portuguese Directorate-General for Energy, and researchers from the INESC Coimbra R&D institute that had consultancy experience in EE and DSM issues. These experts were gathered in workshops to foster crossing perspectives and creativity.

2.3.2 Using SSM to elicit objectives

An SSM intervention typically includes a sequence of activities, each one building on the constructs and insights derived from the preceding ones. A classic formulation is to (1) enter the situation considered problematical, (2) express the problem situation, (3) formulate root definitions of relevant systems of purposeful activities, (4) build conceptual models of the systems named in the root definitions, (5) compare models with real-world activities, (6) define possible changes that are both desirable and feasible, and (7) take action to improve the problem situation. This sequence does not need to be linear, allowing returning to an earlier activity at any moment (Checkland, 1999; Checkland and Scholes, 1990).

The main results of the different tools applied during the first stages, or the 'finding out', were compiled into a rich picture (Fig. 1). In this step, the main actors, their main roles, and concerns were identified. From this identification, the main DMs, who could make use of a new 'system to evaluate the interest of energy efficiency initiatives', were identified as an energy efficiency agency, responsible for promoting energy efficiency on behalf of the society; the energy market regulator for the remaining monopoly markets; the regulated transmission and distribution companies; and the competitive retail companies.

The elicitation of potential objectives for each actor can result immediately from this first stage. For instance, examining the rich picture eased the discussion of the objectives of the agency, namely, concerning its interactions with other actors. This facilitated the identification of objectives such as 'third-party support' (i.e. support from other parties to EE initiatives), which clearly emerged from the interaction with energy service companies (ESCOs) and appliance manufacturers, or policy and budget constraints derived from the interaction with government. Tables 1 and 2 present a preliminary list of objectives that resulted from this study, regarding the entities that operate on behalf of the society (an energy agency and the market regulator) and the entities that are driven mostly by private concerns, the electricity companies, here comprising the transmission, the distribution, and the competitive retail companies.

There are several elicitation questions that can be suggested for discussion during the 'finding out' stage, most of which were used in this study. An initial question is

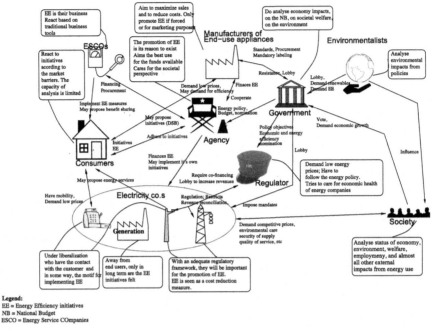

FIG. 1

Rich picture for the energy efficiency case study.

Modified with permission from Neves, L.P., Dias, L.C., Antunes, C.H., Martins, A.G., 2009. Structuring an MCDA model using SSM: a case study in energy efficiency. Eur. J. Oper. Res. 199, 834–845. © Elsevier.

Table 1 Objectives and concerns unveiled during the 'finding out' stage: agency and regulator

Agency	Market regulator
Societal welfare	Societal welfare
Budget constraints	Companies welfare
Policy constraints	Policy constraints
Third-party support	Minimise energy tariffs

Societal objectives common to agency and market regulator

Reduction of environmental impacts
- Emission of atmospheric pollution
- Water pollution
- Endangering of species
- Habitat reduction
- Visual impacts
- Land requirements
- Health effects

Reduction of hazards
Improvement of the quality of service
Reduction of dependence from nonendogenous energy sources
Improvement of domestic comfort and welfare

Table 2 Objectives and concerns unveiled during the 'finding out' stage: electricity companies

Electricity companies
Cost minimisation
To attract and/or keep customers
Regulatory or contractual constraints
Maximise revenues (minimise revenue loss)

simply to ask why the situation is problematical. The answer will typically suggest objectives that are not being met satisfactorily. The identification of relevant actors (clients, would-be problem solvers, 'problem owners', or other) should result from an initial assessment of what are their objectives. Looking at the problem situation from a 'social role' perspective allows inquiring about norms of behaviour and what constitutes good performance in these roles. The discussion about norms or other types of social constraints may show that these exist to meet the objectives of some stakeholder, whereas the values used to deem what is a good performance directly translate into objectives of the role holder. Finally, discussing how power is obtained, manifested, and preserved may unveil political objectives such as fairness, need of accountability for decisions made, or need of a strong leadership (Neves et al., 2009).

The definition of one or more relevant systems is part of another usual SSM stage: formulation of relevant purposeful activity models. In this stage, the objective is the clear definition of a system model to use as a tool for learning. In this study, the choice was to model an evaluation system for EE initiatives, rather than a system of initiatives or the electricity market as a system, although this could have been expanded if needed. The system was defined generally enough to fit each one of the four potential evaluators of EE initiatives, and the following system definition was agreed:

'A system to evaluate the interest of promoting each initiative to foster the efficiency of energy end-uses considering the direct advantages and disadvantages to the promoter, as to other entities involved' (Neves et al., 2009).

This definition clarifies that it is not the initiative that is being evaluated, but 'the interest of promoting' the initiative, which stresses the need to consider the interests of who promotes the initiative. It also acknowledges that the promoter should consider the impacts of the initiative on other entities. Furthermore, it focuses on the efficiency of end uses, thereby excluding efficiency in power generation and transmission.

The root definition led to a 'CATWOE' analysis, another typical element of an SSM process, which aims to identify relevant components of a system according to the meanings that form the CATWOE acronym, as in the following (Neves et al., 2009):

- Customer—The beneficiaries of victims of the transformation performed by the system under analysis. In this case, the beneficiaries include the initiative promoter, the external sponsor if any, the consumers who benefit with the initiative, the society as it concerns to environmental and other benefits, the manufacturers and sellers of equipment, etc., and, as victims, the electricity companies which may reduce sales as a consequence, or the manufacturers of the replaced equipment.

- Actors—Who performs the transformation. In this example, the DM, that is, the promoter of the initiative or someone who has the responsibility of evaluating it, due to some contract. One of the entities referred to previously mentioned.
- Transformation—The objective of the system, normally defined as a transformation of an input in an output, as in initiative with unknown interest → interest known.
- *Weltanschauung*—The worldview that gives a meaning to the transformation: an initiative is implemented only if its advantages overwhelm its disadvantages to the promoter, including the ones resulting from the reactions of other entities affected.
- Owner—Who has the power to initiate or to stop the transformation: the DM or someone at a degree above in the hierarchy (e.g. the government as the power above the energy agency or the market regulator).
- Environment—The background and constraints in which the transformation occurs. For this example, the capability of obtaining relevant data, estimation of initiative success (potential adherence by end-users), budget, international agreements, and directives.

The root definition and the CATWOE analysis revealed other important objectives. As in the 'finding out' stage, a list of questions for debate can be suggested to help unveil objectives in a general case (Neves et al., 2009):

- Customer: Asking what benefits or harm may result to the beneficiaries or victims of an activity and why may these be important to them leads to a previously discussed question: which interests of the problem owners are at stake?
- Actors: Asking, for each role, what would be the difference between a good and a poor performance can unveil objectives.
- Transformation: A relevant question here, although it may overlap the previous one to a great extent, is how to judge the quality or success of the transformation.
- *Weltanschauung*: This 'world view' statement often indicates, directly or indirectly, some objectives or constraints (e.g. the reactions of other affected entities in the *Weltanschauung* lead to think of objectives like minimising interference with other entities or promoting acceptability from other entities).
- Owner: As the owner is the answer to the question 'who could stop this activity?' A relevant question to unveil objectives is 'why would the owner want to stop this activity?' (or less drastically 'why would the owner want to downgrade or upgrade this activity?').
- Environmental constraints: Constraints lead to objectives. Some constraints refer to nondisputed rules or norms that cannot be broken, leading to the objective of complying with these norms (e.g. to keep tap water safe to

drink). Other constraints reveal objectives that may be traded off against other objectives (e.g. to deliver postal packages in less than 36 h or to stay within a budget).

The development of the conceptual model is another stage in which some ideas occur due to the need of thinking about the different activities. The definition of criteria for effectiveness, efficacy, and efficiency that are part of building a conceptual model leads directly to the expression of objectives. Furthermore, it may be possible to look for objectives hidden in the operational activities. Purpose-seeking questions such as 'why is this activity important?' may help unveiling these objectives.

The conceptual model that resulted from the workshops is shown in Fig. 2. Its discussion resulted in new objectives related to the monitoring and control activities, such as maximise success (participation), maximise assessment capability, or minimise risks (e.g. of failing to meet targets).

At the comparison stage, the model in Fig. 2 was compared with cost–benefit analysis and the California Energy Efficiency Policy Manual (CPUC, 2007), under the perspective of different actors to learn about their concerns and objectives. The

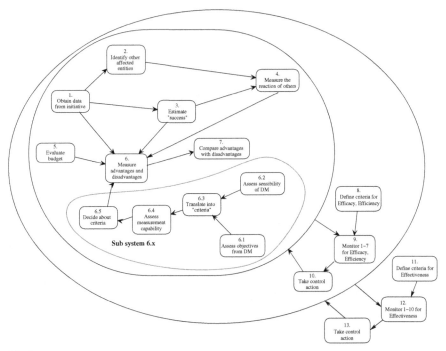

FIG. 2

Conceptual model.

Reproduced with permission from Neves, L.P., Dias, L.C., Antunes, C.H., Martins, A.G., 2009. Structuring an MCDA model using SSM: a case study in energy efficiency. Eur. J. Oper. Res. 199, 834–845. © Elsevier.

need to define monitoring and control activities for efficacy, effectiveness, and efficiency (Checkland, 1989; Friend, 1989) also led to an increased understanding of the problem.

The comparison and debate stage can also be used to uncover objectives. When comparing models with real-world activities, it may be discussed why each discrepancy detected may be considered a negative (or a positive) aspect. Finally, when defining possible changes, the discussion about whether these changes are desirable and feasible, asking 'why/why not' questions, may again reveal new objectives. In the workshops, the current methodologies of assessment were compared with the conceptual model, and their components were analysed (Neves et al., 2009).

SSM was useful to uncover a 'cloud of objectives' for each potential evaluator of EE initiatives, but this cloud still lacked structure. Keeney's value-focused thinking (Keeney, 1992, 1996) 'devices' were then used to expand and refine the list of 'objectives' obtained at the end of the SSM workshops and to define the structure 'hierarchy of fundamental objectives and network of means–ends objectives' that would comply with the set of desirable properties defined by Keeney (essential, controllable, complete, measurable, operational, decomposable, nonredundant, concise, and understandable). The lists of objectives were then subject to an analysis to identify which of them were end objectives and which were means that lead to that end. One of the resulting trees of objectives is illustrated in Fig. 3.

2.3.3 Discussion and concluding remarks

In the intervention described, SSM played a central role in suggesting questions for eliciting a 'cloud of objectives' that each potential evaluator of EE initiatives may pursue. The guidelines provided by Keeney (1992) in the framework of *Value-Focused Thinking* were then used to amend and structure these objectives. Nevertheless, the value focusing was already present during SSM, and the learning that stemmed from SSM was still present when using Keeney's devices; hence, the combined use of these methodologies is richer than a simple sequence.

This study also suggested a list of general elicitation questions to be answered during an SSM intervention (Table 3), aiming at revealing the objectives of a DM in MCDA interventions. The answers to these questions may unveil the objectives of the DM concerning the evaluation of options in the creation or modification of a system.

3 Elicitation in multicriteria decision aiding

3.1 Multicriteria parameter elicitation

The essential goal of MCDA is to help DMs to communicate and reason about a decision, guiding them to increase their insight into the decision situation, and to identify a course of action that can be justified and audited. Therefore, MCDA does not dictate what DMs should choose on supposedly objective grounds. MCDA is a sound methodology in the processes it follows and the way it processes information,

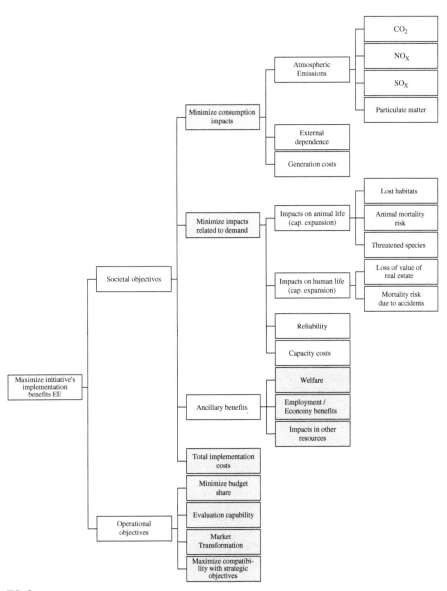

FIG. 3

Objectives for evaluation in the perspective of an energy agency.

Modified with permission from Neves, L.P., Dias, L.C., Antunes, C.H., Martins, A.G., 2009. Structuring an MCDA model using SSM: a case study in energy efficiency. Eur. J. Oper. Res. 199, 834–845. © Elsevier.

Table 3 Guidelines for using SSM to reveal MCDA objectives

Elicitation questions	Contribution of the answer
The finding out stage	
Why is the situation a problem?	Objectives that are not currently being met satisfactorily
What are the objectives of each actor?	Objectives for each actor
What norms of behaviour exist for each actor?	Norms and social constraints can exist to meet the objectives of some stakeholder(s)
What constitutes good performance in each actor's role?	Objectives for the actor
Through what 'commodities' is power manifested, and how are these commodities obtained, preserved, and passed on?	Objectives of a political nature regarding the distribution and the control of power
The system definition stage	
Customer: what benefits or harm may result?	Objectives to maximise or minimise, resp.
Actors: what would define a good role?	Objectives for the actor
What is the *Weltanschauung*?	May entail objectives or constraints (that can also be reframed as objectives)
Owner: why would it stop, downgrade, or upgrade the activity?	Owner's objectives
What are the environmental constraints?	Constraints can be reframed as objectives
By which criteria would effectiveness, efficacy, and efficiency be measured?	Objectives concerning these aspects
Why is this activity *A* important?	Objectives pursued by the activity
The system comparison stage	
Is this difference between the models and the real-word activities a negative or positive aspect and why? (for each debated difference)	Objectives whose degree of achievement would change
Why is/isn't this change desirable? (for each debated change)	Objectives whose degree of achievement would change

Modified with permission from Neves, L.P., Dias, L.C., Antunes, C.H., Martins, A.G., 2009. Structuring an MCDA model using SSM: a case study in energy efficiency. Eur. J. Oper. Res. 199, 834–845. © Elsevier.

but it acknowledges that the evaluation is partially subjective. Indeed, there is often subjectivity in the choice of the alternatives to evaluate, and the choice of the evaluation criteria is always subjective. Other subjective elements are the parameters defining the importance of each criterion (e.g. criteria weights) and other parameters of the aggregation method.

MCDA recognises the legitimacy that the preferences of a DM (possibly an entity or a group) have a bearing on the results. The DM can be, for instance, a consumer choosing a product or a service, a team of managers from a company planning and investment, or a government setting up a new policy. The preference elicitation part of MCDA is a process to set the parameters of the evaluation model that reflect preferences of the DMs. The process usually involves asking questions to the DMs and to translate their answers into information allowing to set, or to constrain, the parameter values. Such questions vary from method to method due to their diverse characteristics and the distinct nature of their parameters, for example, questions to be asked in multiattribute value methods (Morton, 2018) are very different from questions to be asked in ELECTRE methods (Dias and Mousseau, 2018).

Preference elicitation can use a direct or an indirect approach. A direct elicitation will focus the dialogue on the value of the parameters, whereas an indirect elicitation will focus the dialogue on the results the method ought to produce. In the latter case, the DMs make judgements about some alternatives provided as examples (e.g. stating that a is preferred to b or stating that a is illustrative of category C). With this information, an indirect elicitation method infers parameters that reproduce these partial results. The pioneering UTA (*UTilités Additives* in French) method (Jacquet-Lagrèze and Siskos, 2001) uses a ranking of a reference set of alternatives as input to yield an additive value function model that reproduces the ranking provided as close as possible. Indirect elicitation has also been applied in ELECTRE methods (Dias and Mousseau, 2018) to infer parameter values from outranking judgements.

Just because a consumer, a company, or a government agency has legitimacy to make choices, this does not mean that they have to set the evaluation parameters on their own. Often, the DM involves experts or other stakeholders in the elicitation process. There are several ways to elicit preferences from a group. One possibility is to elicit the individual preferences separately so that each one's answers are not influenced by other elements of the group, which can be done using written questionnaires or orally in interviews, for example (Dias et al., 2018a; Myllyviita et al., 2012). Another option is to promote workshops that gather the group to define a model's parameters together, as a single entity (De Brucker et al., 2013). The use of decision support software can be helpful to gather and process the inputs of different stakeholders (Geldermann et al., 2009).

Analysts and policy-makers dealing with energy and environment issues may feel uncomfortable with the subjectivity that underlies a vector of parameter values. Therefore, some studies use not one but several vectors representing different 'profiles' of preferences, each one emphasising different aspects, for example (Roth et al., 2009). One may instead consider a continuous region (with an infinite number of vectors) in the parameter space, whose boundaries reflect the range of preferences considered to be reasonable in face of the existing information. The latter approach is often coupled with stochastic analyses, which simulate random parameter vectors and then present statistics about the results that ensue, as in the recent examples of Dias et al. (2018a, b) and Kirppu et al. (2018).

3.2 Case study—Appraisal of initiatives to promote energy efficiency

As an example of multicriteria preference elicitation approaches for energy decisions, this section briefly overviews an application of ELECTRE to sorting energy efficiency initiatives. The detailed study is presented by Neves et al. (2008).

The purpose of this study was to consider which energy efficiency initiatives are more interesting to be promoted, considering costs and benefits, but using an MCDA approach able to address ancillary effects that are hard to cope with using traditional cost–benefit analysis. The evaluation was formulated as a sorting problem with four predefined categories for the initiatives: The worst category was 'to discard' (C^1); the best category was 'to implement' (C^4); and two intermediate categories were added, 'maybe, tending to no' (C^2) and 'maybe, tending to yes' (C^3). This was defined to allow the evaluation of each alternative on its own merits, rather than in competition against each other, and to allow a finer partition than a yes/no dichotomy. The set of alternatives to be sorted comprised 24 energy efficiency measures selected from the International Database on Energy Efficiency Programs (INDEEP) database made available by the International Energy Agency (Vreuls, 2004).

The MCDA method ELECTRE TRI (Figueira et al., 2013) was chosen because it was built specifically for sorting problems, it accommodates imprecision in performance measurement, and it implements a noncompensatory approach in which an extremely poor performance on one criterion cannot be cancelled out by a very good performance on some other aspect. The parameters that define the ELECTRE TRI evaluation model are the following:

- *Reference profiles*. These are multicriteria performance vectors that separate the categories. Each profile b^k is the upper limit of category C^k and the lower limit of category C^{k+1}.

- *Weights and cutting level*. The weights of the n criteria, denoted $k_1, \ldots, k_n$, indicate their respective importance (each weight k_j can be interpreted as a number of votes allocated to the respective criterion). The cutting level, denoted λ, defines the required majority level to agree that an alternative outranks the lower limit of a category.

- *Indifference and preference thresholds*. An indifference threshold, denoted q_j, is the largest performance difference (usually zero or close to zero) such that the DM does not distinguish between two alternatives on the jth criterion. The preference threshold, denoted p_j, is the smallest performance difference (usually small but greater than q_j) such that the DM is convinced one alternative is better than another one on the jth criterion. In this application, the analysts set the indifference threshold to 1% and the preference threshold to 10% for the quantitative criteria, and both thresholds were set to 0 for the qualitative criteria (i.e. different levels imply a preference).

- *Veto thresholds*. The veto threshold, denoted v_j, is a performance difference so large that it cannot be compensated on other criteria, that is, if an alternative is worse than the reference profile by a difference of v_j or more on the jth criterion, then it cannot outrank the profile even if the performance of the alternative is much better in all other criteria.

Briefly described, the classification process usually proceeds in the following way. Given the parameter values, an alternative can be classified in a category above C^k only if the total weight of the criteria that agree it outranks profile b^k

(those in which its performance is similar or better) is at least the required majority λ, and simultaneously, there is no criterion in which performance is so poor that this outranking is vetoed. The alternative is classified in a category such that it is good enough to outrank its lower limit but not good enough to outrank its upper limit.

Prior to eliciting preferences and using the ELECTRE TRI model to obtain a classification, a structuring process defined the set of evaluation criteria and the perspectives to be involved (see Section 2). According to this process, a total of four separate evaluations were carried out on behalf of different stakeholders: energy agency, market regulator, distribution utilities, and competitive retail companies. For each stakeholder, a set of objectives was structured and reflected into evaluation criteria. For instance, the criteria considered for the energy agency perspective were energy savings (MWh), demand savings (MW), total implementation cost (k€), welfare employment/economy benefits (qualitative), benefits in other resources (qualitative), budget share (k€), evaluation capability (qualitative), market transformation potential (qualitative), and compatibility with strategic objectives (qualitative). Qualitative criteria were evaluated on a scale with levels: very negative, negative, neutral, positive, very positive, and excellent.

The DMs who set the parameters of the evaluation model in this case were a group of five experts in energy efficiency. They were asked individually to perform assessments on the qualitative criteria through a written questionnaire. Following a Delphi process (Delbecq et al., 1986), the answers were summarised, and the questionnaire was applied again until the differences in the responses were considered to be small, using the median of the responses as the final assessment.

The experts were asked to provide also the performance levels corresponding to the upper limit b^1 of the first category, 'to discard' (C^1), and the lower limit b^3 of the best category, 'to implement' (C^4). First, the responses were aggregated to define b^1 and b^3. Then, the reference profile b^2, which separates the two 'maybe' categories, was set equal to the midpoints between b^1 and b^3.

The experts weighted the criteria according to their perceived importance for each of the evaluation perspectives (the different stakeholders). This resulted in five weight vectors for each perspective. Rather than aggregating these five weight vectors, the corresponding classifications were determined, as illustrated in Fig. 4.

To obtain a single-category classification, the IRIS software (Dias and Mousseau, 2003) was used. This software uses linear programming to infer a vector of weights that is as far as possible from violating the classifications provided by the DMs. Hence, for each initiative, a single category was obtained, and this was done for each one of the four evaluation perspectives. The result according to the multiple perspectives could then be summarised using diagrams such as the ones illustrated in Fig. 5. In this figure, the heavier line depicts the classifications using the base data, whereas the lighter lines correspond to a robustness analysis on the input data corresponding to two extreme scenarios: ±10% for all quantitative criteria and ±1 level in the qualitative scales.

Initiatives	Categories			
	C¹	C²	C³	C⁴
a1		■		
a2			1	4
a3		1	4	
a4			2	3
a5		1	3	1
a6		1	4	
a7		1	4	
a8			4	1
a9			1	4
a10			4	1
a11			3	2
a12		3	2	
a13			■	
a14				■
a15				■
a16			■	
a17			4	1
a18				■
a19			2	3
a20				■
a21			■	
a22			2	3
a23			2	3
a24		1	4	

FIG. 4

Classification of 24 energy efficiency initiatives according to five different experts considering the energy agency perspective.

Reproduced with permission from Neves, L. P., Martins, A. G., Antunes, C. H., Dias, L. C., 2008.
A multi-criteria decision approach to sorting actions for promoting energy efficiency. Energy Policy 36(7),
2351–2363. © Elsevier.

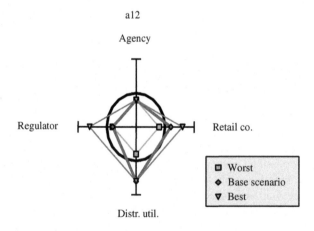

a12

Agency

Regulator

Retail co.

Distr. util.

□ Worst
◆ Base scenario
▽ Best

FIG. 5

Classification of efficiency initiatives under different stakeholder perspectives.

Reproduced with permission from Neves, L. P., Martins, A. G., Antunes, C. H., Dias, L. C., 2008.
A multi-criteria decision approach to sorting actions for promoting energy efficiency. Energy Policy 36(7),
2351–2363. © Elsevier.

3.3 Case study—Evaluation of policies to foster the development of smart grids

A second example of multicriteria preference elicitation for energy decisions consisted in an assessment of policies to foster the development of smart grids, a project included in the R&D Program mandated by the electricity sector regulator in Brazil.

The structuring part of this study led to the choice of the seven objectives to be taken into account (Antunes et al., 2016) and the eight policies to be evaluated (Dantas et al., 2018). A Delphi process was carried out to elicit qualitative assessment of the different policies on each objective (Dantas et al., 2018). Fig. 6 illustrates how the evaluation of the eight policies on a given objective was elicited (level 0 corresponds to no impact; levels +1 to +5 represent slight, moderate, strong, very strong, and extreme positive impact, and negative levels similarly denote a negative impact). The same process was used to elicit information concerning the weights for the different objectives (Fig. 7 illustrates the questions used to elicit the importance of each one). This process involved a group of 28 experts and stakeholders, seven of which from government agencies, state-owned companies, and government scientific advisors (representing a government perspective), plus eight high-level representatives of private companies in the sector (representing a business perspective), and 13 academics and consultants (representing a knowledge perspective) (details in Dantas et al., 2018). This allowed obtaining results for each one of these subsets of respondents and global conclusions considering the set of all these DMs. As in the previous application, the median answers were used for the assessment of how well each policy would perform on each criterion.

Initially, this study aimed at sorting the policies according to four categories: C^1 (uninteresting—the policy would lead to a situation that is equal to or worse than the current situation), C^2 (wait and see—the policy would lead to a situation that is only slightly better than the current situation), C^3 (implement with priority—the policy would clearly contribute to improve the current situation), and C^4 (implement with maximum priority—the policy would clearly contribute to a large improvement to the current situation). The ELECTRE TRI method, presented in the previous section, was used for this purpose. The analysts defined the profiles bounding the categories, setting level +1 as the lower bound for C^2, level +3 as the lower bound for C^3, and level +5 as the lower bound for C^4. These are demanding levels, but the analysts also set an indifference threshold of 0.5 and a preference threshold of 2.5 for all the criteria, so that concordance is 75% if the performance is one level below the profile and concordance is 25% if the performance is two levels below the profile.

Contrary to the previous application, no single vector of weights was inferred or chosen to represent the DM preferences. Rather, the analysis acknowledged the diversity of opinions considering multiple weight vectors that satisfied a reasonable set of constraints derived from the DM's qualitative answers [please see Dias et al. (2018a) for details].

Concerning the veto power of the objectives, the questionnaire elicited opinions taking as a reference the category C^3. First, the DM had to ponder whether a negative

	−5 = extremely negative	−4	−3	−2	−1	0 = no impact	+1	+2	+3	+4	+5 = extremely positive
1—Mandatory Rollout of Smart Meters	O	O	O	O	O	O	O	O	O	O	O
2—Regulatory Changes to Foster Innivation in the Energy Sector	O	O	O	O	O	O	O	O	O	O	O
3—Improvement of Research & Development and Demonstration Projects	O	O	O	O	O	O	O	O	O	O	O
4—Incentive Policies for Promoting Demand-Side Management, Distributed Generation and Storage	O	O	O	O	O	O	O	O	O	O	O
5—Establishing Quality Standards for the Telecommunications Industry	O	O	O	O	O	O	O	O	O	O	O
6—Regulation of New Business Models	O	O	O	O	O	O	O	O	O	O	O
7—Development Plan for Smart Cities	O	O	O	O	O	O	O	O	O	O	O
8—National Development Policy for Smart Grid Industry	O	O	O	O	O	O	O	O	O	O	O

FIG. 6

Qualitative assessment of the impact of different policies regarding one of the objectives.

Reproduced with permission from Dantas, G.A., de Castro, N. J., Dias, L., Antunes, C. H., Vardiero, P., Brandão, R., Rosental, R., Zamboni, L., 2018. Public Policies for Smart Grids in Brazil. Renew. Sust. Energ. Rev. 92, 501–512. © Elsevier.

FIG. 7

Qualitative assessment of the importance of the objectives.

Table 4 Range of classifications and ranking considering all the participants (Policies P1–P8 as described in Fig. 6)

	Worst case	Best case	Minimum credibility of outranking (b^2, b^3)	Rank
P4—DSM/DG/S incentives	C^4	C^4	(1, 0.71)	1st
P2—Regulatory changes	C^3	C^4	(1, 0.61)	2nd
P6—New business models	C^3	C^4	(1, 0.57)	3rd
P7—Smart cities	C^3	C^4	(1, 0.53)	4th
P3—R&D and demonstration	C^3	C^3	(1, 0.47)	5/6th
P8—Smart grid industries	C^3	C^3	(1, 0.47)	5/6th
P5—Telecom standards	C^3	C^3	(0.85, 0.05)	7th
P1—Roll-out smart meters	C^3	C^3	(0.80, 0.05)	8th

impact for one of the objectives would justify not implementing it, even if the impact on all other objectives were positive. Then, the DM would optionally define, for each objective, what impact levels (the same presented in Fig. 6) would be negative enough to discard implementation. For each objective, the median level was used to summarise the answers of the respondents.

Since multiple weighting vectors were accepted, the result is not expected to be a clear-cut classification. In this case, policies P1 (mandatory roll out of smart meters), P5 (definition of mandatory telecommunication quality requirements), and P8 (national development plan of smart grid industries) would always be sorted in C^3, whereas policy P4 (incentives for demand side management, distributed generation, and storage) would always be sorted in C^4. Policies P2 (regulatory changes for technological innovation), P3 (improving R&D and demonstration project schemes), P6 (regulatory changes for new business models), and P7 (smart city development plan) could be sorted in C^3 or in C^4, depending on the weights chosen.

In addition to sorting the policies, it was also considered of interest to obtain a ranking of the policies within the same category by means of a process that would respect the sorting rationale [considering how far each policy was from falling to an inferior category, as described by Dias et al. (2018a)]. In the end, results were obtained considering all the participants (Table 4) and also considering each subset (perspective) separately. The results were rather similar among the perspectives, the main exception being policy P6, which was robustly sorted into C^4 by the government perspective and was robustly sorted into C^3 by the business and knowledge perspectives.

4 Preference elicitation with conjoint analysis
4.1 Conjoint analysis of stated preferences

Consumer preferences data in conjoint analysis are commonly collected through revealed preference (RP) or stated preference (SP) techniques. If a product is already available in the market, RP can be obtained by observing consumers' actual purchase choices. However, if the purpose is to analyse preference data of products that are not yet in the market, for attributes (or criteria) that are not present in existing products or for attribute levels that are beyond the ones that currently exist in the market, it is impossible to collect RP. In such cases, SP are pointed out as the most suitable technique to collect data, because they can be elicited through designed experiments that measure preferences for hypothetical alternatives (Kroes and Sheldon, 1988). SP studies are the most commonly used collection method of preference data, because conjoint analysis is mainly used to assess consumer preferences for products that are planned to be introduced in the market. Additionally, SP have been considered more cost-effective than RP as with a relatively small sample, it is possible to obtain much information (Dagsvik and Liu, 2009; Kroes and Sheldon, 1988).

The development of a conjoint analysis study with SP data usually encompasses six steps [adapted from (Green and Srinivasan, 1990) and (Hanley et al., 2001)]:

1. Selection of attributes

This step consists in identifying the relevant attributes that will characterise the set of products to be analysed. The choice of attributes is not arbitrary, and it needs to account for several aspects, such as selecting a small number of attributes to minimise the estimation efforts, choosing realistic attribute levels and using attributes that are relevant and related to the chosen subject (Sattler and Hensel-Borner, 2007). The selection of attributes can be made through focus groups, interviews, and surveys, or it can be based on previous studies (Hanley et al., 2001).

2. Assignment of levels

After selecting the attributes, a set of levels (specific values) has to be defined for each attribute. This definition should consider several aspects (Hanley et al., 2001; Kotri, 2006) such as being realistic, that is, as close as possible to real product values, and nonlinearly spaced, allowing to capture nonlinear utility functions within attributes.

3. Choice of preference elicitation method

There are three conjoint analysis elicitation methods of preference data: rating scale methods, rank methods, and the choice-based conjoint (CBC) (also known as choice experiment) methods (Louviere, 1988). The first consists in rating a set of products by assigning rates from a predefined scale. In the second method, preference data are elicited through a ranking exercise where a set of products is ordered from the most preferred to the least preferred. The third conjoint method, CBC, consists in asking consumers to choose the most preferred product among several sets of alternatives, where each set comprises two or more multiattribute products (Green et al., 1972; Jain et al., 1979). In CBC, the collection of data, by consisting in simulated purchase decisions, is considered to be more realistic and simple than providing product ratings (Borghi, 2009; Jaeger et al., 2001).

4. Choice of experimental design

A conjoint analysis study can take one of two forms, a full factorial design (or full profile design) or a fractional factorial design. A full factorial design consists in presenting the consumer with all the possible combinations of the attribute levels. A full factorial design therefore entails a tedious and cost-prohibitive process by having consumers considering so many products at once. For instance, if a study comprises five attributes characterised by five levels each (5^5), each consumer would have to assess 3125 different products. This justifies the common use of the second form of experimental design, the fractional factorial design where only a few products are assessed by each consumer. The number of products is computed according to the minimum necessary to estimate efficiently preferences of different attributes on the dependent variable, that is, the SP of purchase of a product (Kotri, 2006; Kuhfeld, 2010).

5. Definition of SP questions

When fractional design is selected in the previous stage, the constructed product combinations have to be grouped before being presented to consumers. As consumers generally have limits on their ability to process information, the number of questions should not be too high or too difficult because it may compromise the acquisition of quality data (Carson et al., 1994). Fatigue or boredom from consumers due to answering to a lot of questions will increase the probability that their answers exhibit high levels of randomness (Day et al., 2012). On the other hand, the design fitness to estimate preferences is assessed according to its efficiency, that is, a measure of the quality of the fractional factorial design. A survey design is efficient when both variance and covariance of the estimated parameters are minimal (Huber and Zwerina, 1996). The process of survey design is highly complex, and therefore, it is very common to use software to perform that design (e.g. Sawtooth and XLSTAT). Considering the aspects that should be taken into account in the definition of the SP questions mentioned earlier, a comfortable number of questions should be defined where it is possible to consider the design requirements to estimate preferences while avoiding overloading consumers with too much information (Johnson and Orme, 2010).

After completing the survey design, preference data can be elicited through face-to-face interviews or online through specific software.

6. Estimation models

When the focus is the analysis of consumer preferences, there are three groups of estimation models regarding the level of aggregation (Moore, 1980): aggregation models, segmented models, and individual-level models. The selection of the model is dependent on the specific goals of each research and on the preference pattern that may emerge in the consumers included in the analysis (Kotler, 2000):

- Aggregation models are more suited when homogeneous preferences are observed, that is, when consumers have similar preferences, or when the research goal is to define a global marketing strategy for a specific market. Examples are logit models, mixed logit models, and probit models.
- Segmented models should be selected when several groups with similar preferences are identified, that is, when clusters of preferences are identified or when a segmentation of the market is sought. For example, the latent class model provides preferences estimations for homogeneous groups of consumers.

- Individual-level models are more appropriate when the variance among consumers is large, that is, when there are high levels of heterogeneity on consumer preferences (Kotler, 2000). CBC hierarchical Bayes (CBC/HB) is an example of a model that computes preferences estimations for each consumer individually.

4.2 Case study—Modelling consumer preferences for electric vehicles (EVs)

To illustrate the application of conjoint analysis techniques to elicit consumer preferences for a product where energy considerations are important, namely, alternative fuel vehicles (AFVs), a case study in Portugal is presented in this section. The purpose of this study was to analyse consumer preferences for electric vehicles (EVs) by identifying the most determinant attributes and the preference structure of preferences.

The number of studies that elicited consumer preferences through conjoint analysis-based techniques applied to the context of purchasing AFVs is extensive. In the last decade, there were more than 60 studies focused on assessing preferences for AFVs through conjoint analysis either focused on AFVs in general (e.g. Hackbarth and Madlener, 2016; Soto et al., 2018; Zhang et al., 2011) or in specific vehicle technologies such as battery electric vehicles (BEVs) (e.g. Eggers and Eggers, 2011; Liao et al., 2018; Lieven, 2015; Parsons et al., 2014), plug-in hybrid electric vehicles (PHEVs) (e.g. Axsen et al., 2016; Krause et al., 2016; Sheldon et al., 2017), and hybrid electric vehicles (HEVs) (e.g. Cirillo et al., 2017; Mau et al., 2008). The main goals of such conjoint analysis preference studies were to estimate vehicle demand, to determine the more important vehicle attributes on future vehicle adoption, and to identify which factors can speed the diffusion of new vehicle technologies.

The present study started with the selection of the appropriate attributes to characterise the vehicle set. A survey was implemented to collect the characteristics most valued by Portuguese consumers in the purchase of a vehicle. Free elicitation was used as a preference elicitation procedure (Steenkamp and van Trijp, 1997), where consumers were asked to name the attributes they considered relevant under the presented context. Focusing on the purpose of differentiating vehicle technologies and considering the frequency with which each attribute was mentioned, purchase price, fuel consumption, range, and CO_2 emissions, in this order, were found to be the most relevant characteristics for consumers (Oliveira and Dias, 2015). The type of engine was added to this list of attributes to distinguish the vehicle technology of each alternative. The attributes are described as follows:

- Type of engine: vehicle technology
- Purchase price: cost to acquire a vehicle, measured in €
- Range: distance that can be driven without fuelling/charging the vehicle, measured in km

- Fuel consumption: cost to drive 100 km, measured in €/100 km
- CO_2 emissions: quantity of CO_2 emissions released to the environment during the usage phase of the vehicle, measured in g/km

To approximate the vehicle purchase scenario with the real Portuguese context, the definition of the attribute values (levels) was based on the characteristics of existent EVs in Portugal. For instance, for the 'type of engine' attribute, five existing technologies in Portugal were considered, namely, BEV, PHEV, HEV, diesel, and gasoline. All the attribute levels are depicted in Table 5.

Among the conjoint analysis elicitation techniques, CBC was selected to elicit consumer preferences in this study due to its already mentioned advantages. Since a full factorial design would comprise $3125 = 5^5$ alternatives of products to analyse, the CBC design followed a fractional factorial design to obtain eight versions of nine questions each, and then, one of these versions was randomly assigned to each consumer. Each question comprised three vehicles to order (rank order questions), that is, to indicate which one was the most and least preferred alternative in each triplet according to his or her preferences. Fig. 8 shows an example of a CBC question.

After the definition of CBC questions, data were collected through face-to-face interviews (105 consumers). The estimation procedure used to analyse the preference data was choice-based conjoint/hierarchical Bayes (CBC/HB) as it allows analysing data at the individual level and presents a high level of predictive validity, that is, the estimated preference structure of each consumer is closer to the consumer stated preferences. CBC/HB is a random effects model that pools the data through an iterative process and provides utilities for each attribute of each consumer allowing to capture consumers heterogeneity (Allenby and Ginter, 1995; Orme, 2009b). The consumer is assumed to choose the alternative that yields the maximum utility. The overall utility of an alternative for a consumer is obtained by adding up the part-worth utilities (for this consumer) corresponding to the attribute levels of that alternative on the multiple attributes (Malhotra, 2008).

The CBC/HB method provides two main outputs concerning each consumer: the part-worth utilities, that is, the utilities corresponding to the different attribute levels, and the relative importance of each attribute for the vehicle purchase decision. Table 6 presents an example of CBC/HB outputs for one of the consumers that

Table 5 Attribute levels for the experimental design

Attribute	Levels
Type of engine	BEV/PHEV/HEV/diesel/gasoline
Price	24,000€/27,000€/30,000€/32,000€/34,000€
Range	150 km/250 km/350 km/900 km/1200 km
Fuel consumption (per 100 km)	2€/4€/6€/8€/10€
CO_2 emissions (per km)	50 g/90 g/110 g/130 g/150 g

Choose the best and worst option according to your preferences:

	Option A	**Option B**	**Option C**
Type of engine	Diesel	BEV	PHEV
Purchase price	27000 €	30000 €	34000 €
Range	1200 Km	150 Km	1200 Km
Fuel consumption	6€/100Km	2€/100Km	4€/100km
CO_2 Emissions	130g/Km	50g/Km	90g/Km
BEST OPTION	☐	☐	☐
WORST OPTION	☐	☐	☐

FIG. 8

Example of a CBC question.

participated in this study. To illustrate the type of preference information that can be obtained based on these outputs, the following observations can be derived from this consumer's preferences:

- PHEV is most preferred type of vehicle, and BEV is the least preferred.
- HEV, diesel and gasoline engine types are similar in terms of preference.
- Among the attribute levels, the highest impacts on preferences are the price increment from 30,000€ to 32,000€, the range increment from 900 to 1200 km, and the fuel consumption increment from 8 to 10€/100 km.
- Fuel consumption and type of engine are the most determinant attributes in a vehicle purchase context.
- The influence of CO_2 emissions in the vehicle purchase is very small.

Fig. 9 presents the distribution of the individual-level preference estimates, where it is possible to observe the homogeneity (or heterogeneity) of preferences for each attribute. For instance, the monetary attributes, purchase price, and fuel consumption present higher heterogeneity, while preferences within range and CO_2 emissions values are more similar across consumers. Regarding the type of engine attribute, BEV has much higher variation of preferences in comparison with the other vehicle technologies.

Considering the relative importance of attributes for all consumers, the most determinant attribute for consumers' decisions is the type of technology, closely followed by fuel consumption and purchase price (Fig. 10).

The outputs obtained through conjoint analysis provide policy-makers with valuable information that can be used to support the design of effective strategies to promote the adoption and diffusion of AFVs in the market. The part-worth utilities forecast how appealing each combination of attribute levels would be, and, for a hypothetical set of choices, statistical techniques can estimate the market share of each option in that set among the sampled population (Huber et al., 1993). For instance, this allows forecasting how the purchase behaviour of the studied population would influence the market share impact of a price subsidy or an increase in battery range due to technological evolution.

Table 6 Example of outputs from CBC/HB for a specific consumer

Attribute	Part-worth utilities	Relative importance
Type of engine		24%
BEV	4.46	
PHEV	10.28	
HEV	8.80	
Gasoline	8.05	
Diesel	8.42	
Purchase price		20%
24,000	10.08	
27,000	9.69	
30,000	8.97	
32,000	6.01	
34,000	5.25	
Range		19%
150	5.99	
250	7.53	
350	7.90	
900	7.92	
1200	10.67	
Fuel consumption		36%
2	10.87	
4	10.05	
6	9.42	
8	7.57	
10	2.09	
CO_2 emissions		1%
50	8.09	
90	8.09	
110	8.00	
130	7.92	
150	7.90	

5 Conclusions

This chapter aimed to explain elicitation of preferences in the context of energy decisions, as part of building a decision model that represents the views of DMs (and possibly other stakeholders) and allows them to perform rational and informed decisions. Knowing the preferences of DMs is essential as energy-related problems usually involve economic, social, environmental, and technical aspects, that is, multiple criteria. Preferences are reflected in the choice of evaluation criteria and the role played by each of the criteria.

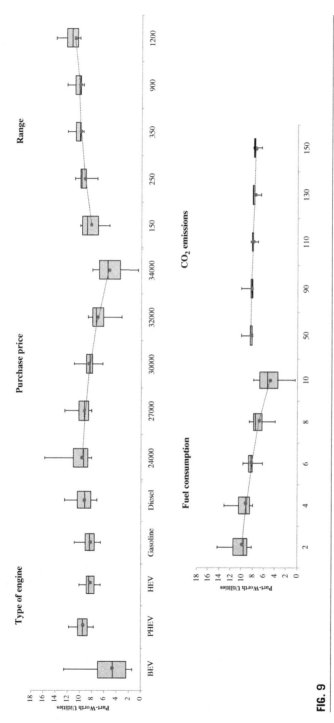

FIG. 9

Preference distribution of each attribute (the dots represent the average utility value of each attribute level, and the dashed line represents the part-worth utility function for each attribute).

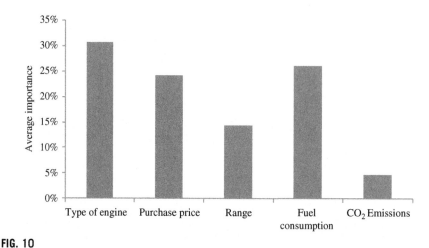

FIG. 10

Aggregated relative importance of attributes.

The different elicitation processes discussed in this chapter can be used in energy-related decision problems, being particularly suited to characteristics such as the presence of multiple stakeholders and multiple objectives. Moreover, we believe these techniques have a huge potential to be combined, yielding improved results over what they can achieve individually.

The set of tools discussed in this chapter has multiple potential uses concerning energy decision-making. When dealing with a decision problem of an individual, a group of individuals, an organisation, or a government, an energy expert can use MCDA elicitation tools to build a preference model for the DMs. Such preference model will help DMs choose their preferred course of action in a way that is consistent with their preferences, based on sound logic, and auditable. Prior to the MCDA elicitation, however, the use of PSM is invaluable to structure the information about the decision situation and its context and in particular to define which are the alternatives and by which criteria they should be evaluated.

In other contexts, an energy expert might be concerned with the preferences of a population, namely, to plan the introduction of new products or new policies. Then, a model of the preferences of an entire population is needed, which can be derived from CA elicitation tools. Again, the prior use of PSM within a focus group can be useful to find out which attributes of concern should be introduced in the CA survey design.

We believe that the basic elements of the tools presented in this chapter are not hard to learn by an energy expert. Their use, however, is as much of an art as it is a science, once we are dealing with sociotechnic processes. The skills required to use these tools improve mainly with the experience in applying them. Thus, the help of analysts with experience in using these tools in different contexts will be beneficial for the energy experts. Experts in preference elicitation processes need not to be experts in energy (nor in healthcare, or finance, or any of the multiple domains where they act as analysts): they should be experts in the process of using these tools. Energy experts can thus count on their guidance while providing themselves the requisite expertise on the energy field. With experience, though, many energy experts will eventually master both types of expertise.

References

Ackermann, F., Eden, C., 2010. Strategic options development and analysis. In: Reynolds, M., Holwell, S. (Eds.), Systems Approaches to Managing Change: A Practical Guide. Springer London, London, pp. 135–190.

Allenby, G.M., Ginter, J.L., 1995. Using extremes to design products and segment markets. J. Mark. Res. 32, 392–403.

Antunes, C.H., Henriques, C.O., 2016. Multi-objective optimization and multi-criteria analysis models and methods for problems in the energy sector. In: Greco, S., Ehrgott, M., Figueira, J. (Eds.), Multiple Criteria Decision Analysis, International Series in Operations Research & Management Science. vol. 233. Springer, New York, pp. 1067–1165.

Antunes, C.H., Dias, L., Dantas, G., Mathias, J., Zamboni, L., 2016. An application of Soft Systems Methodology in the evaluation of policies and incentive actions to promote technological innovations in the electricity sector. Energy Procedia 106, 258–278.

Axsen, J., Goldberg, S., Bailey, J., 2016. How might potential future plug-in electric vehicle buyers differ from current "Pioneer" owners? Transp. Res. Part D: Transp. Environ. 47, 357–370.

Bana e Costa, C.A., Ensslin, L., Corrêa, É.C., Vansnick, J.-C., 1999. Decision support systems in action: integrated application in a multicriteria decision aid process. Eur. J. Oper. Res. 113 (2), 315–335.

Bana e Costa, C.A., Fernandes, T.G., Correia, P.V.D., 2006. Prioritisation of public investments in social infrastructures using multicriteria value analysis and decision conferencing: a case study. Int. Trans. Oper. Res. 13 (4), 279–297.

Bana e Costa, C.A., Lourenço, J.C., Oliveira, M.D., Bana e Costa, J.C., 2014. A Socio-technical Approach for Group Decision Support in Public Strategic Planning: The Pernambuco PPA Case. Group Decis. Negot. 23 (1), 5–29.

Belton, V., Ackerman, F., Shepherd, I., 1997. Integrated support from problem structuring through to alternative evaluation using COPE and V.I.S.A. J. Multi-Criteria Decis. Anal. 6 (3), 115–130.

Belton, V., Stewart, T.J., 2002. Multiple Criteria Decision Analysis: An Integrated Approach. Kluwer, Boston.

Bennett, P., Bryant, J., Howard, N., 1989. Drama theory and confrontation analysis. In: Rosenhead, J. (Ed.), Rational Analysis for a Problematic World. John Wiley and Sons, Chichester.

Bond, S.D., Carlson, K.A., Keeney, R.L., 2008. Generating objectives: Can decision makers articulate what they want? Manag. Sci. 54 (1), 56–70.

Borghi, C., 2009. Discrete choice models for marketing: New methodologies for optional features and bundles. Master thesis, University Leiden, Mathematic Institute.

De Brucker, K., Macharis, C., Verbeke, A., 2013. Multi-criteria analysis and the resolution of sustainable development dilemmas: a stakeholder management approach. Eur. J. Oper. Res. 224 (1), 122–131.

Carson, R.T., Louviere, J.J., Anderson, D.A., Arabie, P., Bunch, D.S., Hensher, D.A., Johnson, R.M., Kuhfeld, W.F., Steinberg, D., Swait, J., Timmermans, H., Wiley, J.B., 1994. Experimental analysis of choice. Mark. Lett. 5, 351–367.

Cerreta, M., Panaro, S., Cannatella, D., 2012. Multidimensional spatial decision-making process: local shared values in action. In: Murgante, B., Gervasi, O., Misra, S., Nedjah, N., Rocha, A.M.A.C., Taniar, D., Apduhan, B.O. (Eds.), Computational Science and Its Applications—ICCSA 2012, Lecture Notes in Computer Science. Springer, Berlin Heidelberg, pp. 54–70.

Checkland, P., 1989. An application of soft systems methodology. In: Rosenhead, J. (Ed.), Rational Analysis for a Problematic World. John Wiley & Sons, Chichester, pp. 71–100.

Checkland, P., 1999. Soft Systems Methodology: A 30 Year Retrospective. John Wiley & Sons, Chichester.

Checkland, P., Scholes, J., 1990. Soft Systems Methodology in Action, second ed. John Wiley & Sons, Chichester.

Cirillo, C., Liu, Y., Maness, M., 2017. A time-dependent stated preference approach to measuring vehicle type preferences and market elasticity of conventional and green vehicles. Transp. Res. A Policy Pract. 100, 294–310.

Coelho, D., Antunes, C.H., Martins, A.G., 2010. Using SSM for structuring decision support in urban energy planning. Technol. Econ. Dev. Econ. 16, 641–653.

CPUC, 2007. Energy Efficiency Policy Manual, 3.1. ed. California Public Utilities Commission. http://docs.cpuc.ca.gov/EFILE/RULINGS/74969.pdf.

Dagsvik, J.K., Liu, G., 2009. A framework for analyzing rank ordered data with application to automobile demand. Transp. Res. A Policy Pract. 43, 1–12.

Dantas, G.A., de Castro, N.J., Dias, L., Antunes, C.H., Vardiero, P., Brandão, R., Rosental, R., Zamboni, L., 2018. Public Policies for Smart Grids in Brazil. Renew. Sust. Energ. Rev. 92, 501–512.

Day, B., Bateman, I.J., Carson, R.T., Dupont, D., Louviere, J.J., Morimoto, S., Scarpa, R., Wang, P., 2012. Ordering effects and choice set awareness in repeat-response stated preference studies. J. Environ. Econ. Manag. 63, 73–91.

Delbecq, A.L., Van de Ven, A.H., Gustafson, D.H., 1986. Group techniques for program planning: a guide to nominal group and Delphi processes. Green Briar Press, Middleton, Wisconsin.

Dias, L.C., Antunes, C.H., Dantas, G., de Castro, N., Zamboni, L., 2018a. A multi-criteria approach to sort and rank policies based on Delphi qualitative assessments and ELECTRE TRI: The case of smart grids in Brazil. Omega 76, 100–111.

Dias, L.C., Morton, A., Quigley, J., 2018b. Elicitation: State of the Art and Science. In: Dias, L.C., Morton, A., Quigley, J. (Eds.), Elicitation—The science and art of structuring judgement. Springer, Cham, pp. 1–14.

Dias, L.C., Mousseau, V., 2003. IRIS: A DSS for Multiple Criteria Sorting Problems. J. Multi-Criteria Decis. Anal. 12, 285–298.

Dias, L.C., Mousseau, V., 2018. Eliciting multi-criteria preferences: ELECTRE models. In: Dias, L., Morton, A., Quigley, J. (Eds.), Elicitation—The Science and Art of Structuring Judgement. Springer, Cham, pp. 349–375.

Eden, C., 1989. Using cognitive mapping for strategic options development and analysis. In: Rosenhead, J. (Ed.), Rational Analysis for a Problematic World. John Wiley and Sons, Chichester.

Eggers, F., Eggers, F., 2011. Where have all the flowers gone? Forecasting green trends in the automobile industry with a choice-based conjoint adoption model. Technological Forecasting & Social Change 78 (1), 51–62.

Eggers, F., Sattler, H., 2011. Preference measurement with Conjoint Analysis: Overview of state-of-the-art approaches and recent developments. GfK Mark. Intell. Rev. 3 (1), 36–47.

Ensslin, L., Dutra, A., Ensslin, S.R., 2000. MCDA: a constructivist approach to the management of human resources at a governmental agency. Int. Trans. Oper. Res. 7 (1), 79–100.

Figueira, J., Greco, S., Roy, B., Slowinski, R., 2013. An overview of ELECTRE methods and their recent extensions. J. Multi-Criteria Decis. Anal. 20, 61–85.

Franco, L.A., Lord, E., 2011. Understanding multi-methodology: Evaluating the perceived impact of mixing methods for group budgetary decisions. Omega 39, 362–372.

Friend, J., 1989. The Strategic Choice Approach. In: Rosenhead, J. (Ed.), Rational Analysis for a Problematic World. John Wiley and Sons.

Friend, J., 2011. The strategic choice approach. In: Wiley Encyclopedia of Operations Research and Management Science. Wiley.

Friend, J., Hickling, A., 2012. Planning Under Pressure, third ed. Routledge.

Geldermann, J., Bertsch, V., Treitz, M., French, S., Papamichail, K.N., Hämäläinen, R.P., 2009. Multi-criteria decision support and evaluation of strategies for nuclear remediation management. Omega 37 (1), 238–251.

Greco, S., Matarazzo, B., Slowinski, R., 2001. Rough sets theory for multicriteria decision analysis. Eur. J. Oper. Res. 129, 1–47.

Green, P.E., Carmone, F.J., Wind, Y., 1972. Subjective evaluation models and conjoint measurement. Behav. Sci. 17 (3), 288–299.

Green, P.E., Krieger, A.M., Wind, Y., 2001. Thirty years of conjoint analysis: reflections and prospects. Interfaces 31 (3), 56–73.

Green, P.E., Srinivasan, V., 1990. Conjoint analysis in marketing: New developments with implications for research and practice. J. Mark, 3–19.

Hackbarth, A., Madlener, R., 2016. Willingness-to-pay for alternative fuel vehicle characteristics: A stated choice study for Germany. Transp. Res. A Policy Pract. 85, 89–111.

Halme, M., Kallio, M., 2011. Estimation methods for choice-based conjoint analysis of consumer preferences. Eur. J. Oper. Res. 214 (1), 160–167.

Hanley, N., Mourato, S., Wright, R.E., 2001. Choice modelling approaches: A superior alternative for environmental valuation? J. Econ. Surv. 15 (3), 435–462.

Hickling, A., 1989. Gambling with frozen fire? In: Rosenhead, J. (Ed.), Rational Analysis for a Problematic World. John Wiley and Sons, Chichester.

Huber, J., Wittink, D.R., Fiedler, J.A., Miller, R., 1993. The effectiveness of alternative preference elicitation procedures in predicting choice. J. Mark. Res. 30 (1), 105–114.

Huber, J., Zwerina, K., 1996. The importance of utility balance in efficient choice designs. J. Mark. Res. 33, 307–317.

Ishizaka, A., Nemery, P., 2013. Multi-criteria Decision Analysis: Methods and Software. Wiley, Chichester.

Jacquet-Lagrèze, E., Siskos, Y., 2001. Preference disaggregation: 20 years of MCDA experience. Eur. J. Oper. Res. 130, 233–245.

Jaeger, S.R., Hedderley, D., MacFie, H., 2001. Methodological issues in conjoint analysis: a case study. Eur. J. Mark. 35 (11), 1217–1237.

Jain, A.K., Mahajan, V., Malhotra, N.K., 1979. Multiattribute preference models for consumer research: a synthesis. In: North American Advances in Consumer Research. vol. 6. pp. 248–252.

Johnson, R., Orme, B.K., 2010. Getting Started With Conjoint Analysis: Strategies for Product Design and Pricing Research. Research Publishers LLC., Madison.

Keeney, R.L., 1992. Value-Focused Thinking. Harvard Press, Cambridge, Massachussets.

Keeney, R.L., 1996. Value-focused thinking: Identifying decision opportunities and creating alternatives. Eur. J. Oper. Res. 92, 537–549.

Keeney, R.L., Raiffa, H., 1993. Decisions With Multiple Objectives–Preferences and Value Tradeoffs. Cambridge University Press, Cambridge & New York.

Keisler, J., 2012. Is Value Focused Thinking a Problem Structuring Method or Soft OR or what? In: Management Science and Information Systems Faculty Publication series. vol. 42.

Kirppu, H., Lahdelma, R., Salminen, P., 2018. Multicriteria evaluation of carbon-neutral heat-only production technologies for district heating. Appl. Therm. Eng. 130, 466–476.

Kotler, P., 2000. Marketing Management, Millennium Edition. Prentice Hall, New Jersey.

Kotri, A., 2006. Analysing customer value using Conjoint Analysis: The example of a packaging company. University of Tartu, Estonia, Tartu working paper.

Krause, R.M., Lane, B.W., Carley, S., Graham, J.D., 2016. Assessing demand by urban consumers for plug-in electric vehicles under future cost and technological scenarios. Int. J. Sustain. Transp. 10 (8), 742–751.

Kroes, E.P., Sheldon, R.J., 1988. Stated preference methods: an introduction. J. Transp. Econ. Policy 22, 11–25.

Kuhfeld, W.F., 2010. Marketing Research Methods in SAS. Experimental Design, Choice, Conjoint and Graphical Techniques. SAS 9.2 Edition.

Liao, F., Molin, E., Timmermans, H., Wee, B.V., 2018. The impact of business models on electric vehicle adoption: A latent transition analysis approach. Transp. Res. A Policy Pract. 116, 531–546.

Lieven, T., 2015. Policy measures to promote electric mobility—a global perspective. Transp. Res. A Policy Pract. 82, 78–93.

Louviere, J.J., 1988. Conjoint analysis modelling of stated preferences: a review of theory, methods, recent developments and external validity. J. Transp. Econ. Policy 22, 93–119.

Luce, R.D., Tukey, J.W., 1964. Simultaneous conjoint measurement: a new type of fundamental measurement. J. Math. Psychol. 1, 1–27.

Malhotra, N., 2008. Marketing Research: An Applied Orientation, fifth ed. Pearson Education, India.

Marttunen, M., Lienert, J., Belton, V., 2017. Structuring problems for multi-criteria decision analysis in practice: a literature review of method combinations. Eur. J. Oper. Res. 263, 1–17.

Mau, P., Eyzaguirre, J., Jaccard, M., Collinsdodd, C., Tiedemann, K., 2008. The "neighbor effect": Simulating dynamics in consumer preferences for new vehicle technologies. Ecol. Econ. 68 (1-2), 504–516.

Mingers, J., Rosenhead, J., 2004. Problem structuring methods in action. Eur. J. Oper. Res. 152, 530–554.

Montibeller, G., Belton, V., Ackermann, F., Ensslin, L., 2008. Reasoning maps for decision aid: an integrated approach for problem-structuring and multi-criteria evaluation. J. Oper. Res. Soc. 59, 575–589.

Moore, W.L., 1980. Levels of aggregation in conjoint analysis: An empirical comparison. J. Mark. Res. 17, 516–524.

Morton, A., 2018. Multiattribute Value Elicitation. In: Dias, L., Morton, A., Quigley, J. (Eds.), Elicitation—The Science and Art of Structuring Judgement. Springer, Cham, pp. 287–311.

Myllyviita, T., Holma, A., Antikainen, R., Lähtinen, K., Leskinen, P., 2012. Assessing environmental impacts of biomass production chains—application of life cycle assessment (LCA) and multi-criteria decision analysis (MCDA). J. Clean. Prod. 29-30, 238–245.

Myllyviita, T., Lähtinen, K., Hujala, T., Leskinen, L.A., Sikanen, L., Leskinen, P., 2014. Identifying and rating cultural sustainability indicators: a case study of wood-based bioenergy systems in eastern Finland. Environ. Dev. Sustain. 16, 287–304.

Netzer, O., Toubia, O., Bradlow, E.T., Dahan, E., Evgeniou, T., Feinberg, F.M., Feit, E.M., Hui, S.K., Johnson, J., Liechty, J.C., Orlin, J.B., Rao, V.R., 2008. Beyond conjoint analysis: advances in preference measurement. Mark. Lett. 19 (3-4), 337–354.

Neves, L.P., Martins, A.G., Antunes, C.H., Dias, L.C., 2008. A multi-criteria decision approach to sorting actions for promoting energy efficiency. Energy Policy 36 (7), 2351–2363.

Neves, L.M.P., Martins, A.G., Antunes, C.H., Dias, L.C., 2004. Using SSM to rethink the analysis of energy efficiency initiatives. J. Oper. Res. Soc. 55, 968–975.

Neves, L.P., Dias, L.C., Antunes, C.H., Martins, A.G., 2009. Structuring an MCDA model using SSM: a case study in energy efficiency. Eur. J. Oper. Res. 199, 834–845.

Oliveira, G.D., Dias, L.C., 2015. Which criteria matter when selecting a conventional or electric vehicle? In: Energy for Sustainability 2015—Sustainable Cities: Designing for People and the Planet, 14-15th May, Coimbra, pp. 1–10.

Orme, B., 2009a. Which conjoint method should I use?. Research paper series, Sawtooth Software.

Orme, B., 2009b. Software for Hierarchical Bayes: Estimation for CBC data. Research paper series, Sawtooth Software.

Parsons, G.R., Hidrue, M.K., Kempton, W., Gardner, M.P., 2014. Willingness to pay for vehicle-to-grid (V2G) electric vehicles and their contract terms. Energy Econ. 42, 313–324.

Rodrigues, T.C., Montibeller, G., Oliveira, M.D., Bana e Costa, C.A., 2017. Modelling multi-criteria value interactions with Reasoning Maps. Eur. J. Oper. Res. 258, 1054–1071.

Rosenhead, J., Mingers, J. (Eds.), 2008. Rational analysis for a problematic world revisited: problem structuring methods for complexity, uncertainty and conflict, 2. ed., repr. ed. Wiley, Chichester.

Roth, S., Hirschberg, S., Bauer, C., Burgherr, P., Dones, R., Heck, T., Schenler, W., 2009. Sustainability of electricity supply technology portfolio. Ann. Nucl. Energy 36 (3), 409–416.

Roy, B., 1991. The outranking approach and the foundations of electre methods. Theor. Decis. 31 (1), 49–73.

Roy, B., 1996. Multicriteria Methodology for Decision Aiding. Kluwer Academic, Dordrecht.

Saaty, T.L., 2008. Decision making with the analytic hierarchy process. Int. J. Serv. Sci. 1 (1), 83–98.

Sattler, H., Hensel-Borner, S., 2007. A comparison of conjoint measurement with self-explicated approaches. In: Gustaffson, A., Herrmann, A., Huber, F. (Eds.), Conjoint Measurement: Methods and Application. Springer, Berlin Heidelberg, pp. 3–30.

Scholz, S.W., Meissner, M., Decker, R., 2010. Measuring consumer preferences for complex products: a compositional approach based on paired comparisons. J. Mark. Res. 47 (4), 685–698.

Sheldon, T.L., Deshazo, J.R., Carson, R.T., 2017. Electric and plug-in hybrid vehicle demand: lessons for an emerging market. Econ. Inq. 55 (2), 695–713.

Soto, J.J., Cantillo, V., Arellana, J., 2018. Incentivizing alternative fuel vehicles: the influence of transport policies, attitudes and perceptions. Transportation, 45 (6), 1721–1753.

Smith, C.M., Shaw, D., 2019. The characteristics of problem structuring methods: a literature review. Eur. J. Oper. Res. 274, 403–416.

Steenkamp, J.-B.E.M., van Trijp, H.C.M., 1997. Attribute elicitation in marketing research: a comparison of three procedures. Mark. Lett. 8 (2), 153–165.

Vreuls, H., 2004. Task 1. INDEEP Final Report, International Energy Agency.

Wolfslehner, B., Vacik, H., 2011. Mapping indicator models: from intuitive problem structuring to quantified decision-making in sustainable forest management. Ecol. Indic. 11 (2), 274–283.

Yoon, K., Hwang, C.-L., 1995. Multiple Attribute Decision Making: An Introduction. Sage Publications.

Zhang, T., Gensler, S., Garcia, R., 2011. A study of the diffusion of alternative fuel vehicles: an agent-based modelling approach. J. Prod. Innov. Manag. 28, 152–168.

Promoting behaviour change

A critical review of energy behaviour change: The influence of context

4.1

Eva Heiskanen, Kaisa Matschoss, Senja Laakso, Eeva-Lotta Apajalahti

University of Helsinki, Helsinki, Finland

1 Introduction

The need for drastic cuts in carbon emissions has intensified the interest in policies for interventions to change people's energy behaviour (Clayton et al., 2016; Dietz et al., 2009; Kok et al., 2011). Yet the question whether interventions work similarly at different times, in different places, and in different social contexts is highly relevant for their ecological validity and transferability (Clayton et al., 2016; Heiskanen et al., 2018). Considering, for example, household energy use for space heating, there are huge variations among countries in physical infrastructures and cultural expectations of thermal comfort (Laakso and Heiskanen, 2017). Intervention studies focusing on individual behaviour in isolation might fail to acknowledge these differences, thus falling short in creating changes in energy use patterns on a larger scale.

The dominance of the individualist perspective prevalent in psychological research on energy use has recently gained a great deal of criticism (Shove, 2010; Keller et al., 2016). Some authors have pointed out that this criticism paints a stylised and simplistic picture of existing intervention research (Whitmarsh et al., 2011), and the debate has continued for several rounds (Wilson and Chatterton, 2011; Shove, 2011). While the debate has raised welcome discussion on the scope and focus of energy conservation policy, it has not—until yet—greatly influenced how policymakers engage with energy users. Sociological critiques of energy use in society have been in many places overshadowed by enthusiasm surrounding behavioural economics and 'nudges' (Jones et al., 2014).

Much of the debate around individualist versus social conceptions of energy conservation has focused on problem definition: whether it is a problem of individual behaviour or of the social organisation of everyday life (Shove, 2011). We are aware of fundamental ontological differences among methodologically individualist and culturalist perspectives but take a different approach to this debate: we start by accepting the problem definition of energy interventions, that of changing individuals'

Energy and Behaviour. https://doi.org/10.1016/B978-0-12-818567-4.00015-6

energy behaviour, but focus on the ecological validity, transferability, and scalability of existing research (Laakso and Heiskanen, 2017; Heiskanen et al., 2018) by exploring how context is reflected in research on behaviour change interventions. The recognition of context dependence has implications for the evaluation of traditional behaviour change interventions (such as convenience, information, feedback, or social influence), which, in the literature, are usually treated as universal and context-independent *mechanisms*, rather than dependent on context to achieve their expected outcomes (Pawson and Tilley, 1997).

Context here refers not only to spatial, geographical, or institutional locations such as particular countries or towns but also to the sets of social rules, norms, and values (e.g. at the workplace) and sets of social relationships pre-existing the introduction of the intervention (following the definition of Pawson and Tilley, 1997). Furthermore, several overlapping contextual aspects are likely to influence the outcomes of intervention mechanisms. If the role of contexts is neglected in the overall evaluation of behavioural interventions, we are at risk of not understanding why interventions result in particular outcomes (Clayton et al., 2016; Laakso, 2019). Since the interventions depend on changing the context of individuals, the feasibility of maintaining and replicating such context changes is also critical for the durability and scalability of interventions in energy behaviours.

In this chapter, we first offer an overview of the existing research evidence from meta-analyses and systematic reviews concerning the most popular behaviour change interventions aiming to support energy conservation in the home while paying special attention to how context is taken into account. We focus on interventions to change habitual energy behaviours (i.e. daily routines rather than one-shot investments, cf. Steg, 2008), because the role of context has been explored less for these kinds of routine behaviours (Šćepanović et al., 2017). We explore the role of three overlapping contexts that influence behaviour change interventions: organisational context, geographical (i.e. national and spatial) context, and practice as context. Through the notion of practice as context, we also attempt to stretch the problem space. As one example, some social influence interventions inevitably extend from the individual to the social sphere by articulating and communicating a particular type of (descriptive) social norm ('what others are doing'). Such interventions are constrained by prevailing social norms in the intervention context (Schubert, 2017). Interventions can also attempt to stretch existing social norms, but then, they need to do this differently. Towards the end of the chapter, we provide suggestions on how to better address—and potentially even change—contextual factors in sustainable energy interventions, using insights from existing cases.

2 Overview of the most popular behaviour change intervention mechanisms

In this section, we investigate the existing research evidence from meta-analyses and review studies concerning various interventions to support energy conservation. Here, interventions are examined as generic and universal set of *mechanisms*

(Pawson and Tilley, 1997) that can be more of less effective in terms of changing behaviour towards desired outcomes. These mechanisms draw on diverse traditions in behavioural science and often even a combination of theoretical bases (Ölander and Thøgersen, 2014; Wilson and Dowlatabadi, 2007). In the following, we review evidence of the effectiveness of these mechanisms while also keeping an eye open for indications of how context might influence outcomes.

The psychological literature (Ones and Dilchert, 2012; Osbaldiston and Schott, 2012) identifies four broad intervention types: convenience/salience, information, monitoring/feedback, and social influence. While different authors use different groupings, Table 1 outlines the various mechanisms commonly understood to fall under each category. We attempt to relate our findings back to the categories in this table in our review.

Several behaviour change mechanisms aim to change the context in which the behaviour occurs (e.g. by making behaviour change easy via a prompt reminding addressees to turn off the light) or to add information to the context, either via persuasion or feedback. Context change is also central to behaviour change strategies drawing on the popular notions of 'nudge' and 'choice architectures', that is, attempts to change the context of decision-making without explicitly forbidding a particular option (Thaler and Sunstein, 2008. The nudge tools are largely similar to those explored in other behaviour change interventions (e.g. Lehner et al., 2016), with the exception of the notion of defaults, that is, attempts to rectify behaviour that is based on the 'status quo bias' (Thaler and Sunstein, 2008), which we have classified under 'convenience/salience'. Social influence is also a popular 'nudge tool', with a focus on *descriptive social norms* (telling people what others do) (Thaler and Sunstein, 2008; Lehner et al., 2016). Other categories of social influence mechanisms include the recruitment of peer influencers (such as 'block leaders' or 'environmental

Table 1 Mechanisms used in behaviour change interventions

Convenience/ salience	Information	Monitoring/ feedback	Social influence
<ul><li>Technical devices and support</li><li>Default settings</li><li>Prompts, checklists, visuals</li></ul>	<ul><li>Persuasion, why-to information</li><li>Education, training, how-to information</li></ul>	<ul><li>Informative feedback</li><li>Rewards, recognition</li></ul>	<ul><li>Peer influencers, modelling</li><li>Public commitments, goal setting, competition, challenge</li><li>Social comparison feedback</li><li>Games, competitions</li></ul>

Modified from Ones, D. S., Dilchert, S., 2012. Environmental sustainability at work: a call to action. Ind. Organ. Psychol. 5(4), 444–466 and Osbaldiston, R., Schott, J. P., 2012. Environmental sustainability and behavioural science: meta-analysis of proenvironmental behaviour experiments. Environ. Behav. 44(2), 257–299.

champions'), the enticement of people to make public commitments or set goals, and the provision of comparative feedback (Abrahamse and Steg, 2013).

Osbaldiston and Schott (2012) conducted a meta-analysis of 87 published studies of 253 research-driven experimental interventions that involved an observed pro-environmental outcome (in recycling, energy conservation in the home, or driving behaviour). It shows that there is quite a large variation in how effective different interventions are in influencing different kinds of behaviours (recycling, water and energy usage, etc.). Table 2 summarises their findings concerning weighted average effect sizes, which is a measure of how much the experimental group differs from the control group. Column 4 expresses the weighted average effect sizes for all interventions and column 5 for home energy conservation interventions. According to Osbaldiston and Schott (2012), an effect size of 0.20, for example, means that the treatment group scored 0.20 standard deviation units higher than the control group. Using the standard normal distribution table, percentages can also be assigned to effect sizes. An effect size of 0.2 means that the intervention increased the desired behaviour by 8%; one of 0.5 that the intervention increased the desired behaviour by 19%; and one of 0.8, an increase of 29% (Osbaldiston and Schott, 2012).

Osbaldiston and Schott's (2012) meta-analysis suggests that interventions have different outcomes depending on the targeted behaviour (e.g. waste recycling vs. home energy use). Moreover, there is variation in how effective different combinations of interventions are, since some combinations (e.g. making it easy combined with prompts or goal setting combined with instructions) outperformed other combinations. The findings concerning the effectiveness of commitments (both alone and in combination with other interventions) are reinforced by a dedicated meta-analysis (Lokhorst et al., 2013) focusing only on studies involving commitment interventions ($n=19$).

As another example, Delmas et al. (2013) conducted a meta-analysis of 156 experimental studies (mostly from the United States) investigating various interventions to promote electricity saving in households (energy-saving tips, audits and consulting, various kinds of feedback, monetary saving information, and monetary incentives). On average, the interventions applied in the studies rendered a saving effect of 7.4%. When modelling the effects in combination with the entire set of interventions, Delmas et al. (2013) could not find (independent) conservation effects from some of the interventions that are often deemed effective, such as individual usage feedback and comparative feedback (though they noted that the sample included few comparative feedback studies). Yet they found that real-time feedback and home energy audits triggered energy conservation. They also found that the effects of the interventions declined over time. Monetary saving information was found to actually increase energy usage, which Delmas et al. (2013) interpret to be because the savings to be obtained from electricity conservation are relatively small in the typical US context.

The focus of the Delmas et al. (2013) meta-analysis is not on contextual factors (i.e. geographical or social contexts), but the study suggests they play a role in the outcomes obtained. The standard deviation of the total effect size across these

Table 2 Selected results from Osbaldiston and Schott's (2012) meta-analysis of experimental pro-environmental behaviour interventions

Category	Mechanism type	Example	Weighted average effect size	
			All interventions	Home energy interventions
Convenience	Making it easy	Providing low-flow showerheads	0.49	n.a.
	Prompts	Reminder to turn off lights when leaving room	0.62	0.00
Information	Justifications	'Why-to' information, for example, how much material is dumped that could be recycled	0.43	0.17
	Instructions	'How-to' information, for example, use blinds to avoid overheating	0.31	0.17
Monitoring	Feedback	For example, monthly electricity billing	0.31	0.28
	Rewards	Cash, coupons, rebates, prizes	0.46	0.45
Social influence	Social modelling	Demonstration of desired behaviour by peers	0.63	0.74
	Cognitive dissonance	'Foot-in-the-door' intervention where participants asked for a small act first	0.94	0.29
	Commitment	Pledge cards	0.40	0.55
	Setting goals	Aiming, for example, 20% reduced power consumption	0.64	0.31

studies is relatively high, 10.0 (compared with the saving effect of 7.4%). Delmas et al. (2013) only explored the impact of two contextual factors: of these, socio-demographic factors (income, education, etc.) were a significant explanatory factor for the observed total effect size and for the individual effects of energy audits, monetary incentives, and individual usage feedback. This can be interpreted to imply that people who are likely to be concerned about high electricity usage, for example due to large bills or environmental concerns, and perhaps people who have the means and opportunity to invest in energy-saving measures are more responsive than average to these interventions. The results show that it might not make sense to examine the impacts of savings, for example, without considering their share in total household expenditure (i.e. the socioeconomic context).

Abrahamse and Steg (2013) have made a separate meta-analysis of 20 studies on social influence instruments: peer influencers, public commitments, modelling and social comparison, and group feedback. They found these mechanisms to be effective in behaviour change, both compared with control groups and to other types of interventions. They found an effect size of 0.35 (i.e. small to medium) compared with control groups and an effect size of 0.22 compared with the other types of interventions tested. Here, the use of neighbourhood 'block leaders' (i.e. volunteer peer influencers providing information and serving as an example their own social networks) was the most effective type of intervention. However, Abrahamse and Steg (2013) found that intervention outcomes depended on the target group: social influence strategies were more effective for employees than for students or hotel guests. They also found personal and face-to-face influence more effective than, for example, social comparison feedback provided via energy bills or electronically. These observations suggest that social influence interventions require a context of interpersonal communication and group identification to reach their full potential.

The meta-analytical evidence suggests that some of the intervention research might have low ecological validity (i.e. results may not be relevant in the actual context of implementation, see Nastasi and Schensul, 2005). One example is the problems observed with one of the psychologists' favourite interventions, commitments. According to Lokhorst et al. (2013), few studies report on how they include or exclude people who refused to make a commitment in their analysis. If only volunteers are included, it is clear that the ecological validity of commitment making (as a policy instrument) is low. Another example can be drawn from Abrahamse and Steg (2013): communicating descriptive social norms (i.e. average energy consumption of neighbours) may backfire if people are already consuming below the communicated norm and start to consume more when they are informed of the norm (see also Diffney et al., 2013). Furthermore, ecological validity is also constrained by the fact that few studies have assessed the long-term outcomes, that is, whether behaviour change persists after the intervention (Abrahamse and Steg, 2013; Delmas et al., 2013).

A similar concern about the persistence of effects can be found in reviews of mobility behaviour change. A systematic review by Arnott et al. (2014) on 27 behavioural interventions to reduce car use (targeted motivation and psychological capability through education and knowledge, incentivisation, or enablement and habit reduction) found no evidence on the significant efficacy of interventions. These findings are supported by a review of 77 behavioural interventions by Graham-Rowe et al. (2011) and Redman et al. (2013) review on 74 studies of public transport improvements. Their results indicate that interventions, such as free travel cards, encourage bus use in the short term, but the effects are not maintained when the incentives are removed. Similarly, economic disincentives can reduce car use, but again, this approach may only be effective while the disincentives remain in place (Graham-Rowe et al., 2011). Accessibility, reliability, and mobility provision must also be provided to sustain the switch from car use (Redman et al., 2013). For example, Fujii and Kitamura (2003) and Thøgersen (2009) conducted studies on the impact of a free travel card on total 396 car users in Kyoto and Copenhagen. In both cases, the intervention succeeded in attracting car users to use public transportation, but in post intervention, the participants failed to use public transportation more than control subjects (Thøgersen, 2009).

Even though experimental research suggests that there are effective interventions, the picture becomes mixed when considering the meta-analytical evidence. Even if there are (in aggregate) effective interventions like convenience, feedback, and peer influencers, few interventions invariably have statistically significant effects on energy consumption, and the variance in outcomes is large. This observation supports our proposition that intervention mechanisms work differently in different contexts (Clayton et al., 2016; Pawson and Tilley, 1997), though relevant aspects of context may be difficult to determine from reported studies. Slightly stronger effects are observed when considering combinations of interventions, which again suggest that diverse interventions serve as contexts for each other (Dahler-Larsen, 2001). It seems reasonable to argue that at least part of the variability is due to the context-dependence of interventions.

3 Critical review of the role of context in behaviour change interventions

As the previous section illustrates, several behaviour change mechanisms aim to change the context in which the behaviour occurs yet often fail to account for other aspects of the context in which the intervention is placed. This is because they often draw on a line of psychological research that aims to uncover universal invariables in human behaviour, rather than more middle-range contextual conceptualisations (Gergen, 1973). This section investigates the evidence on the context dependence of behavioural change attempts. We first present our conceptual framework and then explore the influence of different contextual features on each category of interventions.

3.1 Interventions as context dependent: Realistic evaluation

To conceptualise the role of context, we draw on Pawson and Tilley's (1997) concept of realistic evaluation, which aims to extract generalisable lessons from analyses of interventions. Realistic evaluation explicitly recognises the importance of context in making an intervention work (Pawson and Tilley, 1997). Intervention *outcomes* are thus dependent on both the type of intervention *mechanism* used and the *context* (Pawson and Tilley, 1997). Mechanisms work differently in different contexts, and the context can be part of the reason why a particular initiative or project works or fails (Pawson and Tilley, 1997). Revealing the contextual factors and their role for a certain outcome is important not only for understanding the outcomes within different contexts but also for the transferability and generalisability of interventions (Blamey and Mackenzie, 2007).

While there are necessarily several definitions of context, Pfadenhauer et al. (2017, p. 5) illuminate the context of an intervention: 'Context reflects a set of characteristics and circumstances that consist of active and unique factors, within which the implementation is embedded. As such, context is not a backdrop for implementation, but interacts, influences, modifies and facilitates or constrains the intervention and its implementation. Context is usually considered in relation to an intervention, with which it actively interacts. It is an overarching concept, comprising not only a physical location but also roles, interactions and relationships at multiple levels'. Context thus refers to not only spatial, geographical, or institutional locations (such as particular countries or towns) but also the prior sets of social rules, norms, values, and sets of social relationships pre-existing the introduction of the intervention (Blamey and Mackenzie, 2007; Pawson and Tilley, 1997).

In the following section, we explore three categories of context: organisational context, geographical (i.e. national and spatial) context, and 'practice as context'. Organisational context is highly relevant for energy use, because energy is used for different purposes, for example, in households and at workplaces, and control and authority over energy use ensue from different sources (Staddon et al., 2016). Both workplaces and households may involve individuals who 'behave' in a certain way, but their behaviour is regulated by different considerations. Geographical context seems obvious, considering that energy use is influenced by a range of climatic, policy, and institutional factors and by different traditions within the built environment (Wilhite et al., 1996). Finally, it is likely that the social practices in which the targeted behaviour is embedded are also likely to influence outcomes, since there is variability among intervention outcomes when different types of behaviours (which are part of diverse social practices) are targeted (Osbaldiston and Schott, 2012). Fig. 1 illustrates

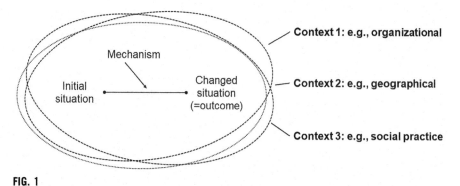

FIG. 1

Pawson and Tilley's (1997) realistic evaluation framework and three types of overlapping contexts (authors' visualisation).

the notion of context-mechanism-outcome combinations, suggesting three types of overlapping contexts that are likely to influence the outcomes of interventions.

This perspective has implications for the evaluation of generic behaviour change interventions, which are usually treated as universal and context independent in the psychological literature. The literature on these interventions thus only refers to *mechanisms* and *not to the mechanism–context combinations*. In the parlance of Pawson and Tilley (1997), such a definition of interventions could be termed interventions as mechanisms (IAM), whereas when we understand the outcomes of interventions as dependent on both mechanism and context, we could speak of interventions in real life (IIRL).

3.2 Organisational context: Energy-saving behaviour at home and at the workplace

Most of the research on behaviour change interventions has been conducted in the context of individuals or private households (Ones and Dilchert, 2012) or in lab settings devoid of context. However, since meta-analyses often fail to distinguish between interventions targeting one individual from interventions influencing several household members, the role of household context remains largely unexplored. In the following, we first consider the social dynamic of household and workplace contexts. We then turn to investigating how organisational context can influence the outcomes of commonly used intervention mechanisms.

3.2.1 The social dynamic of household and workplace contexts

Some studies have assessed the role of household context, often in a qualitative manner to understand the social dynamics underlying interventions. For example, Hargreaves et al. (2010) explored the way in which 15 households used an energy monitoring display they were offered as part of a larger research project aiming to influence energy use via feedback. Among others, the monitoring displays provoked family negotiations, sometimes in the form of deliberation among household members but often in the form of disputes or stealthy avoidance (e.g. a participant's wife hid the display). Similar dynamics were observed by Nyborg (2015) in an analysis of 49 Danish households pioneering smart home devices.

Outcault et al. (2018) investigated the social dynamics underlying the failure of two interventions to promote passive cooling measures (e.g. opening windows, turning off air conditioning, and using fans) in California and Japan. One important reason for this failure turned out to be the decision to engage a single household member without giving advice on whether and how participants should involve the other members of their households, even though most measures had an impact on other household members. Outcault et al. (2018) found that several measures required coordination (i.e. opening windows only when the air conditioning unit was off), which was sometimes successful but more often not. Participants had to resort to imposing measures on other household members, which could lead to acceptance, to adaptation, or to resistance and social disruption. In conclusion, Outcault et al. (2018)

suggested that interventions might avoid some of these problems by including instructions for households on how they could coordinate actions productively (via, e.g., setting an example or engaging in friendly competition rather than dictating and nagging) to avoid social disruption.

The same intervention can work very differently at work and at home. Nye and Hargreaves (2010) present an apt example of how two similar, team-based behaviour change interventions at work (Environment Champions) and at home (EcoTeams), based on group deliberation and feedback auditing waste production and energy use, worked differently at the workplace and in households. The advice and feedback shared within the Environmental Champions teams were interpreted in terms of 'hard data' considered to conform to business understandings of appropriate drivers for behaviour change, whereas the household EcoTeams focused more on sharing 'folk knowledge' defining what green lifestyles mean and how participants could use locally available resources in more environmentally sustainable ways.

Indeed, Nye and Hargreaves (2010) found that the intervention served a completely different purpose in these different contexts. Where the Environmental Champions intervention worked within conventional rules for office conduct so as to renegotiate their meaning in incremental fashion and to encourage co-workers to conform to the new rules, the household EcoTeams served to help households to share social support and to find ways and workarounds to lead a greener lifestyle within the constraints of local systems of provision. This example shows how interventions employing similar mechanisms, such as feedback and social influence, result in different outcomes for energy behaviour at home and at work. At the workplace, the outcomes were confined to incrementally energy-efficient 'good business behaviour' and did not result in radical changes, whereas in households, the interventions gave rise to attempts towards deeper energy savings. This also illustrates the importance of addressing the conventions and norms, which greatly impact on energy behaviours.

3.2.2 Energy behaviour change interventions in organisations

Organisations are collective settings, usually with some kind of hierarchical structure, where shared resources are governed by formal systems of management (Staddon et al., 2016; Ones and Dilchert, 2012; Andrews and Johnson, 2016). A large number of organisations today have made some kind of environmental commitment, almost invariably including commitments to save energy (e.g. St Lawrence, 2004). However, there is limited systematic research on behavioural energy interventions at other organisations than schools and universities. In the following, we draw on a systematic review of 22 behaviour change interventions in the workplace by Staddon et al. (2016) and a review of 84 energy interventions in commercial office buildings by Kosonen and Kim (2017), looking at contextual influence on each category of interventions (cf. Table 1).

Convenience interventions in organisations can consist of checklists, tips, energy dashboards, automation, or increased control over devices (termed 'enablement' and 'environmental restructuring' by Staddon et al., 2016). Convenience measures were part of the top-performing types of intervention combinations in organisations.

Though the independent effect of these types of measures was difficult to deduce from the studies, most of which combined several intervention types, Staddon et al. (2016) pointed out that investments in the workplace might, in addition to convenience, also signal organisational commitment, rather than leaving all the responsibility to employees (see also Young et al., 2015). While convenience is clearly important, it requires good coordination in organisations, since Staddon et al. (2016) also report on a study where the information technology support unit undermined the outcomes of an intervention making it more convenient for employees to turn off their workstations, since the unit wanted workstations on to run updates.

Information was often a part of a package of interventions but alone rendered low impacts (4% reduction), in the studies reviewed by Kosonen and Kim (2017), though moderate outcomes were gained from personalised information campaigns combined with peer influencers (see also Staddon et al., 2016). However, information could be part of a package of measures producing impressive outcomes (31% energy savings), as in a school programme combining assemblies, assignments, signs, posters, and letters to parents, as well as an activism by leaders and staff (Staddon et al., 2016). Importantly, Staddon et al. (2016) point out that some of the limitations of information interventions may be due to the fact that information campaigns are often top-down and disregard the contextualised information that organisation members might have, which might contribute to more effective measures if taken into use (see Foster et al., 2012).

Feedback is today increasingly digital and can be enhanced with games that entail an element of competition. For example, Kosonen and Kim (2017) found visible short-term impacts from 'serious games', such as an online game translating employee's energy use into the health of virtual chickens. On the other hand, Staddon et al. (2016) found that in some cases, personal feedback on energy use attracted little employee interest, suggesting that energy use feedback needs to compete with other workplace priorities. Feedback can also raise potential privacy concerns and anxieties over increased managerial control (Staddon et al., 2016). Staddon et al. (2016) point out that intervention studies have rarely taken into consideration these wider workplace contexts, including how the interventions interact with employees' feelings of responsibility, control, trust, and privacy, or the influence of staff relations and wider organisational pressures. Kosonen and Kim (2017) also point to the practical feasibility of scaling up research-driven feedback interventions in real-life commercial buildings.

Rewards are a much-investigated topic in workplaces and communities (Norton et al., 2015; Young et al., 2015). The outcomes clearly depend on the social dynamics between participants. Incentives could crowd out intrinsic motivation and result in tactical behaviour (Alberts et al., 2016; Staddon et al., 2016), or they might increase intrinsic motivation when they are interpreted as evidence of the participants' competence (Maki et al., 2016). Kosonen and Kim (2017) found that some employees considered some reward (e.g. badges and prizes) trivial, and Staddon et al. (2016) found that none of the highest-performing interventions in their review employed incentivisation. These observations suggest that rewards should treat employees seriously and respectfully and align with existing organisational cultures and priorities.

Social influence interventions are complex to apply in organisations. There are many case studies on the role of leaders and 'environmental champions' in setting an example for other employees and promoting energy-saving efforts within their own social networks (Andrews and Johnson, 2016; Nye and Hargreaves, 2010). Staddon et al. (2016) found that a large share of the relatively successful studies in their review involved peer educators and champions, either as an intended or unintended part of the intervention. Comparative feedback and competition have provided not only some good energy-saving outcomes (Kosonen and Kim, 2017) but also reactance by employees suggesting that intervention designers should carefully consider issues of privacy and the scale on which comparisons are made (Staddon et al., 2016). Another example from Staddon et al. (2016) is the counterproductive behaviour produced by one of the energy-saving games, where users teased their co-workers by turning on other people's workstations to make them lose points. Social influence interventions also highlight the issue of persistence, which might be different for a short-term competition and an organically emerging 'social movement' to save energy within an organisation.

In summary, there is contextual variation in the effectiveness of energy behaviour interventions directed at individuals due to organisational policies and cultures (Ones and Dilchert, 2012). In real organisations, energy savings can be hindered by a lack of personal concern over the consequences of workplace energy use and a lack of common purpose among employees (Andrews and Johnson, 2016). For example, Leygue et al. (2017) found that identification, commitment, and overall well-being in the workplace influenced employees' intentions to engage in energy - saving behaviours, since energy saving at work depends on people wanting to help their organisation. Organisations also consist of smaller work groups, each with their own work styles (including social networks and codes of appropriate behaviour), and it is at this level that work context influences daily energy behaviour in real buildings (Bull and Janda, 2018). Indeed, a notable factor in existing reviews is their focus on how organisational context might hinder the effectiveness of interventions, rather than on how it might help (Janda, 2014), which would require a more fine-grained exploration of organisational context before deciding to apply a particular intervention (Foster et al., 2012; Kosonen and Kim, 2017; Staddon et al., 2016).

3.3 Geographical context

Cross national comparative research of energy interventions is difficult to find, which is strange because national and spatial context is likely to influence the effectiveness of such intervention mechanisms in several ways (Clayton et al., 2016). For example, thermal comfort is experienced differently in different countries (Vávra et al., 2016). Moreover, even within Europe, there are large institutional and infrastructural differences in the extent to which residents can control their space heating in different types of dwellings (Laakso and Heiskanen, 2017). And even within an individual city, as in the case of Detroit, Michigan, different district housing people of different class and ethnicity can experience diverse problems in space heating (Bednar et al., 2017).

Hence, geographical context can be implicated in each of the categories outlined in Table 1 (convenience, information, feedback, and social influence).

Outcomes from increased *convenience* are likely to depend on how *in*convenient energy saving is to start with and what measures would improve convenience in the users' physical and institutional context (Bednar et al., 2017). Increasing convenience in a culturally sensitive way requires close understanding of the context of energy use. For example, Spencer et al. (2015) created design guidelines for increasing the convenience of energy saving in laundry care, based on observations of households in the United Kingdom, India, and Brazil. The authors found very different configurations of laundering in these different country contexts. They concluded that increasing convenience requires understanding *where* laundry is done (e.g. bathroom, specific laundry room, or on the rooftop), what the *touchpoints* of the process are (who does what and when), and what the *perceptions and aspirations of the users* are.

Information-based interventions might entail different behavioural outcomes in different types of cultural contexts. Morren and Grinstein (2016) conducted a meta-analysis of determinants of behaviour based on 66 published studies from 28 countries, looking at studies testing the theory of planned behaviour (TPB), which is often used to guide information-based interventions. They categorised countries according to level of economic development, the UN Human Development Index, and collectivism–individualism (based on the World Values Survey). Since most of the studies are from OECD countries, this categorisation is perhaps not the best, but it rendered some suggestive findings. The study found that attitudes and intentions predict behaviour better in wealthier and individualistic countries (e.g. in Switzerland, the Netherlands, and Denmark) than in more collectivist and less wealthy countries. This is not surprising since the attitude construct is based on assumptions of teleological and individualistic behaviour. In contrast, social norms were found to be equally predictive of behaviour across the diverse contexts.

Mainly *information-based* interventions have been systematically reviewed by Šćepanović et al. (2017), based on 189 reported field studies of residential energy interventions in different geographical contexts. Tenure of the residence was the most influential *physical* context: participants in rented homes (common in some countries and rare in others) were less likely to respond to all kinds of commonly used interventions, particularly audits. In terms of *socioeconomic contexts*, Šćepanović et al. (2017) found that low-income and vulnerable households are likely to respond to interventions differently from other groups, suggesting that many intervention mechanisms may have an implicit middle-class bias. *Cultural factors* were most widely recognised in field interventions: interventions could clash with the cultural context if participants feel that their comfort, lifestyles, and habits are threatened. In terms of *political and governmental factors*, audits were found to gain support from favourable political contexts, whereas several types of interventions were found to be undermined by the political conservativism of participants and by unsupportive energy utilities. Gamification and rewards, communication on energy conservation co-benefits, goal setting, and commitment were found to perform relatively well across political contexts (Šćepanović et al., 2017), though it is not known how participants for games or commitments were selected (i.e. whether they perhaps represent a relatively cosmopolitan group of educated people).

There are also some observations concerning geographical variation in responsiveness to *feedback*. Vine et al. (2013) suggest that cultural context might influence the acceptability of comparative feedback, while Winther and Bell (2018) found that feedback from in-home displays had different implications in the United Kingdom (where electricity is carbon intensive but used for few applications) and in Norway (where electricity use has fewer environmental but greater cost implications due to its use for space heating). Interest in feedback may thus be influenced by the physical set-up of homes and by structural factors in the energy supply industry, such as tariff structures (EEA, 2013). Such factors might underlie the observed differences of the impact of in-home display and informative energy billing pilots, where European pilots outperformed those conducted in the United States (Stromback et al., 2011). Drawing on insights from experienced practitioners from 10 European countries, Backhaus and Heiskanen (2009) summarised some of the context dependencies of feedback: feedback works where people have the opportunity to change, and the information makes sense in terms of participants' prior experience (see also Bird and Legault, 2018).

The applicability of *social influence* interventions is likely to depend on cultural and spatial patterns of social interaction (Bale et al., 2013), that is, 'who influences whom and how' (Backhaus and Heiskanen, 2009). It is also likely to work differently in different cultural contexts: as an example, Culiberg and Elgaaied-Gambier (2016) found that injunctive social norms (i.e. what people's friends think they should do) influenced pro-environmental behaviours in Slovenia, but not in France. Practical experience from different countries (Backhaus and Heiskanen, 2009) suggests that the recruitment of peer influencers works where the participant group is fairly homogeneous and trusts each other, and its use requires a good understanding of the social dynamics in the local context. The cultural context may be equally important for practical applications of *commitments and goal setting*, which require that goals are accepted in the cultural context while being measurable (Backhaus and Heiskanen, 2009). Thus, Moloney et al. (2010) recommend that behaviour change interventions should be tailored to specific communities and systems of provision, they should recognise why participants might develop commitment to participate, and they should build on an understanding of how interventions influence participants' sense of agency.

The relevance of spatial context is highlighted by intervention studies related to mobility, which often focus on a specific area and/or journeys, aim to encourage people in reducing private car use, and include a combination of mechanisms targeting behaviour change. In their systematic review of 22 studies, Ogilvie et al. (2004) assessed the effectiveness of various interventions in shifting people from using cars towards walking and cycling in urban areas. They found some evidence that targeted behaviour change programmes (engaging a motivated group of volunteers by providing them materials such as self-help packages, leaflets, timetables, and free bus trial tickets) resulted in a shift of about 5% of all trips. However, they did not find evidence of the independent effect of publicity campaigns, engineering measures, or other non-personalised interventions. Similarly, Yang et al. (2010) concluded that it is unclear whether addressing attitudes and perceptions about cycling alone would yield increases in cycling, since a lack of supportive infrastructure might limit willingness to take up cycling, particularly in areas without an established cycling culture (see also Arnott et al., 2014).

In the same vein, Pucher et al. (2010) found positive associations between interventions and levels of cycling in their review of 139 studies on cycling interventions (such as changes in infrastructure and public transport, education and marketing programmes, bicycle access programmes, and legislative measures). However, this was most clearly the case in communities that implemented a set of strategies to increase bicycling (infrastructure provision and pro-bicycle programmes, supportive land use planning, and restrictions on car use), indicating that a comprehensive approach produces a greater impact than individual, uncoordinated measures. However, there was considerable variation in estimated impacts. They found no direction of causality, which left it open whether infrastructural or policy changes led to increased cycling levels, or vice versa, and whether the same infrastructure provision, programme, or policy would work in different contexts (Pucher et al., 2010).

Mobility behaviour change interventions have also demonstrated the potential power of a change in spatial context, that is, people moving from one place to another, as an opportunity to change mobility patterns (Graham-Rowe et al., 2011; Laakso, 2017; Schäfer et al., 2012; Thompson et al., 2011; Thøgersen, 2009). For example, when people move to a new locality, support such as free public transport tickets in conjunction with tailored advice has served to promote modal shift towards public transport. Habit disruptions have been found to provide an important 'window of opportunity' to change behaviour, but the extent of such change is likely to be limited without adjustments to the cultural and structural factors (e.g. Laakso, 2017; Thøgersen, 2012).

3.4 'Practice as context': Challenging shared routines and conventions

Several authors argue that the mechanisms used in energy behaviour interventions should be considered in the context of the broader household or workplace practices in which they attempt to intervene (Hargreaves, 2018; Nye and Hargreaves, 2010; Staddon et al., 2016). The core of this argument is that social life can be considered to consist of sets of 'scripts' or practices that enable and constrain people in performing daily activities: such as workplace codes of behaviour or collective expectations concerning cleanliness and thermal comfort (Hargreaves, 2011; Shove, 2003). These practices are shaped by systems of provision (such as appliances and the infrastructures they are connected to), embodied capabilities, and routines and shared understandings of appropriate and 'normal' behaviour (Laakso, 2019; Reckwitz, 2002; Shove and Walker, 2014). While there are differences in how individuals perform practices of, for example, keeping clothes clean, there is also a common base of similar socially shared routines, values and technologies that makes people in similar contexts behave rather similarly. For example, almost everyone in wealthy Northern countries today launders their clothes, or has them laundered, several times a month—in contrast to much more infrequent washing days in history (Shove, 2003).

The notion of social practice overlaps with other contexts. It is obvious that workplaces consist of different practices than homes and that conventions of cleanliness, thermal comfort, and other elements that constitute energy-using activities in the home vary between countries (Wilhite et al., 1996). It might nevertheless be useful for practitioners to think about the practice as a context when planning and implementing interventions on energy use. Most of the practice-based literature has advocated for broad, societal changes to reverse unsustainable consumption patterns (Shove, 2014), but there is also an emerging research stream on interventions in practices (see Spurling et al., 2013), often based on existing intervention mechanisms combined with more open-ended experimentation.

Looking back at the behaviour change intervention categories (Table 1), the notion of increased *convenience of energy saving* is, on the surface, rather compatible with a social practice approach. The current conventions of energy use build on historically shaped systems of provision that have made a particular mode of highly intensive energy use convenient and easy. Should we thus make more energy-saving practices more convenient and wasteful practices more inconvenient, as argued by, for example, Spurling et al. (2013)? The social practice approach also calls into question existing expectations about comfort and convenience, such as the increasingly widespread expectation of stable indoor temperatures brought about by the advance of central heating (Luo et al., 2016). From a social practice approach, it has been argued that current attempts to promote energy efficiency as convenient can serve to *reify* existing energy intensive standards and expectations (Shove, 2018). From this perspective, outcomes from a sustainability perspective can remain modest if the attempt is to only change behaviours incrementally without challenging rising expectations towards comfort and convenience. Hence, from a practice as context perspective, practitioners should carefully consider 'what is made convenient' and thus normalised.

Can interventions challenge existing understandings of convenience and comfort? Several studies drawing on the concept of 'adaptive comfort' have attempted to do so. Studies have tested the ways in which people respond to new, more adaptive practices of thermal comfort, such as heating the person rather than the entire room (Kuijer, 2014; Sahakian, 2014; Strengers, 2014). They have also drawn on 'natural experiments', such as blackouts or lower-than-average temperatures, to explore how established conceptions of normality are challenged by ruptures in everyday life (Jalas et al., 2017). Cross-cultural analyses of different thermal comfort practices (Wilhite et al., 1996) have raised interest in exploring generative tensions between expectations of comfort and convenience in different countries (Laakso and Heiskanen, 2017).

The promotion of adaptive practices has also been explored in the context of reducing domestic hot water use, drawing on a combination of several conventional behaviour change interventions adapted to the context of showering and bathing practice. For example, based on an analysis of bathing practices in the Netherlands, Japan, and India, Kuijer and De Jong (2011) conceived a new energy-conserving bathing practice called splash, 'a shift from flowing to contained water'[a]. The Washlab intervention by Davies et al. (2015) was based on an adaptive water scenario, among others, where information and communication technologies were employed to communicate fluctuations in water availability linked with variable water charges, social benchmarking information, and suggestions for appropriate forms of washing, alongside automatic adjustments to flows of water-using devices (Davies et al., 2015). This technological component would be complemented by the promotion of adaptive washing skills including the use of splash or gel washing to achieve cleanliness in times of low water flow. The participants testing the splash concept not only experienced the new way of bathing as rewarding, effective, and relaxing but also reported discomfort, mainly because of the incompatibility of this way of bathing with bathroom designs (Kuijer and De Jong, 2011). In the Washlab follow-up study, participants reported that they had continued taking shorter showers and using devices (such as timers and water metres) to save water. In contrast, households did not continue incorporating novel products (such as dry cleaning or 2-in-1 products) in their routines (Davies et al., 2015). There thus appear to be limits to how far expectations can be 'stretched'.

The practice context surrounding and arising from *feedback* has been explored in several studies. As an example, Foulds et al. (2017) explored the practices surrounding the online energy feedback tool (iMeasure) and its users; the householders studied saw themselves as being 'monitors' of energy, instead of 'managers' (or 'savers') of energy. Strengers (2013) argues that while conventional approaches to energy consumption feedback often help householders identify wasteful or unnecessary energy use, in so doing, they can even serve to legitimise the remaining bulk of energy demand as normal and necessary (see also Hargreaves et al., 2010, 2013). Even when motivated to make changes in their energy-related behaviour by feedback and other measures, people might feel unable to compromise the existing levels of comfort and cleanliness (Nye and Hargreaves, 2010). In this way, while potentially generating small savings for individual households, expectations of even more energy intensive lifestyles remain unchallenged (Shove and Walker, 2014). It has also been argued that users might struggle to connect the use of energy to the set of practices it enables; instead, 'practice feedback' has been suggested as a way to overcome the deficits of traditional energy feedback in situating energy use within the broader dynamics of everyday life and social practices (Strengers, 2013).

---

[a] 'Splash' included a basin containing warm water, a seated position, and a ritual of splashing water over the body with a scoop (Kuijer and De Jong, 2011, p. 4).

The notion of broader social norms and conventions has obvious connections to the *social influence* interventions in psychological research, which aim to influence individual behaviour by articulating and communicating social norms (via, e.g., peer influencers, providing descriptive social norm information, modelling appropriate behaviour, or eliciting public commitments). This set of interventions depends on *differences* between the behaviour of the targeted individuals and the set of norms or behaviours communicated. Two iconic examples are Goldstein et al. (2008) study of hotel guests who are informed that 75% of other guests reuse their towels and Nolan et al. (2008) intervention to promote cooling by fans rather than air conditioning by informing subjects that 77% of households in their county use fans. But where do the 'better' norms come from? In the long term, it would be unethical to make social norm claims that are invalid (Schubert, 2017); hence, for example, the use of descriptive social norm information is rather limited to subjects who do not (yet) conform to prevailing norms of 'good behaviour'. Yet perhaps there are interventions that can stretch existing social norms.

One example of an intervention that aimed to stretch existing social norms is the Jeans Challenge experiment launched by Jack (2013). In Melbourne, 31 people made a commitment to wear the same pair of jeans for 3 months without washing them, thus challenging the social norms on cleanliness. Some participants experimented with alternative strategies to keep their jeans fresh without using water, energy, or chemicals, but many did not, preferring to keep wearing the jeans without specific cleanliness efforts, developing routines and meanings that incorporated wearing the jeans without washing them. A Facebook group for participants established a context for and provided peer support in challenging the accepted norm that washing is better than not washing. The participants reported that they had also started to wash their other clothes less frequently (Jack, 2013). This is thus an example of a social influence intervention that did not articulate and communicate an existing social norm, but rather served to stretch and transform norms, at least within the community of participants.

Another example are the HomeLabs interventions studied by, for example, Devaney and Davies (2017) and Doyle (2014) in which the challenge was to disrupt the norms associated with the intertwined household practices that shape actual moments of, for example, food consumption and heating. The findings highlight that while uniform interventions (such as weekly targets and guidelines, tools, infographics, and expert advice as well as peer support via social media) were provided to each of the participating households, the reactions to, and impacts of, those interventions varied across and even within households. Interventions were experienced differently as they entered novel situations with specific social relations and dynamics created by diverse household structures, life stages, and family contexts (Devaney and Davies, 2017). Devaney and Davies (2017) also stress the social dynamics of change, where some members of the households became important drivers of practice change in the home. Household dynamics and daily practices thus create a context within which the interventions work.

Following the insights from the aforementioned projects on interventions in practice, in the ENERGISE project[b], approximately 300 households in eight European countries were challenged to change their daily practices related to space heating and washing laundry (Laakso et al., 2017). The households were provided with common challenges of setting their indoor temperature to 18°C and reducing their laundry wash cycles by half, or they could define their own challenges that better fitted in their life situations. The intervention was hence based on deliberation and setting a challenge to disrupt the status quo in homes, to study how households adapt their practices to a new situation, and to question social norms related to cleanliness and comfort. The participants reported that the challenge has led them to wash fuller cycles and to contemplate before washing a piece of clothing, whether it actually does need washing or is there another way to prolong its time of use before washing it. However, for most participants, it was still important that the clothes used outside home were clean and tidy. In terms of heating, some participants found it possible to adapt to colder temperatures (and even considered previous temperatures too hot) and learned that instead of maintaining stable indoor temperature, keeping, for example, the bedroom cooler actually made them sleep better. However, it was also found that a centralised infrastructure for heating in some countries makes it difficult to adjust the heating in individual apartments. There are thus limits to how far even practice-based interventions can 'stretch' prevailing normalities that are embedded in established infrastructures and institutions.

4 Conclusions: Addressing context in research and practice

This chapter has illustrated the role of the context in energy behaviour interventions. In reality, there are multiple contexts that simultaneously impact the outcomes of an intervention. The outcomes of behaviour change interventions in mobility, for example, depend on the spatial context (e.g. the availability of public transportation), organisational context (e.g. expectations towards car ownership at work), and the daily practices interlinked with mobility (e.g. family dynamics and work or leisure practices). The relation between interventions and contexts is thus complex, and as this chapter has illustrated, for the initiatives to work at a specific time and in a specific place, these contextual factors need to be addressed. Contextual factors are relevant for implementation, but they gain added relevance when considering the

[b] ENERGISE is a research initiative of 10 European research partners to achieve a greater scientific understanding of the social and cultural influences on energy use in households. Funded under the European Union's H2020 Research and Innovation programme under grant agreement number 727642 for 3 years (2016–19), ENERGISE develops, tests, and assesses options for a bottom-up transformation of energy use in households and communities across Europe. The research presented here has received funding from the European Union's H2020 Research and Innovation Programme under grant agreement number 727642. The sole responsibility for the content of this paper lies with the authors. Online: http://www.energise-project.eu/.

transferability or scalability of the interventions, which is crucial for policies aiming to make household energy use more sustainable on a wider scale.

Table 3 summarises our findings, showing how intervention mechanisms are dependent on organisational, geographical, and practice contexts to reach their expected outcomes. The organisational context highlights the role of the spatial organisation of tasks performed, the relationships among members and the lines of communication within the organisation, be it a household, workplace, or student dormitory. An important message from this line of research is that *social influence is always present*, whether or not it is a planned part of the intervention: members of the household or organisation may support or undermine each other's energy -saving behaviours through their own behaviour. While social influence interventions can be effective, social norms or existing social relations may also prevent changes if they are not experienced as appropriate in context. This is particularly important when the interventions target socially sensitive aspects of energy use, such as those related to notions of cleanliness.

The research on geographical context highlights the diversity of physical and cultural environments, having implications for what would be helpful convenience measures, what type of information is appreciated, how feedback is translated into action, how social influence interventions can draw on existing social networks, and what kinds of goals and commitments have cultural resonance. It is also important to bear in mind that the geographical contexts can vary not only between but also within nations, such as in urban and rural areas. While inspiration can be gained from interventions in other places, dedicated research is required to ensure that interventions fit their physical and cultural context.

Lastly, our review of 'practice as context' shows the tensions between interventions and the context of socially shared practices they attempt to change: What is made convenient, what feedback is used for, and what types of social norms are communicated. The challenge is to address and disrupt the prevailing normalities that steer daily behaviour and patterns related to energy use, such as those of showering, washing laundry, or managing heat at home. Deliberation with participants in interventions supports the development of 'innovations in practice', such as questioning the given assumptions related to 'appropriate' indoor temperature and finding new ways to keep warm. A focus on practice as context also allows the intervener to consider the wider set of practices and their interlinkages, as well as the potential for the altered practices to spread.

Intervention studies should attend to ecological validity and practical feasibility if they are to be genuinely policy relevant. Otherwise, for example, interventions relying on information and commitments may end up engaging only those people who are already keen to conserve energy, whereas they might not be meaningful or persuasive for those who do not find saving energy equally urgent. Alongside a good knowledge of psychological mechanisms, practitioners need a good understanding of context, which they can gain by engaging with participants on the ground. Studies that explore failed interventions, like the study by Outcault et al. (2018), could be particularly valuable in creating sensitivity towards contextual

Table 3 Summary of the influence of contextual factors on energy intervention outcomes

Interventions as mechanisms (IAMs)	Interventions in real life (IIRL)		'Practice as context'
	Organisational context	Geographical context	
Convenience/salience: – Technical devices and support – Default settings – - Prompts, checklists, visuals	Spatial organisation of task environment Lines of communication within the household/workplace Relations of trust and authority within the household/workplace	Compatibility with/ability to enhance existing physical and institutional environment	Extent to which existing expectations towards convenience are challenged versus normalised
Information: – Persuasion, why-to information – Education, training, how-to information	Communication/cooperation/resistance within the household/workplace In workplaces: depends on employees' willingness to help their team or organisation	Individualist versus collectivist culture Need to relate to existing infrastructure and institutions (which also influence lines of communication) Some messages (e.g. co-benefits) appear to work across contexts	Awareness of routinised and invisible daily practices Individuals' capacity and communities' capacity to challenge shared social Conventions and overcome structural obstacles in the physical environment
Monitoring/feedback: – Informative feedback – - Rewards, recognition	Communication/cooperation/resistance within the household/workplace Relation between participants influences interpretation of feedback/reward Ways in which new tools are appropriated (e.g. intended use vs use for teasing co-workers)	Acceptability of different types of feedback Ability to make sense of feedback in terms of prior experience Ability to act on feedback	Types of practices supported by feedback, for example, monitoring versus management Extent to which feedback challenges versus normalises existing expectations about energy consumption
Social influence: – Peer influencers, modelling – Public commitments, goal setting, competition, games – - Social comparison feedback	Existing level of consumption vis-à-vis communicated norm Social networks and relations within households and organisations	Existing social networks and relations between peer influencers and targets Cultural sensitivity to norms and acceptance of goals and competition	What types of norms are communicated Whether there are opportunities to 'stretch' existing norms around everyday practices

factors among practitioners and other research users. On the other hand, context need not be only a hindrance, but indeed part of why a highly successful intervention works (Janda, 2014), as was illustrated by a comprehensive school programme that gave rise to an energy-conserving social movement within the organisation (Staddon et al., 2016).

Our analysis of the literature on interventions in context has shown that when working in the field—in real-life conditions—the mechanism that an intervention aims to test or use cannot be separated from its context. Articulating the context dependence of outcomes helps to highlight under which contextual condition-specific mechanisms are likely to produce expected outcomes, thereby also indicating that such outcomes are not to be expected unless those contextual conditions are met. If there is a lack of 'fit' between intervention mechanism and context, one might consider adapting the mechanism. An alternative approach is highlighted by our examples of 'practice as context', where the intervention might also be targeted towards changing the wider context beyond making a simple behaviour more convenient or communicating an established norm. Such research might contribute to 'ecological validity' in a broader sense, since some energy behaviour interventions have been criticised for not attending to the most environmentally significant consumption (Clayton et al., 2016; Wynes et al., 2018).

The extent to which such contextual shaping originating from small-scale interventions is durable, transferrable, and scalable is still an open question. However, our observations provide some inspiration for practitioners to not only consider context when designing interventions but also consider whether interventions might explicitly and actively 'stretch' the boundaries of prevailing expectations and norms surrounding energy use in households and organisations. However, this requires looking beyond the confines of energy use alone, into broader aspects of context change.

References

Abrahamse, W., Steg, L., 2013. Social influence approaches to encourage resource conservation: a meta-analysis. Glob. Environ. Chang. 23 (6), 1773–1785.

Alberts, G., Gurguc, Z., Koutroumpis, P., Martin, R., Muûls, M., Napp, T., 2016. Competition and norms: a self-defeating combination? Energy Policy 96, 504–523.

Andrews, R.N., Johnson, E., 2016. Energy use, behavioural change, and business organizations: reviewing recent findings and proposing a future research agenda. Energy Res. Soc. Sci. 11, 195–208.

Arnott, B., Rehackova, L., Errington, L., et al., 2014. Efficacy of behavioural interventions for transport behaviour change: systematic review, meta-analysis and intervention coding. Int. J. Behav. Nutr. Phys. Act. 11 (1).

Backhaus, J., Heiskanen, E., 2009. Rating expert Advice on How to Change Energy Behaviour. Research Note 2. Changing Behaviour Project, European Commission.

Bale, C.S., McCullen, N.J., Foxon, T.J., Rucklidge, A.M., Gale, W.F., 2013. Harnessing social networks for promoting adoption of energy technologies in the domestic sector. Energy Policy 63, 833–844.

Bednar, D.J., Reames, T.G., Keoleian, G.A., 2017. The intersection of energy and justice: modeling the spatial, racial/ethnic and socioeconomic patterns of urban residential heating consumption and efficiency in Detroit, Michigan. Energy Build. 143, 25–34.

Bird, S., Legault, L., 2018. Feedback and behavioral intervention in residential energy and resource use: a review. Curr. Sustain./Renew. Energ. Rep. 5 (1), 116–126.

Blamey, A., Mackenzie, M., 2007. Theories of change and realistic evaluation: peas in a pod or apples and oranges? Evaluation 13 (4), 439–455.

Bull, R., Janda, K.B., 2018. Beyond feedback: introducing the 'engagement gap' in organizational energy management. Build. Res. Inf. 46 (3), 300–315.

Clayton, S., Devine-Wright, P., Swim, J., Bonnes, M., Steg, L., Whitmarsh, L., Carrico, A., 2016. Expanding the role for psychology in addressing environmental challenges. Am. Psychol. 71 (3), 199.

Culiberg, B., Elgaaied-Gambier, L., 2016. Going green to fit in—understanding the impact of social norms on pro-environmental behaviour, a cross-cultural approach. Int. J. Consum. Stud. 40 (2), 179–185.

Dahler-Larsen, P., 2001. From programme theory to constructivism on tragic, magic and competing programmes. Evaluation 7 (3), 331–349.

Davies, A.R., Lavelle, M.J., Doyle, R., 2015. Washing HomeLabs: Longitudinal Impact Study. High Level Findings. CONSENSUS Project. Available at: http://www.consensus.ie/wp/sample-page/papers-reports/.

Delmas, M.A., Fischlein, M., Asensio, O.I., 2013. Information strategies and energy conservation behaviour: a meta-analysis of experimental studies from 1975 to 2012. Energy Policy 61, 729–739.

Devaney, L., Davies, A.R., 2017. Disrupting household food consumption through experimental HomeLabs: outcomes, connections, contexts. J. Consum. Cult. 17 (3), 823–844.

Dietz, T., Gardner, G.T., Gilligan, J., Stern, P.C., Vandenbergh, M.P., 2009. Household actions can provide a behavioural wedge to rapidly reduce US carbon emissions. Proc. Natl. Acad. Sci. 106 (44), 18452–18456.

Diffney, S., Lyons, S., Valeri, L.M., 2013. Evaluation of the effect of the Power of One campaign on natural gas consumption. Energy Policy 62, 978–988.

Doyle, R., 2014. Heating. In: Davies, A.R., Fahy, F., Rau, H. (Eds.), Challenging Consumption. Pathways to a More Sustainable Future. Routledge, Oxon/New York, pp. 112–134.

EEA, 2013. Achieving energy efficiency through behaviour Change: What Does It Take? Copenhagen, European Environment Agency. Available at: http://www.eea.europa.eu/publications/achieving-energy-efficiency-through-behaviour. (Accessed February 11, 2019).

Foster, D., Lawson, S., Wardman, J., Blythe, M., Linehan, C., 2012. Watts in it for me?: design implications for implementing effective energy interventions in organisations. In: Proceedings of the SIGCHI Conference on Human Factors in Computing Systems. ACM, pp. 2357–2366.

Foulds, C., Robinson, R., Macrorie, R., 2017. Energy monitoring as a practice: investigating use of the iMeasure online energy feedback tool. Energy Policy 104 (2017), 194–202.

Fujii, S., Kitamura, R., 2003. What does a one-month free bus ticket do to habitual drivers? An experimental analysis of habit and attitude change. Transportation 30 (1), 81–95.

Gergen, K.J., 1973. Social psychology as history. J. Pers. Soc. Psychol. 26 (2), 309.

Goldstein, N.J., Cialdini, R.B., Griskevicius, V., 2008. A room with a viewpoint: using social norms to motivate environmental conservation in hotels. J. Consum. Res. 35 (3), 472–482.

Graham-Rowe, E., Skippon, S., Gardner, B., et al., 2011. Can we reduce car use and, if so, how? A review of available evidence. Transp. Res. A Policy Pract. 45 (5), 401–418.

Hargreaves, T., 2011. Practice-ing behaviour change: applying social practice theory to pro-environmental behaviour change. J. Consum. Cult. 11 (1), 79–99.

Hargreaves, T., 2018. Beyond energy feedback. Build. Res. Inf. 46 (3), 332–342.

Hargreaves, T., Nye, M., Burgess, J., 2010. Making energy visible: a qualitative field study of how householders interact with feedback from smart energy monitors. Energy Policy 38 (10), 6111–6119.

Hargreaves, T., Nye, M., Burgess, J., 2013. Keeping energy visible? Exploring how householders interact with feedback from smart energy monitors in the longer term. Energy Policy 52, 126–134.

Heiskanen, E., Laakso, S., Matschoss, K., Backhaus, J., Goggins, G., Vadovics, E., 2018. Designing real-world laboratories for the reduction of residential energy use: articulating theories of change. Gaia 27 (1), 60–67.

Jack, T., 2013. Nobody was dirty: intervening in inconspicuous consumption of laundry routines. J. Consum. Cult. 13 (3), 406–421.

Jalas, M., Hyysalo, S., Heiskanen, E., Lovio, R., Nissinen, A., Mattinen, M., Rinkinen, J., Juntunen, J.K., Tainio, P., Nissilä, H., 2017. Everyday experimentation in energy transition: a practice-theoretical view. J. Clean. Prod. 169, 77–84.

Janda, K.B., 2014. Building communities and social potential: between and beyond organizations and individuals in commercial properties. Energy Policy 67, 48–55.

Jones, R., Pykett, J., Whitehead, M., 2014. The geographies of policy translation: how nudge became the default policy option. Environ. Plan. C: Gov. Policy 32 (1), 54–69.

Keller, M., Halkier, B., Wilska, T.A., 2016. Policy and governance for sustainable consumption at the crossroads of theories and concepts. Environ. Policy Gov. 26 (2), 75–88.

Kok, G., Lo, S.H., Peters, G.J.Y., Ruiter, R.A., 2011. Changing energy-related behaviour: an intervention mapping approach. Energy Policy 39 (9), 5280–5286.

Kosonen, H.K., Kim, A.A., 2017. Advancement of behavioural energy interventions in commercial buildings. Facilities 35 (5/6), 367–382.

Kuijer, L., 2014. Implications of Social Practice Theory for Sustainable Design. PhD Dissertation, Technical University of Delft.

Kuijer, L., De Jong, A.M., 2011. Practice theory and human centered design: a sustainable bathing example. In: Nordic Design Research Conference 2011, Helsinki.

Laakso, S., 2017. Giving up cars—the impact of a mobility experiment on carbon emissions and everyday routines. J. Clean. Prod. 169, 135–142.

Laakso, S., 2019. Experiments in everyday mobility: social dynamics of achieving a sustainable lifestyle. Sociol. Res. Online 24, 235–250. e-pub ahead of print, https://doi.org/10.1177/1360780418823222.

Laakso, S., Heiskanen, E., 2017. Good Practice Report: Capturing Cross-Cultural Interventions. Deliverable 3.1 of ENERGISE, European Network for Research, Good Practice and Innovation for Sustainable Energy. Online: http://energise-project.eu/deliverables.

Laakso, S., Heiskanen, E. Matschoss, K., 2017. ENERGISE Living Labs Background Report. ENERGISE—European Network for Research, Good Practice and Innovation for Sustainable Energy, Deliverable 3.2.

Lehner, M., Mont, O., Heiskanen, E., 2016. Nudging—a promising tool for sustainable consumption behaviour? J. Clean. Prod. 134, 166–177.

Leygue, C., Ferguson, E., Spence, A., 2017. Saving energy in the workplace: why, and for whom? J. Environ. Psychol. 53, 50–62.

Lokhorst, A.M., Werner, C., Staats, H., van Dijk, E., Gale, J.L., 2013. Commitment and behaviour change: a meta-analysis and critical review of commitment-making strategies in environmental research. Environ. Behav. 45 (1), 3–34.

Luo, M., de Dear, R., Ji, W., Bin, C., Lin, B., Ouyang, Q., Zhu, Y., 2016. The dynamics of thermal comfort expectations: the problem, challenge and implication. Build. Environ. 95, 322–329.

Maki, A., Burns, R.J., Ha, L., Rothman, A.J., 2016. Paying people to protect the environment: a meta-analysis of financial incentive interventions to promote proenvironmental behaviours. J. Environ. Psychol. 47, 242–255.

Moloney, S., Horne, R.E., Fien, J., 2010. Transitioning to low carbon communities—from behaviour change to systemic change: lessons from Australia. Energy Policy 38, 7614–7623.

Morren, M., Grinstein, A., 2016. Explaining environmental behaviour across borders: a meta-analysis. J. Environ. Psychol. 47, 91–106.

Nastasi, B.K., Schensul, S.L., 2005. Contributions of qualitative research to the validity of intervention research. J. Sch. Psychol. 43 (3), 177–195.

Nolan, J.M., Schultz, P.W., Cialdini, R.B., Goldstein, N.J., Griskevicius, V., 2008. Normative social influence is underdetected. Personal. Soc. Psychol. Bull. 34 (7), 913–923.

Norton, T.A., Parker, S.L., Zacher, H., Ashkanasy, N.M., 2015. Employee green behaviour: a theoretical framework, multilevel review, and future research agenda. Organ. Environ. 28 (1), 103–125.

Nyborg, S., 2015. Pilot users and their families: inventing flexible practices in the smart grid. Sci. Technol. Stud. 28 (3), 54–80.

Nye, M., Hargreaves, T., 2010. Exploring the social dynamics of proenvironmental behaviour change. A Comparative Study of Intervention Processes at Home and Work. J. Ind. Ecol. 14 (1), 137–149.

Ogilvie, D., Egan, M., Hamilton, V., et al., 2004. Promoting walking and cycling as an alternative to using cars: systematic review. BMJ 329, 763.

Ölander, F., Thøgersen, J., 2014. Informing versus nudging in environmental policy. J. Consum. Policy 37 (3), 341–356.

Ones, D.S., Dilchert, S., 2012. Environmental sustainability at work: a call to action. Ind. Organ. Psychol. 5 (4), 444–466.

Osbaldiston, R., Schott, J.P., 2012. Environmental sustainability and behavioural science: meta-analysis of proenvironmental behaviour experiments. Environ. Behav. 44 (2), 257–299.

Outcault, S., Sanguinetti, A., Pritoni, M., 2018. Using social dynamics to explain uptake in energy saving measures: lessons from space conditioning interventions in Japan and California. Energy Res. Soc. Sci. 45, 276–286.

Pawson, R., Tilley, N., 1997. Realistic Evaluation. Sage, London.

Pfadenhauer, L.M., Gerhardus, A., Mozygemba, K., Lysdahl, K.B., Booth, A., Hofmann, B., Rehfuess, E., 2017. Making sense of complexity in context and implementation: the Context and Implementation of Complex Interventions (CICI) framework. Implement. Sci. 12 (1), 21.

Pucher, J., Dill, J., Handy, S., 2010. Infrastructure, programs, and policies to increase bicycling: an international review. Prev. Med. 50, 106–125.

Reckwitz, A., 2002. Toward a theory of social practices: a development in culturalist theorizing. Eur. J. Soc. Theory 5 (2), 243–263.

Redman, L., Friman, M., Gärling, T., Hartig, T., 2013. Quality attributes of public transport that attract car users: a research review. Transp. Policy 25, 119–127.

Sahakian, M., 2014. Keeping Cool in Southeast Asia: Energy Consumption and Urban Air-Conditioning. Palgrave Macmillan, p.265.

Šćepanović, S., Warnier, M., Nurminen, J.K., 2017. The role of context in residential energy interventions: a meta review. Renew. Sust. Energ. Rev. 77, 1146–1168.

Schäfer, M., Jaeger-Erben, M., Bamberg, S., 2012. Life events as windows of opportunity for changing towards sustainable consumption patterns? J. Consum. Policy 35 (1), 65–84.

Schubert, C., 2017. Green nudges: do they work? Are they ethical? Ecol. Econ. 132, 329–342.

Shove, E., 2003. Comfort, Cleanliness and Convenience: The Social Organization of Normality. Berg, Oxford/New York.

Shove, E., 2010. Beyond the ABC: climate change policy and theories of social change. Environ Plan A 42 (6), 1273–1285.

Shove, E., 2011. On the difference between chalk and cheese—a response to Whitmarsh et al's comments on "Beyond the ABC: climate change policy and theories of social change". Environ Plan A 43 (2), 262–264.

Shove, E., 2014. Putting practice into policy: reconfiguring questions of consumption and climate change. Contemp. Soc. Sci. 9 (4), 415–429.

Shove, E., 2018. What is wrong with energy efficiency? Build. Res. Inf. 46 (7), 779–789.

Spencer, J., Lilley, D., Porter, S., 2015. The opportunities that different cultural contexts create for sustainable design: a laundry care example. J. Clean. Prod. 107, 279–290.

Spurling, N. J., McMeekin, A., Southerton, D. et al., 2013. Interventions in Practice: Reframing Policy Approaches to Consumer Behaviour. Executive Summary of the SPRG report, 'Interventions in Practice'. Available at: http://www.sprg.ac.uk/uploads/sprg-report-sept-2013.pdf.

Shove, E., Walker, G., 2014. What is energy for? Social practice and energy demand. Theory Cult. Soc. 31 (5), 41–58.

St Lawrence, S., 2004. Review of the UK corporate real estate market with regard to availability of environmentally and socially responsible office buildings. J. Corporate Real Estate 6 (2), 149–161.

Staddon, S.C., Cycil, C., Goulden, M., Leygue, C., Spence, A., 2016. Intervening to change behaviour and save energy in the workplace: a systematic review of available evidence. Energy Res. Soc. Sci. 17, 30–51.

Steg, L., 2008. Promoting household energy conservation. Energy Policy 36 (12), 4449–4453.

Strengers, Y., 2013. Smart Energy Technologies in Everyday Life: Smart Utopia? Palgrave Macmillan, Basingstoke.

Strengers, Y., 2014. Smart energy in everyday life: are you designing for resource man? Interactions 21, 24–31.

Stromback, J., Dromacque, C., Yassin, M. H., 2011. The Potential of Smart Meter Enabled Programs to Increase Energy and Systems Efficiency: A Mass Pilot Comparison Short Name: Empower Demand. Vaasa ETT.

Thaler, R., Sunstein, C., 2008. Nudge: Improving Decisions about Health, Wealth, and Happiness. Yale University Press, New Haven CT.

Thøgersen, J., 2009. Promoting public transport as a subscription service: effects of a free month travel card. Transp. Policy 16 (6), 335–343.

Thøgersen, J., 2012. The importance of timing for breaking commuters' car driving habits. In: Warde, A., Southerton, D. (Eds.), The Habits of Consumption. Studies Across Disciplines in the Humanities and Social Sciences, vol. 12. Helsinki Collegium for Advanced Studies, Helsinki, pp. 130–140.

Thompson, S., Michaelson, J., Abdallah, S., Johnson, V., Morris, D., Riley, K., Simms, A., 2011. 'Moments of Change' as Opportunities for Influencing Behaviour. A research report completed for the Department for Environment, Food and Rural Affairs by nef (the new economics foundation), DERFA, London. Available at: http://orca.cf.ac.uk/43453/.

Vávra, J., Peters, V., Lapka, M., Craig, T., Cudlínová, E., 2016. What shapes the temperatures of living rooms in three European regions? Sociální studia/Social Studies 12 (3).

Vine, D., Buys, L., Morris, P., 2013. The effectiveness of energy feedback for conservation and peak demand: a literature review. Open J. Energ. Effi. 2 (1), 7–15.

Whitmarsh, L., O'Neill, S., Lorenzoni, I., 2011. Climate change or social change? Debate within, amongst, and beyond disciplines. Environ Plan A 43 (2), 258–261.

Wilhite, H., Nakagami, H., Masuda, T., Yamaga, Y., Haneda, H., 1996. A cross-cultural analysis of household energy use behaviour in Japan and Norway. Energy Policy 24 (9), 795–803.

Wilson, C., Chatterton, T., 2011. Multiple models to inform climate change policy: a pragmatic response to the 'beyond the ABC'debate. Environ Plan A 43 (12), 2781–2787.

Wilson, C., Dowlatabadi, H., 2007. Models of decision making and residential energy use. Annu. Rev. Environ. Resour. 32, 169–203.

Winther, T., Bell, S., 2018. Domesticating in-home displays in selected British and Norwegian households. Sci. Technol. Stud. 31 (2), 19–38.

Wynes, S., Nicholas, K.A., Zhao, J., Donner, S.D., 2018. Measuring what works: quantifying greenhouse gas emission reductions of behavioural interventions to reduce driving, meat consumption, and household energy use. Environ. Res. Lett. 13 (11), 113002.

Yang, L., Sahlqvist, S., McMinn, A., Griffin, S., Ogilvie, D., 2010. Interventions to promote cycling: systematic review. BMJ 341.

Young, W., Davis, M., McNeill, I.M., et al., 2015. Changing behaviour: successful environmental programmes in the workplace. Bus. Strateg. Environ. 24 (8), 689–703.

Urban low-carbon futures: Results from real-world lab experiment in Berlin

4.2

Fritz Reusswig, Wiebke Lass, Seraja Bock,

Potsdam Institute for Climate Impact Research (PIK), Potsdam, Germany

1 Introduction: The experimental dispositive of climate policy

By the end of 2018, 184 parties of the *United Nations Framework Convention on Climate Change* (UNFCCC) have ratified the *Paris Agreement* (PA), negotiated in late 2015, entering into force on 4 November 2016. Its central aim is to strengthen the global response to the threat of climate change by keeping a global temperature rise this century well below 2°C above preindustrial levels and to pursue efforts to limit the temperature increase even further to 1.5°C. The Paris Agreement requires all parties to put forward their best efforts through national contributions and to strengthen these efforts in the years ahead. This includes requirements that all parties report regularly on their emissions and on their implementation efforts. A rapid and full decarbonisation of the world's economies until 2050 will be needed to achieve the 1.5°C goal in 2100 (IPCC, 2018).

The Paris Agreement has been criticised for its reliance on voluntary state action, the so-called *Nationally Determined Contributions* (NDCs) (see Benveniste et al., 2018). A positive side aspect of the NDCs is the fact that by their introduction, a broad range of policy instruments—including carbon taxes—is put on the table. Before Paris, nations have strived for binding targets and favored reputed market-based instruments, such as emissions trading (Kyoto instruments). To put it more succinctly, Paris has broken the neoliberal spirit that has dominated climate policy since Kyoto.

However, it is also obvious that national pledges are not sufficient. According to the *Climate Action Tracker* (cf. https://climateactiontracker.org/), a nongovernmental and nonprofit expert platform on tracking climate action by nations globally, current policies of nation states will most probably lead to an increase of global mean temperature of about 3.4°C above preindustrial levels until 2100; if indicated national intentions (pledges) are included, the increase will be about 3.2°C. Germany's climate

Energy and Behaviour. https://doi.org/10.1016/B978-0-12-818567-4.00016-8

419

policy, for example, is rated 'insufficient', which is better than 'critically insufficient' (e.g. the United States or Russia), or 'highly insufficient' (e.g. China or Canada), but still not compliant with the PA. If no significant further action is taken until 2030, the 1.5°C goal will not be met (UNEP (United Nations Environment Programme), 2018).

Another implication of the PA is that other actors come into play alongside the nation states. While international climate policy is still built on nation states and state associations (such as the EU), many more actors across the globe are engaged in many kinds of actions in order to mitigate against anthropogenic climate change. The *United Nations Climate Change Secretariat*, which is observing these actions, recently counted about 9000 cities, around 240 states and regions below the central state level, and more than 6000 businesses in 120 countries that have been setting carbon goals, implementing low-carbon policies or individual actions largely independent of nation states (UNCCS (United Nations Climate Change Secretariat), 2018). Some authors even assume that US non-nation–state actors such as federal states or cities do have the potential under the Paris Agreement to actively fight against the attempt of the Trump Administration to obstruct the Agreement's goal, for example, by withdrawing from the PA (Cooper, 2018).

Especially *cities* have attracted the attention of the research community, both as state actors at the lower end of the federal political hierarchy and as 'containers' or 'stages' for all kinds of voluntary local climate action, many of which driven by local businesses, nongovernment organisations, and individual citizens. Most of these actions have chosen an experimental design, many of them in the institutional form of an urban living lab (ULL). For example, Bulkeley and Castán Broto (2013) have done a survey of climate change experiments taking place in 100 global and megacities, spanning from policy over sectoral experiments to spatially limited urban transition labs. They do not conceive experiments in a formal scientific sense, but rather defined them as "purposive interventions in which there is a more or less explicit attempt to innovate, learn or gain experience" (Bulkeley and Castán Broto, 2013, p. 363). ULLs have received a particular attention by both transdisciplinary and transition research-oriented approaches (cf. Bulkeley et al., 2018a, b; Castán Broto, 2017; Engels and Walz, 2018; Evans and Karvonen, 2014; Evans et al., 2016; Menny et al., 2018; Rogga et al., 2018; Schneidewind et al., 2018; Voytenko et al., 2016; Wolfram and Frantzeskaki, 2016). Jensen et al. (2017) have collected more than 1,000 ULL projects (sustainable energy consumption change initiatives) in 30 European countries, and Jensen et al. (2018) have tried to systematise them.

The upswing of these urban labs is no coincidence. The modern city can be seen as the paradigm of an experimental turn. Long before Park (1929) conceived the city as a field of experimentation for humanity and Chicago as its paradigm, the European city of the 19th century emerged as a real laboratory for the development and testing of new sociotechnical systems (e.g. infrastructures for water supply and sanitation, electricity, gas, waste, and transport), including the corresponding business and governance models (cf. for the case of Berlin: Monstadt, 2007; Moss, 2014). Barthel et al. (2014) have shown how the modern city has unfolded as interplay between household changes and sociotechnical changes at the city level and how this interplay enabled the population densities and city sizes of today. The modern city can thus be understood largely as the *product* of sociotechnical experiments, not just as the spatial *container* in which they take place.

This short historical reminder should prevent us from the false conclusion that urban or real-world labs would be a completely new format in the urban context. In addition, the idea of innovation is built-in into the 'DNA' of modern societies, at least when it comes to new products and services. Experimental and lab settings have also become a common feature for the codevelopment of sustainable energy solutions at the user-producer interface (Dupont et al., 2019; Liedtke et al., 2015).

We would define experiments in this context as inclusive, practice-based, and challenge-led initiatives, which are designed to promote system innovation through social learning under conditions of uncertainty and ambiguity (Sengers et al., 2016). In this perspective, the term 'experiment' does not only refer to the transfer of a scientific research concept to social contexts but also reflect the 'abstract radicalism' of climate policy together with a reflexive turn in modern society policy making in general (Boström et al., 2017; Gross and Krohn, 2005). What distinguishes the recent upswing of labs and experiments—at least in our interpretation—is the coincidence of two facts:

- *Climate policy* in itself does have an *experimental disposition*. This results from its abstract radicalism. Climate policy is radical, because the rates and the timing of an almost complete decarbonisation of the whole energy system are a huge an unprecedented challenge.[a] And while other 'radicalisms' (e.g. in social policy) often lack a precise quantification of their goals, climate policy can exactly quantify its goal functions (e.g. in terms of temperature or emission goals or timing). On the other hand, this quantitative radicalism remains abstract when it comes to the concrete measures that have to be taken: Should we invest in direct electric cars, should we rather invest in hydrogen cars, or would it be more feasible to run a downsized car-sharing fleet with biofuels? Should we prescribe the passive house standard for every new building, or can we manage to generate enough renewable space heating, so that lower energy standards for buildings can be realised, and so forth? Many scenarios to reach climate neutrality by 2050 reveal that there are various kinds of options that can be followed (internal consistency provided), the choice of which depends on further criteria (such as costs or social acceptability) and cannot be derived from the quantitative climate goals alone. These options come along with uncertainties with respect to their feasibility, and some of them impose choices (or compromise) between environmental efficiency and social justice.
- *Climate policy* needs to be *fast*. Despite of being abstract and thus open to political debate, climate policy must reach effects very quickly, if 2050 goals should be met. Waiting until, say, 2045 is no good idea, especially

---

[a] When it comes to greenhouse gas emissions, the energy system has to be complemented by land use and food issues, transport issues are included anyway, the same holds for the building sector.

when it comes to larger-scale investments into 'slow' objects/processes (such as building codes, retrofitting, and power plants). There are many path dependencies and carbon lock-ins that have to be overcome, and external forces are often needed (Unruh, 2002).

In short: climate change is a "super wicked problem", and these kinds of problems comprise four key features: time is running out, those who cause the problem also seek to provide a solution, the central authority needed to address them is weak or nonexistent, and irrational discounting occurs that pushes responses into the future (Levin et al., 2012). This leads many analysts to believe that one-shot 'big bang' policies were needed, requiring behavioural change by *all* relevant populations *immediately* (see Lazarus, 2009). Not only many climate scientists but also some activists think that way. This policy paradigm, as understandable as it might be, runs the risk of failing to garner adequate support or, in those rare cases where such policies are adopted, is likely to produce societal resistance or conflicts that hamper implementation and compliance.

In order to avoid both failures—wait until all uncertainties and problems have been resolved or put all one's eggs in one basket immediately—moving to an experimental policy mode seems the only feasible way out. Triggering and enabling sociotechnical innovations; conditioning the political system to learning processes; and taking advantage of the multi-layered character of the system, norm activation, and coalition building with relevant societal actors in order to broaden support over time are some of the key features here (see Levin et al., 2012). We will come back to this point at the final section of this chapter.

What is important here is to consider the 'experimental turn' in sustainability transition research not as a 'fashion' (which it in fact *also* is), but as an adequate and promising answer to the specific character of climate change policy as a super wicked problem.

When it comes to energy-related behaviour, most of the labs tend to influence *consumer* behaviour by various forms of experimental interventions. The role of the *citizen* as a political subject has received less attention, although a focus on the citizen would allow for linking small-scale spatial or sectoral experiments to larger-scale—for example, city wide—transformations. In this chapter, we will report from a real-lab experiment in Berlin, Germany, in which both roles—the consumer and the citizen—of participating individuals have explicitly been addressed. This double perspective is especially important when it comes to issues like evaluating the outcomes of real labs (Luederitz et al., 2017), learning (Singer-Brodowski et al., 2018), upscaling (Niamir et al., 2018), social innovations (Godin, 2015; Howaldt et al., 2018), and system-wide transformations towards low-carbon societies (EC (European Commission), 2017; Heiskanen et al., 2018; WBGU (German Advisory Council on Global Change), 2011).

This chapter focuses on another format of the 'experimental turn' in sustainability transition research: *real-world laboratories* (RwLs). RwLs have attracted quite some attention both from scientists and from funding agencies in Germany recently (cf. GAIA, 2018). RwLs can be defined as research-oriented, long-term, transformative, and transdisciplinary change initiatives focusing on sustainability in a specific context (Heiskanen et al., 2018; Beecroft and Parodi, 2016). They are characterised by a normative framing (as they aim to contribute to a more sustainable development) and by real-world problems as a starting point (e.g. reducing CO_2 emissions of private households). But they do so by mobilising (transdisciplinary) scientific knowledge and methods, usually in a coproduction setting with real-world actors. They follow an experimental approach by initiating and studying real-world interventions. They usually aim at scientific learning and at empowering practical agents at the same time, and they want to ultimately contribute to real-world changes by reflection and diffusion processes, explicitly taking into account the context factors they are operating in/with (Hilger et al., 2018).

However, the potential for emissions reduction of individuals/private households is substantial. According to an analysis of 73 countries, Hertwich and Peters (2009) assess the private household share of total carbon footprints to be 72%. Ivanova et al. (2016) have analyzed 43 countries (mostly from the EU) and give an estimate of 65% for the carbon footprint share of private households.[b] These data include direct and indirect emissions, the latter being 'incorporated' in the products, and services individuals buy and use (life cycle analysis). Given the increasingly complex and global character of production chains and trade, it is clear that especially indirect CO_{2eq} emissions are both hard to retrieve and often difficult to change.[c]

But do individuals actually take on that responsibility? Indeed, consumers are often largely unaware of their indirect energy use (Abrahamse et al., 2007). Would they change their behaviours if they had credible and readily usable information on their carbon footprints? Do they support more stringent climate policies if they manage to reduce—or rather not? In a recent study (IGES (Institute for Global Environmental Strategies, Aalto University, and D-mat ltd.), 2019) per capita, consumption-based targets of greenhouse gas emissions scenarios compatible with the Paris Agreement temperature goal (1.5°C) have been calculated. According to the results, reductions to $2.5\,t\;CO_{2eq}$ in 2030 and to 0.7 t in 2050 have to be achieved.[d] Are households aware of the dimension of the challenge?

Research designs that are bound to status quo or incremental change purposes are most probably not suited to answer these questions. Much greater emphasis is needed on approaches that can accelerate learning and actions that lead to transformations

---

[b] The *German Statistical Office* (Destatis (Statistisches Bundesamt), 2018) estimates a 67.5% share of private households in Germany's total emissions (2014, life cycle-based assessment).

[c] It is important to note that life cycle-based assessments attribute all product-related emissions to final consumption, and thus, the consumer, whether or not the individual, can directly influence this footprint, for example, by choosing a more efficient egg cooker ($92\,g\;CO_2$) instead of an open pot ($221\,g\;CO_2$) to cook an egg. But when the egg enters the household, $908\,g\;CO_2$ have been already generated due to production and distribution processes (PCF, 2009, cf. Goodall, 2007). It is thus necessary to distinguish assignment from responsibility.

[d] If human carbon sink technologies are included (such as carbon capture and storage, CCS), the figures change to 3.2 t (2030) and 1.5 t (2050) (IGES (Institute for Global Environmental Strategies, Aalto University, and D-mat ltd.), 2019).

towards a low-carbon, resilient, and sustainable future (Fazey et al., 2018; Stern et al., 2016). And unlike most economists assume, incentives to reduce energy demand or carbon emissions go well beyond making the energy saving alternatives more attractive, financially or otherwise. The presence of behavioural barriers can diminish the potential for energy and emission savings by anywhere between 63% and 80% (Niamir et al., 2018) and thus needs to be addressed as well.

Real-world labs can be seen as a way to explore the possibilities of ambitious household action but can also help to reveal barriers to action at a very practical level. And they offer the opportunity to have a closer look at the interplay—including both the possible synergies and the conflicts—between the respective roles of consumers and citizens. RwL results can thus help to inform policy makers with some deeper insights into the effect of energy and climate policy instruments on the household level.

The rest of this chapter is organised as follows: in the next section (2), we will describe the project goal methods and major interventions of our (*Klimaneutral leben in Berlin*) KLIB real-world urban lab, followed (3) by a short overview of its major results and outputs. Furthermore, the household members in their role as citizens are explored. In doing so, an ethical dimension (4) and a political dimension (5) are discussed. The final section (6) will draw some conclusions and give some policy recommendations.

2 Research design

It is a well-established finding in environmental and energy-related social science research that private households can play an important role in saving energy and reducing CO_2 emissions (e.g. Dietz et al., 2009; Steg, 2009; Stern et al., 2016) and that contextual factors as well as individual lifestyles are influencing not only the current energy use and emissions but also the willingness and capacity to save energy and reduce emissions (Bin and Dowlatabadi, 2005; Lutzenhiser, 1993; Reusswig, 2009, 2010).

After having helped to design a decarbonisation strategy until 2050 for the city of Berlin as a whole[e], the Potsdam Institute for Climate Impact, scientists wanted to focus on the role of the individual citizen in urban climate policies, the birth of the project idea of KLIB ('Climate Neutral living in Berlin'). The project was funded by the German Federal Ministry for the Environment (BMU) as a nationwide flagship project on innovative solutions for climate protection, running from 2017 to 2018, with a 1-year real-lab phase with private households participating as its core.[f]

The main goal of KLIB was to find out whether private households could be motivated to reduce their personal carbon footprint in a 1-year real-world experiment, utilising a variety of intervention instruments. Which intervention work, which ones don't? What factors influence the adoption of reduction strategies, what are the barriers to this adoption? At the same time, we wanted to learn more about the interplay between the respective roles of the consumer and the citizen. 'Real world'

---

[e] See the documents at https://www.berlin.de/senuvk/klimaschutz/politik/index_en.shtml.
[f] Grant No. 03KF0062.

refers to the fact that neither technological equipment nor (material) incentives have been provided to participants. They had to operate under existing market and policy conditions.[g]

The recruitment of the KLIB real lab took place via a project website, newspaper ads, and the stakeholder media channels. It was clear from the outset that people with an interest in climate policy in general and personal action on it in particular would more probably be willing to participate in the 1-year experiment. 208 Berlin households showed interest, but after more detailed information about the experimental setting (1-year lab, weekly tracking), 170 households did fill out the baseline survey in late 2017. The household size distribution differed slightly from the Berlin average distribution, with less one person households (32% in the project vs 54.6% in the city) and more 4+ person households (21% in the project vs 8.1% in the city). More families with children and more people living in shared flats have been participating in the project. With respect to income, KLIB households do have a higher than average share of middle-income classes, as both the lower and the upper end of the Berlin household net income distribution is underrepresented. However, it is interesting to note that 19.4% low-income households had chosen to participate.

The continuous weekly tracking was a challenge, and the project team expected a drop-out rate of about 50%. The real performance of KLIB households was, in fact, better. From 152 active participants, 72 (47.4%) did continuously track their behaviour over the full period, that is, 52 weeks with almost no lacks. Twenty-three households (15.1%) had minor, 46 (30.3) major data gaps, and 11 households (7.2%) missed the tracking right from the start. All households, independently of their CO_2 tracking behaviour, are nevertheless regarded as participants of the experiment, since drop out is seen as a reaction to the project setting and performance and will be researched in a later stage. The current chapter is based on the analysis of the 72 households that did track continuously.

The experimental character of the project was defined by a couple of interventions.

(1) As a precondition for active participation, households did have to complete a first annual version of the carbon tracker in order to get their *baseline values* for the year 2017, the preceding year of the experiment. This feedback can be considered the first intervention tool.

(2) As mentioned, the *carbon tracker*—an online tool—was used on a weekly basis, with weekly reminders sent by e-mail.[h] Households entered their consumption and other activities (such as car driving or eating less meat) and received their own footprint, a comparison with all other participants and a comparison with the German average footprint (currently at

---

[g] This is a clear difference to experiments where participants are equipped with innovative technologies and try them out. Real-world market and policy constraints are largely excluded.

[h] In addition to these weekly reminders, we were reminding reluctant participants via phone after longer breaks (e.g. holidays), a more personal (but also time-consuming) intervention that did in fact significantly improve participation rates.

11.63 tons CO_{2eq}^{-year}). The—anonymous—results were displayed on the project website (cf. http://klimaneutral.berlin).[i]

(3) On the same website, people could communicate on issues related to the project in a project-wide public discussion *forum*, one of our major peer-to-peer learning spaces.

(4) A closed *Facebook* group has been established serving the same purpose but with a special offer to the preferred social media channel of KLIB households.

(5) The biweekly *newsletter* of KLIB was distributed to all households, not only providing the seasonal tips but also addressing climate policy issues of the day, for example, articles on driving bans for diesel cars, technological alternatives to kerosene, or political debate about the shutdown of lignite power plants in Germany. This intervention tool was a key element in addressing the citizen, not only the consumer. A short version was sent to all participants via e-mail as a teaser, with longer versions provided at the website—another way of motivating people to repeatedly visit the website.

(6) The KLIB team provided many practical *low-carbon tips* on the website in all domains under research, ranging from small (e.g. substitute exotic functional food items by more regional alternatives) to big points (e.g. alternatives to air travel trips and change to green energy providers). The sequence of these tips did follow a seasonal pattern, increasing the everyday life fit of this intervention tool.

(7) In addition to the website tips, the project actively offered *products and services* of the KLIB stakeholder network, consisting of 20 Berlin-based businesses or NGOs. For example, *Berliner Stadtwerke* (https://berlinerstadtwerke.de/) was involved: a public utility company that distributes locally generated renewable energy, or *Greenpeace Energy*, a subsidiary company to Greenpeace Germany (https://www.greenpeace-energy.de). Other examples include *BioCompany*, offering organic food (https://www.biocompany.de/), or *Florida Eis*, a local ice cream company with a renewable-based production and distribution line https://www.floridaeis.de/). Of particular importance has been *BUND* Berlin (https://www.bund-berlin.de/), an environmental NGO, offering professional energy consulting at home. The consultants met the participants at their homes and could address various ways of energy saving and green energy substitution in a tailored way.

(8) During a couple of *live meetings* with the whole group information could be disseminated and questions asked, but the most important goal of these meetings has been community-building among participants. This intervention tool was especially important with respect to the social character of the

---

[i] The (protected) website was designed in a very person-oriented way, with colored balloons symbolising the current emissions, a gray balloon in the background symbolising the 2017 baseline. People could upload photographs in order to personalise their entries, and every week a new household was portrayed with some key features such as the quarter in Berlin where they lived, family size, or hobbies. This personalised approach was inspired by the insight that stories and storytelling as a way of understanding, communicating, and influencing others (Moezzi et al. 2017).

innovation. As one workshop participant did put it: 'For me it is important to know whether I am the only "crazy person" trying to reduce my carbon footprint, or if there are others in my city too'.

(9) The *mass media* response to KLIB was surprisingly high, given the small number of participants. Two factors turned out to be crucial here: (a) Mass media were particularly attracted by the high degree of personalisation. While climate change mitigation usually comes along as a highly political and at the same time abstract issue, in KLIB, it did have 'a face'—people like you and me; (b) as most households wanted to reduce their footprints on a very concrete basis, mass media could report on success stories, not on the highly problematic side of climate change (such as disastrous impacts or failing policies). As all media activities have been reported on the website and as many households did actively cooperate with the media, this element can be seen as a final intervention 'tool'.[j]

The *social* character of social innovations was primarily met by interventions 3, 4, 8, and 9, while the *innovation* character was primarily addressed interventions 1, 2, 6, and 7. Intervention no. 5 (newsletter) covered both aspects. It is important to highlight that it was not only innovative *products* (e.g. gas from wind or a plug-in solar application for the balcony) or *services* (a weekly carbon tracker) but also innovative *practices* that were brought to the attention of participants. Political practices, such as participating in a political demonstration (e.g. against coal mining), were explicitly included.

In addition to a weekly monitoring, the KLIB team did undertake two surveys in order to better understand the sample and to answer the research questions. A shorter one took place right before the beginning of the real-lab phase in late 2017 the second and larger one during the lab phase in summer 2018.

In their assessment of interventions designed to reduce residential energy use, Heiskanen et al. (2018) have identified five different types of instruments: (1) need-based tailored support (+); (2) pioneering practices (+++); (3) challenges, competition, and game (+); (4) learning by doing (+); and (5) peer-to-peer learning (++).[k]

A control group has not been set up for KLIB, as one would expect for classical experimentation or intervention studies, rendering a control for Hawthorne effect seemingly impossible. Avoiding Hawthorne effect is difficult for a quasiexperimental setting like in KLIB. This well-known effect refers to situations where participants behave differently because they are being observed, as in the original experiments at the Western Electric plant in the late 1920s, measuring the effects of altered working conditions (e.g. improved lighting) on productivity.[l] In our context, households were

---

[j] KLIB thus fits quite well into a rather narrow definition of urban laboratories as formalised settings, '(1) involving a specific setup of instruments and people that (2) aims for the controlled inducement of changes and (3) the measurement of these changes' (Karvonen and van Heur, 2014, p. 383).

[k] In brackets the relevance for KLIB: +++ highly relevant, ++ moderately relevant, + less relevant, 0 not relevant at all.

[l] Reanalysis of the original Hawthorne data has shown that the name-giving effect has been massively overstated and the remaining subtle Hawthorne effects in 'its presence and magnitude depend heavily on economic and psychological features of the environment that can only be understood with further theoretical and empirical modelling' (Lewitt and List, 2009, p. 4; cf. Jones, 1992).

explicitly invited to change their behaviour due to the participation in the lab, and a close monitoring (i.e. observation) was an integral part of the study design.

In a recent overview paper on the state and perspectives of energy social science, the authors find that 'the laboratory setting of most true experiments is rather artificial' and advise scientists 'to employ a "natural" or "quasiexperimental" research design' (Sovacool et al., 2018, p. 22). A real-world lab cannot isolate 100 households for a year from their social and/or geographical contexts—quite the opposite: remaining in these contexts is a precondition for obtaining valid results. With respect to the effects of the regular carbon tracker feedback and the footprint results, the German average served as the control group.

3 Carbon footprint results

From many studies, we know that a general concern about anthropogenic climate change and its consequences is common to vast majorities in Germany and other countries. Fifty-two percent of all Germans find it very important to reach climate neutrality by 2050 (1.5°C Paris Agreement goal), but only 16% hold it to be probable that we can in fact achieve it (BMUB/UBA, 2017, p. 38). There are many reasons for this scepticism also on an individual level due to the following:

- missing or hard-to-acquire product-/service-related carbon footprint information
- a lack of expert feedbacks on footprint issues and related uncertainties
- a lack of repeated motivational inputs
- a lack of peer-to-peer exchange on the issue and related problems associated with reducing one's own footprint

KLIB tried to overcome these deficits by its study design, enabling households to track the CO_2 consequences of their consumer behaviour on a weekly basis, giving both on- and offline expert support, motivating people on a weekly basis (and by other incentives, such as visits of partners, e.g. the climate-friendly ice cream producer in summertime), and enabling on- and offline peer-to-peer exchange. Nevertheless, the KLIB setting could not—and did not intend to—remove everyday barriers to the adoption of low-carbon solutions.

The overall carbon footprint results from the 1-year experiment are the following (cf. Fig. 1).

The KLIB households had started on 1 January 2018, with a baseline carbon footprint of 8.754 kg per capita (or 24.6% less than the German average[m]), and after 1 year of participation in the real-world lab, they ended up with an average of

---

[m] Data on the German average footprint are taken from the Federal Environment Agency (UBA) and its online carbon calculator, based on annual tracking data (cf. http://www.uba.co2-rechner.de/de_DE/). The private company KlimAktiv, which runs the UBA calculator, has been developing both the KLIB baseline calculator and the weekly carbon tracker. Using the same methodology and data sources, German per capita footprints are thus comparable with KLIB footprints.

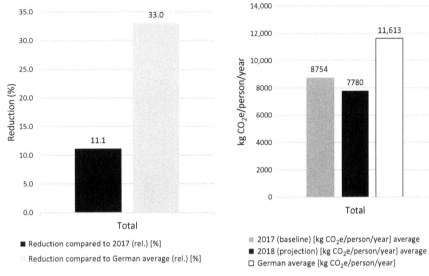

FIG. 1

Baseline (2017) and real-lab overall performance (2018) results of KLIB households in comparison with German average both in total (kg/cap) (*left*) and in relative numbers (percent) (*right*).

7.780 kg/cap, which is 11.1% less than their 2017 baseline and 33% less than the German average. Although the project has been attracting people with a (slightly) better-than-average carbon footprint, the KLIB households managed to reduce their average carbon footprint from 8754 (2017, SD=3648) to 7780 kg (2018, SD=3180), which is 11.3% less than in 2017. While some households managed to reduce by about 40%, others reported increases by up to 30%. The relatively distinctive difference of the initial footprint to the German average reflects two different facts:

(1) There is a *self-selection bias* of participants that led to a sample of people with a higher-than-average general interest in climate change and carbon footprint reduction. Many actions to reduce the personal carbon footprint (e.g. switching to green electricity) had already been taken by many participants. The recruitment process was done via a press release and via the customer magazines of KLIB stakeholders like Bio Company (Berlin's biggest organic food retailer, https://www.biocompany.de/) or *BürgerEnergie Berlin* (a citizen organisation that engages for a remunicipalisation of the city power grid, https://www.buerger-energie-berlin.de/). This led to a slight 'green bias' of the sample.[n]

---

[n] It is worth noting that 'green' consumers (at the attitudinal level) often also are higher income consumers, and income is a major driver of both consumption and emissions (Kleinhückelkotten et al., 2016; Pichler et al., 2017). This relativises the self-selection effect.

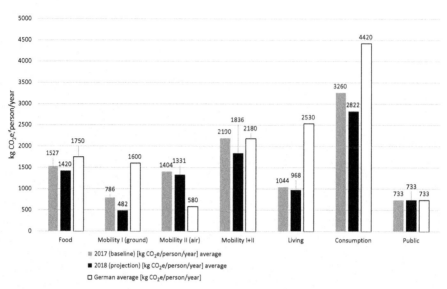

FIG. 2

Baseline (2017) and real-lab sectoral performance (2018) results of KLIB households in comparison with German average both in total (kg per capita).

(2) There is also an *urban bias* of the real-lab experiment. Big cities do have a lower share of single detached homes, for example, which lead to a higher per capita energy consumption, and cities do provide a much better public transport system, reducing the need for individual car use. The same holds for the availability of organic food stores and other facets of a higher urban density, indicating that urban form, urban lifestyles, and time use of households are linked (Gill and Moeller, 2018; Jones and Kammen, 2014; Sovacool and Brown, 2010).[o]

Having a closer look at the results by sectors, we find that participants reduced less in food (7%) and living (7.2%) than in mobility (16.2%) and (other) consumption (13.4%) sectors (cf. Fig. 2).

The survey data and the entries into the KLIB forum reveal some of the reasons behind these differences. Many of the participating households had already changed

---

[o] According to Pichler et al. (2017), the per capita greenhouse gas footprint for Berlin is 8.9 tons (2015). This figure includes direct- and indirect consumption-based emissions but excludes public emissions (e.g. for public lightning, the police, or firefighters). In KLIB, these latter emissions (0.733 t) have been included as well. If we add them to the footprint reported in Pichler et al. (2017), we end up by a per capita footprint of 9.633 tons. If we relate the KLIB households opening account to these figures, our households did start with about 10% less emissions than the Berlin average. This relativises the self-selection bias of the KLIB sample, while the urban structure bias would become more important.

their diets to less meat and dairy intensive meals. The same holds for living-related emissions: about 80% of the participants had switched to green electricity before the project, and the home size of the sample was 17% lower than the German average. If we split mobility into air travel and ground travel, we find that KLIB households did not achieve much of a reduction in the former domain (5.3%), but performed much better in the latter (38.7%). Air travel is the only domain in which KLIB households did perform worse than the German average right from the outset: while the average German has an annual footprint of 580 kg in this domain, KLIB households started with an initial value of 1404 kg, almost three times higher. This reflects the rather high share of middle-income households together with a relative high share of scientists and self-employed occupations.[p]

A more detailed look at the households that increased their carbon footprints during the project reveals that for the majority of them, air travel was the major driver. A personal interview from one participant may illustrate this point:

We have performed quite well this year in most domains. For example, it was a great help to have an energy consultant at our home, providing us with a lot of tailor-made tips. Next to light bulb substitution, which I had anticipated to be a possibility, he identified or fridge (cooling and freezing combination) to be the major driver of electricity consumption at home. We substituted it by a highly efficient new device and could immediately see how the tracker numbers went down! Together with many other small things (e.g. in the food domain), we've been quite successful and, yes, rather proud of being a 'good KLIB household'. All the more surprising was the fact that we jumped upward when our son went to New Zealand this year for a school exchange. This 'ate up' all the other savings—and left us a little frustrated. I didn't know how important air travel is! (Male participant, 55 yrs, married, and a father of two kids).

Ground mobility and other consumption have, on the other hand, been domains with very marked reduction rates. With respect to ground mobility, the urban context of the KLIB real-world lab is important: shifting from car use to public transport or bicycle—the two activities performed by the vast majority of participants—is much easier in an urban agglomeration (cf. for Reckien et al., 2007).

Various studies have reported that energy savings due feedback range from 3% to 20% (e.g., Abrahamse et al., 2005; Darby, 2006; Ehrhardt-Martinez et al., 2010; Fischer, 2008; and Harries et al., 2013). The reported variations may be attributable to differences in study design as feedback has taken a variety of forms (e.g., marketing campaigns or electronic communications) using diverse study groups (ranging from self-selected volunteers to random population samples).[q] The CO_2 savings achieved

---

[p] While many KLIB households did not fly at all, the statistical measure 'average' (mean) is very sensitive to a few high values. The median for flying is only 390 kg. This includes only private trips (e.g. holidays), while business air travel trips were excluded as being in the responsibility of the employer.

[q] The methodological robustness and rigour of studies does also play a role, with more rigorous studies reporting lower savings (Buchanan et al., 2015; Delmas et al., 2013).

by the KLIB experiment can thus be assessed to be quite substantial. IGES (Institute for Global Environmental Strategies, Aalto University, and D-mat ltd.) et al. (2019) have calculated that achieving the Paris Agreement goal of 1.5°C will need a total reduction of individual footprints of 80%–93% by 2050, with annual reduction rates of 8%–12%. The KLIB result is situated in the upper range of that bracket und can be said to be in line with global climate policy targets—provided that KLiB results could be replicated.

4 Household members as citizens I: The ethical dimension

The numbers that we have presented so far may look impressive in comparison with other intervention studies. But do private household members do have a moral obligation to reduce their carbon footprints? The social sciences can provide answers to questions like 'What motivates individuals?' or 'Do people act in favor of a collective good if they assume their own efficacy to be near zero?' But whether or not people as individual consumers do have a moral obligation to make particular consumption choices cannot be answered by empirical behavioural studies. The social sciences do tell us that social norms are an untapped tool that may be useful for encouraging households to reduce and/or shift their electricity consumption (cf. Horne and Kennedy, 2017). But whether or not people ought to have social or personal norms is another question—a philosophical question in the end.

The rather young philosophical discipline of consumer ethics is a controversial domain (Young, 2011). Nevertheless, both utilitarians and Kantians mostly argue in favor of individual moral duties—albeit for different reasons. Utilitarians focus on the outcomes of (aggregated) individual action. Anthropogenic climate change affects the ecosystems worldwide and thus, among other things, the living conditions of (future) generations, and it creates direct harm to people (property, health, and lives).[r]

> *A number of individuals who do not yet constitute a collectivity (…) can be held morally collectively responsible for a harm which has been caused by the predictable aggregation of individual actions.*
>
> **Cripps (2011, p. 174).**

This even holds if no individual had the *intention* to harm someone and even if each individual contribution is marginal with respect to harm. Marginal harm is nonzero harm (Kagan, 2011), and millions of individual marginal contributions sum up to a substantial share of global harm (Nolt, 2011).

Duties to reduce harm do exist whether or not others cooperate: if I reduce my annual footprint by 1 kg and you increase yours by 1 kg, we are off with 0 kg as a

---

[r] This not only holds for some distant future, where the major effects of anthropogenic climate change will unfold in unprecedented ways (cf. meltdown of planetary ice caps and consecutive sea level rise) but also for current events (such as the very dry and hot European summer 2018) to the degree that they can, at least in part, attributed to climate change.

total effect. This is of course a *worse* outcome as compared with full cooperation, with both of us reducing by 1 kg, resulting in a total of 2-kg reduction. But it is still a *better* result than the 2 kg increase that would have resulted in our combined noncooperation, that is, you *and* me increasing our respective footprints by 1 kg. As the harm that results from climate change is caused by our *combined* emissions, the noncooperation of one party is no excuse for a second party to not acting. While free-riding problems clearly exist at a psychosocial level, this *empirical* evidence does not automatically translate into an *ethical* disqualification of *partial* cooperation. Some authors even postulate supererogatory duties of individuals in case of marginal harm due to very small shares of responsibility (Hohl and Roser, 2011), and even if no common institutional framework for solutions does exist (Cripps, 2013; Young, 2011). This individual responsibility is 'caused' by the carbon footprint, that is, the individual share in collective harm, and moderated by the ability to act—the first might be termed 'blame responsibility', while the former can be called 'task responsibility' (Stilz, 2011). But also, authors who are more influenced by Kant's deontological ethics do support individual obligations in face of collective action problems. The concept of individual integrity is very important in this context: even if others do not act in favor of the good, we owe it to our own integrity, which is intricately linked to our moral self-concepts (Korsgaard, 2009; Williams, 1993).

The KLIB project was interested not only in what the consumer could achieve in direct (personal) carbon footprint reduction during 1 year but also in the question of how this endeavour would relate to the attitudes and behaviours of the citizens involved. In doing so, we were guided by the idea of a holistic concept of 'climate citizenship', encompassing both the consumer and the (political) citizen.[s] One way of conceptualising climate citizenship in our context was to ask for the motivation to participate in the real-world lab. If one clusters the answers to this open question, the following picture emerges (cf. Fig. 3).

Consumer-related motives are the primary ones: interest in one's own carbon footprint and its improvement yield support rates from 75% to 85%. Trying out new behavioural patterns (43%) or pure curiosity (19.5%) can also be a motive here, just as to contribute to climate protection in general (72%) or to motivate others for climate protection (25%). To be a good example for one's own children (18%) can be seen as a mixed motive, addressing both the in-home performance of kids, as well as to influence the next generation, an issue of intergenerational justice. These results encourage us in assuming that even projects (experiments) with a clear focus on the consumer and on 'private' action must not forget to address the citizen, that is, both ethical and political facets of people.

---

[s] In the literature, this recently emerged concept has various, often competing definitions, mainly due to the political theory background one wishes to give it. A main tension runs between liberal and communitarian theoretical approaches (cf. Symons and Karlsson, 2018) or practices (cf. Keenis, 2016). The European Union supports the idea, but an analysis of the relevant documents reveals that climate citizenship is considered mainly as a private sphere activity performed by individuals and consists mainly in small adjustments in daily life (Vihersalo, 2017). In our terminology, this kind of understanding refers to the consumer, not the citizen.

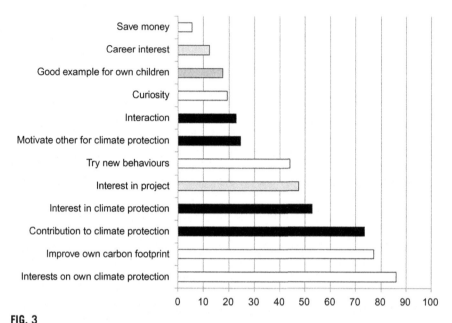

FIG. 3

Motivation of households to participate in KLIB project (in %, multiple answers possible). White boxes, consumer; black boxes, citizen; gray boxes, mixed or other.

While the motivation question is focusing on the willingness to participate in the first place, the question about the effectiveness of the KLIB intervention tools is focusing on the impacts of the experimental setting on the participants. As mentioned previously (Section 2), there were nine interventions in use. Fig. 4 lists them at the bottom line. We have been measuring the subjective (self-reported) impact of these interventions on three facets of the participants: (1) their carbon footprint, (2) the overall climate consciousness (awareness of the problem, awareness of my own contribution to it), and (3) social engagement (e.g. talking to others, participating in public action, and voting attitudes).

The result shows that—in the eyes of the participants—both the baseline survey and the weekly carbon tracking have had high impacts not only on the carbon footprint (which could be expected) but also on climate consciousness (40% 'agree') and on social engagement. In particular, it is the tracker that has had a significant impact (18.5%), which can be attributed to the constant, repeated nature of having to fill out the tracker input boxes. The forum turns out to be more helpful for footprint issues (18%), as we had expected, indicating that peer-to-peer learning is an important feature of the real-world lab. The website tips (19%) have been important for reducing the carbon footprint, but the biweekly newsletter was even more so (29.6%). This supports again the suggestion that constant, repeated information is needed to sustain individual efforts over time. Another result reveals that with respect to climate consciousness, live meetings (29.5%) have been trumping social media (Facebook in our case) (5.1%) by far—even in times of a growing relevance of virtual interactions.

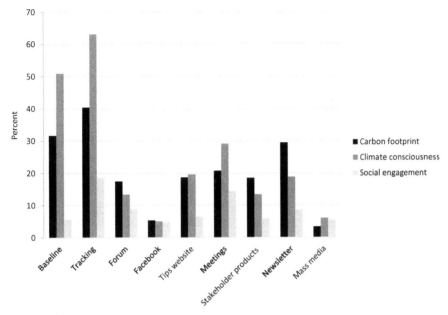

FIG. 4

Impact of intervention formats (*x*-axis) on participants' carbon footprint, climate consciousness, and social engagement (*y*-axis: percentage 'agree').

We cannot tell how this relation will evolve over time, but so far, there is clear evidence that face-to-face communication remains crucial for information exchange and community building. This once again highlights the personalised approach of KLIB. Mass media coverage of KLIB was very high as compared with other projects we have been doing, most of them addressing much larger sociospatial units (such as the city of Berlin as a whole). Nevertheless, the participants stated that this fact had only little influence on their lab performance.

5 Household members as citizens II: The political dimension

KLIB has demonstrated that consumers can reduce their individual carbon footprint at degrees required to meet the Paris Agreement goals. At the same time, it is a very plausible assumption that the KLIB households will not be able to repeat this success without modifying the political boundary conditions of household consumption.[t]

---

[t]   One indicator for this assumption is the fact that a positive correlation (Pearson's $R=0.5$) holds between the baseline footprint 2017 and individual total reduction results (2018): The lower a household's carbon footprint at the beginning of the lab, the smaller the total reductions achieved after 1 year. This suggests that there are (systemic) limits to individual reduction strategies and that the overall 11% reduction of KLIB in 2018 could not easily be repeated in subsequent years.

Following this line of thought, attempts to resolve the climate problem by consumer (household) action alone have received strong criticism. Akenji (2014) for example, argues that without taking the social, economic, and political contexts into account focusing on consumer choices and their sustainable transformations (often termed 'green consumerism') runs the risk of 'consumer scapegoatism'. Grunwald (2010) rejects appeals to individual consumer action as a 'privatisation of sustainability' and calls for political action instead.

This sounds very reasonable in the first place. But what exactly do we mean by 'politics', and how does citizen or consumer action relate to it? Grunwald (2010), for example, criticises the 'privatisation' of sustainability issues, which he sees as intrinsically political in nature. He follows a liberal understanding of the political sphere, which must not be confounded with 'private' issues, such as questions of consumption or the good life in general. Whatever the political dealing with more sustainable consumption patterns might mean, a direct moral or political influence on individual (consumption) preferences cannot be a meaningful part of it. In addition to this decidedly liberal political conception, Grunwald is also attached to a system theory approach to society: different subsystems (such as the economy or law) follow their own logic with subsystem-specific codes, and individuals can only try to influence the codes of these subsystems. In combination with widespread planning scepticism (Van Assche and Verschraegen, 2008), this leads to a very conservative approach that allows only for incremental changes in a political system largely determined by its own path dependency.

Against the background of our project results, we would like to propose a different view of the political system and the role of the individual in it. This other view, first of all, calls into question the strict separation between the private and the political sphere. Yes, consumption choices are morally or politically indifferent in the first place, but only to the extent that the cumulative externalities of individual consumption decisions do not affect the rights of others—current or future generations. Exactly, this precondition is often violated in modern mass-consumer societies, not only with respect to climate change (Princen et al., 2002).[u] Even within the liberal paradigm of strictly separating private and public, governments have found it necessary to intervene in private consumption decisions, for example, bans on tobacco or weapons in many countries show.

In addition, private consumption preferences and decisions are often much less 'private' as the argument assumes. Next to the lack of internalised environmental losses, governments around the world grant industries and private households environmentally harmful subsidies, for example, for fossil fuels or nonsustainable agricultural practices—in Germany alone €57 billion in 2012.[v] This means that environmentally harmful 'private' consumption choices are to some degree driven or enabled by nonsustainable policies.

---

[u] Jaeggi (2016) even argues that 'private' forms of life can be subject to public moral reasoning to the degree that they incorporate (follow and exemplify) ethical norms and that public criticism of private forms of life can at least follow the principle of internal critique.

[v] Cf. https://www.umweltbundesamt.de/en/environmentally-harmful-subsidies#textpart-1 (Accessed 12 December 2018).

Grunwald advises individuals to get involved politically, but he seals off their consumer preferences against any criticism or political intervention. This can even be interpreted as an implicit encouragement for nonsustainable lifestyles. In any case, the question is raised how individual consumers, protected against any public criticism or critical moral considerations, should be willing to opt for public policies that restrict their nonsustainable but free consumer choices. What would motivate a SUV driving, flying-a-lot, heavily meat consuming climate sceptic to ask for political regulation constraining his or her lifestyle? Or, put more generally, how is the link between the individual carbon footprint of consumers and their attitudes to climate politics as citizens?

In the context of the KLIB project, we have been trying to answer this question by testing the so-called crowding-out hypothesis. This hypothesis tries to specify the *interplay* between the citizen and the consumer. It states that for a given individual, household behaviour and policy support are (partial) substitutes rather than complements. This hypothesis was confirmed by survey experiments with more than 14,000 participants in Japan following the shutdown of the Fukushima power plant in 2011 (Werfel, 2017). [w] Translated into the KLIB context, the rationale behind this (partial) substitution effect would be: if individuals themselves can effectively reduce CO_2 emissions by low-carbon consumer and household behaviour, they see no need for further government action, and thus, there is no support for it by the citizen. We tested this hypothesis by correlating two variables: achieved individual reductions versus support for a carbon tax. Our results show no significant correlation (Pearson's $R = 0.08$), so that we have to reject the crowding-out hypothesis for our (small!) sample. We have also asked whether participants think that der German government was doing enough for climate protection, should do more, or should do less. The result reveals a clear ceiling effect: independent of the individual emission changes over time, virtually all participants supported stronger government action. While this might be influenced by the self-selection bias of our sample, it still indicates that—within this sample— no crowding-out takes place and even people who 'do their lot' insist in stronger government action.

We found, however, a weak negative correlation ($R = -0.29$) between baseline emissions and the support for a carbon tax, leading us to the hypothesis that particularly low-emission households are in favor of stronger government action. Interviews we had with these households reveal the rationale behind that: low-emission households take strong action themselves (e.g. completely switching off the fridge in winter and travelling to Spain by train instead of flying) that often are associated with higher subjective costs (money, time, and social appreciation) exactly because the sociotechnical boundary conditions still favor carbon-intensive modes of consumption.

---

[w] Willis and Schor (2012) have found a positive correlation between conscious consumption and the support for political action.

In many areas, I have been climate friendly for about 35 years, and I am equally committed both professionally and privately to bring about the required change in the framework conditions (e.g. in Green Budget Germany)[x]. My carbon footprint currently stands at 7.2 t per year. I cannot be persuaded that it would be up to me to improve it further, as long as the German average is 11.6 t and as long as hundreds of thousands of top emitters in this country are located somewhere above 40 tons—not to mention some who passively or actively inhibit or prevent the change in the framework conditions (Male participant, 61 years., single household).

The plea for stricter government action is thus motivated by a concern not only for more climate effective regulation but also for more social justice by creating a level playing field for all.[y]

Developing an attitude towards climate politics is not the only option that climate citizens do have. There is a wide array of possible fields of action:

- *Cheap talk.* Individuals can simply communicate with relatives, friends, neighbours, colleagues etc. on the issue of climate change and on their participation in the real-world lab[z].
- *Public voice.* People can put forward their viewpoints and ideas in all kinds of public and social media.
- *Public engagement.* People can perform visible acts of public engagement, for example, by participating in demonstrations or by supporting environmental NGOs.
- *Voting behaviour.* People can express their proclimate preferences by voting for the respective parties or candidates in political elections.

The link between consumer and citizen is a complex issue and cannot be easily mapped by linear quantitative correlations (Atkinson, 2015). Attempts to measure the connection between these two roles often miss an important point: how do individuals themselves perceive this link (De Moor, 2017)? An appropriate way to get hold of both the complexity and the subjective perception of this link is to talk to people. In KLIB, we have asked participants about their perceptions of synergies and conflicts of these two roles. The dominant feedback was that most people do accept an individual consumer responsibility for climate change mitigation—after all, KLIB was based on voluntary participation—but that there are clear limits to what individual consumers can do. It is here where the citizen has to step in.

---

[x] *Green Budget Germany* (http://www.foes.de/home/?lang=en) was founded in 1994 as a nongovernmental organisation. It focuses on all elements of market-oriented ecofiscal policy: environmental taxation, emission trading, removal of environmentally harmful subsidies, promotion of renewable energies, and green growth.

[y] See Jenkins et al. (2018) for an assessment of the role of energy justice in global sociotechnical transitions.

[z] 'Cheap talk' refers to the widespread attitude that (mere) communication is easy to have but inefficient, for example, as compared to actually 'do' something. Elinor Ostrom has shown empirically how exactly 'cheap talk' can help to protect common pool resource from overuse (see references in Ostrom, 2010). Experimental economic studies too can show on a quantitative level that this assumption holds (cf. Palfrey et al., 2017).

Participation in KLIB has given us many helpful hints for saving. But it has also shown where the scope of individual actions ceases and where the need for systemic change by politics begins. Clearly, no climate-friendly lifestyle is possible without the latter. A law for CO_2-neutral products must be established—to protect the climate and resources and to relieve the consumer! It is actually an imposition to shift climate protection to the consumers. Responsibility should be taken by the political system so that it creates the conditions for a climate-friendly lifestyle for everyone. KLIB has made it even clearer to me that political action is absolutely necessary. And, of course, we're part of that system too. As a result of KLIB, I decided to become a member of the Green Party. My eldest daughter has been joining the Fridays for Future demonstrations recently, and we have supported it—although she missed a school day (Female participant, 35 years, married, three children).[aa]

This statement clearly shows that participants have been aware of the limits of their own personal carbon footprint reduction possibilities and that the need for more ambitious climate policies is perceived. As the previous quote also shows, participants with a preference for these climate policies see themselves as 'part of the system', and deliberately act as climate citizens.

We have asked households for their preferences of particular climate policies, offering to them a number of options extracted from the current political debate in Germany (cf. Fig. 5).

Interestingly, a carbon tax (66.7%) receives more support than improved EU emission trading (47.4%), and natural carbon sinks (64.9%) are preferred over carbon capturing and storage (CCS) (7%). Although many households did complain about a lack of information on carbon footprints, this policy option (50.9%) was trumped by the support for an immediate coal exit (56.1%) and the removal of environmentally harmful subsidies (93.1%), the top priority in our sample.[ab]

KLIB households did reduce their footprints as consumers, but they also felt their responsibility as citizens, both in an ethical and a political dimension. In our view, this result clearly rejects the assumption that climate change has to be tackled *either* by the consumer *or* by the citizen, as some scholars (implicitly) state. Both roles are needed; the will and the capability to perform them are given, although not every member of the KLIB group did do so in a uniform way. Research projects that address the individual consumer should—this at least is our conclusion—try to find ways for engaged citizens to express their views, preferences, and political actions as well.

[aa] *Fridays for Future* (https://fridaysforfuture.de) organises regular demonstrations of young people for a stringent climate policy. The international movement is inspired by a school strike of Greta Thunberg, a 16-year old Swedish climate activist. In Germany, there are currently (March 2019) 155 active local groups. The movement started in late 2018, and many young members of the KLIB households did participate in these initial demonstrations.

[ab] This last result is surprising to some degree. There was a vivid public debate on a rapid coal exit in Germany during the real-lab phase 2018, which has also been discussed in the KLIB forum. No such debate had been taking place on harmful subsidies.

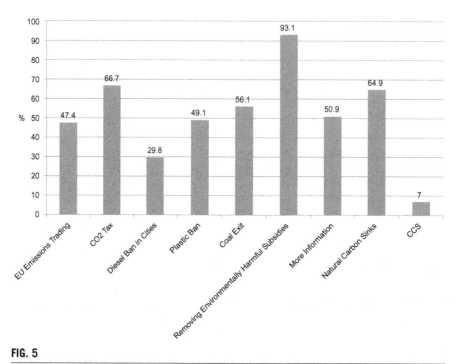

FIG. 5
Support of selected climate policy options by KLIB households in percent.

6 Conclusions and policy recommendations

In this chapter, we have reported some results from an urban real-world lab in Berlin, Germany, with about 100 households trying to reduce their carbon footprints over a 1-year period. Designed as an intervention study, the households have been provided with a core feedback instrument, the carbon tracker, enabling them to trace their weekly life cycle-based CO_{2eq} emissions in the domains of food, mobility, housing energy, other consumption, and public infrastructure. Other intervention instruments included a website, a newsletter, an online forum, face-to-face group gatherings, and at-home visits by professional energy consultants.

The households managed to reduce their carbon footprints by 11.3% within that year, with a significant variance. This result is in line with the implications of the Paris Agreement for private households, namely, an annual reduction of emissions by about 10% (IGES (Institute for Global Environmental Strategies, Aalto University, and D-mat ltd.) et al., 2019). But while KliB managed to achieve this goal in a 1-year real-lab experiment, the Paris goals would require this result to be replicated *every* year until 2050 and by *all* consumers.

We have shown that the chosen intervention instruments did have different effects, with a more technical device—the carbon tracker—influencing also climate awareness and a more social intervention—face-to-face gatherings—influencing the motivation to act as climate conscious citizens. We could reject the crowding-out hypothesis—consumer action being a substitute to an active citizen role—confirmed in a study on Japanese households (Werfel, 2017) as KLIB households have shown an independent relation between both roles, with slight positive correlations at the lower end of total personal CO_2 emissions. We have highlighted the citizen role by focusing on ethical and political aspects of the KLIB household performance, supporting a 'holistic' approach to individuals instead of addressing consumers and citizens separately.

Recent years have seen a significant boom in experimental research and the laboratory approach in the context of sustainability (cf. Bulkeley et al., 2018b; GAIA, 2018; Howaldt et al., 2018; Kivimaa et al., 2017; Marvin et al., 2018; Sengers et al., 2016). The KLIB project adds a small delta to these studies. The experiments involved serve various purposes, including niche creation, market creation, societal problem solving, and spatial planning. Some experiments can have significant outcomes that change the discourse, whereas others facilitate the emergence and diffusion of new technologies or direct changes in the built environment. In the case of KLIB, changes in consumption and everyday practices could be achieved by appropriate feedbacks and by addressing not only the consumer but also the citizen. A new technological feature—the carbon tracker—was a centrepiece for this success. However, this technological device displayed its supportive effects only due to its embedding in expert feedbacks, social learning, norm building, and a political discourse.

So, a first policy recommendation would be that the energy policy mix should aim at encouraging and facilitating mutual learning processes for consumers, both with respect to knowledge and motivation. Accompanying information and policy instruments that change values have the potential to greatly contribute to the effectiveness of the more conventional policy approach. Future work may focus on testing the interplay of information and economic policies (subsidies and taxes), calling for a combination of experimental and modelling approaches. Technical and economic biases need to be overcome; the energy system needs to be conceived as a complex sociotechnical system (cf. Eyre et al., 2018; Miller et al., 2015).

Given the wealth of lab/experimental approaches and outcomes, a second policy recommendation may be that a systematic deliberate combination of different types of experiments, with each contributing slightly different aspects to the processes, is needed. Only a few experimental studies have systematically looked at the question of what upscaling would mean in terms of energy efficiency and CO_2 reduction effects. Methodologically, this can be achieved by combining intervention/case study designs with agent-based modelling (cf. Niamir et al., 2018). These studies can show quantitatively what others have demonstrated on a more qualitative basis: the aggregate effect of small, bottom-up lifestyle changes can reduce the carbon and overall environmental footprint of societies significantly (Chicca et al., 2018).

As it would still be naïve to believe that even more and combined social experiments and local labs would 'change the world', there is a clear need to better link the 'niches' of these experiments with the 'mainstream' of energy and climate policies, that is, the regime and landscape levels.

Turnheim et al. (2018) have convincingly argued that the broad range of experimental settings and labs that we have seen in recent years—many of them in urban contexts—need to have an impact on climate governance at various levels. One of them is the city (Wolfram and Frantzeskaki, 2016; Wolfram et al., 2016), and civil society actors can contribute substantially due to the various roles they can play here (Frantzeskaki et al., 2016; Kronsell and Mukhtar-Landgren, 2018). But we need *learning* cities, that is, cities that are able to institutionalise processes of knowledge accumulation, application, and reflexive modification in various contexts (Campbell, 2009). The deliberate attempt of public leaders and managers to convene the relevant actors, facilitate collaboration and cocreation, and catalyse the development and realisation of innovative ideas needs to be supplemented with persistent attempts to build a strong 'innovation culture' in public organisations (Dobni, 2008). Creating an innovation culture involves recruiting and nurturing creative talent, enhancing diversity and mobility, and encouraging staff members to use their professional knowledge to generate and test new ideas. It also involves challenging a zero-error culture and the detailed rules and regulations and demotivating performance measurement systems that prevent innovation (Torfing and Ansell, 2017). Finally, it involves attempts to create flatter and more flexible organisations with clear mission objectives and strong leadership to breach administrative silos.

Deducing rapid changes in CO_2-relevant everyday practices from single cases, overstating the probability of society-wide transformations from biased samples, and adhering to normative biases of experiments are some of the most common pitfalls in case-based sustainability transition studies (Haxeltine et al., 2017). The KLIB research team is fully aware of them. One key remedy against these pitfalls is a critical analysis of the sociotechnical and governance boundary conditions in which real labs are embedded (cf. for the case of Berlin: Reusswig and Lass, 2017). Path dependencies, interests, and the power of incumbent actors must be taken into account in order to give a realistic assessment of the sustainability transition impact of small-scale experiments (Fligstein and McAdam, 2012).

The fourth and final recommendation would be to utilise the outputs of experimental projects like KLIB to more actively support the idea of climate citizenship and to define it in a less 'privatist' way as the EU does. Anderson et al. (2017) have shown that the strength of national government domestic climate policies in democracies crucially depends upon the public opinion towards climate change and the degree of public support on renewable energies. Cao and Ward (2017) have shown that domestic climate policy and transnational climate governance networks are complements, not substitutes for moving global climate policies ahead. This also means that citizens via their governments do have an indirect influence on global climate policies. In a world of growing right-wing populism and their often detrimental energy and climate policy agendas (Schaller and Carius, 2019), this is may be the most important policy message: ambitious climate policy goals (such as the 1.5°C goal of the Paris Agreement) are by no means mere outcomes of detached elite policy processes (at the European level 'Brussels' would be the appropriate symbol) but find active support among consumer and citizens 'out there'.

Most social scientists in the field of consumer activation have asked for more stringent political action in order to support greener lifestyles. However, for many of them, the political system remains a black box, often detached from the world of consumers and citizens. A closer look at the political system is needed, and thus a closer cooperation with political scientists can be recommended. Especially in democracies, where consumers are voters, the political system is very sensitive to changes in attitudes and behaviours of the electorate, and not only majorities are important here. The success of populism in many European countries, for example, does not only rely upon their election results and the representation in parliaments, it is also felt indirectly, with other parties serving some issues and discourses of these parties, despite of their (still) relatively small electoral support. A closer look at the influence potential of civil society on the policy cycle and on changing advocacy coalitions (Sabatier, 1988; Sabatier and Jenkins-Smith, 1993) shows that the political system is not a closed system. And it is not set to incremental changes in fixed sectors. Parag and Janda (2014) highlighted the ways in which 'middle actors' (such as professionals, real-estate actors, and congregations) may have the agency and capacity to bring about change because of their position in energy systems and the knowledge and connections they can dispose of. Proactive consumer and citizen efforts can both profit form and support these middle actors, thus increasing pressures on the political system. If real-world experiments can contribute to these insights, then they are really helpful tools for a transformative (social) science.

References

Abrahamse, W., Steg, L., Vlek, C., Rothengatter, T., 2005. A review of intervention studies aimed at household energy conservation. J. Environ. Psychol. 25 (3), 273–291.

Abrahamse, W., Steg, L., Vlek, C., Rothengatter, T., 2007. The effect of tailored information, goalsetting, and tailored feedback on household energy use, energy-related behaviors, and behavioral antecedents. J. Environ. Psychol. 27 (4), 265–276.

Akenji, L. (2014) Consumer scapegoatism and limits to green consumerism. J. Clean. Prod., 63, 13-23. doi:https://doi.org/10.1016/j.jclepro.2013.05.022 (Accessed 12 November 2018).

Anderson, B., Böhmelt, T. & Ward, H. (2017) Public opinion and environmental policy output: a cross-National analysis of energy policies in Europe. Environ. Res. Lett., 12 (11), 114011. doi:https://doi.org/10.1088/1748-9326/aa8f80 (Accessed 17 November 2018).

Atkinson, L., 2015. Locating the politics in political consumption: a conceptual map of four types of political consumer identities. Int. J. Commun. 9, 2047–2066.

Barthel, S., Oswalt, P., Schmidt, A., von Mende, J., 2014. Privater Haushalt und städtischer Stoffwechsel. Eine Geschichte von Verdichtung und Auslagerung—Berlin 1700-1930. Archplus 47 (218), 92–103.

Beecroft, R., Parodi, O., 2016. Reallabore als Orte der Nachhaltigkeitsforschung und Transformation. Einführung in den Schwerpunkt. Technikfolgenabschätzung—Theorie und Praxis 25 (3), 4–8.

Benveniste, H., Boucher, O., Guivarch, C., Le Treut, H., Criqui, P., 2018. Impacts of nationally determined contributions on 2030 global greenhouse gas emissions: uncertainty analysis and distribution of emissions. Environ. Res. Lett. 13, 014022.

Bin, S., Dowlatabadi, H., 2005. Consumer lifestyle approach to US energy use and the related CO_2 emissions. Energy Policy 33 (2), 197–208.

BMUB/UBA (Federal Ministry for the Environment/Federal Environmental Agency) (Eds.), 2017. Umweltbewusstsein in Deutschland 2016. Ergebnisse einer repräsentativen Bevölkerungsumfrage, Berlin, Dessau, BMUB/UBA.

Boström, M., Lidskog, R., Uggla, Y., 2017. A reflexive look at reflexivity in environmental sociology. Environ. Sociol. 3 (1), 6–16.

Buchanan, K., Russo, R., Anderson, B., 2015. The question of energy reduction: the problem(s) with feedback. Energy Policy 77, 89–96.

Bulkeley, H., Castán Broto, V., 2013. Government by experiment? Global cities and the governing of climate change. Trans. Inst. Br. Geogr. 38 (3), 361–375.

Bulkeley, H., Castán Broto, V., Edwards, G.A.S. (Eds.), 2018a. An Urban Politics of Climate Change: Experimentation and the Governing of Socio-Technical Transitions. Routledge, London/New York.

Bulkeley, H., Marvin, S., Voytenko Palgan, Y., McCormick, K., Breitfuss-Loidl, M., Mai, L., von Wirth, T. & Frantzeskaki, N. (2018b) Urban living laboratories: conducting the experimental city? Eur. Urban Region. Stud. Available from: https://journals.sagepub.com/doi/full/10.1177/0969776418787222 (Accessed 2 December 2018).

Campbell, T., 2009. Learning cities: knowledge, capacity and competitiveness. Habitat Int. 33 (2), 195–201.

Cao, X., Ward, H., 2017. Transnational climate governance networks and domestic regulatory action. Int. Interact. 43 (1), 76–102.

Castán Broto, V., 2017. Urban Governance and the Politics of Climate change. World Dev. 93, 1–15.

Chicca, F., Vale, B., Vale, R. (Eds.), 2018. Everyday lifestyles and sustainability. In: The Environmental Impact of Doing the Same Things Differently. Routledge, London.

Cooper, M., 2018. Governing the global climate commons: the political economy of state and local action, after the U.S. flip-flop on the Paris Agreement. Energy Policy 118, 440–454.

Cripps, E., 2011. Climate change, collective harm and legitimate coercion. Crit Rev Int Soc Pol Phil 14 (2), 171–193.

Cripps, E., 2013. Climate Change and the Moral Agent. Individual Duties in an Interdependent World. Oxford University Press, Oxford.

Darby, S., (2006) The Effectiveness of Feedback on Energy Consumption: A Review for DEFRA of the Literature on Metering, Billing and Direct Displays. Environmental Change Institute, University of Oxford, Oxford. Available from: https://www.eci.ox.ac.uk/research/energy/downloads/smart-metering-report.pdf (Accessed 7 November 2018).

De Moor, J., 2017. Lifestyle politics and the concept of political participation. Acta Politica 52 (2), 179–197.

Delmas, M.A., Fischlein, M., Asensio, O.I., 2013. Information strategies and energy conservation behavior: a meta-analysis of experimental studies from 1975 to 2012. Energy Policy 61, 729–739.

Destatis (Statistisches Bundesamt) (2018) Umweltökonomische Gesamtrechnungen. Direkte und indirekte CO_2-Emissionen in Deutschland 2005-2014. Wiesbaden, Statistisches Bundesamt. Available from: https://www.destatis.de/DE/Publikationen/Thematisch/UmweltoekonomischeGesamtrechnungen/CO2EmissionenPDF_5851305.pdf;jsessionid=3A0E39548626E12E48CFA01C7BC0B961.InternetLive1?__blob=publicationFile (Accessed 24 November 2019).

Dietz, T., Gardner, G.T., Gilligan, J., Stern, P.C., Vandenbergh, M.P., 2009. Household actions can provide a behavioral wedge to rapidly reduce U.S. carbon missions. Proc. Natl. Acad. Sci. U.S.A. 106 (44), 18452–18456.

Dobni, C., 2008. Measuring innovation culture in organizations. Eur. J. Innov. Manag. 11 (4), 539–559.

Dupont, L., Mastelic, J., Nyffeler, N., Latrille, S., Seulliet, E., 2019. Living labs as a support to trust for co-creation of value: application to the consumer energy market. J. Innov. Econ. Manag. 28 (1), 53–78.

EC (European Commission) (2017) KI-01-17-409-EN-N. Social Innovation as a Trigger for Transformations. The Role of Research. Brussels, European Commission. Available from: https://ec.europa.eu/research/social-sciences/pdf/policy_reviews/social_innovation_trigger_for_transformations.pdf (Accessed 14 November 2018).

Ehrhardt-Martinez, K., Donnelly, K. A. & Laitner, J. A., (2010) Advanced Metering Initiatives and Residential Feedback Programs: A Meta-Review for Household Electricity-Saving Opportunities. American Council for an Energy-Efficient Economy. Report Number E105. Available from: https://www.smartgrid.gov/files/ami_initiatives_aceee.pdf (Accessed 12 November 2018).

Engels, A. & Walz, K. (2018) Dealing with multi-perspectivity in real-world laboratories. Experiences from the transdisciplinary research project Urban transformation laboratories. *Gaia,* 27 (S1), 39-45. Available from: https://www.wiso.uni-hamburg.de/fachbereich-sowi/professuren/engels/archiv/engelswalz-gaia-s1-2018.pdf (Accessed 12 November 2018).

Evans, J., Karvonen, A., 2014. 'Give me a laboratory and I will lower your carbon footprint!'—Urban laboratories and the governance of low-carbon futures. Int. J. Urban Reg. Res. 38 (2), 413–430.

Evans, J., Karvonen, A., Raven, R., 2016. The Experimental City. Routledge, London/New York.

Eyre, N., Darby, S. J., Grünewald, P., McKenna, E. & Ford, R. (2018) Reaching a 1.5°C target: socio-technical challenges for a rapid transition to low-carbon electricity systems. Phil. Trans. R. Soc. A, 376 (2119). Available doi https://doi.org/10.1098/rsta.2016.0462 (Accessed 10 December 2018).

Fazey, I., et al., 2018. Ten essentials for action-oriented and second order energy transitions, transformations and climate change research. Energy Res. Soc. Sci. 40, 54–70.

Fischer, C., 2008. Feedback on household electricity consumption: a tool for saving energy? Energy Effic. 1 (1), 79–104.

Fligstein, N., McAdam, D., 2012. A Theory of Fields. Oxford University Press, New York/Oxford.

Frantzeskaki, N., et al., 2016. Elucidating the changing roles of civil society in urban sustainability transitions. Curr. Opin. Environ. Sustain. 22, 41–50.

GAIA, 2018. Special issue: labs in the real world: advancing transdisciplinarity and transformations. Ecol. Perspect. Sci. Soc. 27 (S1).

Gill, B., Moeller, S., 2018. GHG emissions and the rural-urban divide. A carbon footprint analysis based on the German official income and expenditure survey. Ecol. Econ. 145 (C), 160–169.

Godin, B., 2015. Innovation Contested. The Idea of Innovation Over the Centuries. Routledge, Milton Park.

Goodall, C., 2007. How to live a low-carbon life. In: The Individual's Guide to Stopping Climate Change. Earthscan, London and Sterling, VA.

Gross, M., Krohn, W., 2005. Society as experiment: sociological foundations for a self-experimental society. Hist. Hum. Sci. 18 (2), 63–86.

Grunwald, A., 2010. Wider die Privatisierung der Nachhaltigkeit: Warum ökologisch korrekter Konsum die Umwelt nicht retten kann. Gaia 19 (83), 178–182.

Harries, T., Rettie, R., Studley, M., Burchell & K., Chambers, S., 2013. Is social norms marketing effective? A case study in domestic electricity consumption. Eur. J. Mark. 47 (9), 1458–1475.

Haxeltine, A., Pel, B., Wittmayer, J., Dumitru, A., Kemp, R. & Avelino, F. (2017) Building a middle-range theory of transformative social innovation. Theoretical pitfalls and methodological responses. Eur. Public Social Innov. Rev., 2 (1), 59-77. doi: 10.31637/epsir.17-1.5 (Accessed 10 November 2018).

Heiskanen, E., Laakso, S., Matschoss, K., Backhaus, J., Goggins, G., Vadovics, E., 2018. Designing real-world laboratories for the reduction of residential energy use. Articulating theories of change. Gaia 27 (S1), 60–67.

Hertwich, E.G., Peters, G.P., 2009. Carbon footprint of nations: a global, trade-linked analysis. Environ. Sci. Technol. 43 (16), 6414–6420.

Hilger, A., Rose, M., Wanner, M., 2018. Changing faces—factors influencing the role of researchers in real-world laboratories. Gaia 27 (1), 138–145.

Hohl, S., Roser, D., 2011. Stepping in for the polluters? Climate justice under partial compliance. Analyse Kritik 2011 (2), 477–500.

Horne, C., Kennedy, E.H., 2017. The power of social norms for reducing and shifting electricity use. Energy Policy 107, 43–52.

Howaldt, J., Kaletka, C., Schröder, A. & Zirngiebl, M. (2018) Atlas of Social Innovation—New Practices for a Better Future. Sozialforschungsstelle/TU Dortmund University, Dortmund. Available from: https://www.socialinnovationatlas.net/ (Accessed 30 November 2018).

IGES (Institute for Global Environmental Strategies, Aalto University, and D-mat ltd.) et al. (2019) 1.5-Degree Lifestyles: Targets and Options for Reducing Lifestyle Carbon Footprints. Technical report. Hayama, Institute for Global Environmental Strategies, Aalto University and D-mat ltd. Available from: https://pub.iges.or.jp/pub/15-degrees-lifestyles-2019 (Accessed 10 December 2018).

IPCC (2018) Global Warming of 1.5°C. An IPCC Special Report on the Impacts of Global Warming of 1.5°C Above Preindustrial Levels and Related Global Greenhouse Gas Emission Pathways, in the Context of Strengthening the Global Response to the Threat of Climate Change, Sustainable Development, and Efforts to Eradicate Poverty. Summary for Policy Makers. [Online] Available From: http://www.ipcc.ch/report/sr15/ (Accessed 5 December 2018).

Ivanova, D., Stadler, K., Steen-Olsen, K., Wood, R., Vita, G., Tukker, A., Hertwich, E.G., 2016. Environmental impact assessment of household consumption. J. Ind. Ecol. 20 (3), 526–536.

Jaeggi, R., 2016. Critique of Forms of Life. Harvard University Press, Cambridge, MA.

Jenkins, K., Sovacool, B.K., McCauley, D., 2018. Humanizing sociotechnical transitions through energy justice: an ethical framework for global transformative change. Energy Policy 117 (C), 66–74.

Jensen, C.L., Goggins, G., Fahy, F., Grealis, E., Vadovics, E., Genuse, A., Rau, H., 2018. Towards a practice-theoretical classification of sustainable energy consumption initiatives: insights from social scientific energy research in 30 European countries. Energy Res. Soc. Sci. 45, 297–306.

Jensen, C. L. et al (2017) Catalogue of Existing Good Practice Examples of Programmes and Interventions. ENERGISE—European Network for Research, Good Practice and Innovation for Sustainable Energy, Deliverable No 2.1., Aalborg University. Available from: http://energise-project.eu/sites/default/files/content/WP2%20Deliverable%202%201%20FINAL.pdf (Accessed 25 November 2018).

Jones, S.R., 1992. Was there a Hawthorne effect? Am. J. Sociol. 98 (3), 451–468.

Jones, C., Kammen, D.M., 2014. Spatial distribution of U.S. household carbon footprints reveals suburbanization undermines greenhouse gas benefits of urban population density. Environ. Sci. Technol. 48, 895–902.

Kagan, S., 2011. Do I make a difference? Philos Public Aff 39 (2), 105–141.

Karvonen, A., van Heur, B., 2014. Urban laboratories: experiments in reworking cities. Int. J. Urban Reg. Res. 38 (2), 379–392.

Keenis, A., 2016. Ecological citizenship and democracy: communitarian versus agonistic perspectives. Environ. Polit. 25 (6), 949–970.

Kivimaa, P., Hildén, M., Huitema, D., Jordan, A., Newig, J., 2017. Experiments in climate governance—a systematic review of research on energy and built environment transitions. J. Clean. Prod. 169, 17–29.

Kleinhückelkotten, S., Neitzke, H.P. & Moser, S. (2016) Repräsentative Erhebung von Pro-Kopf-Verbräuchen natürlicher Ressourcen in Deutschland (nach Bevölkerungsgruppen). UBA Texte 39/2016. Umweltbundesamt, Dessau. Available from: https://www.umweltbundesamt.de/sites/default/files/medien/378/publikationen/texte_39_2016_repraesentative_erhebung_von_pro-kopf-verbraeuchen_natuerlicher_ressourcen.pdf (Accessed 1 October 2018).

Korsgaard, C., 2009. Self-Constitution: Agency, Identity, and Integrity. New York, Oxford University Press, Oxford.

Kronsell, A., Mukhtar-Landgren, D., 2018. Experimental governance: the role of municipalities in urban living labs. Eur. Plan. Stud. 26 (5), 988–1007.

Lazarus, R.J., 2009. Super wicked problems and climate change: restraining the present to liberate the future. Cornell Law Rev. 94, 1153–1234.

Levin, K., Cashore, B., Bernstein, S., Auld, G., 2012. Overcoming the tragedy of super wicked problems: constraining our future selves to ameliorate global climate change. Policy. Sci. 45, 123–152.

Lewitt, S.D., List, J.A., 2009. Was There Really a Hawthorne Effect at the Hawthorne Plant? An Analysis of the Original Illumination Experiments. National Bureau of Economic Research (NBER), Working paper 15016, NBER, Cambridge/MA.

Liedtke, C., Baedeker, C., Hasselkuß, M., Rohn, H., Grinewitschus, V., 2015. User-integrated innovation in sustainable living labs: an experimental infrastructure for researching and developing sustainable product service systems. J. Clean. Prod. 97, 106–116.

Luederitz, C., et al., 2017. Learning through evaluation: a tentative evaluative scheme for sustainability transition experiments. J. Clean. Prod. 169, 61–76.

Lutzenhiser, L., 1993. Social and behavioural aspects of energy use. Annu. Rev. Energy Environ. 18, 247–289.

Marvin, S., Bulkeley, H., Mai, L., McCormick, K., Voytenko Palgan, Y. (Eds.), 2018. Urban Living Labs: Experimenting With City Futures. Routledge, Milton Park/New York.

Menny, M., Voytenko Palgan, Y., McCormick, K., 2018. Urban living labs and the role of users in co-creation. Gaia 27 (S1), 68–77.

Miller, C.A., Richter, J., O'Leary, J., 2015. Socio-energy systems design: a policy framework for energy transitions. Energy Res. Soc. Sci. 6, 29–40.

Moezzi, M., Janda, K.B., Rotmann, S., 2017. Using stories, narratives, and storytelling in energy and climate change research. Energy Res. Soc. Sci. 31, 1–10.

Monstadt, J., 2007. Urban governance and the transition of energy systems: institutional change and shifting energy and climate policies in Berlin. Int. J. Urban Reg. Res. 31 (2), 326–343.

Moss, T., 2014. Socio-technical change and the politics of urban infrastructure: managing energy in Berlin between dictatorship and democracy. Urban Stud. 51 (7), 1432–1448.

Niamir, L., Filatova, T., Voinov, A., Bressers, H., 2018. Transition to low-carbon economy: assessing cumulative impacts of individual behavioral changes. Energy Policy 118, 325–345.

Nolt, J., 2011. How harmful are the average American's greenhouse gas emissions? Ethics Policy Environ. 14 (1), 3–10.

Ostrom, E., 2010. A Polycentric Approach for Coping With Climate Change. Background Paper to the World Development Report 2010. Policy Research Working Paper 5095, The World Bank, Washington, D.C.

Palfrey, T., Rosenthal, H., Roy, N., 2017. How cheap talk enhances efficiency in threshold public goods games. Games Econ. Behav. 101, 234–259.

Parag, Y., Janda, K.B., 2014. More than filler: middle actors and socio-technical change in the energy system from the 'middle-out'. Energy Res. Soc. Sci. 2014 (3), 102–112.

Park, R.E., 1929. The city as a social laboratory. In: Smith, T.V., White, L.D. (Eds.), Chicago: An Experiment in Social Science Research. University of Chicago Press, Chicago, pp. 1–19.

PCF (Product Carbon Footprint Pilot Project Germany), 2009. Product Carbon Footprinting—The Right Way to Promote Low Carbon Products and Consumption Habits? Experiences, Findings and recommendations from the Product Carbon Footprint Pilot Project Germany. Thema 1, Berlin.

Pichler, P.-P., Zwickel, T., Chavez, A., Kretschmer, T., Seddon, J., Weisz, H., 2017. Reducing urban greenhouse gas footprints. Nat. Sci. Rep. 7 (1), 14659.

Princen, T., Maniates, M., Conca, K. (Eds.), 2002. Confronting Consumption. The MIT Press, Cambridge, MA.

Reckien, D., Ewald, M., Edenhofer, O., 2007. What parameters influence the spatial variations in CO_2 emissions from road traffic in Berlin? Implications for urban planning to reduce anthropogenic CO_2 emissions. Urban Stud. 44 (2), 339–355.

Reusswig, F., 2009. Consuming Nature. Modern Lifestyles and Their Environment. Potsdam University, Habilitation Thesis.

Reusswig, F., 2010. Sustainability transitions through the lens of lifestyle dynamics. In: Lebel, L., Lorek, S., Daniel, R. (Eds.), Sustainable Production and Consumption Systems. Knowledge, Engagement and Practice. Springer, Dordrecht/Heidelberg/London/New York/Berlin, pp. 39–60.

Reusswig, F., Lass, W., 2017. Urbs laborans: Klimapolitische Realexperimente am Beispiel Berlins. In: Böschen, S., Groß, M., Krohn, W. (Eds.), Experimentelle Gesellschaft. Das Experiment als wissensgesellschaftliches Dispositiv. Nomos, Baden-Baden, pp. 311–340.

Rogga, S., Zscheischler, J., Gaasch, N., 2018. How much of the real-world laboratory is hidden in current transdisciplinary research? Gaia 27 (S1), 18–22.

Sabatier, P., 1988. An Advocacy Coalition Framework of policy change and the role of policy-oriented learning therein. Policy. Sci. 21 (2-3), 129–168.

Sabatier, P.A., Jenkins-Smith, H.C., 1993. Policy Change and Learning. An Advocacy Coalition Approach. Westview Press, Boulder, CO.

Schaller, S., Carius, A., 2019. Convenient Truths. Mapping Climate Agendas of Right-Wing Populist Parties in Europe. Adelphi, Berlin.

Schneidewind, U., Augenstein, K., Stelzer & F., Wanner, M., 2018. Structure matters: real-world laboratories as a new type of large-scale research infrastructure. A framework inspired by Giddens' structuration theory. Gaia 27 (S1), 12–17.

Sengers, F., Wieczorek, A. J. & Raven, R. (2016) Experimenting for sustainability transitions: a systematic literature review. Technol. Forecast. Social Change doi https://doi.org/10.1016/j.techfore.2016.08.031 (Accessed 20 November 2019).

Singer-Brodowski, M., Beecroft, R., Parodi, O., 2018. Learning in real-world laboratories. A systematic impulse for discussion. Gaia 27 (S1), 23–27.

Sovacool, B.K., Axsen, J., Sorrell, S., 2018. Promoting novelty, rigor, and style in energy social science: towards codes of practice for appropriate methods and research design. Energy Res. Soc. Sci. 45, 12–42.

Sovacool, B.K., Brown, M.A., 2010. Twelve metropolitan carbon footprints: a preliminary comparative global assessment. Energy Policy 38, 4856–4869.

Steg, L., 2009. Promoting household energy conservation. Energy Policy 36 (12), 1449–1453.

Stern, P.C., Janda, K.B., Brown, M.A., Steg, L., Vine, E.L., Lutzenhiser, L., 2016. Opportunities and insights for reducing fossil fuel consumption by households and organizations. Nat. Energy 2016 (1), 16043.

Stilz, A., 2011. Collective responsibility and the state. J. Polit. Philos. 19 (2), 190–208.

Symons, J., Karlsson, R., 2018. Ecomodernist citizenship: rethinking political obligations in a climate-changed world. Citizsh. Stud. 22 (7), 685–704.

Torfing, J., Ansell, C., 2017. Strengthening political leadership and policy innovation through the expansion of collaborative forms of governance. Public Manag. Rev. 19, 37–54.

Turnheim, B., Kivimaa, P., Berkhout, F. (Eds.), 2018. Innovating Climate Governance. Moving Beyond Experiments. Cambridge University Press, Cambridge.

UNCCS (United Nations Climate Change Secretariat) (2018) Yearbook of Global Climate Action 2018. Marrakech Partnership. Bonn, UNCCS. Available from: https://unfccc.int/sites/default/files/resource/GCA_Yearbook2018.pdf (Accessed 25 November 2019).

UNEP (United Nations Environment Programme) (2018) Emissions Gap Report 2018. Nairobi, UNEP. Available from: https://www.unenvironment.org/resources/emissions-gap-report-2018 (Accessed 14 November 2019).

Unruh, G.C., 2002. Escaping carbon lock-in. Energy Policy 30 (4), 317–325.

Van Assche, K., Verschraegen, G., 2008. The limits of planning: Niklas Luhmann's systems theory and the analysis of planning and planning ambitions. Plan. Theory 7, 263–283.

Vihersalo, M., 2017. Climate citizenship in the European Union: environmental citizenship as an analytical concept. Environ. Polit. 26 (2), 343–360.

Voytenko, Y., McCormick, K., Evans, J., Schliwa, G., 2016. Urban living labs for sustainability and low carbon cities in Europe: towards a research agenda. J. Clean. Prod. 123, 45–54.

WBGU (German Advisory Council on Global Change) (2011) World in Transition—A Social Contract for Sustainability. Berlin, WBGU. Available from: https://www.wbgu.de/fileadmin/user_upload/wbgu.de/templates/dateien/veroeffentlichungen/hauptgutachten/jg2011/wbgu_jg2011_en.pdf (Accessed 21 November 2019).

Werfel, S.H., 2017. Household behaviour crowds out support for climate change policy when sufficient progress is perceived. Nat. Clim. Chang. 7, 512–515.

Williams, B., 1993. Shame and Necessity. University of California Press, Berkeley, Los Angeles, London.

Willis, M.M., Schor, J.B., 2012. Does changing a light bulb lead to changing the world? Political action and the conscious consumer. Ann. Am. Acad. Polit. Social Sci. 644 (2012), 160–190.

Wolfram, M. & Frantzeskaki, N. (2016) Cities and systemic change for sustainability: prevailing epistemologies and an emerging research agenda. Sustainability, 8 (2), 144. doi https://doi.org/10.3390/su8020144 (Accessed 22 November 2019).

Wolfram, M., Frantzeskaki, N., Maschmeyer, S., 2016. Cities, systems and sustainability: status and perspectives of research on urban transformations. Curr. Opin. Environ. Sustain. 22, 18–25.

Young, I.M., 2011. Responsibility for Justice. Oxford University Press, New York.

Further reading

Castán Broto, V. & Bulkeley, H. (2013) A survey of urban climate change experiments in 100 cities. Glob. Environ. Chang., 23 (1), 92—102. doi:https://doi.org/10.1016/j.gloenv-cha.2012.07.005 (Accessed 7 November 2018).

Dalton, R.J., 1988. Citizen Politics in Western Democracies: Public Opinion and Political Parties in the United States, Great Britain, West Germany, and France. Chatham House Publishers, New York.

Overview of the European Union policies to promote more sustainable behaviours in energy end-users

4.3

Paolo Bertoldi[a]

European Commission, Joint Research Centre, Ispra, Italy

1 Introduction

Energy is needed for several useful services such as providing heating, cooling and lighting, mobility, food preparation, water, and drive power. With the progress of humankind and with technological development, energy use has increased due to the need and desire for additional and faster services such as television, mobile phones, computers, internet, and fast trains. Energy services were initially provided by renewable energy, in particular biomass (but also hydropower and wind power as well as animal power). With the start of the industrial revolution, biomass was gradually replaced by fossil fuels, starting with coal. At the same time, thermal and electrical machines were introduced and therefore the concept of energy efficiency, that is, the ratio of output useful service (e.g. amount of lighting) and the input energy (e.g. kilowatt-hour or oil supplied to the lamp). Gradually, due to technological progress and the cost of energy, the efficiency of energy production and energy end-use has increased (Ruzzenenti and Bertoldi, 2014). However, this has not resulted in an absolute energy reduction nor in a per capita energy reduction, due to the economic expansion, more production of goods, and more services requested by end-users. Raising gross domestic product (GDP) and living standards in both developed countries and developing countries has increased the demand for energy services in all the sectors of the economy, thus increasing global energy demand. To meet the increasing energy demand, large investments in energy supply side have been necessary for extraction, conversion, and distribution to end-users of energy, in particular for electricity. Very often, end-use efficiency could have been implemented at reduced or

---

[a] Disclaimer: The views expressed are purely those of the authors and may not in any circumstances be regarded as stating an official position of the European Commission.

Energy and Behaviour. https://doi.org/10.1016/B978-0-12-818567-4.00018-1

similar costs than supply side solutions, with reduced emissions, pollution, and other cobenefits (Fawcett and Killip, 2019), such as lower impact on health, job creation, and enhanced security of supply.

The indicator normally used for the efficiency of a nation or region is the energy intensity, that is, the ratio between the energy consumption and GDP. Over time, energy efficiency has improved, and energy intensity has been reduced (also due to the growth of GDP), in particular in developed countries, although the energy consumption has increased. In the EU, energy consumption peaked in 2006 and then has decreased till 2014, when it started to grow again (Tsemekidi Tzeiranaki et al., 2018). Similarly, indicators for energy efficiency in industrial production are the ratio between input energy and the physical output production.

Burning fossil fuels, to meet energy services requested by end-users (e.g. heating, cooling, cooking, lighting, and mobility), has many drawbacks. First of all, fossil fuel consumption results in CO_2 emissions. CO_2 is the main greenhouse gas responsible for the global warming, which intensified at the time of the industrial revolution and still continues with the average global Earth surface temperature being 1 °C higher today than in the preindustrial era (IPCC, 2018; Bertoldi, 2018). Secondly, other pollutants, for example, NO_X and particulate matter (PMs), are also produced in the combustion of fossil fuels, which deteriorate local air quality and affect the indoor environment. In 1988, the United Nations General Assembly defined climate change a 'common concern of mankind' (United Nations, 1998). The scientific community formally established the Intergovernmental Panel on Climate Change (IPCC) in 1988. The United Nations Conference on Environment and Development held in Rio de Janeiro in June 1992 established the United Nations Framework Convention on Climate Change (UNFCC) and the subsequent regular Conference of the Parties (COPs) (Gupta, 2010). The Kyoto Protocol (UNFCCC, 2007), an international treaty in the frame of the UNFCCC, was adopted in December 1997 and entered into force in February 2005. Under the Kyoto Protocol Annex I, countries (i.e. 38 industrialised countries) committed to binding greenhouse gases (GHG) emission targets to be achieved between 2008 and 2012 (in the first commitment period) compared with a base year (for most of the countries set at 1990) (Maamoun, 2019).

The Paris Agreement at COP 21 in December 2015 is a major step forward in climate change negotiation and global engagement to limit global warming. The Paris Agreement aims at limiting the increase of global average temperature to 'well below 2°C above preindustrial levels and pursuing efforts to limit the temperature increase to 1.5°C above preindustrial levels' (UNFCCC, 2015), with the 'aim to reach global peaking of GHG as soon as possible' and 'achieve a balance between anthropogenic emissions by sources and removal by sinks of GHG in the second half of this century' (UNFCCC, 2015). Since the Rio Conference, energy efficiency and energy demand reductions have been identified as a key mitigation option by several IPCC Assessment Reports and UNFCCC documents, protocols, and international agreement (Bertoldi, 2018).

Another important point is that conventional energy sources (in particular fossil fuels) are limited and found only in some specific geographical locations. Many countries around the world do not have reserves of fossil fuels, but have to import them, thus being exposed to the fluctuation of the market and possible unforeseen disruptions. The provision of energy reserves, supply contracts, and indigenous energy sources contribute to the security of energy supply. Energy efficiency and the reduction of energy demand are a key component of security of supply, by reducing the need for energy imports. The societal goal to reduce energy consumption started in 1973 with the first oil crisis generated by the oil embargo by the members of the Organisation of Arab Petroleum Exporting Countries and then by the second oil crisis in 1979 (Grossman, 2015; Rüdiger, 2014). Oil was used for vehicles and in many European nations for heating and power generation. Policy makers started to adopt policies to reduce oil usage in vehicles, buildings, industry, power generation, and equipment to mitigate the higher energy prices. To reduce energy consumption, policies have focused on the promotion and adoption of efficient technologies and on changing user's energy behaviour, both at the private/personal and organisation level (Geller et al., 1994).

The reduction of energy consumption through energy efficiency technologies and end-user behaviour is of key importance for the mitigation of climate change and the security of energy supply. In addition, as highlighted in several articles and reports, energy efficiency offers many cobenefits (Fawcett and Killip, 2019; IEA, 2014). However, there are several barriers, presented in detail in the next section, that prevent the full adoption of energy efficiency technologies and positive behaviour change. Therefore, policies to overcome these barriers have been implemented.

This chapter has the following structure: Section 2 introduces the concept of barriers to energy efficiency; Section 3 presents and classifies the policies needed to remove, reduce, and overcome the identified barriers to energy efficiency and energy conservation; Section 4 presents and discusses, in more detail, the policies which, in the author's view, are best suited to induce end-user behaviour change, with some of these policies not yet implemented; finally, Section 5 presents an historical overview the EU main policies to promote energy efficiency and how these policies have taken into account end-users' behaviour change. Conclusions and recommendations are provided at the end of the chapter.

2 Barriers to energy efficiency[b]

Following the oil crisis in OECD countries in the 1970s, energy efficiency started to be considered as an important option to promote energy security (Blumstein et al., 1980). Scientists identified that a large unexploited energy saving potential was still available and showed that with the implementation of energy-efficient technologies the same useful service could be obtained with less energy input. Scientists identified the 'energy efficiency gap', defined as the untapped energy efficiency potential

---

[b] This section presents the most common barriers to investments in energy efficiency and behaviour change in relation to energy consumption.

or the difference between the cost-optimal level of energy efficiency and the level of energy efficiency actually implemented by end-users (Hirst and Brown, 1990; Jaffe and Stavins, 1994). Moreover, scientists also identified barriers to investments and proposed the adoption of energy efficiency policies and policy packages to overcome these barriers (Hirst and Brown, 1990; Howarth and Andersson, 1993). Barriers to energy efficiency have been investigated and categorised by different scholars in different categories. Reddy (1991) classified the barriers in the following categories, related to the different actors: (i) consumer related, (ii) equipment manufacturer related, (iii) utility related, (iv) financial institution related, and (v) government related. Weber (1997) proposed the following classification of barriers based on different activities and structures: (i) institutional barriers, (ii) market barriers, (iii) organisational, and (iv) behavioural barriers. Sorrell et al. (2000) proposed a taxonomy of barriers in three categories: (i) neoclassical, (ii) behavioural, and (iii) organisational. Reddy (2002) categorised the barriers combining the two earlier classifications into (i) financial-economic, (ii) technical, (iii) awareness and information, (iv) institutional-organisational, (v) regulatory, and (vi) personnel and behavioural barriers. Sorrell et al. (2011) identified the following barriers with an economic focus: (i) risk, (ii) imperfect information, (iii) hidden costs, (iv) access to capital, (v) split incentives, and (vi) bounded rationality. More recently, Cagno et al. (2012) regrouped barriers combining the type of actors, the function, and the economics into (i) market, (ii) government/politics, (iii) technology/service suppliers, (iv) designers and manufacturers, (v) energy suppliers, (vi) capital suppliers, (vii) organisational, (viii) economic, (ix) behavioural, (x) competence, and (xi) awareness. Almost all the classification schemes presented earlier identify end-user behaviour as a key barrier, both at the individual and organisational level. Barriers are also presented and discussed in other chapters of this book (see, e.g. Arning et al., Mundaca et al., Palm and Thollander and Reusswig et al.)

To remove, reduce, or overcome the aforementioned barriers, a number of energy efficiency policies and policy packages (Hirst and Brown, 1990; Brown, 2001; Lucon et al., 2014; IPCC, 2018; Kern et al., 2017; Bertoldi, 2018) have been adopted by regional and national governments or regional organisations (e.g. European Union). The following section presents the classification of most common energy efficiency policies, as proposed by different scholars, with some of them targeting end-user behaviour.

3 Classification of energy efficiency policies

As indicated by several authors, since late 1970s, energy efficiency policies have been adopted to overcome the energy efficiency gap and the barriers described in the previous section. This section introduces the main energy efficiency policies adopted over the years, with some of them targeting end-users' behaviour. As in many other public policy sectors and, in particular, in environmental policy, there is no single policy (or policy measure) able to overcome all the barriers. Because one policy may address one or more barriers and a single barrier may be addressed by more than one policy, a range of policies are needed. For example, end-user behaviour may be seen as one barrier, but there are various ways to address it through a combination of policies and policy measures.

Similar to the barriers described earlier, energy efficiency policy measures have been classified in different ways by different authors. Based on Opschoor et al.'s (1994) categorisation of environmental policies, Markandya et al. (2014) classified energy efficiency policies in three broad categories: command and control (e.g. including building codes and appliances standards), price instruments (e.g. taxes, subsides, tax deductions, credits, permits, and tradable obligations), and information instruments (labels, audits, smart meters, and information). Shen et al. (2016) follows the three-category classification with the following breakdown of policy instruments: mandatory administration instruments, economic incentive instruments, and voluntary scheme instruments. He further subdivides these three categories in three further categories: law, regulation, and code and standards; subsidies, tax, and loan incentives; and R&D, certification and labels, and government services. Sterner and Robinson (2018) organise policies in four overarching categories: price type (taxes, subsidies, and fees), rights with quota (tradable permits, property rights, and certificates), quantity-type regulation (efficiency standards, bans, etc.), and informational/legal (information disclosure, voluntary agreements, etc.).

In its energy efficiency policy database,[c] the International Energy Agency (IEA) classifies energy efficiency policies in seven categories: behavioural measures; economic instruments; information and education; policy support; regulatory instruments; research, development, and deployment (RD&D); and voluntary approaches, with many categories subdivided in several subcategories—for example, regulatory instruments include codes and standards, mandatory audits, and supplier's obligation schemes.

The Mesures d'Utilisation Rationnelle de l'Energie (MURE) database[d] proposes the following classification of policies for the household sector at a disaggregated level. Seven broad classes are identified: two normative/regulatory classes (one normative and one informative), financial, fiscal and tariffs, information/education and training, voluntary type of measures, and cross-cutting measures (mainly energy or CO_2 taxation). The seven categories are further divided into 38 subcategories.

The EU Energy Efficiency Directive of 2012, in Article 7, introduces some policy instruments that EU Member States (MSs) can adopt as alternative to energy efficiency obligation schemes (considered in the directive as a policy instrument), which are energy or CO_2 taxes; financing schemes and instruments or fiscal incentives; regulations or voluntary agreements; standards and norms for products, buildings, and vehicles; energy labelling schemes; and training and education.

---

[c] See https://www.iea.org/policiesandmeasures/energyefficiency/.

[d] The MURE database contains energy efficiency policy measures introduced by EU Member States. More information is available at http://www.measures-odyssee-mure.eu/, Accessed 12/07/2019.

Based on the earlier presented scheme and on the evaluation of EU MSs National Energy Efficiency Action Plans by the European Commission Joint Research Centre, the author proposes the following classification of policy measures (Table 1, Bertoldi and Economidou, 2018) with eight categories. Whilst most of the policies in the succeeding text have an impact on end-user behaviour, some policy measures aim specifically to change end-user behaviour in relation to energy consumption (these policies are indicated in italics).

Both at national and local level (e.g. regions and cities), different policies and combination of policies are adopted to remove barriers to energy efficiency investments and to change end-user's behaviour. Very often, different policy measures,

Table 1 Categorisation of policy measures.

Regulatory	Building codes, minimum energy performance standards (MEPS) for new and existing buildings, energy efficiency standards for appliances and equipment, refurbishment obligations, procurement regulations, phase-out of inefficient equipment. *Mandatory energy labelling*
Financial and fiscal	Grants/subsidies, preferential loans, tax incentives, *energy/carbon taxation. Feed-in tariff*
Information and awareness	General information, *information campaigns, information centres, energy audits, energy labelling schemes,* governing by example, information exchange, *awareness campaigns,* demonstration programmes, *energy consumption feedback, smart meters and smart billing*
Qualification, training, and quality assurance	Professional training, training courses, vocational education, quality standards
Market based	Incentives facilitating third-party financing/ESCOs, energy efficiency obligation schemes (EEOSs), white certificates,[a] incentives for the producers of innovative technologies, technology deployment schemes. *Personal carbon allowances*
Voluntary action	*Voluntary certification and labelling programmes,* voluntary and negotiated agreements
Infrastructure investments	Investments in transportation infrastructure (e.g. railways and road networks), energy infrastructures (e.g. generation plants, electrical grid, substations, and local distribution); smart meter roll-out
Other	Other measures that do not fall under one of the aforementioned categories

[a] *Energy efficiency obligations coupled with a trading system for energy efficiency measures resulting in certified energy savings (tradable white certificates). Obligations can be coupled with various trading options: trading of certified energy savings, trading of eligible measures without formal certification, or trading of obligations (Bertoldi and Rezessy, 2008).*
Based on Bertoldi P., Economidou M., 2018, EU me mber states energy efficiency policies for the industrial sector based on the NEEAPs analysis, in ECEEE Industrial Summer Study Proceedings, 2018-June, pp. 117–127, Available at https://www.eceee.org/library/conference_proceedings/eceee_Industrial_Summer_Study/2018/1-policies-and-programmes-to-drive-transformation/eu-member-states-energy-efficiency-policies-for-the-industrial-sector-based-on-the-neeaps-analysis/.

forming a policy package (Lucon et al., 2014; Kern et al., 2017), are designed and introduced with some coherence and are aiming at overcoming, eliminating, or reducing one or more barriers. In relation to changing end-user behaviour, information and awareness measures are often the most used, but more and more economic/fiscal and market-based policy instruments are also used (Kern et al., 2017).

However, it is important to highlight that the majority of these policies are designed to improve the technical efficiency of the equipment or systems and not to encourage more efficient behavioural practices. Examples of these policies are incentives to buy a new efficient appliance with no limitation on the size of the appliance or to remove the old appliance (old refrigerators or old TV may remain in use in the house) or incentives to buy a more efficient boiler, which, by reducing the heating cost, may induce the household to have a higher indoor temperature (this is the so-called direct rebound effect (Lebot et al., 2004)). There are even some policy measures, which may, unintentionally, result in or encourage higher energy consumption. For example, efficiency standards may be more favourable to larger equipment due to methodological issues in determining the efficiency metrics used in the policy measure.[e] Also, building performance standards or certificates, as introduced in the EU by the Energy Performance of Buildings Directive (EPBD),[f] which are based on the ratio between energy consumption and the buildings' area over 1 year, do not provide the information on the total building consumption.[g] The risk for the user is that energy labels or building certificates encourage the purchase of not only more efficient but also larger equipment or buildings. To conclude, these examples of policies (e.g., energy labels or building certificates), as currently formulated, risk encouraging end-users to the purchase of not only more efficient but also larger equipment or larger buildings (Shove and Moezzi, 2002; Lebot et al., 2004; Ruzzenenti and Bertoldi, 2014).

4 Policies to address energy and behaviour: Energy conservation, sufficiency, and life style change

To meet the EU[h] energy and GHG targets for 2030 and to reach carbon neutrality by 2050 and contributing to limiting the maximum temperature increase to 1.5 °C

---

[e] In the EU, the efficiency metrics for refrigerators was defined as the consumption per unit of volume and larger refrigerators with a more favourable surface to volume ratio have a better energy label. Therefore, energy end-users may be encouraged to buy larger refrigerators instead of smaller ones.

[f] Directive *2002*/91/EC of the European Parliament and of the Council of 16 December *2002* on the Energy Performance of Buildings.

[g] An example is a building with a surface of $100\,m^2$ and specific consumption of $150\,kWh/year$ (with an hypothetical energy class C) resulting in a total annual consumption of $15,000\,kWh$ and a more "efficient" building with a surface of $200\,m^2$ and specific consumption of $100\,kWh/year$ (with an hypothetical energy class B) resulting in a total annual consumption of $20,000\,kWh$.

[h] Whilst the present chapter is focused on the EU energy efficiency policies, the policies described in this section are or could be adopted and implemented by countries outside the EU. Many of these countries have committed to large reductions in GHG emissions in their NDCs in the frame of the Paris Agreement.

by the end of the century, as foreseen in the Paris Agreement, there is the need to complement energy efficiency technology deployment with an effective change in the behaviour of consumers and organisations to ensure that real energy savings are achieved and overall energy consumption in the EU is reduced. From a policy perspective, this means coupling technological change with promoting measures to limits to growth, such as energy conservation and sufficiency. If *energy efficiency* is reducing energy input whilst keeping energy services constant, *energy conservation* is often thought of as reducing energy input by reducing energy services (Labanca and Bertoldi, 2018). Neither efficiency nor conservation has a normative limit. Thomas et al. (2019) define *energy sufficiency* as 'a strategy aiming at limiting and reducing the input of technically supplied energy towards a sustainable level'. In an energy sufficiency scenario, energy input is reduced, whilst utility/technical service changes in quantity or quality, provided that energy services are still 'sufficient' for the basic needs of the individual. To achieve this end, effective policies that trigger behaviour change related to energy conservation and energy sufficiency have to be designed and adopted (IPCC, 2018). This section describes seven different types of policies, highlighting which have been implemented and which are new. In the author view, these are the most suited policies to change the end-user's behaviour and have been indicated as such in Table 1. These policies address information and communication campaigns, energy consumption feedback, energy labelling, energy and carbon taxes, personal carbon allowances, building carbon allowances, and energy saving feed-in tariffs.

4.1 Information and communication campaigns

Information campaigns have been a common type of policy adopted by national and local governments to change end-user behaviour. Since the oil crisis in the 1970s, there have been information campaigns trying to stimulate consumers to reduce energy consumption based on different societal goals. Goals include security of energy supply (e.g. reducing dependence on sources of fossil fuels,[i] reducing peak electricity to avoid blackouts,[j] and improving local air quality[k]), and climate change. Information campaigns have taken several different forms from general advertising campaigns on the need to save energy to specific and tailored information provided to change specific types of end-uses; for example, residential end-users may be provided with information on how to cook efficiently or teenagers to switch off game consoles after use.

The impact of information campaigns has been well analysed in Diffney et al. (2013) who provide a complete literature review on the topic. Although some authors

---

[i] Several campaigns where launched in the 1970s following the first oil shock.
[j] Campaigns to reduce peak power to avoid blackouts were introduced in Brazil and California.
[k] In several cities, internal combustion engine vehicles cause very high level of air pollutants; citizens are informed of this and encouraged to avoid using private vehicles for a specific period.

report large savings in some specific advertisement campaigns (Reiss and White, 2008), most of the authors agree that the effect of both targeted and general advertisement and persuasion campaigns has a short lifetime and the effects tend to decrease over time (Simcock et al., 2014; Diffney et al., 2013; Reiss and White, 2008).

The meta-analysis carried out by Delmas et al. (2013) analysed the results of 156 different types of information campaign, including the provision of individual feedback (see section on feedback in the succeeding text), experiments involving 525,479 participants in 59 peer-reviewed studies, dating from 1975 to 2012. Overall, information campaigns were effective, and participants reduced their energy use by an average of 7.4%. Comparing different strategies, the researchers showed that energy audits and consultation, when individuals are informed about their own energy use and given advice on how to lower their consumption, were the most effective (Delmas et al., 2013). Under this strategy, consumers reduced their energy use on average by 13.5%. The next best approach identified by Delmas et al. (2013) was providing individuals with comparisons with their peers' energy use (this is often called 'social comparison', and it is classified as 'provision of feedback'); this approach reduced consumption by 11.5%. Surprisingly, the authors show that strategies that provided information on money savings or provided monetary incentives (e.g. payments for reduced energy usage) actually resulted, on average, in an increase in energy use by the participants. The authors suggest that this may happen because, if other altruistic reasons (such as environmental concerns) are not considered, many participants may find the potential savings or remunerations too small and not a sufficient reason to conserve energy. Provision of accurate energy consumption information to end-users has been one of the policies pursued by the EU as described in the section on EU policies.

Delmas et al. (2013) also noted that rates of energy use slowly increased in longer studies. This finding suggests that, in particular, general information campaigns may not have a sustained effect and the authors call for further research on the impact of repeated campaigning to recapture the attention and engagement of the end-users, avoiding the risk of creating some fatigue.

Delmas et al. (2013) concluded that 'non-monetary, information-based strategies can be effective at reducing overall energy use in controlled experimental studies'. This is an important finding, because it suggests that information and feedback programmes targeting conservation through behavioural change should be considered alongside with efforts to reduce energy consumption through technological improvements (Delmas et al., 2013).

When designing information measures, it is important to be aware of the psychological and social approaches to human decisions and behaviour. Here, we present a short summary of the main theories, although in other chapters of the book, such as Abrahamse and Schuitema, and Heiskanen et al., these concepts and theories are presented in more detail. The behaviour model developed by Ajzen and Fishbein (Ajzen, 1991; Ajzen, 1988) for the understanding of the individual behaviour, such as energy-related conduct and decisions, assumes there is a direct and rational relationship between behaviour and intentions and it places attitude in front of intentions, as a background for it. Intentions are derived from beliefs that develop according to background factors.

A second model introduced by Egmond et al. (2005), similar to that of Ajzen and Fishbein, is focused on the determinants. The model is based on the PRECEDE-PROCEED model of Green and Kreuter (2005), which also places the intentions as the central input for human behaviour, but on planning interventions to change individual behaviour. Intentions to save energy were found to be formed by 'predisposing factors'. Three classic approaches are distinguished in the literature:

- the price-based approach—save money.
- the environmental approach—save the planet.
- the social approach—be a good citizen.

The *price-based approach* to energy conservation alone is not successful in engaging and promoting a real change in energy end-user's behaviour (Allcott, 2011).

To overcome the problems faced by the price-based approach, the *environmental approach was proposed*. This approach relies on raising awareness on climate change and making people take personal environmental action seriously (Allcott, 2011).

The social norm approach integrates the social norms (referred to the perception of what is commonly done in a situation) as the basis for information and awareness measures on energy behaviour (Gifford, 2011; Schultz, 2007). This is based on the fact that people are social beings and respond to group norms. Social norms have a huge power to influence proenvironmental behaviour (Griskevivius, 2008; Nolan et al., 2008). They not only spur but also guide end-users' action in direct and meaningful ways.

In addition, from the literature analysed, researchers and practitioners have recommended that people need to be inspired, to be engaged, and to have fun when receiving the message (how people feel about a given situation often has a potent influence on their decisions) (Slovic and Peters, 2006). The message needs to be carefully selected and kept as simple as possible focusing on the following: entertain, engage, embed, and educate (Owen and Dewick, 2015).

4.2 Energy consumption feedback

In recent years, amongst the actions to induce energy savings through the understanding of the importance of consumer behaviour (from individual psychological and social norm points of view), researchers, utilities, and policy makers have focused their attention on energy consumption feedback. The use of this mechanism has also been enhanced by the diffusion of smart meters and the internet, but in some cases, it is still based on traditional means as energy billing. As indicated later in the text, energy billing and advanced meters are an important area of EU policy for energy savings and energy efficiency. The impact of energy consumption feedback is discussed in detail in other chapters of this book (Abrahamse and Schuitema and Heiskanen et al.). Here, we present a short summary.

There are two types of feedback: indirect and direct. Under these two broad categories of feedback, it is possible to identify some subcategories, allowing different types of interaction and response from the energy providers and energy users. Table 2 proposes a classification of feedback systems.

Zangheri et al. (2019) evaluated 123 studies on residential feedback system impact and concluded that feedback could contribute to a household's energy consumption reduction in the range of 5%–10%, in a realistic perspective. Zangheri highlighted that energy feedback works better when it is tailored to the specific household characteristics and it is presented in a clear, engaging way and complemented by advices on saving energy. In addition, the high and regular frequency delivery of feedback was found to be also specifically effective.

4.3 Energy labelling

The provision of information on energy efficiency and energy consumption using equipment (e.g. a refrigerator and a car) or energy-related equipment and material (e.g. a window and insulation materials) has been one of the most common policies adopted by countries around the world since the 1980s.[1] The most traditional manner to provide information to potential buyers is through a label attached to equipment at the point of sale. Labels could be of mandatory nature, that is, introduced by legislation and with the obligations to be displayed on equipment. Labels can be also of voluntary nature, that is, each manufacturer decides whether to participate in a labelling scheme and display the label. An example of a well-known voluntary labelling scheme is the US EPA Energy Star,[m] which covers both buildings and equipment (for more information, see Peters in this book). In addition, some energy labels provide more detailed information on equipment energy efficiency, such as annual energy consumption, size, and other important environmental impacts (e.g. water consumption), as in the case of the EU energy label or the US Energy Guide[n] label, providing only energy consumption, or the Australian Energy Rating Label.[o] Most of the labels also provide a simple classification to help end-users to easily identify the most efficient equipment. For example, in the EU, equipment is divided in seven classes according to the efficiency (A to G, with A being the most efficient class), whilst in Australia a star system is adopted, with the most efficient equipment having more stars. Other labels are a 'one class only' efficiency label, that is, only efficient equipment meeting certain level of energy performance can have the label, the best example of this label is the US Energy Star label. Labelling has also been extended

---

[1] It started in the US and EU and then spread to many other OECD and developing countries. More information can be found at the following website https://clasp.ngo/. The labelling policies in the EU are described in the section on EU energy efficiency policies.

[m] For more information, see https://www.energystar.gov/products?s=mega.

[n] For more information, see https://www.energy.gov/energysaver/appliances-and-electronics/shopping-appliances.

[o] For more information, see http://www.energyrating.gov.au/.

Table 2 Types of feedback.

Type of feedback	Subtype of feedback	Medium	Type of information	Communication
Indirect feedback	Standard billing	Paper	– Historical energy consumption – Historical comparison	One-way communication
	Enhanced billing	– Paper – Electronic environment (e-bill)	– Energy consumption, rewards – Energy efficiency advice – Social comparison – Historical comparison	One-way communication
Direct Feedback	Direct feedback with IHD	– In-house display – Web environment	– Real-time information – Social comparison – Historical comparison	One-way communication
	Direct with connected devices	– In-house Display – Web environment – Smart meter	– Real-time information – Appliance disaggregation – Social comparison – Historical comparison	Two-way communication

to cars (e.g. in the EU) and to buildings (e.g. US Energy Start label for buildings or the EU Energy Performance Certificate adopted under the EPBD) (see later for a description of the EPBD).

4.4 **Energy and carbon taxes**

Energy and/or carbon taxes are a key energy efficiency policy in all the classifications of energy efficiency policy instruments discussed earlier in the chapter, including the JRC classification based on EU policies and EU Member States' policies. The energy and/or carbon taxes are also a key policy at EU Member States[p] level. The energy and/or carbon taxes are also explicitly mentioned in Article 7 of the 2012 Energy Efficiency Directive (EED) amongst the alternative measures that Member States can adopt to fulfil the Article 7 target. This section presents the key rationale behind the adoption of an energy and/or carbon taxes.

There is an extensive literature on the impact of price signals on energy consumption, even if not all end-users would respond to a price increase (it depends on the price elasticity of the demand for energy) (Kettner-Marx and Kletzan-Slamanig, 2018). Energy taxation is a well-known energy efficiency and fiscal policy, often introduced more to raise revenue rather than discourage consumption.

Energy or carbon taxation is a very powerful policy instrument used in some countries around the world.[q] Energy and CO_2 taxation is also limiting the rebound effect[r] as the tax would penalise additional consumption triggered by the rebound (Peng et al., 2019, Font Vivanco et al., 2016; Freire-González, 2020). It is quite complex to define an optimum level of taxation (also taking into account the variation of energy prices) to achieve the desired level of consumption reduction or CO_2 emission reduction (Metcalf and Weisbach, 2013). High energy prices tend to reduce the energy consumption, particularly in less affluent households, and thus, particular care is needed to avoid unintended effects such as energy poverty. As for any energy efficiency policy distributional effect and equity considerations, these have to be carefully considered and, when necessary, mitigated (Borozan, 2019). The carbon tax revenues collected by governments could be used for supporting investments in energy efficiency and clean energy technologies. Hence, the introduction of a carbon tax can be neutral or even positive to the economy, as investments in clean technologies generate additional revenues. In addition, in the long term, a carbon/energy tax could gradually replace the tax on labour reducing the labour cost (e.g. the example of the German Eco-tax), thus helping to create additional jobs in the economy. This is known in literature as double divided (Jaeger and Dividend, 2013; Freire-González and Ho, 2019).

---

[p] Since the 1990s, there is a proposal to introduce in the EU a uniform energy tax, but this has never been agreed amongst Member States.

[q] For example, Sweden, France, Finland, South Africa, and Argentina. See for more information https://carbonpricingdashboard.worldbank.org/map_data. It is also important to notice that a carbon price can be set also through market-based instruments such as emission trading schemes.

[r] For a presentation and discussion of the rebound effect, see Ruzzenenti and Bertoldi (2014).

An energy or carbon tax could be a valid alternative to quota systems, such as cap and trade (e.g. the EU emission trading system), which will also raise the fossil fuel energy price or the personal carbon allowances, as discussed in the next section.

Taxes could also be used to penalise inefficient behaviour and favour the adoption of efficient behaviour and technologies. As example, taxes are already used in some jurisdictions to promote energy efficiency in cars by linking the annual road tax for each vehicle based on the CO_2 emissions, resulting in higher running costs for inefficient vehicles. In addition to avoid the rebound effect caused by more efficient vehicles, the road tax could also be based on the kilometres driven by the vehicles, which is now feasible with modern technology such as GPS. Vehicles with very low emissions could be incentivised, by lowering the car purchase tax or VAT, and at the same time penalise vehicles with high emissions. This measure would promote the efficient vehicles. This possible additional use of taxes would be compatible with the carbon tax on the fuel as described earlier.

Similarly, the building/property tax (at the time of purchase and the annual property tax) could be partly or totally based on the CO_2 emissions of the buildings. In the EU, this could be based on the Building Energy Performance Certificate under the EPBD or based on real emissions due to energy consumption (metered data).

4.5 Personal carbon allowance

Personal carbon allowances have been proposed and investigated by several authors (Fleming, 1997; Ayres, 2007; Hillman, 1998; Raux et al., 2015; Bristow et al., 2010; Fawcett, 2010; Starkey, 2012a,b; Burgess, 2016; Raux and Marlot, 2005, Fawcett and Parag, 2017). This policy was also analysed in detail by the UK government but then discharged as a new policy instrument due to its complexity and other possible drawbacks (House of Commons, 2007-08). As other market-based instruments, it is based on a cap set by governments, the cap being then apportioned to all the participants, in this case physical persons, which have to meet their annual target.

In practice, the government sets the amount of emissions that a person can emit based on his/her energy consumption (house, transport fuel, air-travel, etc.—in principle, it could also include the food purchased and/or the carbon content of goods purchased). The scheme will allocate (free allocation, but allowances could also be auctioned) to each person (or household) her/his carbon budget for the year. Each time there are energy expenditures (e.g. gas bill or fuel for the car at the petrol station), the amount of CO_2 emissions linked to that purchase will be deducted from the annual budget of carbon allowances. To make the system more flexible, the institution that administers the scheme could have some reserve allowances that could be purchased should any person need more or allow the trade of allowances between people (persons saving energy could have a surplus of allowances and would be able to sell them in the market, whilst people needing more energy than the ones allocated would need to buy additional allowances). In the case of allowance scarcity, the personal allowances would result in higher fuel costs, similarly to a carbon tax;

would avoid or minimise the rebound effect; and would add a carbon price to every energy purchase. Although the accounting technology for the personal carbon allowances is available (via smart cards, smart phones, internet, etc.), the system could be quite challenging in its setup and information and training of end-users. Finally, in common with many other environmental policies, the distributional effects have to be assessed carefully as this policy instrument may favour well-off people able to purchase additional carbon allowances or install technologies that reduce their carbon emissions.

4.6 Building carbon allowances

The scheme earlier described for the personal carbon allowances could also be applied to both residential and nonresidential buildings, that is, assigning a yearly amount of CO_2 emissions or kilowatt-hour (energy consumption) per building per year. The advantages would be less complex than personal allowances as buildings have metered or billed energy sources (e.g. gas, electricity, delivered heat, and heating oil), with on-site renewable generation and sustainable biomasses not being included. The scheme could allocate the emission allowances to each individual building and thus stimulate investments in energy efficiency and renewable energies and energy savings resulting from behaviour actions (e.g. lowering thermostat temperature) by building occupant or landlords (the allowance could be split between landlord and tenant to take into account the split incentive barrier).[s] This scheme would only address the building emissions, whilst personal carbon allowances could, in principle, also address other sectors. For commercial buildings, some policies similar to this already exist, for example, the UK CRC Energy Efficiency Scheme[t] or the Tokyo Metropolitan Carbon and Trade Scheme (Bertoldi et al., 2013a) even though the UK scheme is working more as an energy tax. There could be a strong synergy between the innovative policies described in this chapter, such as property tax based on carbon emissions, building carbon allowances, and feedback systems, all based on the metered or billed energy consumption.

4.7 Energy savings feed-in tariff

Rather than trying to 'discourage' consumption (and inefficiency) with an additional energy tax and get through the complexities of trying to define an optimum level of taxation, public money (or money raised through a small wire or consumption charge, as in the case of white certificates schemes) can be used to reward and give incentives to energy saved, as a result of technology implementation and/or as a

---

[s] For example, tenants may be only responsible for electricity due to appliances (assuming the tenant is selecting and owning the appliances) whilst landlord being responsible for heating.

[t] More information available at https://www.gov.uk/government/collections/crc-energy-efficiency-scheme.

result of energy conservation (resulting also from behaviour change) (Eyre, 2013; Bertoldi et al., 2013b; Neme and Cowart, 2012). This can be seen as a core feature of a possible energy saving feed-in tariff (ES FiT). Unlike investment grants, which reward consumers based on the size of their investment in energy efficiency technologies, a FIT rewards end-users based on the operational performance of their investment or behaviour change in terms of energy savings.[u]

With an ES FiT, the consumers would benefit from an additional financial incentive on top of the monetary savings resulting from reduced energy bills. This is why an ES FiT can be seen as a performance-based subsidy, whereby actions undertaken by end-users—both in terms of investment in energy efficiency technology and in terms of behavioural action and energy conservation—are remunerated based on the savings achieved, possibly differentiated by the type of action or by end-use sector. In terms of design, the ES FiT could be either based on the actual amount of saved kilowatt-hour of electricity or cubic metre of gas (referred hereinafter as quantity-based FiT, e.g. based on the actual quantity of savings) or based on a fixed threshold achieved (target-based FIT).

In the case of quantity-based FiT, the subsidy can be awarded based on saved amount of energy compared with a predefined and agreed energy consumption (ex post and based on meter reading) with or without adjustment for climatic and other 'external' conditions.

In the case of a target-based FiT, the FiT subsidy can be awarded contingent upon the reduction of the amount of consumed energy by a certain amount (target) or on reaching a certain threshold. It is based on the energy consumption as indicated on the energy bill with possible normalisation for exogenous factors such as occupancy levels. A target-based FiT uses data that are compiled and regularly communicated to the end-user via the bill. Billing on the basis of actual consumption shall be performed frequently enough to enable customers to regulate their own energy consumption.

As for personal carbon allowance, this policy is not yet proposed in EU legislation nor implemented in any MSs.

5 European Union policies for sustainable behaviours in energy end-use

This section analyses how the historical development, implementation, and impact of EU energy efficiency[v] policies have addressed energy conservation, energy sufficiency, and end-user behaviour. It separates the material into two phases: an early

---

[u] Normalising energy savings to account for autonomous savings, which occurred without any action on the side of the consumer (e.g. reduced occupancy levels of times).

[v] As already illustrated in the introduction, energy efficiency policies are those policies that aim at improving the technical efficiency of energy consuming equipment, systems, and buildings and aim at reducing energy demand through end-user behaviour change. The end-user behaviour change can be focused both on adopting more efficient equipment, systems, and buildings or to better manage their usage, as well as at reducing the energy use through limitation of services and goods. This latter concept is also identified as energy sufficiency.

phase (1970–2005) and a recent phase (2005–18). The pivotal change between these two phases is marked by the 2006 Action Plan, which introduced for the first time the 20% energy reduction by 2020.

5.1 **The early phase (1970–2005)**

Following the oil crises in 1974, the European Council adopted resolutions promoting energy savings with the goal of reducing energy consumption by 1985 by 15% below the January 1973 estimation (Council Resolution of 17 December 1974, OJ C 153/2) and by introducing a target for energy intensity and adopting energy policies including energy pricing policies (Council Resolution of 9 June 1980, OJ C 149/1). In 1986, the concept of energy efficiency emerged as target, shifting the attention from energy savings to energy efficiency, by introducing a new target for energy intensity, which should then be reduced by 20% by 1995 (Council Resolution of 16 September 1986, OJ C 241/1). In 1987, the communication entitled 'Towards a continuing policy for energy efficiency in the European Community' (COM(1987) 233 final) proposed 14 energy efficiency measures to Member States to help achieve the 1995 target. Seven policies out of the 14 policies recommended in the communication are related to the provision of consumer information, seen as essential to trigger investments in energy efficiency in a period of low oil price.

As already indicated, in 1990, the climate change issue started to emerge, and in the same year, the European Council of Environment and Energy Ministers agreed on 29 October 1990 on stabilising total CO_2 emissions in year 2000 at the 1990 levels.

The 'SAVE' Directive of 1993 (Council Directive 93/76/EEC) was one of the first EU 'mandatory' policies to recognise the role of consumer information and awareness to change end-users' behaviour. In particular, this was implemented through mandating Member States to implement programmes for (i) the certification of buildings with the description of the building energy characteristics to provide to the consumer information on the energy efficiency level of the buildings and (ii) on the billing of heating, air-conditioning, and hot water to be based on actual consumption in such a way to trigger the adoption of measures to reduce energy consumption. Moreover, the directive also requested that the building occupants should be able to regulate their own consumption of heat, cold or hot water. Unfortunately, the directive language was not strong enough, and several Member State implemented the directive is a loose manner.[w]

Another important EU initiative to promote consumer awareness and behaviour change was the energy labelling of domestic appliances,[x] which was established

---

[w] Although the Directive used a mandatory language "Member States shall draw up and implement programmes", the Directive did not impose any technical requirement or methodologies to Members States as in the future EPBD and EED, hence the implementation of this Directive varied a lot between Member States.

[x] For the list of existing labelling regulation and products labelled visit: https://ec.europa.eu/info/energy-climate-change-environment/standards-tools-and-labels/products-labelling-rules-and-requirements/energy-label-and-ecodesign/energy-efficient-products_en.

in 1992 with the adoption of the Energy Labelling Framework Directive (Council Directive 92/75/EEC). This framework directive introduced the concept of EU harmonised labelling, leaving the design of the label to the implementing directive for specific domestic appliances. The majority of the energy labels introduced under this framework directive adopted a similar design, with seven efficiency classes: A to G, with A the most efficient class (in green colour) and G the less efficient (in red colour). The design of the label was established through a real testing of different label layouts and classification schemes with a sample of consumer to identify the most effective design to attract the consumer attention and stimulate them to the right purchase decision.

In 1998, the commission adopted the communication on energy efficiency (Energy Efficiency in the European Community—Towards a Strategy for the Rational Use of Energy—(COM(1998) 246 final), 29.04.1998), which set out a community strategy for energy efficiency. The strategy called for Member States to promote awareness on energy efficiency, with particular attention to increasing consumer information on cost-effective energy efficiency opportunities. The Council endorsed the strategy and proposed an indicative target of 1 percentage point per year improvement in energy intensity until the year 2010. The following commission 'Action Plan to improve Energy Efficiency in the European Community' (COM(2000) 247 final) adopted in year 2000 stated the importance of changing behaviour regarding mobility, to foster more energy-efficient transport modes such as public transport. In addition, more attention was placed on consumer information, in particular through a new EU campaign focusing the attention of consumers and other stakeholders on energy efficiency and valorising its environmental and other benefits (i.e. cobenefits).

The EU adopted legislation on technical requirements for existing and new buildings (EPBD) and the major buildings components, for example, heating and cooling systems under the Energy Performance of Buildings Directive (EPBD)[y] (Directive 2002/91/EC), and individual energy using equipment under the Ecodesign Directive[z] (Directive 2005/32/EC). The Ecodesign Directive sets mandatory energy performance requirements, that is, (low) level of energy consumption. At the same time, the EU and national Member States policy makers identified the need to speed up investments in energy efficiency. The EPBD introduced amongst other technical provisions the energy performance certificate (EPC) for buildings (Amecke, 2012; Li et al., 2019). The EPC, similarly to energy label for appliances, should help end-users to buy or rent property with low energy consumption. In many Member States, the EPC adopted an A to G scale, with A being the most efficient class of buildings.

---

[y] The EPBD has been amended in 2010 (2010/31/EU) and in 2018 (2018/844/EU). For more information see https://ec.europa.eu/energy/en/topics/energy-efficiency/energy-performance-of-buildings#content-heading-0.

[z] Directive 2005/32/EC of the European Parliament and of the Council of 6 July 2005 establishing a framework for the setting of ecodesign requirements for energy-using products and amending Council Directive 92/42/EEC and Directives 96/57/EC and 2000/55/EC of the European Parliament and of the Council. For more information see https://ec.europa.eu/growth/industry/sustainability/ecodesign_en.

The EPC has to be displayed in advertisement and be provided to potential buyers and renters. Amecke (2012) found that the impact of the EPC on residential building buyers was rather limited. This was confirmed by Hårsman et al. (2016).

5.2 **The recent phase (2005–18)**

To boost the role and contribution of energy efficiency to the three pillars of EU energy policy, the European Commission presented in 2005 a Green Paper (COM (2005) 265 final) on 'Energy Efficiency or Doing More With Less', introducing suggestions for new and additional energy efficiency policies at the horizontal level, for example, energy taxation and national energy efficiency action plans. For the first time, an EU official policy paper recognised that 'improving energy efficiency is a broad term. In this Green Paper, it covers, firstly, a better use of energy through improvements in energy efficiency and, secondly, energy saving through changes in behaviour'. Moreover, the Green Paper stated that 'Energy saving in the overall sense also derives from a change in consumer behaviour. This means, for example, a policy of making public transport more attractive and thereby encouraging car users to take the bus or train instead; or educating people on how to reduce heat losses from their house, notably through correct use of the thermostats'. The Green Paper highlighted the role of information campaigns to change consumer behaviour with the launch of the Sustainable Energy Europe 2005–08 campaign. It was designed to bring about a genuine change in behaviour by the main players. The Green Paper also recognised the role of energy and carbon taxes to change the behaviour.

The European Commission published an Energy Efficiency Action Plan 2006 (COM(2006) 545 final). The communication proposed an overall realistic energy saving target of 20% for 2020, to be achieved through new measures and the strengthening of existing policies. A number of measures were introduced in the action plan to change the consumer behaviour. The action plan stated that 'the efficient use of energy requires factors that motivate, facilitate and reinforce rational and responsible behaviour. Institutional capacity, awareness, and clear, credible and accessible information on energy-using technologies and techniques are important predisposing elements for rational market behaviour. Education and training are required for all stakeholders, and information technology is vital' (Directive 2005/32/ EC). The Ecodesign Directive sets mandatory energy performance requirements, that is, (low) level of energy consumption. At the same time, the EU and national Member States policy makers identified the need to speed up investments in energy efficiency. The Energy Saving Directive (ESD) was adopted in 2006 (Directive 2006/32/EC).

The ESD introduced an indicative energy saving target for Member States of 9% by 2016; amongst many measures to promote a market for energy efficiency, including establishing financing schemes for investments in energy efficiency and removing barriers to energy services, the ESD requested Member States to ensure that end-users were provided with competitively priced individual metering and informative billing that showed their actual energy consumption. As far as possible, bills must be based on actual energy consumption. As already indicated, enhanced billing and real consumption metering are powerful mechanisms to promote energy savings and behaviour change.

In March 2007, EU leaders committed Europe to become a highly energy-efficient, low carbon economy and agreed on energy and climate targets, known as the '20-20-20' targets, by setting three key objectives for 2020: (i) a 20% reduction in EU greenhouse gas emissions from 1990 level, (ii) raising the share of EU energy consumption produced from renewable resources to 20%, and (iii) improving energy efficiency to achieve a 20% savings on EU primary energy consumption (Conclusions of the European Council of 8 and 9 March 2007). It is important to notice that the 'energy efficiency target' is, in reality, an 'energy saving' target as the target is expressed as maximum energy primary and final energy consumption for the EU[aa] and could be reached in several manners, including energy reduction resulting from behaviour and sufficiency. The targets were enacted through the climate and energy package in 2009. The climate and energy package is a set of binding legislation, which aims to ensure the European Union meets its ambitious climate and energy targets for 2020.

To provide a legal basis to the 2007 energy efficiency target, the Energy Efficiency Directive (EED) (Directive 2012/27/EU) was adopted in October 2012. The directive quantifies the 20% energy efficiency target for 2020 defined in the climate and energy package and establishes a common framework of legally binding measures for the promotion of energy efficiency in the Member States to reach the target by 2020. The EED requires Member States to set indicative national energy efficiency targets and legally binding measures to help the EU reach its 20% energy efficiency target. In particular, all EU Member States are required to implement policy measures that improve energy efficiency.

In addition to the primary and final energy reduction targets, the EED introduced some binding national measures, that is, measures the Member States have to adopt, such as legal obligations to establish energy saving obligation schemes (Art. 7), mandatory energy audits (Art. 8), provision of metering and billing consumer information (Art. 9–11), renovation of public buildings (Art. 5), promotion of energy services (Art. 19), enabling demand response (Art. 15), programmes to change consumer behaviour (Art. 12), and national strategies for the building renovation (Art. 4). In particular, Article 7 mandated Member States to set up Energy Efficiency Obligation Schemes (EEOSs), through which distributors and/or retail energy sales companies have to deliver new annual energy savings, equivalent to 1.5% of the annual energy sales, amongst final energy consumers (Fawcett et al., 2019). As an alternative policy option, Member States can achieve the same energy savings through the implementation of alternative policy measures or a combination of alternative measures and EEOSs; these measures include carbon taxes, energy labelling, and information programmes. The Article 7 energy savings must be delivered in the period from 2014 to 2020.

[aa] Based on 20% reduction compared with a fixed baseline calculated in 2006 with the PRIMES model.

The ESD metering and billing measures were reinforced in the EED (Art. 9–11) covering gas, electricity, and the supply of heat. Of particular interest is Article 12 on 'Consumer Information and Empowering Programme'. The article states that Member States must adopt 'a range of instruments and policies to promote behavioural change which may include: (i) fiscal incentives; (ii) access to finance, grants or subsidies; (iii) information provision; (iv) exemplary projects; (v) workplace activities'. In addition, Member States shall identify and implement 'ways and means to engage consumers and consumer organisations during the possible roll-out of smart meters through communication of (i) cost-effective and easy-to-achieve changes in energy use and (ii) information on energy efficiency measures'.

In addition, the ESD and the EED requested Member States to prepare National Energy Efficiency Action Plans (NEEAPs). The NEEAPs are national strategic documents intended to present how a Member State introduces new policies and programmes (or relying on existing ones) to reach the 9% ESD energy saving target. NEEAPs under the EED were intended to present which policies and measures Member States were adopting and implementing to reach their own 2020 targets. NEEAPs under the ESD were submitted in 2008 and 2011, whilst NEEAPs under the EED were submitted in 2014 and 2017.

In 2014, the commission proposed a new set of climate and energy targets for 2030:

- A 40% GHG reduction compared with 1990
- A 27% contribution of renewable sources in the total energy consumption
- A 27% improvement in energy efficiency compared with the 2006 baseline used for the 2020 target

This set of targets represents the EU National Determined Contribution to the Paris Agreement. During subsequent discussions of the legislation to achieve the 2030 targets in the EU Council and Parliament, the renewable target was increased to 32% and the energy efficiency target to 32.5%.

Following the commission's proposal for an update to the EED and the proposal for an extension of the Art. 7 EEOSs till 2030, the revised EED[ab] entered into force on 24 December 2018.

The EU has recently adopted a communication on the long-term 2050 climate action strategy[ac] calling for carbon neutrality by 2050. The strategy highlights the central role of EE in the EU decarbonisation process and calls for an energy consumption reduction of 50% compared with 2005. This reduction is primarily expected in the building sector.

[ab] Directive (EU) 2018/2002 of the European Parliament and of the Council of 11 December 2018 amending Directive 2012/27/EU on energy efficiency. PE/54/2018/REV/1, OJ L 328, 21.12.2018. pp. 210–230. Available online: https://eur-lex.europa.eu/legal-content/EN/TXT/?uri=uriserv:OJ.L_.2018.328.01.0210.01.ENG.

[ac] 28/11/2018—COM (2018) 773—A Clean Planet for all—A European strategic long-term vision for a prosperous, modern, competitive, and climate neutral economy, Available at https://eur-lex.europa.eu/legal-content/EN/TXT/?uri=CELEX:52018DC0773.

It is also important to highlight that the EU policies on energy efficiency described earlier have been supported through the financing of pilot projects in the frame of the SAVE programme initially (since 1992) and then the Intelligent Energy Europe programme in a second phase.[ad] More recently, the EU Horizon 2020[ae] supports research, demonstration, and market uptake of energy-efficient technology projects including projects to change end-users' behaviour.[af] Very soon, the new EU R&D Horizon Europe programme will be launched.

6 Conclusions

Several barriers prevent end-users—at both the individual and group levels—to make the best investments in energy efficiency and discourage them from adopting practices, life styles, and work styles that favour energy conservation and/or sufficiency. This low adoption of efficient technologies and sufficient practices results in the 'energy efficiency gap'. Several policy and policy packages have been adopted and implemented by governments at local, national, and EU level to avoid, overcome, or reduce the barriers. In general, most of the policies implemented around the world focus on the 'technical' efficiency of products, systems, or buildings, including changing the end-user behaviour towards investments in energy-efficient solutions. Additional policy measures have targeted consumer information, through a variety of campaigns, training, and media in different end-use sectors, including energy consumption feedback. Carbon pricings and energy and carbon taxes are also important policy tools to induce behaviour change depending on end-users' price elasticity. Only very recently, policies to change the end-users' energy consumption through energy conservation and sufficiency measures have emerged. This chapter presented and discussed some of these innovative policies, such as personal carbon allowances, building carbon allowances, and energy saving feed-in tariffs. It is highlighted that some of these policies are not yet implemented.

Focusing on the EU policies, since the 1980s, following the oil price shocks, the reduction or the moderation of energy demand has been a priority for the EU energy policy to increase energy security and mitigate climate change. Early EU policies targeted the reduction of energy consumption growth or the improvement of energy intensity. Some of the main policy measures in the EU, that is, the EPBD and the Ecodesign Directive, target the 'technical' efficiency through energy performance standards in buildings through the improvement of technical equipment (e.g. HVAC and lighting) and increased building insulation and in energy-related end-use equipment in the case of the ecodesign. The behavioural component has mainly focused on changing the purchase behaviour of consumers towards more efficient equipment

[ad] For more information, visit: https://ec.europa.eu/easme/en/section/energy/intelligent-energy-europe.

[ae] For more information, visit: https://ec.europa.eu/easme/en/horizon-2020-energy-efficiency.

[af] As example of projects focused on end-users' behaviour are the CONSEED (https://www.conseed-project.eu/), PENNY (http://www.penny-project.eu/), and COBHAM (http://www.cobham-erc.eu/about/) projects. More projects can be identified in the CORDIS database (https://cordis.europa.eu/).

with the implementation of the mandatory energy labelling scheme for appliances and later with the energy performance certificate for buildings introduced by the EPBD. The second action in EU energy policy to change end-user's behaviour has been through the provision of frequent information on real energy consumption, starting from the heating consumption bill to be based on metered consumption rather than on estimated consumption. More recently, the EU policy makers have been focusing on additional measures to trigger a behaviour change in the Energy Services Directive and in particular in the Energy Efficiency Directive, for example, with the implementation of smart meters and smart billing for electricity, gas, and heat consumption and the provision of feedback to end-users and to match energy bills with real consumption. The Energy Efficiency Directive also reinforced the need of targeted and effective consumer information to trigger consumer behaviour change.

Also, the major EU R&D financial programme H2020 has introduced several calls and has supported several projects to better understand the dynamics of consumer behaviour from an economic, social, and psychological point of view.

An important evidence of the impact of EU and Member States policies is that in 2017 total final energy consumption in the EU is lower than the peak reached in 2006, showing that policies have contributed to this reduction amongst other factors such as the 2008 economic crisis and technology development (Tsemekidi Tzeiranaki et al., 2019). The commission has recently proposed a carbon neutrality target for 2050. However, the current set of policies is not enough to reach the ambitious climate targets for 2050, and the goal of the Paris Agreement to stabilise temperature increase well below 2 °C.

In the author's view, the EU energy efficiency and energy conservation policies must increase their focus on effective policies inducing a behaviour change in relation to energy consumption. Some of the innovative policies presented and discussed in this chapter offer a great potential although they have not yet been implemented. This new attention to effective policies for energy conservation, behaviour change, and sufficiency will contribute to enable the EU to reach carbon neutrality by 2050 and the goal of the Paris Agreement.

References

Ajzen, I., Kuhl, J., 1988. From intentions to actions: A theory of planned behaviour. In: Beckmann, J. (Ed.), Action-Control: from cognition to behaviour. Springer, Heidelberg, Germany, pp. 11–39.

Ajzen, I., 1991. The theory of planned behaviour. Organizational Behaviour and Human Decision Processes 50 (2), 179–211.

Allcott, H., 2011. Social norms and energy conservation. J. Public Econ. 95 (2011), 1082–1095.

Amecke, H., 2012. The impact of energy performance certificates: a survey of German home owners. Energy Policy 46, 4–14.

Ayres, R., 2007. Environmental market failures: Are there any local market-based corrective mechanisms for global problems? Mitig. Adapt. Strateg. Glob. Chang. 1, 289–309.

Bertoldi, P., 2018. The Paris Agreement 1.5°C goal: what it does mean for energy efficiency? In: Proceedings of 2018 ACEEE Summer Study on Energy Efficiency in Buildings. American Council for an Energy-Efficient Economy, Washington, DC.

Bertoldi, P., Rezessy, S., 2008. Tradable white certificate schemes: Fundamental concepts. Energy Efficiency 1, 237. https://doi.org/10.1007/s12053-008-9021-y.

Bertoldi, P., Labanca, N., Rezessy, S., Steuwer, S., Oikonomou, V., 2013a. Where to place the saving obligation: Energy end-users or suppliers? Energy Policy 63, 328–337.

Bertoldi, P., Rezessy, S., Oikonomou, V., 2013b. Rewarding energy savings rather than energy efficiency: Exploring the concept of a feed-in tariff for energy savings. Energy Policy 56, 526–535.

Bertoldi, P., Economidou, M., 2018. EU member states energy efficiency policies for the industrial sector based on the NEEAPs analysis. In: ECEEE Industrial Summer Study Proceedings. pp. 117–127. 2018-June. Available at: https://www.eceee.org/library/conference_proceedings/eceee_Industrial_Summer_Study/2018/1-policies-and-programmes-to-drive-transformation/eu-member-states-energy-efficiency-policies-for-the-industrial-sector-based-on-the-neeaps-analysis/.

Blumstein, C., Krieg, B., Schipper, L., York, C., 1980. Overcoming social and institutional barriers to energy conservation. Energy 5 (4), 355–371.

Borozan, D., 2019. Unveiling the heterogeneous effect of energy taxes and income on residential energy consumption. Energy Policy 129, 13–22.

Bristow, A., Wardman, M., Zanni, A., Chintakayala, P.K., 15 July 2010. Public acceptability of personal carbon trading and carbon tax. Ecol. Econ. 69 (9), 1824–1837.

Brown, M.A., 2001. Market failures and barriers as a basis for clean energy policies. Energy Policy 29, 1197–1207. 00067-2.

Burgess, M., October 2016. Personal carbon allowances: a revised model to alleviate distributional issues. Ecol. Econ. 130, 316–327.

Cagno, E., Worrell, E., Trianni, A., Pugliese, G., 2012. Dealing with barriers to industrial energy efficiency: an innovative taxonomy. In: Proceedings of ECEEE 2012 Summer Study on Energy Efficiency in Industry.

Delmas, M.A., Fischlein, M., Asensio, O.I., 2013. Information strategies and energy conservation behaviour: A meta-analysis of experimental studies from 1975 to 2012. Energy Policy 61, 729–739.

Diffney, S., Lyons, S., Malaguzzi Valeri, L., November 2013. Evaluation of the effect of the power of one campaign on natural gas consumption. Energy Policy 62, 978–988.

Egmond, C., Jonkers, R., Kok, G., December 2005. A strategy to encourage housing associations to invest in energy conservation. Energy Policy 33 (18), 2374–2384.

Eyre, N., January 2013. Energy saving in energy market reform—The feed-in tariffs option. Energy Policy 52, 190–198.

Fawcett, T., November 2010. Personal carbon trading: a policy ahead of its time? Energy Policy 38 (11), 6868–6876.

Fawcett, T., Killip, G., 2019. Re-thinking energy efficiency in European policy: Practitioners' use of 'multiple benefits' arguments. J. Clean. Prod. 210, 1171–1179.

Fawcett, T., Parag, Y., 2017. Personal Carbon Trading. https://www.taylorfrancis.com/books/e/9781849776721.

Fleming, D., 1997. Tradable quotas: Using information technology to cap National Carbon Emissions. Eur. Environ. 7, 139–148.

Font Vivanco, D., Kemp, R., van der Voet, E., 2016. How to deal with the rebound effect? A policy-oriented approach. Energy Policy 94, 114–125.

Freire-González, J., 2020. Energy taxation policies can counteract the rebound effect: analysis within a general equilibrium framework. Energy Efficiency . in press.

Freire-González, J., Ho, M.S., 2019. Carbon taxes and the double dividend hypothesis in a recursive-dynamic CGE model for Spain. Economic Systems Research 31 (2), 267–284.

Geller, H., DeCicco, J., Laitner, S., Dyson, C., 1994. Twenty years after the embargo US oil import dependence and how it can be reduced. Energy Policy 22 (6), 471–485.

Gifford, R., May–Jun 2011. The dragons of inaction: Psychological barriers that limit climate change mitigation and adaptation. American Psychologist 66 (4), 290–302.

Green, L., Kreuter, M., 2005. Health Program Planning: An Educational and Ecological Approach, 4th ed McGraw-Hill, New York, NY.

Griskevivius, V., 2008. Social norms: An underestimated and underemployed level for managing climate change. IJSC 3, 5–13.

Grossman, P.Z., 2015. Energy shocks, crises and the policy process: a review of theory and application. Energy Policy 77, 56–69.

Gupta J. A History of International Climate Change Policy Volume Vol. 1, September/October 2010, WIREs Climate Change

Hårsman, B., Daghbashyan, Z., Chaudhary, P., 2016. On the quality and impact of residential energy performance certificates. Energy and Buildings 133, 711–723.

Hillman, M., 1998. Carbon budget watchers. Town and Country Planning 67, 305. (special issue on climate change).

Hirst, E., Brown, M., 1990. Closing the efficiency gap: Barriers to the efficient use of energy. Resour. Conserv. Recycl. 3, 267–281.

House of Commons, 2007-08 Environmental Audit Committee, Personal Carbon Trading, Fifth Report of Session 2007–08, Available at https://publications.parliament.uk/pa/cm200708/cmselect/cmenvaud/565/565.pdf

Howarth, R.B., Andersson, B., 1993. Market barriers to energy efficiency. Energy Econ. 262–272.

IEA, 2014. Capturing the Multiple Benefits of Energy Efficiency. ISBN 978-92-64-22072-0 International Energy Agency, Paris. Available at: http://www.iea.org/bookshop/475Capturing_the_Multiple_Benefits_of_Energy_Efficiency.

IPCC Special Report on Global Warming of 1.5 °C, 2018, Available at https://www.ipcc.ch/sr15/

Jaeger, W.K., Dividend, D., 2013. In: Shogren, J.F. (Ed.), Encyclopedia of Energy, Natural Resource, and Environmental Economics. Elsevier, Waltham, pp. 37–40.

Jaffe, A.B., Stavins, R.N., 1994. The energy-efficiency gap: What does it mean. Energy Policy 22 (10), 804–810.

Kern, F., Kivimaa, P., Martiskainen, M., 2017. Policy packaging or policy patching? The development of complex energy efficiency policy mixes. Energy Res. Soc. Sci. 23, 11–25.

Kettner-Marx C., Kletzan-Slamanig D., Energy and Carbon Taxes in the EU Empirical Evidence with Focus on the Transport Sector, WIFO Working Papers, No. 555, February 2018, Available at https://www.econstor.eu/handle/10419/179309

Labanca, N., Bertoldi, P., 2018. Beyond energy efficiency and individual behaviours: Policy insights from social practice theories. Energy Policy 115, 494–502.

Lebot B., Bertoldi P., Harrington P., Consumption versus efficiency: Have We Designed the Right Policies and Programmes? In Proceedings of 2004 ACEEE Summer Study on Energy Efficiency in Buildings, American Council for an Energy-Efficient Economy, Washington, DC.

Li, Y., Kubicki, S., Guerriero, A., Rezgui, Y., 2019. Review of building energy performance certification schemes towards future improvement. Renew. Sust. Energ. Rev. 113, 109244.

Lucon, O., Ürge-Vorsatz, D., Zain Ahmed, A., Akbari, H., Bertoldi, P., Cabeza, L.F., Eyre, N., Gadgil, A., Harvey, L.D.D., Jiang, Y., Liphoto, E., Mirasgedis, S., Murakami, S., Parikh, J., Pyke, C., Vilariño, M.V., 2014. Buildings. In: Edenhofer, O., Pichs-Madruga, R., Sokona, Y., Farahani, E., Kadner, S., Seyboth, K., … Minx, J.C. (Eds.), Climate Change 2014: Mitigation of Climate Change. Contribution of Working Group III to the Fifth Assessment Report of the Intergovernmental Panel on Climate Change. Cambridge University Press, Cambridge, United Kingdom and New York, NY, USA.

Maamoun, N., 2019. The Kyoto protocol: Empirical evidence of a hidden success. Journal of Environmental Economics and Management 95, 227–256.

Markandya, A., Labandeira, X., Ramos, A., 2014. Policy Instruments to Foster Energy Efficiency. Available at https://ideas.repec.org/p/efe/wpaper/01-2014.html.

Metcalf, G.E., Weisbach, D., 2013. Carbon taxes. In: Shogren, J.F. (Ed.), Encyclopedia of Energy, Natural Resource, and Environmental Economics. Elsevier, Waltham, pp. 9–14.

Neme C., Cowart R. Energy Efficiency Feed-in-Tariffs: Key Policy and Design Considerations. Regulatory Assistance Project. April 2012.

Nolan J., Wesley Schultz P., Cialdini R., Goldstein N., Griskevicius V., 2008, Normative social influence is underdetected, Personality and Social Psychology Bulletin, Vol 34, Issue 7, pp. 913–923, 2008.

Opschoor, J.B., de Savornin Lohman, A.F., Vos, H.B., 1994. Managing the Environment: the Role of Economic Instruments. Organization for Economic Press.

Owen, P., Dewick, P., 2015. How effective is a games-centric approach in changing student eco behaviours? Research Evaluation Report.

Peng, J., Wang, Y., Zhang, X., He, Y., Taketani, M., Shi, R., Zhu, X., 2019. Economic and welfare influences of an energy excise tax in Jiangsu province of China: A computable general equilibrium approach. Journal of Cleaner Production 211, 1403–1411.

Raux, C., Marlot, G., 2005. A system of tradable CO2 permits applied to fuel consumption by motorists. Transport Policy 255–265.

Raux, C., Croissant, Y., Pons, D., March 2015. Would personal carbon trading reduce travel emissions more effectively than a carbon tax? Transportation Research Part D: Transport and Environment 35, 72–83.

Reddy, A.K.N., 1991. Barriers to improvements in energy efficiency. Energy Policy 19 (10), 953–961.

Reddy, S.B., 2002. Barriers to the Diffusion of Renewable Energy Technologies. Monograph, Centre for Energy and Environment UNEP, Denmark.

Reiss, P., White, M., Autumn 2008. What changes energy consumption? Prices and public pressures. RAND Journal of Economics 39 (3), 636–663.

Rüdiger, M., 2014. The 1973 oil crisis and the designing of a Danish energy policy. Historical Social Research/Historische Sozialforschung 39 (4(150)), 94–112.

Ruzzenenti, F., Bertoldi, P., 2014. Energy conservation policies in the light of the energetics of evolution. In: Labanca, N. (Ed.), Complex Systems and Social Practices in Energy Transitions; Green Energy and Technology. Springer, Cham, Switzerland.

Schultz, P.W., 2007. The constructive, destructive, and reconstructive power of social norms. Psychological Science 18 (5).

Shen, L., He, B., Jiao, L., Song, X., Zhang, X., 2016. Research on the development of main policy instruments for improving building energy-efficiency. Journal of Cleaner Production 112 (Part 2), 1789–1803.

Shove, E., Moezzi, M., 2002. What do standards standardise? In: Proceedings of 2002 ACEEE Summer Study on Energy Efficiency in Buildings, American Council for an Energy-Efficient Economy, Washington, DC.

Simcock, N., et al., February 2014. Factors influencing perceptions of domestic energy information: Content, source and process. Energy Policy 65, 455–464.

Slovic, P., Peters, H., 2006. Risk perception and affect. Current Directions in Psychological Science 15 (6), 322–325.

Sorrell, S., Scleich, J., Scott, S., O'Malley, E., Trace, F., Boede, U., Ostertag, K., Radgen, P., 2000. Reducing barriers to energy efficiency in private and public organisations. Report to the European Commission, in the framework of the Non-Nuclear Energy Programme JOULE III, Brighton.

Sorrell, S., Mallett, A., Nye, S., 2011. Development Policy, Statistics and Research Branch. Barriers to industrial energy efficiency: A literature review. Working paper 10/2011 United Nations Industrial Development Organisation.

Starkey, R., 2012a. Personal carbon trading: A critical survey: Part 1: Equity. Ecological Economics 73 (15), 7–18.

Starkey, R., 2012b. Personal carbon trading: A critical survey Part 2: Efficiency and effectiveness. Ecological Economics 73, 19–28.

Sterner, T., Robinson, E., 2018. Selection and design of environmental policy instruments. In: Handbook of Environmental Economics. 1st ed. 4. Available https://www.elsevier.com/books/handbook-of-environmental-economics/dasgupta/978-0-444-53772-0.

Thomas, S., Thema, J., Brischke, L., et al., 2019. Energy sufficiency policy for residential electricity use and per-capita dwelling size. Energy Efficiency 12, 1123. https://doi.org/10.1007/s12053-018-9727-4.

Tsemekidi Tzeiranaki, S., Bertoldi, P., Diluiso, F., Castellazzi, L., Economidou, M., Labanca, N., Serrenho, T., Zangheri, P., 2019. Analysis of the EU residential energy consumption: trends and determinants. Energies 12, 1065.

Tsemekidi Tzeiranaki, S., Bertoldi, P., Labanca, N., Castellazzi, L., Ribeiro, S.T., Economidou, M., Zangheri, P., 2018. Energy consumption and energy efficiency trends in the EU-28 for the period, 2000–2016, EUR 29473 EN. Publications Office of the European Union, Luxembourg.

UNFCCC, 2007. The Kyoto Protocol. Available at: https://unfccc.int/process/the-kyoto-protocol/history-of-the-kyoto-protocol/text-of-the-kyoto-protocol.

UNFCCC, 2015. The Paris Agreement. Available at: https://unfccc.int/process-and-meetings/the-paris-agreement/the-paris-agreement.

United Nations, 1998. General Assembly of 6 December 1988. A/RES/43/53, Available at https://www.un.org/documents/ga/res/43/a453.htm3r0

Weber, L., 1997. Some reflections on barriers to the efficient use of energy. Energy Policy 25 (10), 833–835.

Zangheri, P., Serrenho, T., Bertoldi, P., 2019. Energy savings from feedback systems: a meta-studies review. Energies. article in press.

Further reading

Brischke, L.-A., Lehmann, F., Leuser, L., Thomas, S., Baedeker, C., 2015. Energy sufficiency in private households enabled by adequate appliances. In: ECEEE 2015 Summer Study Proceedings. ECEEE, Stockholm, pp. 1571–1582.

Levy, J., 2003. Applications of prospect theory to political science. Synthese 135, 215.

Parry, I., 2013. Green tax design in the real (second-best) world. In: Shogren, J.F. (Ed.), Encyclopedia of Energy, Natural Resource, and Environmental Economics. Elsevier, Waltham, pp. 161–168.

A brief history of behaviour in US energy programs: Landscape, integration, and future opportunities

4.4

Jane S. Peters,
Opinion Dynamics Corporation, Portland, OR, United States

1 Introduction

The United States has been addressing end-user energy behaviour since the mid-1970s first in response to the Organization of Petroleum Exporting Countries (OPEC) oil embargo in 1973 and natural gas shortages in the later 1970s and most recently in response to concerns for climate change. The size of the economy and the focus on private sector solutions over government solutions to problems have led to a different environment than that found in other countries. Initial efforts for energy savings emanated from the Federal government in the 1970s but shifted by the end of the 1980s to state legislatures and state public services/utility commissions and investor-owned utilities. There is, as a result, no single locus of energy behaviour efforts.

In this chapter, the author's experience of working as an evaluator of energy programs since 1982 sheds light on the evolution of the landscape in which energy programs have evolved in the United States. The three types of energy programs that have dominated the United States from the 1980s to today in 2019 (energy efficiency, demand management and pricing, and market transformation) may not be the dominant programs in the future. These three program types are currently responding to an increasing interest and commitment to integrating behavioural solutions into energy programs to increase program savings without increasing program costs.

There are emerging program efforts as the need to reduce greenhouse gas emissions brings programs that offer distributed generation and beneficial electrification options to end users. The lessons learned from the efforts to integrate behavioural solutions into energy programs offer suggestions of how to ensure the new energy programs are as effective as possible and that regulators and legislatures structure policies to ensure new programs will include behavioural solutions to optimise program effectiveness.

Energy and Behaviour. https://doi.org/10.1016/B978-0-12-818567-4.00017-X

The chapter is divided into two sections: The first section addresses the landscape in which energy programs operate and its evolution since the 1970s. The first section also introduces the various governmental and private sector actors and their roles in the development and implementation of programs and in engaging end users about their energy use behaviours. The second section presents the three most common types of energy programs—energy efficiency, demand management and pricing, and market transformation—and then reviews how energy behaviour solutions have been integrated into the programs in each category. Based on this historical review, in the conclusions I discuss the remaining opportunity for integration of behaviour into energy programs and what types of policy changes may be needed to achieve greater integration of behaviour into current and future energy programs.

2 The US energy program landscape

Prior to discussing energy behaviour change programs in the United States, it is important to understand the context in which energy behaviour is addressed though energy programs. Energy behaviour is typically defined as the behaviour of the end user, the person who uses energy to accomplish a task. It is also defined as occurring on the demand side versus the supply side of the energy distribution process or behind the meter where the distribution utility has limited access.

An energy end user might be a homeowner or tenant in an apartment who uses a clothes washer or an air conditioner. An end user can also be the workers in a business who use computers, printers, and other energy-using equipment and the owner of the business who buys the computers, printers, and energy-using equipment. The decision-making processes of the business owner, the worker, and the homeowner or tenant are different yet all affect how much energy is being demanded from the energy supply distribution system. Similarly, the salesperson in an electronics store, their manager, and the company buyer for the store are not end users. Yet, their actions affect the decisions end users make. Therefore, changing retailer selling behaviour can influence end-user energy using behaviour.

Energy behaviours are addressed by many different types of energy programs with many different strategies and tactics. The end-user energy behaviours of interest in this chapter are many those addressed by different types of distributed energy resource (DER) programs, which encompass energy efficiency, renewables, cogeneration, battery storage, and demand management and pricing options. Further, the energy program landscape includes private and public sector actors and activities that occur within every level of government and in multiple sectors of the economy.

Considering efforts to influence energy behaviour, the federal government and states play a significant role, with state legislatures and public utility and service commissions directly influencing energy behaviour on nearly a daily basis both for end users and for those that interact with end users. Local cities and counties are establishing policies that influence different aspects of energy behaviour and across the economy nongovernmental organisations (NGOs) and trade associations representing companies and workers in the energy industry also act to influence energy behaviour (Berg et al., 2018; Sciortino et al., 2011).

2.1 Evolution of the US energy program landscape

This section provides an overview of the evolution of the US energy program landscape beginning in the 1970 and culminating in 2019, at the time of this discussion. The broad scope of the US energy program landscape began to emerge following the 1973 oil embargo by the OPEC. At that time, the US Federal Government, under the leadership of Presidents Gerald Ford and Jimmy Carter, began efforts to reduce energy consumption in buildings and transportation, typically employing information campaigns to exhort citizens to conserve energy by changing thermostat setting, wearing sweaters inside, and weatherising homes and businesses (see Ford Message to Congress on Energy 1977[a]; Carter Energy Speech 1977[b]).

In response to President Carter's request for energy legislation, the US Congress initiated energy program efforts of which two focused on end-user energy: Residential Conservation Service (RCS), funding audits to households, begun in 1979, and the Low-Income Weatherization Assistance Program (WAP), funding audits and direct weatherization services, begun in 1976. While the Weatherization Assistance Program is still in operation in 2019, the Residential Conservation Service program was cancelled in the mid-1980s after a recognition that too few households were getting audits and too few audits were leading to home weatherization (Hirst, 1986; Walker et al., 1985). Substantial research by social scientists in the late 1970s and early 1980s demonstrated that information-only campaigns were not effective at changing behaviour (Geller et al., 1982). The failure of the Residential Conservation Service to generate the amount of savings Congress hoped for is not surprising. However, at the time, it was unexpected.

In California, the focus of the regulatory structure shifted to support energy reductions rather than energy sales. As Brownstein (2009) tells in his Atlantic Monthly article, first, the legislature enacted appliance efficiency standards in 1977, and then, in 1978, the California Public Utility Commission (CPUC) instituted a policy known as 'decoupling' that made it possible for the utilities to earn a profit on reduced sales. This policy has been a key driver of utility investment in energy efficiency programs ever since. Utilities in California continue to support energy efficiency programs because they earn money on not selling electricity (Brownstein, 2009).

The Northwest Power and Conservation Council (NPCC), then termed the Northwest Power Planning Council (NPPC), was established as a result of federal legislation in late 1980. The NPCC created the first effort to plan for optimal investment in power plants using the principle of 'least-cost planning'. Least cost planning

---

[a] Ford message to Congress on Energy: https://www.presidency.ucsb.edu/documents/special-message-the-congress-energy.

[b] Carter Energy Speech: https://www.presidency.ucsb.edu/documents/the-energy-shortage-statement-announcing-initiatives-deal-with-the-shortage.

is where the costs of demand reduction through conservation and efficiency improvements are compared directly to the costs of supply side investments in energy generation (NPPC, 1983). This approach has been a key policy framework for investment in energy efficiency by US utilities.

Throughout the 1980s, US Federal Government funding focused on low-income and residential weatherization programs through the national Weatherization Assistance Program and a regional program by the US Department of Energy Bonneville Power Administration. Studies of the impacts of these programs showed that energy savings occurred with the installation of weatherization materials (White and Brown, 1990). Local utilities often facilitated greater program participation by offering rebates, which encouraged their customers to make weatherization investments (Nadel and Geller, 1996).

In the late 1980s, following several years of electric power system brownouts in the Northeastern United States, the region's investor-owned utilities asked their commissions for approval to build more power plants. In response, NGOs sought to apply least-cost planning principles in the Northeast so that utilities would invest first in energy use reduction, before building power plants. One NGO, the Conservation Law Foundation, argued in front of Northeast public service and utility commissions that the region should adopt the Northwest approach; Commissions in Massachusetts and Connecticut were the first Northeastern states to require that utilities invest first in energy reduction through conservation and demand management efforts (New England Conservation Act Coalition, 1993; Smeloff and Asmus, 1997).

Lutzenhiser (1993) in his seminal review of energy research noted the approach taken in the US energy programs uses 'A physical–technical–economic model (PTEM) of energy consumption'. Behaviour, he noted, is "secondary to building thermodynamics and technology efficiencies…". Yet, because research consistently demonstrates that energy savings occurred with the PTEM approach, policy makers overwhelmingly require PTEM solutions in efforts that use government- and utility-provided financial support to offset the cost of improving the energy efficiency of citizens' homes and buildings (Mazur-Stommen and Farley, 2013).

In addition, PTEM programs dominate the energy program landscape because of the cost-effectiveness requirements that emerge from least-cost planning (National Action Plan for Energy Efficiency, 2008). The least-cost planning process sets up an economic comparison between demand and supply side resources. If demand resources are lower cost than the next increment of supply, termed the avoided cost, then the demand resources are more cost-effective than building a supply resource. In most cases, this cost-effectiveness is carried to the 'measure' level, which is that each specific item installed must be cost-effective (Energy Trust of Oregon, 2019; National Action Plan for Energy Efficiency, 2008; Woolf et al., 2012).

Energy efficiency spending for the 20th century by US utilities peaked in 1993 (Smeloff and Asmus, 1997; Sciortino et al., 2011). By 1997, energy efficiency program funding had dropped substantially as commissions and utilities focused their efforts on restructuring (Molina et al., 2010). In the West and Northeast, market transformation efforts became more dominant among the remaining efficiency expenditures. Market transformation introduced non-PTEM approaches to energy efficiency programs such as a longer horizon for adoption (10–20 years) and focus of activities on market suppliers and providers not end users.

The focus on these approaches was abbreviated, because the restructuring efforts in California famously led in 2001 to market manipulation and power shortages. The CPUC ordered the utilities to return to PTEM approaches. A few years later, 2005 saw a return to pre-1995 funding levels, which continued to grow (Sciortino et al., 2011). This revival of funding for PTEM programs led to an increase in energy efficiency programs. By 2010, 34 of the 50 public service and utility commissions required investor-owned utilities in their states to pursue integrated resource planning for electricity and, in 17 states, for natural gas (Barbose et al., 2013). Many public utilities offered energy efficiency programs by this time, and nearly all these programs relied largely on a PTEM approach. Further, these PTEM programs had to meet cost-effectiveness requirements, which meant that the utility be able to trace the location and purchase of each measure to account for and demonstrate the purchaser would not have made the purchase without the utility program (Arimura et al., 2011).

The Great Recession that began in 2008, led to congress enacting the American Recovery and Reinvestment Act of 2009 (ARRA). As a result of ARRA, local governments received substantial funding from the federal government to offer energy efficiency programs. Ninety-nine percent of the $16.8 billion funding provided to the US DOE office of Energy Efficiency and Renewable Energy went to fund: the Weatherization Assistance program for low-income households, the State Energy Program that supported state energy programs, and the Energy Efficiency and Conservation Block Grants that supported a wide range of programs (Pew Center, 2009). One program funded was the Better Building Neighbourhood Program (BBNP) that funded 41 local jurisdictions and not-for-profit organisations to provide whole-building upgrade programs to their constituencies (Research Into Action, 2015). In other efforts, local governments such as New York City, Chicago, San Francisco, Seattle, and Portland, Oregon, increased their investment in energy programs to further reduce greenhouse gas emissions through benchmarking requirements.[c]

During this same period, there emerged a resurgence of interest in non-PTEM approaches to energy. The Behaviour Energy and Climate Change conference, founded in 2007, brought together many researchers from academia and from the energy program landscape of market actors to discuss behaviour change strategies that could further advance energy reduction.

2.2 **US energy program market actors**

In this section, I discuss the different market actors that are engaged in delivering energy programs and energy behaviour change. A variety of actors are involved in the formation of policy that influences the program environment; these include federal,

---

[c] See https://www.buildingrating.org/jurisdictions for a map of jurisdictions with benchmarking ordinances.

state, and local legislative bodies, state and local commissions, and external or intervening parties—usually NGOs that advocate in the policy process.

The US Federal Government's role typically is a supporting one, contributing to energy program development by facilitating a more efficient and effective markets. The US Department of Energy (DOE) operates to bring new energy products and services to market through programs that support research and development, technology transfer, and start-ups, among other activities. The DOE has supported the development of certifications that ensure knowledge is available in the market place, such as the effort to establish Superior Energy Performance certification.[d] The US Environmental Protection Agency (EPA) and DOE jointly developed the ENERGY STAR label to aid residential and nonresidential end users in identifying higher-efficiency products in the marketplace.[e]

Numerous public and private sector actors take on a variety of roles and responsibilities in delivering energy efficiency programs in the United States. These market actors typically fall into one of the following categories: program administrator, implementer and support contractor, evaluator, or trade ally.

Program administrators are the organisations that directly oversee the development and implementation of a variety of energy programs. While not making up the largest number of market actors, program administrators are the key drivers of a variety of energy program activities since they manage the allocation of funds.

Some states act as program administrators and offer programs to their citizens that facilitate investment in more efficient or clean energy solutions. Like some federal efforts, these state efforts include tax credit programs, loan programs, and (less frequently) service programs such as for low-income citizens or target programs for schools (Schweitzer and Tonn, 2005; Sciortino et al., 2011). In a few states, the state government or public service/utility commission oversees a third-party organisation who acts as a program administrator; these states include Vermont, Wisconsin, Oregon, Delaware, and the District of Columbia.

Since the mid-1980s, the largest number of program administrators managing the largest dollar volume of programs is electric and gas utilities (Berg et al., 2018). Typically, these programs are mandated by legislative action or by a public service/utility commission decision as part of their utility proceedings. Some utilities do all steps of program implementation from program strategy, planning, and design to the installation of measures in homes and businesses. While this was common in the 1980s, since then, utility efforts have generally tended to work with implementation contractors and local trade allies to implement programs. Using implementation contractors instead of utility staff and leveraging the skills and contacts of local trade allies is the most effective way to operate programs and enables utilities to scale programs up and down more rapidly than they could while using their own staff.

---

[d] See USDOE Superior Energy Performance website: https://www.energy.gov/eere/amo/superior-energy-performance.

[e] See the ENERGY STAR website: https://www.energystar.gov/about/history/major-milestones.

Local engagement in energy efficiency program administration is expected to increase with the rise of community choice aggregations (CCAs). A CCA can be formed by local government or group of local governments for the purpose of aggregating their community members' purchase of energy supply. In most cases, the CCA chooses to bulk purchase a higher level of renewable energy supply than provided by their local utility company. According to Lean Energy US, eight states have laws that permit CCAs as of summer 2019, and another five states have such legislation under consideration.[f] While the CCAs can purchase energy supply, they do not distribute the energy to their community, which remains in the control of the local utility company. However, some CCAs in California, such as Marin Clean Energy,[g] are now offering energy efficiency, electrification, and renewable energy programs as part of their services. This expansion of program administrators could lead to new approaches in the coming years.

The remaining energy efficiency market actors work with program administrators and include the following:

Implementers and their support contractors are businesses that have the 'boots on the ground' delivering services to customers including all types of distributed energy resources and beneficial electrification. As noted previously, program administrators can be implementers using their own staff to reach out to customers, conduct assessments, and help customers get the services, but most implementers are private sector companies. Examples of companies offering implementation services are CLEAResult,[h] Frontier Energy,[i] and Franklin Energy,[j] which specialise in energy program implementation; ICF,[k] DNV GL,[l] Leidos,[m] and Lockheed Martin,[n] which include program implementation among many other services; and many smaller firms. A professional organisation, the Association of Energy Services Professionals,[o] provides a certification for program managers who work for either program administrators or implementation firms.

Evaluators are businesses and organisations that provide services to program administrators of all types to assess delivery and implementation effectiveness, and the impacts and cost efficiency of energy programs. Evaluators are an objective third party to the program administrator and implementer so that they are able to provide an unbiased assessment. There is no certification at this time for evaluators, but one is being developed by the DOE.

---

[f] Lean Energy US, http://leanenergyus.org/.

[g] Marin Clean Energy, https://www.mcecleanenergy.org/.

[h] CLEAResult, https://www.clearesult.com/.

[i] Frontier Energy, https://frontierenergy.com/.

[j] Franklin Energy, https://www.franklinenergy.com/.

[k] ICF, https://www.icf.com/.

[l] DNV GL, https://www.dnvgl.com/.

[m] Leidos, https://www.leidos.com/.

[n] Lockheed Martin Corporation, https://www.lockheedmartin.com/en-us/index.html.

[o] Association of Energy Services Professionals, https://www.aesp.org/.

Trade allies are the largest group of market actors in the energy program landscape. These include the design and construction industry of architects, engineers, and building and general contractors. It also includes manufacturers and vendors of energy equipment, as well installers and maintenance contractors for energy-using equipment such as lighting, heating, ventilating, and cooling; appliances such as washers, dryers, audiovisual and refrigeration; and drive power equipment such as motors, fans, and blowers. Although many of these market actors focus on standard equipment, energy programs have sought to train and engage with these market actors to ensure they all are aware of and able to sell, install, and service the most efficient options.

3 Types of US energy programs

In this section, I discuss the three most significant types of energy programs through which program administrators deliver energy behaviour change. First, I introduce the three types of program and then discuss each in more depth in the following subsections addressing each of the three types of programs: energy efficiency programs, demand management and pricing programs, and market transformation programs.

There are programs that do not fit these three categories. For instance, distributed energy programs seek to increase the installation of end user-sited solar, wind, cogeneration equipment and, recently, storage systems. Another recently emerging program type is transportation electrification, which focuses on increasing adoption of electric vehicles and developing the infrastructure to support electric vehicle charging. These new programs have a shorter history than the three highlighted in this section.

Another program type that will not be discussed independently is codes and standards programs. Codes and standards programs focus on changing regulations and standards at a state or national level, thus changing the default conditions under which energy behaviours occur. Typically, the implementation of a change in a code or a standard is considered a penultimate step towards full market transformation; thus, in this discussion are part of market transformation programs.

- **Energy efficiency programs** are designed to facilitate improvement of the energy efficiency of buildings, usually following the PTEM model and directly affecting the equipment installed in buildings. These programs are offered to households, commercial, industrial, agricultural, and governmental end users and commonly use a combination of education, financial incentives, and technical advice to increase the purchase and installation of energy efficiency products and services.
- **Demand response and energy pricing programs** are designed to encourage end-use customers to shift, stagger, or otherwise change when they use energy and how much they use at targeted times. These are behaviour change programs and typically use a combination of pricing and education to encourage behaviour change.

- **Market transformation programs** seek to change the rate of adoption of various energy products and services by reducing barriers to market adoption. Market transformation programs rarely use financial incentives to end users, instead relying primarily on changing the structure of the market such as by developing a certification so that trained people are easier to identify, training sales people about how to sell efficient equipment, training or leveraging design charrettes to expand efficiency knowledge among the design community, and using incentives at the wholesale level to buydown the cost of a product at the retail level.

The following sections provide some examples of how behaviour is being integrated into each of the three most common program types.

3.1 Integration of behaviour in US energy efficiency programs

US energy efficiency programs traditionally did not admit to a focus on behaviour, in large part because the program administrators—utilities primarily—have been restrained by their commissions from using advertising to influence energy behaviour (Meyer, 1978). Nonetheless, most energy efficiency programs use marketing and outreach to influence behaviour including strategies like framing, prompts, credible messaging, and gifts. These approaches often were simply aspects of program marketing and advertising efforts and lacked direct intention to change behaviour other than to gain engagement with customers and encourage them to make a specific purchase decision or be more conserving in their use of electricity or natural gas.

Tempchin and LeBlanc's (1992) paper at the American Council for an Energy Efficiency Economy (ACEEE) Summer Study discussed this focus on marketing single programs and technologies arguing for the development of a national social marketing campaign. Their argument was that single focus marketing efforts needed a broader market campaign to facilitate greater engagement with utility energy efficiency programs. Further, with the focus on single program marketing, the marketing efforts themselves were typically secondary to the product.

A 1997 study explored the effectiveness of a marketing campaign that sought to shift residential end users' attitudes and encourage them to purchase efficient water heaters and low flow shower heads and to adjust their thermostats seasonally (Peters et al., 1998). At the time, the ENERGY STAR label had only been in development and use for 5 years, so the utility used mass market billboards and radio along with point-of-sale messaging about preferred products.

Ajzen's (1991) planned behaviour model provided a framework to assess whether the campaign influenced end users' intentions to perform these behaviours and confirmed that it did. The paper met with peer recognition that the approach had promise, yet criticism focused on several factors. Lacking an easy way to confirm the purchased products were indeed energy efficient, the commission staff response was that self-reports could not be trusted (and indeed, subsequent studies have shown that such self-reports are subject to Type 1 and Type 2 errors).[p] Further, the study faced validity concerns because there was no premeasurement of intentions, only postmeasurement. This meant that the utility could not justify savings claims.

The PTEM approach coupled with rebates had been sufficiently effective from the 1980s through 2000s, but there were signs that this might change. Between 2009 and 2014, the cost of saved energy for some programs was increasing (Billingsley et al., 2014). At the same time, natural gas was becoming the primary peak electricity fuel, and natural gas prices were declining, leading to lower avoided costs and the potential that fewer energy efficiency program could meet cost-effectiveness requirements (Tsai and Upchurch, 2016). Seeing the need for major reductions in energy use to meet climate goals, administrators and implementers wondered if it was possible to get residential and nonresidential end users to make more behaviour changes.

In 2008, a new behaviour program was launched by start-up OPower (now part of Oracle Utilities) for the Sacramento Municipal Utility District (Integral Analytics et al., 2012); many utilities continue to offer the program today, and it is the most common behaviour program for energy efficiency (Dougherty et al., 2015; Mazur-Stommen and Farley, 2013; Sussman and Chikumbo, 2016) with estimated savings in 2018 of over 20 TWh (Gheorghiu, 2018). The program acquires residential energy consumption reduction using feedback and social norms. Each month, OPower sends a Home Energy Report (HER) to utility end users. The HER, in letter format, includes a comparison of the end user's energy consumption over the past 30 days with that of others in their neighbourhood with similar homes. The HER offers recommended actions to reduce energy consumption. OPower program effectiveness is measured using a randomised control trial (RCT) experimental design, comparing the consumption of HER recipients (usually numbering 30,000–50,000) with a sample of similar end users (usually 10,000–20,000) who do not receive a HER.

The RCT design, as the evaluation approach, is the heart of the program and is key to utility and public commission acceptance of the assessed savings. Compared with marketing and education programs that use only self reported behaviours for evaluation, such as described in Case study 1, the RCT design is considered the gold standard in research design, because "it reduces bias and provides a rigorous tool to examine cause-effect relationships" (Hariton and Locascio, 2018). Recent studies show that HER programs have saved between 1% and 3% based on evaluations across multiple utilities though savings typically decay after 2 or 3 years, without continued delivery of the HER as households revert back to previous habits (Khawaja and Stewart, 2014).

---

[p] A Type 1 error occurs when the results provide a *false positive* and the null hypothesis is rejected when the null hypothesis should be that there is no difference, and a Type 2 error occurs when the results provide a *false negative* and the null hypothesis is not rejected, when it should be. In a study for the Energy Center of Wisconsin, now Slipstream, 'of those who answered "yes" or "no" to the question "is your unit high efficiency", 56% gave a response verified to be accurate and 44% gave a response determined to be inaccurate. Thus, amongst all purchasers, 36% had an accurate understanding of whether their unit was energy efficient' (Research Into Action, Inc. and Opinion Dynamics Corporation, 2002).

Other behaviour programs surfaced yet the HER program success set in motion several decisions by the CPUC to define behaviour-based programs as programs that use experimental design and information comparing households with similar residences (Karlin et al., 2016). This narrow definition, however, did not set well with energy efficiency professionals and other market actors who wished to have more program options for behaviour program and who see many more ways to integrate behaviour strategies into energy efficiency programs (see Samiullah in Ignelzi et al., 2013).

Two papers in 2013 (Gonzales et al., 2013; Ignelzi et al., 2013) discuss the variety of behaviour strategies that could be incorporated into programs or used to develop new behaviour programs, yet only two of these (social norms and feedback) meet the CPUC definition. Some of the behavioural interventions that can be used in energy efficiency programs are noted later:

1. Commitment
2. Feedback
3. Follow-through
4. Framing
5. In-person interaction
6. Monetary incentives
7. Gifts or rewards
8. Social norms
9. Use of heuristics such as sunk cost bias, default or status quo bias, a bias towards anticipation of negative outcomes, or loss aversion

These behavioural interventions can be combined or used individually to enhance an existing program or to become part of a new program design to affect behaviour. Typically, they are not the key component of a program, for instance, the OPower HER programs discussed earlier include feedback and social normative messaging as the dominant intervention strategies. Variations of the HER program design have used goal setting, commitments, competitions, and games (Ehrhardt-Martinez et al., 2010).

Following the work of Mazur-Stommen and Farley (2013), Dougherty et al. (2015) group programs together that use combinations of the interventions such as noted earlier to elicit behaviour change. The three types of groupings, which Dougherty labels types of families, are cognition family, which typically use messaging and information; the calculus family, which typically anticipates that end users will make economically rational decisions; and the social interaction family, which use human interactions, online interactions, or reciprocity to induce behaviour change (Mazur-Stommen and Farley, 2013). In their 2015 work, from a base of 170 programs, Dougherty and her colleagues identify 58 programs across these three family groupings that have evaluated savings. These families of behaviour change programs are not restricted to energy efficiency programs but include demand management and pricing programs and market transformation programs.

Ignelzi et al. (2013) generated some examples of behaviour strategies that could be integrated into existing program designs:

- For energy audit programs: set audit recommendations in the context of other households to activate social norms and seek a commitment to an action plan of recommendations.
- For appliance rebate programs: supply follow-up communications using emails or sending stickers that prompt energy reducing behaviours after a customer receives an appliance rebate.
- For upstream lighting buydown program: develop social marketing campaigns for different target audiences, for instance households with children in school could be responsive to the benefit of higher quality lighting.

Messaging and outreach have always been a part of energy efficiency programs in the United States as noted by Tempchin and LeBlanc (1992), but recent lessons from cognitive psychology have led to more care in how the opportunity is framed, in seeking commitment as part of participation, in use of experimentation, and in using social interaction and follow-through to gain longer engagement with energy efficiency (Sussman and Chikumbo, 2016). One innovative approach taken by AEP Ohio was to use evaluation findings of the nonenergy impacts important to businesses. The program team at AEP used the nonenergy impacts to craft messaging for their outreach to customers and training for the outreach team to help close the deal with nonresidential end users considering investments in energy upgrades (Stevens et al., 2019).

Cool Choices is one of several organisations that have focused on communities and workplaces using a gamification approach (Grossberg et al., 2015; Vine and Jones, 2015). Cool Choices developed a digital platform that helps businesses and communities track their actions and estimate their emissions reductions. One project involved generating behaviour change solutions from engagement with a variety of community members across 18 Northeastern Wisconsin counties (Cool Choices, 2019). As part of the process, 1350 community members generated ideas for how their community leaders could facilitate greater behaviour change across a variety of resource related areas, one of which was energy, and also reported their own actions. Using the Cool Choices platform, the reported actions resulted in annual reduction of 2500 metric tons of carbon emissions, and the suggestions resulted in 17 recommendations for things community leaders could do to facilitate community members improving their resource management.

Programs with online portals now use AB testing in which the web users are divided randomly into A group or B group and provided different web pages to test response such as click through or time viewing the page. This AB testing approach is used to adjust the messaging and learn what leads to greater response. One program was able to increase response rate and the size of their recruitment funnel over a short 2-week period by adjusting their messaging through the AB testing approach (confidential communication with a program manager in 2012). This type of integration of messaging with experimental design can be highly effective in improving the PTEM approach.

A commitment to integration of behaviour into PTEM-based programs is growing. But much work remains. This is evident most in looking at the oldest type of energy program: energy audits and assessments. A study conducted during the ARRA funding period used a Seattle City Light utility energy audit to study whether an asset-based audit would have appeal to homeowners (Ingle et al., 2012). The asset-based audit could provide homeowners with a score that compared their home with other homes in Seattle and made recommendations for what upgrades could improve their asset score. Ingle and his colleagues found that the asset score was of less interest to the homeowners than the specific tailored recommendations made by the auditors. The audit program used no additional behavioural strategies other than providing specific recommendations, and there were upgrades still left to be done after 8 months post audit.

Seeking to improve energy audits has been a focus of Reuven Sussman at American Council for and Energy Efficiency Economy. In his 2017 study, Sussman and Chikumbo (2016) review how behaviour can improve energy efficiency programs. The study focuses on energy assessments, 32 years after the Residential Conservation Service was suspended. The authors used an online experiment to test six message-framing strategies that assessors could use when discussing upgrades. Several recommendations emerged from this study similar to those noted in Ignelzi et al. (2013).

In 2019, Sussman attended to assessments again. Using content analysis, expert interviews, eye tracking studies, and a survey of assessment recipients, Sussman and his colleagues identified behavioural best practices that would improve the assessments presentation including more social interaction (face-to-face presentations and personalised information) and use of social norms and metaphorical language in the assessment report (Sussman et al., 2019). Their work demonstrates that, as the backbone of the energy upgrade process, using behavioural insights to improve assessments could have large effects on energy savings.

3.2 Integration of behaviour in US demand management and pricing energy programs

Demand management and pricing programs are often considered behaviour change programs because the intention is solely to change the time and way that residential and nonresidential end users use energy in their homes and buildings. Designers of these programs believe that the price signal is what changes end-user behaviour. Yet, because these programs use time-varying prices and end users only learn of the effect when they receive their utility bill, these programs necessitate upfront messaging to inform customers about the different prices and price periods. Evidence suggests most customers understand these rates and make corresponding behaviour changes, with the exception of some lower-income customers who appear to have less understanding, which limits their ability to respond (Nexant and Research Into Action, Inc., 2018).

Messaging typically occurs in the form of notifications by email, phone, or text message and in the form of prompts that can be placed at locations to help customers remember when to turn an air conditioner on or when to stagger their equipment use. These prompts take the form of magnets, static clings, stickers, or just printed cards. Careful consideration of the prompt content using behaviour principles is helpful; however, it is unlikely that all or even most US program administrators engage in such careful thinking across the variety of programs they offer.

Most studies of demand management and pricing programs focus on load impacts. These studies use complex econometric models to disentangle the effects of the pricing signal from weather and other exogenous variables. These studies show evidence that demand management and pricing programs have substantial and significant effects on energy use. Allcott (2011) conducted a study of the first real-time pricing program for residential end users in which just under 700 customers volunteered for the program. From the 700, about 100 were assigned as control, and the remaining nearly 600 were recruited to the real-time pricing program. Alcott found that end users had sufficient price elasticity to respond to the price signals just as nonresidential customers do.

There is less research conducted on why these behaviour changes occur. Does the price signal or the messaging have the greater influence on behaviour? The price increase is typically experienced in a bill increase that lags usage by more than 30 days, and so, it is unable to affect first-month consumption. Only the messaging that a price increase is occurring can influence that. Messaging is a key part of the *Beat the Peak* community game developed by Minnesota Valley Electric Cooperative and now used by other electric cooperatives through the National Rural Electric Cooperative Association[q] to inspire customers to reduce their use during peak periods (Grossberg et al., 2015).

Messaging may influence a household over time as it adjusts to the price signal and receives lagged feedback from the bill. A study for San Diego Electric and Gas Company (SDG&E) of their peak time rebate (PTR) program found that the largest load impacts occurred for those customers who had SDG&E notify them through their MyAccount or had signed up for text messaging so that they were alerted when the peak period occurred that would result in a rebate if they reduced their use (Forster et al., 2013).

A study conducted with residents of the Pecan Street project in Austin, Texas, demonstrated that text and price together can have a substantial effect in a critical peak pricing program (Royal and Zwirn 2018). Three groups of customers only received a behaviour nudge: either a simple text notifying of the peak event, a simple email notifying of the event, or a notification text with a suggested action. One group received a text notifying of the event and the kWh price that would be applied to usage during the peak period, which was a tenfold increase over standard rate. The three groups with a behaviour nudge reduced consumption by 6%, while the group with the nudge and pricing information reduced use by 29% (Royal and Zwirn 2018; Royal and Rustamov, 2018).

---

[q] See the National Rural Electric Cooperative Association website for current program activities https://www.electric.coop/delaware-electric-beat-the-peak/.

The price signal may also have a differential effect depending on household factors, as was observed across the first- and second-year surveys of participants in the Opt-In Time-of-Use Pilot in California (Nexant and Research Into Action, Inc., 2018). In the Opt-In Pilot, over 55,000 utility customers for the three investor-owned utilities in California were recruited to the pilot and then randomly assigned to control or treatment conditions. The utilities provided communications about the rates to those in the treatment conditions to help them know how to respond during peak periods. Over 80% responded to the two surveys, and across all income and climate zones, treatment participants reported taking actions to reduce their use during peak in both the first and second year of the pilot. Notably, however, low-income customers and customers in hotter climate zones reported continuing more actions in year two than did those in moderate climates or those with high and moderate incomes (Nexant and Research Into Action, Inc., 2018). These results suggest that higher bills in the first-year motivated greater persistence of behaviour change for those where the effects were greatest.

Demand management and pricing programs have consistently demonstrated behaviour change by end users as noted earlier, though the drivers that result in the behaviour changes are less well understood; it appears that when end users know that a peak price will occur, they do respond.

3.3 Integration of behaviour in US market transformation energy programs

Market transformation programs use interventions with market actors upstream from the end user, rather than individually focused interventions with end users such as most energy efficiency programs use. As most behaviour interventions are individually focused, there is a limited palette available to market transformation program administrators.

Tempchin and LeBlanc's (1992) paper and presentation at the American Council for an Energy Efficiency Economy (ACEEE) Summer Study discussed this focus on marketing single programs and technologies arguing for the development of a national social marketing campaign. No such national campaign evolved for overall energy efficiency. The American Council for an Energy Efficient Economy (ACEEE) documented a variety of mass market public relations campaigns (Egan and Brown, 2001), including ENERGY STAR, that did surface during the 1990s. None, however, had very robust measurement though some relied on external polls to suggest that changes in overall attitudes were influenced by their campaigns.

The Consortium for Energy Efficiency (CEE) began implementing an annual survey to track awareness of the ENERGY STAR label in 2000. The survey has shown that the Environmental Protection Agency (EPA) and DOE campaigns to increase awareness of the label led to over 90% awareness of the label after 25 year (U.S. EPA, 2017). By 2016, 45% of households responding to the ENERGY STAR awareness survey knowingly purchased an ENERGY STAR labelled product during the 12 months prior to the survey (U.S. EPA, 2017).

The EPA and DOE have cooperated on aspects of the ENERGY STAR program for many years. EPA oversees the labelling of energy-efficient appliances and equipment, and DOE sets standards for appliances and equipment through a standard-setting process. This combination of behaviour strategies—the label as a clear message of what is the highest energy-efficient product in its category and standards setting a default for energy efficiency—has been powerful in the marketplace. ENERGY STAR products are present among all white goods, most electronics, and many business equipment categories.[r]

Market transformation organisations such as the Northwest Energy Efficiency Alliance (NEEA) and Northeast Energy Efficiency Partnership (NEEP) take a regional perspective on the premise that markets are larger than utility service territories or states. However, since many markets tend to be national or international, even a regional initiative may have too limited an effect for a savings claim.

Measurement of program effects in market transformation programs typically relies on developing a program theory of change, developing a logic model of the expected behaviour changes (outcomes), and identifying market progress indicators (outputs and outcomes) that can be tracked within the target market of the program. The type of behaviour changes sought from market transformation program activities include changes in awareness, adoption of new skills and knowledge, demonstration projects providing better estimates of savings and better assessment of installation challenges, and changes in market adoption of the targeted product or service (York et al., 2017).

Among the successful market transformation efforts, the increased adoption of energy-efficient televisions and of resource-efficient clothes washing machines demonstrates how targeting market actor behaviour works (York et al., 2017). The increase in energy-efficient televisions relied on an increase in the ENERGY STAR standard for televisions (changing the default), consumer messaging (framing), and incentives to retailers to promote and sell the efficient televisions (financial reward). Targeting incentives to the midmarket actors led to increased availability of energy-efficient televisions as retailers who had more energy-efficient televisions to sell would get a larger incentive amount (NEEA, 2014; Research Into Action, Inc. and Apex Analytics, LLC, 2015). The increased adoption of resource-efficient clothes washing machines relied on similar practices: increasing the standard for ENERGY STAR qualifying equipment and targeting manufacturers and retailers with incentives to offer resource-efficient washing machines. In the Pacific Northwest, NEEA made these investments between 1997 and 2001. By 2004, manufacturers were offering rebates to customers, and by 2011, 100% of market share for clothes washing machines sold in the Pacific Northwest was in the three most stringent ENERGY STAR standards (NEEA, 2013).

---

[r] See the ENERGY STAR website: https://www.energystar.gov/about/history/major-milestones.

Case study 1

The Building Operator Certification (BOC) program trains facility staff who operate the energy systems in commercial buildings. These building operators attend eight 1-day classes that are designed to increase skills and knowledge about energy systems. At the end of the course, operators earn Level I BOC certification. Additional training leads to Level II and Level III BOC certification.

The evaluation of Level I program savings focuses on a set of key behaviours that are at the core of the training program. A random sample of BOC graduates are surveyed about 6 months after obtaining certification and asked about their energy system operating behaviours. In some cases, a subsample of engineering site visits occurs to observe operator behaviour and review maintenance records. The questions are behaviourally based with each behaviour disaggregated into components that elicit answers from the certified operator about behaviours implemented with greater specificity and accuracy than questions designed around the aggregate behaviour.

The Building Operator Certification program is a key example of the challenge of measuring savings in the regulatory environment of energy efficiency. This program does not meet the definition of a behaviour program in California. However, it does fall under the category of indirect impacts in the California Evaluation Protocols (The TecMarket Works Team, 2006), those that result from information, education, or outreach. Since, the buildings that operators manage have such high energy consumption that the effect of targeted behaviours cannot be measured at the meter. However, across multiple third-party evaluations in different jurisdictions, using this indirect impact approach, evaluators have estimated program savings to fall within similar ranges, leading many regulatory commissions to accept the savings estimates for this behaviour change program (Northwest Energy Efficiency Council, 2016).

Case study 2

Behavioural demand response (BDR) is a program where the utility uses messages to notify customers of the need to reduce load. There is no incentive or penalty for compliance. The messages typically rely on some form of social comparison and may include suggested actions. Customers may be targeted based on geographic location or usage characteristics, or all customers can be included.

Pacific Gas and Electric (PG&E) conduced a 2-year pilot BDR using a randomised control trial with over 75,000 customers residing within the areas serviced by 31 capacity constrained substations (Schellenberg and Brummer, 2017; Thayer et al., 2016). With no incentives or penalties, the key questions PG&E sought to answer were whether there were measurable load impacts from the BDR intervention for each event day and, if so, whether the effects continued.

The BDR pilot provided an email or automated-phone message to selected customers the day before (day-ahead) the anticipated event. This day-ahead notice usually stated that the customer could join with their neighbours in reducing energy use. While not all BDR programs provide tailored postevent feedback, the PG&E program provided postevent feedback within 4–7 days of the event and noted how each household compared with their neighbours (Schellenberg and Brummer, 2017; Thayer et al., 2016).

The experimental design was a two-by-two design to test the effectiveness of the BDR message for HER and non-HER customers. The design randomly assigned HER customers and non-HER customers to both the BDR control and the BDR treatment groups. The analysis found that the BDR intervention for non-HER customers resulted in an average peak reduction of 2.4% in 2015 and 2.9% in 2016; for HER customers, the average peak reduction was 1.8% in 2015 and 1.7% in 2016. These results demonstrated that BDR is effective in curtailing residential peak demand over a 2-year period. Further, the results indicated that HER customers are somewhat less responsive to the BDR message (Thayer et al., 2016). The BDR pilot did not also require an additional impact evaluation as it was a demand response effort and no energy efficiency savings were claimed.

Case study 3

Despite the use of randomised control/treatment design for the Home Energy Reports (HER) program, California regulators require an impact evaluation to confirm the savings claims. According to the evaluation findings report for the California efficiency portfolio for 2013–15, HER programs provided 60% of residential sector savings (CPUC, 2018). This conclusion is based on the impact evaluations CPUC-contracted evaluators conducted for each of the HER programs offered by the California investor-owned utilities.

The evaluation of the HER programs had four research questions: (1) Did the randomisation process produce a balanced sample design for new waves of participants? Was the sample design balance maintained for existing waves after attrition? (2) What are the energy and demand savings for each HER wave? (3) How much energy and demand savings can be jointly reported by both rebate and HER programs? (4) What are the final energy and demand savings for each HER wave? (DNV GL, 2019).

The data required to conduct the impact evaluation included monthly billing data for 12 pre- and 12 postprogram months, program tracking data for HER customers who participated in a utility rebate program, online survey data on efficient light bulb uptake for HER customers (because bulb costs were reduced through an upstream program), and hourly consumption data for HER customers pre- and postprogram summers (DNV GL, 2019).

For the three investor-owned utilities, the impact evaluation found significant savings for each utility for electricity, for gas, and for each annual wave of HER participants. However, while the estimated savings are significant, the percent electricity savings varies by wave. For instance, Pacific Gas and Electric (PG&E) HER savings per household was 1.5% in the 2011 beta wave and declined to less than 0.5% in the 2016 wave. PG&E gas savings per household also varied, beginning at about 0.9% in the 2011 beta wave, dropping to 0.2% in wave 4, and then growing to 0.8% in 2016 (DNV GL, 2019).

In California and in many other states, the impact evaluation conducted for regulatory review defines the accepted program savings, not the results that emerge from the randomised control trial implemented by the HER company.

4 Conclusions on expanding the use of behaviour strategies in US energy program implementation

Dougherty et al. (2015) cite the results of a workshop convened in California in 2014 to address the definition of behaviour programs. The workshop, an outgrowth of the whitepaper by Ignelzi et al. (2013), sought to obtain agreement on what changes were needed to achieve an expanded definition of behaviour programs in California. That definition was proposed to the CPUC Energy Division by the California investor-owned utilities in July 2014 (Southern California Edison, 2014). The recommended revisions were to include evaluation techniques such as quasiexperimentation in addition to randomised control trials and to permit other behaviour interventions in addition to social norms and feedback such as commitment, follow-through, in-person interactions, rewards, or gifts. This proposed revision has not been adopted or rejected by the CPUC.

Table 1 notes my assessment of the frequency with which various energy efficiency programs in the United States use behaviour intervention strategies and what

Table 1 Assessment of frequency of use and opportunity for use of behaviour intervention strategies.

Behaviour intervention	Frequency relative to opportunity	Opportunity
Commitment	Low	Whenever technical assistance or an audit is provided, the provider could set up an action plan and request a commitment to a timeline to complete the recommendations.
Feedback	Medium	The HER and customer bills provide feedback though it is delayed; smart meters can be used to provide immediate feedback on usage, but few tools exist for customers to use and interpret the data; building upgrade program participants could receive feedback on the results of their participation.
Follow-through	Low	Auditors, technical assistance providers, program administrators, and call centres could conduct follow-up calls to ask how customers are using or responding to information provided.
Framing	Medium	Messaging is often used as a behaviour change intervention; framing messages to address heuristics of sunk cost and fear of loss could be used more effectively.
In-person interaction	Medium	Many programs have in-person interactions with technical assistance providers, auditors, training of retail staff, etc.; this cost can seem high but has high reward in changing behaviour.
Monetary incentives	High	Monetary incentives in the form of rebates are used for energy efficiency, demand management, and pricing programs, yet limited research has been conducted to assess the amount of incentive needed so as not to over compensate; incentives may act more as 'pointing' to the right product than being an economic signal.
Gifts or rewards	Medium	Gifts of light bulbs and aerators have been common; rewards and recognition could be provided more often.
Social norms	High	Social comparison is used in HER; lawn signs have been used effectively in local community programs; social comparison and modelling could be used in public relations stories and media efforts.
Heuristics (sunk cost bias/fear of loss/default bias)	Medium	Countering or leveraging heuristics presents a large opportunity for behaviour change; sometimes, this is in messaging as noted earlier and sometimes in setting up program designs so that customers default into the preferred setting or a program rather than having to volunteer.

opportunities might be considered to use them more. Most of these strategies have had low or medium usage over the past 40 years, and even since the push to add behaviour began in the mid-2000s, there remain many opportunities for adding behaviour interventions to programs. The two most used strategies, social norms and feedback, have expanded primarily due to the HER programs described earlier.

The future of behaviour in energy efficiency programs in the United States is promising, but there remains limited activity. The CPUC requirements that behaviour programs use experimental design and focus on social comparison and feedback have yet to be relaxed. In California and those places that follow California policy lead, the need to meet cost-effectiveness requirements at the measure level remains in place and affects program administrator willingness to invest in behaviour strategies.

Adding behavioural interventions to existing programs is occurring, yet it is usually too costly or too difficult to use an experimental approach to test the effectiveness of these changes, other than with online portals or programs. Currently, the Behaviour, Energy, and Climate Change Conference is the most active source of discussion in the United States on the use of behaviour interventions to reduce energy consumption. Though the conference does publish some papers, most presenters choose only to provide a presentation at the conference, making it difficult to draw upon that literature when developing new programs or ascertaining lessons learned.

The refinement and increased adoption of new behind-the-meter energy technologies such as electric vehicles, solar and wind generating equipment, energy storage equipment, combined heat and power, and microturbines is leading to a new set of energy programs. Both the numbers of market actors and their roles and opportunities will continue to expand. The notion of a prosumer where the end users becomes a producer and retailer of energy suggests future energy end users will know much more about energy than those of today. The groundwork for that transition has been laid by the programs of today and the recent past. Currently, the expectations for measurement and evaluation for these behind-the-meter technologies are still evolving. The 2019 International Energy Program Evaluation Conference, in Denver, Colorado, includes many papers addressing evaluations of the behind-the-meter technologies.[s]

Using good behavioural strategies to facilitate the adoption of these new technologies and the technologies and programs that we are already familiar with is important. The key challenge will be the expansion of what state public service/utility commissions accept as behaviour programs and as satisfactory methods to demonstrating the effects of behaviour programs. While experimental design should remain an option and even a preferable option, other approaches such as quasiexperimentation and sound evaluation methods with careful consideration of the effects and how that drives the research design are necessary to enable program administrators to willingly expand their use of behaviour strategies for interventions and overall program designs.

---

[s] The International Energy Program Evaluation Conference is held biannually in different locations across the United States. The 2019 conference occurred in Denver, in August 2019. https://www.iepec.org/.

References

Ajzen, I., 1991. The theory of planned behavior. Organ. Behav. Hum. Decis. Process. 50, 179–211.

Allcott, H., 2011. Rethinking real-time electricity pricing. Resour. Energy Econ. 33 (4), 820–842.

Arimura, T.H., Li, S., Newell, R.G., Palmer, K., 2011. Cost-effectiveness of electricity energy efficiency programs. (Report Number 17556). Available at: https://www.nber.org/papers/w17556. (Accessed May 2019).

Barbose, G.L., Goldman, C.A., Hoffman, I.M., Billingsley, M., 2013. The Future of Utility Customer-Funded Energy Efficiency Programs in the United States: Projected Spending and Savings to 2025. Available at: https://emp.lbl.gov/sites/default/files/lbnl-5803e.pdf. (Accessed May 2019).

Berg, W., Nowak, S., Relf, G., Vaidyanathan, S., Junga, E., DiMascio, M., Cooper, E., 2018. The 2018 State Energy Efficiency Scorecard. American Council for An Energy Efficiency Economy, Washington, DC. Available at: https://aceee.org/sites/default/files/publications/researchreports/u1808.pdf. (Accessed May 2019).

Billingsley, M.A., Hoffman, I.M., Stuart, E., Schiller, S.R., Goldman, C.A., LaCommare, K., 2014. The program administrator cost of saved energy for utility customer-funded energy efficiency programs. (Report Number LBNL-6595E). Available at: https://emp.lbl.gov/publications/program-administrator-cost-saved. (Accessed May 2019).

Brownstein, R., 2009. The California Experiment. The Atlantic Monthly. October 2009. Available at: https://www.theatlantic.com/magazine/archive/2009/10/the-california-experiment/307666/. (Accessed May 2019).

California Public Utilities Commission, 2018. Energy efficiency portfolio report. CPUC, San Francisco, CA. Available at: https://www.cpuc.ca.gov/uploadedFiles/CPUCWebsite/Content/About_Us/Organization/Divisions/Office_of_Governmental_Affairs/Legislation/2018/13-15%20Energy%20Efficiency%20Report_Final.pdf. (Accessed June 2019).

Cool Choices, 2019. Help Us Do the Right Things! Strategies for Expanding Sustainable Practices in Communities. Available at: https://coolchoices.com/insights/. (Accessed May 2019).

DNV GL, 2019. Impact evaluation report: Home energy reports—Residential program year 2016. California Public Utilities Commission. CALMAC ID: CPUC190.01. Available at: http://www.calmac.org/publications/CPUC_Group_A_Res_2016_HER_finalCALMAC.pdf. (Accessed June 2019).

Dougherty, A., Henderson, C., Dwelley, A., Jayaraman, M., 2015. Energy Efficiency Behavioral Programs: Literature Review, Benchmarking Analysis, and Evaluation Guidelines. Available at: http://mn.gov/commerce-stat/pdfs/card-report-energy-efficiency-behavorial-prog.pdf. (Accessed May 2019).

Egan, C., Brown, E., 2001. An analysis of public opinion and communication campaign research on energy efficiency and related topics. (Report Number A013). Available at: https://aceee.org/research-report/a013. (Accessed May 2019).

Ehrhardt-Martinez, K., Donnelly, K.A., "Skip" Laitner, J.A., 2010. Advanced Metering Initiatives and Residential Feedback Programs: A Meta-Review for Household Electricity-Saving Opportunities. Report E105. American Council for and Energy-Efficiency Economy, Washington, DC.

Energy Trust of Oregon, 2019. Determining When Efficiency Is the Best Energy Buy for All Utility Customers. Available at: https://www.energytrust.org/wp-content/uploads/2016/11/GEN_FS_CostEffectiveness.pdf. (Accessed May 2019).

Forster, H., Moran, D., Gettig, B., 2013. What's driving the CART in behavior-based demand response? In: International Energy Program Evaluation Conference, Chicago, 2013. Available at: http://www.iepec.org/conf-docs/conf-by-year/2013-Chicago/023.pdf#page=1. (Accessed May 2019).

Geller, E.S., Winett, R.A., Everett, P.B., 1982. Preserving the Environment: New Strategies for Behavior Change. vol. 2. Pergamon Press, New York, NY191–197.

Gheorghiu, I., 2018. Oracle Utilities Boasts Industry-Leading 20 TWh of Savings for Consumers. Available at: https://www.utilitydive.com/news/oracle-utilities-opower-boasts-industry-leading-20-twh-of-savings-for-cons/540202/. (Accessed May 2019).

Gonzales, P., Peters, J., Spahic-McClaren, M., Dunn, A., Forster, H., 2013. Integrating more behavior change strategies into a portfolio. In: Association of Energy Services Professionals Conference, Orlando, 2013.

Grossberg, F., Wolfson, M., Mazur-Stommen, S., Farley, K., Nadel, S., 2015. Gamified energy efficiency programs. (Report Number B1501). Available at: https://aceee.org/research-report/b1501. (Accessed May 2019).

Hariton, E., Locascio, J., 2018. Randomised controlled trails-the gold standard for effectiveness research. BJOG 125 (13), 1716. Available at: https://www.ncbi.nlm.nih.gov/pmc/articles/PMC6235704/. (Accessed June 2019).

Hirst, E., 1986. Communications on energy Review of the US Residential Conservation Service. Energy Policy, April 1986, 164–166.

Ignelzi, P., Peters, J., Dougherty, A., Randazzo, K., Dethman, L., Lutzenhiser, L., 2013. Paving the Way for a Richer Mix of Residential Behavior Programs. Available at: http://www.calmac.org/publications/Residential_Behavior_White_Paper_5-31-13_FINAL.pdf. (Accessed May 2019).

Ingle, A., Moezzi, M., Lutzenhiser, L., Hathaway, Z., Lutzenhiser, S., Van Clock, J., Peters, J., Smith, R., Heslam, D., Diamond, R., 2012. Behavioral Perspectives on Home Energy Audits: The Role of Auditors, Labels, Reports, and Audit Tools on Homeowner Decision-Making. Available at: https://eetd.lbl.gov/sites/all/files/publications/ingle-lbnl-5715e.pdf. (Accessed May 2019).

Integral Analytics (Integral Analytics, Inc.), Building Metrics, Inc., and Sageview, 2012. Home energy report program impact & persistence evaluation report. Available at: http://www.oracle.com/us/industries/utilities/smud-home-energy-rpt-program-3631971.pdf. (Accessed May 2019).

Karlin, B., Lupkin, L., Ford, R., Bunten, A., Forster, H., Zaval, L., 2016. California Behavioral Definition: Review and Recommendations. . Prepared for: Pacific Gas and Electric.

Khawaja, M.S., Stewart, J., 2014. Long-run savings and cost-effectiveness of home energy report programs. Available at: https://cadmusgroup.com/papers-reports/long-run-savings-cost-effectiveness-home-energy-report-programs/. (Accessed May 2019).

Lutzenhiser, L., 1993. Social and behavioral aspects of energy use. Annu. Rev. Energy Environ. 18, 247–289. https://doi.org/10.1146/annurev.eg.18.110193.001335.

Mazur-Stommen, S., Farley, K., 2013. ACEEE field guide to utility-run behavior programs. (Report Number B132). Available at: https://aceee.org/research-report/b132. (Accessed May 2019).

Meyer, M.A., 1978. Advertising by public utilities as an allowable expense for ratemaking: assault on management prerogative. Valparaiso Univ. Law Rev. 13, 87–126. Available at: http://scholar.valpo.edu/vulr/vol13/iss1/3. (Accessed May 2019).

Molina, M., Neubauer, M., Sciortino, M., Nowak, S., Vaidyanathan, S., Kaufman, N., Chittum, A., 2010. The 2010 State Energy Efficiency Scorecard. American Council for An Energy Efficiency Economy, Washington, DC. Available at: https://aceee.org/sites/default/files/publications/researchreports/e107.pdf. (Accessed May 2019).

Nadel, S., Geller, H., 1996. Utility DSM: what have we learned? Where are we going? Energy Policy 24 (4), 289–302. Washington, DC: American Council for An Energy Efficiency Economy.

National Action Plan for Energy Efficiency, 2008. Understanding Cost-Effectiveness of Energy Efficiency Programs: Best Practices, Technical Methods, and Emerging Issues for Policy-Makers. . Energy and Environmental Economics, Inc. and Regulatory Assistance Project.

NEEA (Northwest Energy Efficiency Alliance), 2013. ENERGY STAR® Efficient Clothes Washers: Success Story. Available at: https://neea.org/img/uploads/neea-previously-funded-initiative-washers.pdf. (Accessed March 2019).

NEEA (Northwest Energy Efficiency Alliance), 2014. Energy-Efficient Televisions: Success Stories. Available at: https://neea.org/img/documents/neea_pfi_tvs.pdf. (Accessed March 2019).

New England Conservation Act Coalition, 1993. Power to Spare: A Plan for Increasing New England's Competitiveness Through Energy Efficiency. New England Energy Policy Council, Boston.

Nexant and Research Into Action, Inc., 2018. California statewide opt-in time-of-use pricing pilot final report. Available at: http://www.cpuc.ca.gov/WorkArea/DownloadAsset.aspx?id=6442457172. (Accessed May 2019).

Northwest Energy Efficiency Council, 2016. Energy Savings for the Building Operator Certification (BOC©) Program. Seattle, WA. Available at: https://www.theboc.info/wp-content/uploads/2017/02/BOC-Energy-Savings-FAQ-2.0-web.pdf. (Accessed June 2019).

NPPC (Northwest Power Planning Council), 1983. Northwest Conservation and Electric Power Plan. NPPC, Portland. Available at: https://www.nwcouncil.org/sites/default/files/1983Plan.PDF. (Accessed May 2019).

Peters, J., Seiden, K., Baggett, S., Morander, L., 1998. Changing Consumer Attitudes to Energy Efficiency: Midterm Results From an Advertising Campaign. ACEEE Summer Study on Energy Efficiency in Buildings, Pacific Grove, CA, 1998 ACEEE, Washington, DC.

Pew Center (Pew Center on Global Climate Change), 2009. U.S. Department of Energy's Recovery Act Spending. Available at: https://www.issuelab.org/resources/11536/11536.pdf. (Accessed May 2019).

Research Into Action, 2015. Evaluation of the better buildings neighborhood program final synthesis report. vols. 1–6. U.S. Department of Energy, Washington, DC. Available at: https://www.energy.gov/eere/better-buildings-neighborhood-program/accomplishments. (Accessed May 2019).

Research Into Action, Inc. and Apex Analytics, LLC, 2015. Television initiative MPER #4. (Report Number E15-316). Available at: https://neea.org/img/uploads/television-initiative-mper-4.pdf. (Accessed May 2019).

Research Into Action, Inc. and Opinion Dynamics Corporation, 2002. Appliance Sales Tracking: 2001 Residential Survey. Prepared for the Energy Center of Wisconsin, now Slipstream. Madison WI. Available at: https://www.seventhwave.org/sites/default/files/234-1.pdf. (Accessed May 2019).

Royal, A., Rustamov, G., 2018. So small pecuniary incentives motivate residential peak energy reductions? Experimental evidence. Appl. Econ. https://doi.org/10.1080/00036846.2018.14189508.

Royal, A., Zwirn, M., 2019. How to harness human nature in a heatwave. Resources for the Future. Online Magazine. 198ІSummer 2018. Available at:https://www.resourcesmag.org/archives/how-to-harness-human-nature-in-a-heatwave/. (Accessed June 2019).

Schellenberg, J., Brummer, W., 2017. Evaluation of PG&E's Two-Year Behavioral Demand Response Study. Presented at 35th Peak Load Management Association Conference, Nashville. TN. Available at:https://www.peakload.org/assets/35thConf/8BrummerSchellen berg.pdf. (Accessed June 2019).

Schweitzer, M., Tonn, B.E., 2005. An Evaluation of State Energy Program Accomplishments: 2002 Program Year. ORNL/CON-492 Oak Ridge National Laboratory, Oak Ridge, TN. Available at: https://www.researchgate.net/publication/237363938/download. (Accessed May 2019).

Sciortino, M., Neubauer, M., Vaidyanathan, S., Chittum, A., Hayes, S., Nowak, S., Molina, M., Sheppard, C., Jacobson, A., Chamberlin, C., Mugica, Y., 2011. The 2011 state energy efficiency scorecard. (Report Number E115). Available at: https://aceee.org/research-report/e115. (Accessed May 2019).

Smeloff, E., Asmus, P., 1997. Reinventing Electric Utilities: Competition, Citizen Action, and Clean Power. Island Press, Washington, DC.

Southern California Edison, 2014. California IOUs Behavior Definition Proposed to CPUC. Available at: https://library.cee1.org/system/files/library/11659/CA_IOUs_Behavior_Definition_Proposed_to_CPUC.pdf. (Accessed May 2019).

Stevens, N., Billing, B., Murakami, S., 2019. The bottom line and energy efficiency: how non-energy impacts improve the bottom line and create targeted messages addressing industry specific pain points. In: International Energy Program Evaluation Conference, Denver, 2019.

Sussman, R., Chikumbo, M., 2016. Behavior change programs: Status and impact. (Report Number B1601). Available at: https://aceee.org/research-report/b1601. (Accessed May 2019).

Sussman, R., Chikumbo, M., Miller, N., 2019. After the audit: Improving residential energy efficiency assessment reports. (Report Number B1901). Available at: https://aceee.org/research-report/b1901. (Accessed May 2019).

Tempchin, R.S., LeBlanc, W.J., 1992. Maximizing Energy Efficiency Investments Through Communication Campaigns: How to Make Reddy Kilowatt® the Moral Equivalent of Smokey the Bear. Available at: https://aceee.org/files/proceedings/1992/data/papers/SS92_Panel6_Paper30.pdf#page=1. (Accessed May 2019).

Thayer, D., Brummer, W., Smith, B.A., Aslin, R., Cook, J., 2016. Is behavioral energy efficiency and demand response really better together? In: Proceedings of the American Council for and Energy Efficient Economy Summer Study on Energy Efficiency in Buildings. Asilomar, CA. Available at: https://aceee.org/files/proceedings/2016/data/papers/2_1222.pdf. (Accessed June 2019).

The TecMarket Works Team, 2006. California Energy Efficiency Evaluation Protocols: Technic, Methodological, and Reporting Requirements for Evaluation Professionals. Prepared for the California Public Utilities Commission. April 2006. Available at: https://www.cpuc.ca.gov/WorkArea/DownloadAsset.aspx?id=5212. (Accessed June 2019).

Tsai, K., Upchurch, J., 2016. Natural Gas Prices in 2016 Were the Lowest in Nearly 20 Years. Available at: https://www.eia.gov/todayinenergy/detail.php?id=29552. (Accessed May 2019).

U.S. EPA (EPA Office of Air and Radiation, Climate Protection Partnerships Division), 2017. National Awareness of ENERGY STAR® for 2016: Analysis of 2016 CEE Household Survey. Available at: https://www.energystar.gov/about/publications_resources. (Accessed May 2019).

Vine, E., Jones, C.M., 2015. A Review of Energy Reduction Competitions: What Have We Learned? Available at: https://uc-ciee.org/behavior-decision-making/1/721/99/nested. (Accessed May 2019).

Walker, J.A., Rauh, T.N., Griffin, K., 1985. The residential conservation service program. Ann. Rev. Energy 10, 285–315.

White, D.L., Brown, M.A., 1990. Electricity Savings Among Participants Three Year After Weatherization in Bonneville's 1986 Residential Weatherization Program. ORNL/CON-305 Bonneville Power Administration, Portland, OR.

Woolf, T., Steinhurst, W., Malone, E., Takahashi, K., 2012. Energy Efficiency Cost-Effectiveness Screening: How to Properly Account for 'Other Program Impacts' and Environmental Compliance Costs. Available at: https://www.raponline.org/knowledge-center/energy-efficiency-cost-effectiveness-screening/. (Accessed May 2019).

York, D., Bastian, H., Relf, G., Amann, J., 2017. Transforming energy efficiency markets: Lessons learned and next steps. (Report Number U1715). Available at: https://aceee.org/research-report/u1715. (Accessed May 2019).

Relevant international conferences, programmes, and journals

This appendix presents potentially useful resources in the field of Energy and Behaviour, including the most relevant international conferences (Table 1), programmes (Table 2), and journals (Table 3).

Table 1 Relevant international conferences including the 'Energy and Behaviour' topic.

Conference	About	Location	Frequency	Link
ACEEE Summer Study on Energy Efficiency in Buildings	Attracts professionals to discuss cutting-edge technologies, strategies, and programmes for reducing energy use and addressing climate impacts. Associated ACEEE conferences include the 'Summer Study on Energy Efficiency in Industry'; the 'National Conference on Energy Efficiency as a Resource'; 'Conference on Health, Environment, and Energy'; and the 'Energy Efficiency Finance Forum' See panels on 'Human Dimensions', 'Social Dimensions', and 'Capturing Savings through Behavior'	The United States	Started in 1980, held biennially	aceee.org/conferences
eceee Summer Study	Delivers comprehensive evidence-based information on all aspects of energy efficiency and contribute to design future policies Has hosted a panel on 'Dynamics of Consumption' since 1999	Europe	Started in 1993, held biennially	www.eceee.org/summerstudy/
Behavior, Energy & Climate Change Conference (BECC)	Focus on understanding human behaviour and decision-making and using that knowledge to accelerate the transition to a low-carbon future	The United States	Started in 2007, held annually	beccconference.org
Behavior, Energy & Climate Change Conference (BECC) Japan	Focus on understanding human behaviour and decision-making and using that knowledge to accelerate the transition to a low-carbon future	Japan	Started 2014, held annually	seeb.jp
European Conference on Behaviour and Energy Efficiency (BEHAVE)	Brings researchers and practitioners involved in end-use energy efficiency to share recent research, new technological developments, and best practices on understanding and influencing behaviour related to energy efficiency https://www.zhaw.ch/storage/hochschule/ueber-uns/veranstaltungen/behave-2018-conference-schedule.pdf	Europe	Started in 2009, held ad hoc, in various locations: Maastricht 2009, Helsinki 2012, Oxford 2014. Coimbra 2016, Zurich 2018	Active links: www.cres.gr/behave/Maastricht.htm https://www.sitra.fi/en/events/behave-conference-energy-efficiency-and-behaviour-2012/ www.inescc.pt/behave2016

Conference	Description	Location	Timeline	Website
International Conference on Energy Research and Social Science	Aims to exploring the nexus of energy and society	Europe, the United States	Started in 2017, held biennially	www.elsevier.com/ events/conferences/ international-conference-on-energy-research-and-social-science/ about
International Energy Program Evaluation Conference (IEPEC)	Provides a forum for presenting, critiquing, and discussing evaluations of energy programmes. Targets energy programme implementers; evaluators of those programmes; local, state, national, and international representatives; and academic researchers involved in evaluation	The United States, Europe	Started in 1984, held biennially	www.iepec.org
International Energy Policy & Programme Evaluation Conference (IEPPEC)	Builds a global community of people involved in evaluating energy policies who will work together to improve the quality and effectiveness of energy policy	Europe, Asia Pacific, Australia	IEPPEC started in 2014. Name change in 2018. Held annually	www.energy-evaluation.org
Sustainable Built Environment (SBE) conferences	Disseminates innovative policies and developments in the field of sustainable urban environment to a broad international audience of specialists in policy, design, construction, and operation of buildings and related infrastructure	Worldwide	Started in 1997, held triennially	www.sbe-series.org/

Table 2 Relevant programmes related with the 'Energy and Behaviour' topic.

Program	Description	Link
ACEEE Behaviour and the Human Dimensions of Energy Use	The American Council for an Energy-Efficient Economy (ACEEE) is a nonprofit organisation, acting as a catalyst to advance energy efficiency policies, programmes, technologies, investments, and behaviours	https://aceee.org/portal/behavior
ASHRAE Multidisciplinary Group on Occupant Behaviour in Buildings	The American Society of Heating, Refrigerating and Air-Conditioning Engineers (ASHRAE) has a Multidisciplinary Task Group on Occupant Behaviour in Buildings to study and integrate occupant behaviour insights in ASHRAE research, guidelines, codes and standards, handbooks, and policies	https://www.ashrae.org/technical-resources/technical-committees/section-mtg-multidisciplinary-task-groups
European Energy Network	Voluntary network of European energy agencies that aims at promoting sustainable energy good and best practice, having a working group on behaviour change	http://enr-network.org/
IEA Demand Side Management Technology Collaboration Programme	Cooperative energy technology programme within the framework of the International Energy Agency (IEA). Develops and promotes tools and information on demand-side management and energy efficiency Most relevant tasks: • Task 23: The Role of Customers in Delivering Effective Smart Grids • Task 24 Phase I and Phase II: Behaviour Change • Task 25 Phase I and Phase II: Business Models • Hard-to-Reach Energy Users in the Residential & Commercial Sectors	http://www.ieadsm.org/iea-demand-side-management-programme/
IEA Energy in Buildings and Communities Programme	International energy research and innovation programme in the buildings and communities field within the framework of the International Energy Agency (IEA). Enables collaborative R&D projects among its 24 member countries Most relevant projects: • Annex 53—Total Energy Use in Buildings: analysis and evaluation methods • Annex 66—Definition and simulation of occupant behaviour in buildings • Annex 79—Occupant-centric building design and operation	www.iea-ebc.org

Table 3 Overview of international journals with publications on topics related with 'Energy and Behaviour'.

Journal	About	Link
Annual Review of Environment and Resources	Publishes authoritative reviews of significant topics within environmental science and engineering, including ecology and conservation science, water and energy resources, atmosphere, oceans, climate change, agriculture and living resources, and human dimensions of resource use and global change. Available since 1976	https://www.annualreviews.org/journal/energy
Applied Energy	Publishes original papers, review articles, technical notes, and letters to the editor in the areas of energy conversion and conservation, the optimal use of energy resources, analysis and optimization of energy processes, mitigation of environmental pollutants, and sustainable energy systems. Available since 1975	https://www.journals.elsevier.com/applied-energy
Buildings & Cities	Publishes high-quality research and analysis on the interplay between the different scales of the built environment. Buildings, blocks, neighborhoods, cities, national building stocks, and infrastructures. It focuses on built environment policy, practices and outcomes, and the range of economic, environmental, political, social, and technological issues occurring over the full life cycle. Available since 2019	https://www.buildingsandcities.org
Building and Environment	Publishes original research papers and review articles related to building science, urban physics, and human interaction with the indoor and outdoor built environment. Available since 1976	https://www.journals.elsevier.com/building-and-environment
Building Research & Information	Publishes original research, information papers, book reviews, and evidence-based commentaries that acknowledge the complex and interrelated nature of the built environment while focusing on the building. Available since 1973	https://www.tandfonline.com/loi/rbri20
Cities	Publishes research on aspects of urban planning and policy, including urban adaptation to climate change, urban governance, smart cities and regions, liveability and quality of life, and the complexities of creating sustainable cities. Available since 1983	https://www.journals.elsevier.com/cities
Climate Policy	Publishes research and analysis on all aspects of climate change policy, including adaptation and mitigation, governance and negotiations, policy design, implementation and impact, and the full range of economic, social, and political issues at stake in responding to climate change. Available since 2001	https://www.tandfonline.com/toc/tcpo20/current
Ecological Economics	Publishes research that fosters the understanding of the interfaces and interplay between 'nature's household' (ecosystems) and 'humanity's household' (the economy) and integrates elements of ecological science, economics, and the analysis of values, behaviours, cultural practices, institutional structures, and societal dynamics. Available since 1989	https://www.journals.elsevier.com/ecological-economics

Continued

Table 3 Overview of international journals with publications on topics related with 'Energy and Behaviour'—cont'd

Journal	About	Link
Energy	Publishes research in mechanical engineering and thermal sciences, with a strong focus on energy analysis, energy modelling and prediction, integrated energy systems, energy planning and energy management, and related topics such as energy conservation, energy efficiency, biomass and bioenergy, renewable energy, electricity supply and demand, energy storage, energy in buildings, and on economic and policy issues. Available since 1976	https://www.journals.elsevier.com/energy
Energy and Buildings	Publishes articles with new research results and new proven practice aimed at reducing the energy needs of buildings and improving indoor environment quality. Available since 1977	https://www.journals.elsevier.com/energy-and-buildings
Energy Economics	Publishes research related to the exploitation, conversion and use of energy, markets for energy commodities and derivatives, regulation and taxation, forecasting, environment and climate, international trade, development, and monetary policy. Available since 1979	https://www.journals.elsevier.com/energy-economics
Energy Efficiency	Publishes new and original work related to energy efficiency, energy savings, energy consumption, energy sufficiency, and energy transition in all sectors across the globe. Since 2008	https://link.springer.com/journal/12053
Energy & Environment	Publishes contributions that exploit the dialogue between the social sciences as energy demand and supply are observed and analysed with reference to politics of policy-making and implementation. Available since 1990	https://journals.sagepub.com/home/eae
Environmental Research Letters	Publishes interdisciplinary research on environmental science across all components of the Earth system, that is, land, atmosphere, cryosphere, biosphere and hydrosphere, and exchanges between these components. Available since 2006	https://iopscience.iop.org/journal/1748-9326
Energy Policy	Publishes research addressing the policy implications of energy supply and use from their economic, social, planning, and environmental aspects. Available since 1973	https://www.journals.elsevier.com/energy-policy
Energy Research & Social Science	Publishes original research and review articles examining the relationship between energy systems and society, namely, by looking beyond the dimensions of technology and economics to include social and human elements of energy systems. Available since 2014	https://www.journals.elsevier.com/energy-research-and-social-science
Environment and Planning	Publishes research on human geography and comprises 'A) Economy and Space', 'B) Urban Analytics and City Science', 'C) Politics and Space', and 'D) Society and Space'. Available since 1969	https://www.eandponline.org

Journal	Description	URL
Environmental and Resource Economics	Publishes research that applies the economic theory and methods to environmental issues and problems. Available since 1991	https://www.springer.com/economics/environmental/journal/10640
Environmental Science & Technology	Publishes advances, trends, and challenges in environmental science, technology, and policy, promoting an interdisciplinary understanding in the environmental field. Available since 2004	https://pubs.acs.org/journal/esthag
Environment and Behavior	Publishes international and interdisciplinary perspectives on the relationships between environments and human behaviour. Available since 1969	https://journals.sagepub.com/home/eab
Environmental Politics	Publishes theoretical and empirical aspects of environmental politics, such as political parties, social movements, NGOs, and campaigns; the design, negotiation, and implementation of environmental policy; and environmental political thought. Available since 1992	https://www.tandfonline.com/loi/fenp20
European Journal of Operational Research	Publishes original papers that contribute to the methodology of operational research and to the practice of decision-making. Available since 1977	https://www.journals.elsevier.com/european-journal-of-operational-research
European Urban and Regional Studies	Publishes research addressing processes of urban and regional development in Europe. Available since 1994	https://journals.sagepub.com/home/eur
GAIA: Ecological Perspectives for Science and Society	Publishes state-of-the-art environmental research and current solutions to environmental problems. Available since 1991	https://www.ingentaconnect.com/content/oekom/gaia
Global Environmental Change	Publishes research that advance knowledge about the human and policy dimensions of global environmental change. Available since 1990	https://www.journals.elsevier.com/global-environmental-change
IEEE Power & Energy Magazine	Publishes research on advanced concepts, technologies, and practices associated with all aspects of electric power from a technical perspective in synergy with nontechnical areas such as business, environmental, and social concerns. Available since 2003	https://ieeexplore.ieee.org/xpl/RecentIssue.jsp?punumber=8014
IEEE Transactions on Smart Grid	Publishes research on smart grid that relates to, arises from, or deliberately influences energy generation, transmission, distribution, and delivery. Available since 2010	https://ieeexplore.ieee.org/xpl/RecentIssue.jsp?punumber=5165411

Continued

Table 3 Overview of international journals with publications on topics related with 'Energy and Behaviour'—cont'd

Journal	About	Link
IEEE Transactions on Sustainable Energy	Publishes research on sustainable energy that relates to, arises from, or deliberately influences energy generation, transmission, distribution, and delivery. Available since 2010	https://ieeexplore.ieee.org/xpl/RecentIssue.jsp?punumber=5165391
International Journal of Energy Research	Publishes papers, review articles, and short communications in the area of novel energy systems and applications, aiming at better efficiency, cost improvements, more effective resource use, improved design and analysis, and reduced environmental impact and hence leading to better sustainability. Available since 1977	https://onlinelibrary.wiley.com/journal/1099114x
International Journal of Sustainable Energy	Publishes research on the science and engineering of sustainable energy systems, such as energy efficiency and conservation, and technological and educational aspects of sustainable energy. Available since 1982	https://www.tandfonline.com/loi/gsol20
International Journal of Sustainable Transportation	Publishes new and innovative ideas on sustainable transportation research in the context of environmental, economic, social, and engineering aspects and current and future interactions of transportation systems and other urban subsystems. Available since 2007	https://www.tandfonline.com/loi/ujst20
International Journal of Urban and Regional Research	Publishes theoretical empirical developments in urban and regional research. Available since 1977	https://onlinelibrary.wiley.com/journal/14682427
International Transactions in Operational Research	Publishes works advancing the understanding and practice of Operational Research and Management Science internationally. Available since 1994	https://onlinelibrary.wiley.com/journal/14753995
Journal of Cleaner Production	Serves as a platform for addressing and discussing theoretical and practical cleaner production, encompassing environmental, and sustainability issues in corporations, governments, education institutions, regions, and societies. Available since 1993	https://www.journals.elsevier.com/journal-of-cleaner-production

Journal	Description	URL
Journal of Industrial Ecology	Publishes research on topics such as material and energy flow studies, technological change, dematerialization and decarbonization, life cycle planning, design and assessment, design for the environment, extended producer responsibility, eco-industrial parks, and product-oriented environmental, policy eco-efficiency. Available since 1997	https://onlinelibrary.wiley.com/journal/15309290
Journal of Consumer Policy	Publishes research on the behaviour of consumers and producers, namely, the impact of new technologies, market regulation and deregulation, internationalization, consumers in less affluent societies, environmental and gender issues, public sector products and services, consumer organizations and agencies, product safety and liability, and interaction of consumption, work, and leisure. Available since 1977	https://link.springer.com/journal/10603
Journal of Consumer Research	Publishes research that describes and explains consumer behaviour. Empirical, theoretical, and methodological articles spanning fields such as psychology, marketing, sociology, economics, communications, and anthropology are featured in this interdisciplinary journal. Available since 1974	http://www.ejcr.org/
Journal of Economic Psychology	Publishes research that improves the understanding of behaviour, in particular, psychological aspects of economic phenomena and processes. Available since 1981	https://www.journals.elsevier.com/journal-of-economic-psychology
Journal of Environmental Psychology	Publishes research addressing the interrelationships between people and their surroundings, including built; social, natural, and virtual environments; the use and abuse of nature and natural resources; and sustainability-related behaviour. Available since 1981	https://www.journals.elsevier.com/journal-of-environmental-psychology
Journal of Environmental Economics and Management	Publishes theoretical and empirical papers devoted to specific natural resource and environmental issues, namely, analysis of environmental policy, environmental behaviour and responses to regulation, and topics of energy economics related to the environment. Available since 1974	https://www.journals.elsevier.com/journal-of-environmental-economics-and-management
Journal of Environmental Economics and Policy	Publishes topics ranging from applications of economic theory and methods to the appraisal and evaluation of environmental issues and the economic evaluation of environmental policy. Available since 2012	https://www.tandfonline.com/toc/teep20/current
Journal of Multi-Criteria Decision Analysis	Publishes papers addressing mathematical, theoretical, algorithmic, and behavioural aspects of MCDA/MCDM, describing case studies or a series of real applications, problem structuring, interaction with decision-makers, and interpretation of results. Available since 1992	https://onlinelibrary.wiley.com/journal/10991360

Continued

Table 3 Overview of international journals with publications on topics related with 'Energy and Behaviour'—cont'd

Journal	About	Link
Journal of Public Economics	Publishes contributions on the problems of public economics, with particular emphasis on the application of modern economic theory and methods of quantitative analysis. Available since 1972	https://www.journals.elsevier.com/journal-of-public-economics
Journal of the Operational Research Society	Publishes original research papers that cover the theory, practice, history, or methodology of operational research. Available since 1950	https://www.tandfonline.com/toc/tjor20/current
Local Environment: The International Journal of Justice and Sustainability	Publishes critical research and practical experience on sustainability planning, policy, and politics in relation to theoretical, conceptual, and empirical studies at the nexus of equity, justice, and the local environment. Available since 1996	https://www.tandfonline.com/toc/cloe20/current
Nature Climate Change	Publishes significant and cutting-edge research on the nature, underlying causes or impacts of global climate change and its implications for the economy, policy, and the world at large. Available since 2011	https://www.nature.com/nclimate/
Nature Energy	Publishes research related with the provision of energy, from the generation and storage of energy, to its distribution and management, the needs and demands of the different actors, and the impacts that energy technologies and policies have on societies. Available since 2016	https://www.nature.com/nenergy/
Nature Human Behaviour	Publishes research of outstanding significance into any aspect of individual or collective human behaviour. Available since 2017	https://www.nature.com/nathumbehav/
Organization and Environment	Publishes research on the management of organisations and its implications for the sustainability and flourishing of the social, natural and economic environment in which they act. Available since 1987.	https://journals.sagepub.com/home/oae
Renewable and Sustainable Energy Reviews	Publishes review papers, original research, case studies, new technology analyses, and expert insights that support the transition to a low-carbon future, including technosocioeconomic aspects of renewable and sustainable energy. Available since 1997	https://www.journals.elsevier.com/renewable-and-sustainable-energy-reviews
Renewable Energy	Publishes research on the various topics and technologies of renewable energy systems and components. Available since 1991	https://www.journals.elsevier.com/renewable-energy

Journal	Description	URL
Science	Publishes papers that are most influential in their fields or across fields and that will significantly advance scientific understanding and present novel and broadly important data, syntheses, or concepts. Available since 1880	https://science.sciencemag.org/
Sustainability	Publishes reviews, regular research papers, communications, and short notes on environmental, cultural, economic, and social sustainability of human beings, which provides an advanced forum for studies related to sustainability and sustainable development. Available since 2009	https://www.mdpi.com/journal/sustainability
Sustainable Cities and Society	Publishes cross-cutting, multidisciplinary research on environmentally sustainable and socially resilient cities. Available since 2011	https://www.journals.elsevier.com/sustainable-cities-and-society
Technological Forecasting and Social Change	Publishes research on the methodology and practice of technological forecasting and future studies as planning tools as they interrelate social, environmental, and technological factors. Available since 1970	https://www.journals.elsevier.com/technological-forecasting-and-social-change
Theory, Culture & Society	Publishes original research and review articles in the social and cultural sciences. Available since 1982	https://journals.sagepub.com/home/tcs
Transport Policy	Publishes research reflecting the concerns of policymakers in government, management strategists in industry, and the public at large, providing independent, original, and rigorous analysis to understand how policy and strategy decisions have been made, monitor their effects, and suggest how they may be improved, in modes: air, maritime, urban, intercity, domestic, and international transport. Available since 1993	https://www.journals.elsevier.com/transport-policy
Transportation Research	Publishes research on passenger and freight transportation modes, including the dimensions of policy analysis; formulation and evaluation; planning; interaction with the political, socioeconomic, and physical environment; and design, management, and evaluation of transportation systems. Comprises 'Part A, Policy and Practice'; 'Part B, Methodological'; 'Part C, Emerging Technologies'; 'Part D, Transport and Environment'; and 'Logistics and Transportation Review'. Available since 1967	https://www.sciencedirect.com/journal/transportation-research https://www.sciencedirect.com/journal/transportation-research
Urban Studies	Publishes submissions that further the understanding of the urban condition and the rapid changes taking place in cities and regions across the globe, whether from an empirical, theoretical, or a policy perspective. Available since 1964	https://journals.sagepub.com/home/usj

Index

Note: Page numbers followed by *f* indicate figures, *t* indicate tables, and *b* indicate boxes.

Printed in the United States
By Bookmasters